The Comparative Study of the Planets

NATO ADVANCED STUDY INSTITUTES SERIES

*Proceedings of the Advanced Study Institute Programme, which aims
at the dissemination of advanced knowledge and
the formation of contacts among scientists from different countries*

The series is published by an international board of publishers in conjunction
with NATO Scientific Affairs Division

A	Life Sciences	Plenum Publishing Corporation
B	Physics	London and New York
C	Mathematical and Physical Sciences	D. Reidel Publishing Company Dordrecht, Boston and London
D	Behavioural and Social Sciences	Sijthoff & Noordhoff International Publishers
E	Applied Sciences	Alphen aan den Rijn and Germantown U.S.A.

Series C – Mathematical and Physical Sciences

Volume 85 – The Comparative Study of the Planets

The Comparative Study of the Planets

*Proceedings of the NATO Advanced Study Institute
held at Vulcano (Aeolian Islands), Italy, September 14-25, 198*

edited by

ANGIOLETTA CORADINI

and

MARCELLO FULCHIGNONI

*Reparto di Planetologia, Istituto di Astrofisica Spaziale
del C.N.R., Rome, Italy.*

D. Reidel Publishing Company

Dordrecht : Holland / Boston : U.S.A. / London : England

Published in cooperation with NATO Scientific Affairs Division

Library of Congress Cataloging in Publication Data

NATO Advanced Study Institute (1981: Isola Vulcano, Italy)
    The comparative study of the planets.

    (NATO advanced study institutes series. Series C, Mathematical and
physical sciences ; v. 85)
    Includes indexes.
        1.   Planets–Congresses.   2.   Solar system–Congresses.
I.   Coradini, Angioletta.   II.   Fulchignoni, Marcello.   III.   Title.
IV.   Series.
QB600.N37   1981   523.2   82-3658
AACR2
ISBN-13: 978-94-009-7812-6   e-ISBN-13: 978-94-009-7810-2
DOI: 10.1007/978-94-009-7810-2

Published by D. Reidel Publishing Company
P.O. Box 17, 3300 AA Dordrecht, Holland

Sold and distributed in the U.S.A. and Canada
by Kluwer Boston Inc.,
190 Old Derby Street, Hingham, MA 02043, U.S.A.

In all other countries, sold and distributed
by Kluwer Academic Publishers Group,
P.O. Box 322, 3300 AH Dordrecht, Holland

D. Reidel Publishing Company is a member of the Kluwer Group

TABLE OF CONTENTS

PART II: PHYSICS OF THE SOLAR SYSTEM

PART III: GEOLOGY OF PLANETARY BODIES

PREFACE

The volume gathers the prominent works by participants in the NATO Advanced Study Institute, "The Comparative Study of the Planets", which was held at Vulcano (Aeolian Islands) from September 14 to September 25, 1981.

The book is intended to be a landmark for all those inter-ested in the problems disclosed through the close-up exploration of planets and satellites, either for professional reasons or simply for scientific knowledge. The topics dealt with concern all methodologies by which the members of the Solar System have been studied and the audience to whom the volume is addressed is, in addition to the experts, mainly graduate and post-graduate students attending courses in Earth Sciences, Physics and Astro-nomy.

The aim we intend to achieve in editing this volume is to offer to planetologists an overview of the present state of knowledge, using the comparative study of the planets as a basis on which to build the themes treated. We think it is one of the few publications of this type, which includes all the subjects concerning the study of a planet. We hope we have succeeded in conveying through the book the message which was clearly expressed by the lecturers of the ASI in Vulcano: what counts most at this stage of planetary research is to be able to single out the fun-damental problems and choose the appropriate set of data to use in approaching these problems.

For this reason the book is divided in three sections: the first section deals with the general problems that concern the Solar System as a whole, while the other two sections concern specific problems of the components of the Solar System, from the point of view of planetary physics and planetary geology, respectively.

The information contained in the book is the result of ten years of study on the nature of the planets. It is a primary synthesis, necessarily incomplete, of the ideas that have been

A.Coradini and M. Fulchignoni (eds.), The Comparative Study of the Planets, ix–x.
Copyright © 1982 by D. Reidel Publishing Company.

discussed and that still enliven meetings among experts. As it is
a type of research that cannot be separated from an interdisciplin-
ary approach, one can sometimes find it difficult to tackle the
the different ways and, perhaps, styles in which researchers with
different backgrounds face problems. One of the aims of the book
is to help the "language" spoken by each to become more familiar
to all in order to promote an ever increasing co-operation among
researchers operating in different fields.

Angioletta Coradini and
Marcello Fulchignoni
Directors, NATO
Advanced Study Institute
Vulcano, Italy

ACKNOWLEDGEMENTS

The island of Vulcano, in the Aeolian Archipelago, was a perfect setting in which to carry out the NATO Advanced Study Institute. Besides offering its beauties of nature for our moments of relaxation, it also represented a natural laboratory for the compared study of volcanic phenomena.

Two days were dedicated to guided field trips to the largest active volcanoes (Stromboli and Vulcano) and to the main eruptive and effusive formations on the various islands.

The field trips were sponsored by the Guardia di Finanza (Coast Guard) in Messina and the Consiglio Nazionale delle Ricerche (National Research Council) who kindly offered the boats that were used. We would like to thank Captain Gentile and the crew members of the Guardia di Finanza for the exquisite hospitality they offered us on board. We would also like to express our appreciation to Captain Scotto and crew members of the motor-boats supplied by the C.N.R., who actively contributed to the success of the field trips.

We also wish to express our most sincere thanks to the mayor of Lipari, Prof. Carnevale, and to the councillor, Mr. Natoli, for their help in organizing the A.S.I. and for having brightened one of the evenings by sponsoring a traditional show with Aeolian dances and songs.

A special thanks goes to Rita Cellini, secretary to the A.S.I., who for almost a year, solved the problems related to the organization of the A.S.I. with skill, always smiling: thank you above all for having had the patience to deal with the two directors of the A.S.I., not always capable of smiling.

Angioletta Coradini and Marcello Fulchignoni

A.Coradini and M. Fulchignoni (eds.), The Comparative Study of the Planets, xi.
Copyright © 1982 by D. Reidel Publishing Company.

PART I

ORIGIN, EVOLUTION AND DYNAMICS OF
THE BODIES IN THE SOLAR SYSTEM

SOME REMARKS ON THE FORMATION OF TERRESTRIAL PLANETS

A.Coradini,C.Federico(*)and G.Magni

I.A.S.Reparto di Planetologia,viale dell'Università 11 – 00185 Roma
(*) also at Istituto di Geologia dell'Università degli Studi di Perugia

ABSTRACT

The process of terrestrial planet formation in a low mass protoplanetary disk is examined.Emphasis is placed on the role played by the gas during coagulation and settling of grains to the central plane of the system.The effects of turbulence are also taken into account.Gravitational instability in a gas-grain thin disk is shown to be a possible mechanism of planetesimal formation.
The theory by which terrestrial planets formed by accumulation of planetesimals in a gas free protoplanetary swarm is briefly reviewed.

1.   INTRODUCTION

It is believed that the Solar System was formed at the end of the gravitational collapse of an interstellar cloud of gas and dust.Theoretical studies of the collapse of a cloud of about 1 $M_\odot$ show the formation of a dense hydrostatic core surrounded by an extended envelope of infalling matter.Depending on the role played respectively by the solid component and by the gas, different evolutionary patterns of the planet formation have been proposed.In the first case,the terrestrial planets and the cores of the Jovian planets are formed through the hyerarchical accumulation of planetesimals with growing sizes.In the second case,gaseous protoplanets are at once formed,and their subsequent dynamical

3

A.Coradini and M. Fulchignoni (eds.), The Comparative Study of the Planets, 3–24.
Copyright © 1982 by D. Reidel Publishing Company.

and internal evolution leads to the formation of the
Jovian and terrestrial planets,partially or completely
removing their gaseous envelope.Giant gaseous protoplanets,formed as result of gravitational instabilities
in the gas component,need a massive protoplanetary ne-
bula.It is not yet clearly understood how a dissipative
mechanism can disperse a large amount of matter.More-
over the exact timing of the Sun formation,the inward
motion of the protoplanets and their thermal evolution
need at present more careful investigations(1,2).

     In this paper emphasis will be placed on the role
of the solid component in the formation of the terrestrial planets.The gas present in a low mass protosolar nebula affects less and less the evolution of the
solid bodies as their mass increases.In fact the gas-
particle interaction strongly affects settling,coagu-
lation of grains and onset of gravitational instabilities in a thin dust disk.On the contrary the evolution
of the swarm of planetesimals can be studied in a gas
free medium.

2.   STRUCTURE OF THE DISK

     Current ideas about the first phases of the for-
mation of planetary system state that an accretion disk
forms,fed by the infalling matter of the residual part
of the protosolar nebula,while the young Sun starts to
evolve at the center of the disk.However bi-and tridimensional numerical models of the collapse of a protostellar cloud(3,4)have not been able to follow the evolution of the nebula up to the stage of formation of
the accretion disk,because the strong dynamical exchanges
of heat,radiation and angular momentum require small
temporal steps and very long computer time.

     The total amount of mass in the disk M_D has two li-
mits.A lower limit is obtained adding to the present
mass of the planetary system the amount of volatiles to
restore the solar composition.Thus M_D lies in the range
$0.01 - 0.07$ $M_\odot$ (5),the uncertainties in mass being due
to uncertainties in the structure of the giant planets.
The upper limit must be estimated taking into account
the efficiency of the possible mechanisms for the ejection of mass from the disk,such as T-Tauri wind and
dissipative processes active in the disk.However this
method results in a poor estimate as a consequence of
the poor physical understanding of these processes.Ca-
meron and Pine(7) assume a total mass of about 2 $M_\odot$ for
the protosolar nebula,but other authors(5,7)believe that
a value less than 1.1 $M_\odot$ is more resonable.

If the disk mass is much less than the mass of the Sun, it is clear that its structure, i.e. surface and volume density and temperature distributions, strongly depends on the poorly known last evolutive phases of the young Sun. Assuming that the formation of the planets takes place in a timescale shorter than the timescale of mass and angular momentum dissipation in the disk, the structure of the disk surface density can be obtained simply taking as mass of the disk the lower limit estimated above. The temperature distribution can be deduced under the hypothesis of high opacity in the gas of the disk(8). If hydrostatic vertical equilibrium is assumed, it results(9)

$$\sigma = 2700 \, r_{au}^{-1.72} \; ; \; \rho = 1.28 \, 10^{-9} \, r_{au}^{-2.72} \; ; \; T = 600 \, r_{au}^{-1} \qquad (1)$$

$$M_D = 0.01 \, M_\odot$$

where r is the radial distance in Astronomical Units. For comparison we report the main disk parameters, assumed by other authors.
Cameron and Pine(6):

$$\sigma = 10^5 r_{au}^{-1} \; ; \; \rho = 7.5 \, 10^{-9} \, r_{au}^{-1.1} \; ; \; T = 2000 r_{au}^{-0.8}$$

$$M_D = M_o$$

Weidenschilling(5):

$$\sigma = 6 \, 10^3 r_{au}^{-1.5} \; ; \; \rho = 3.2 \, 10^{-9} \, r_{au}^{-2.5} \; ; \; T = 600 r^{-1}$$

$$M_D = 0.05 \, M_\odot$$

Safronov(7):

$$\sigma = 10^3 \text{ for } r_{au} \leq 5.2 \; ; \; \sigma = 10^3 r_{au}^{-2} \text{ for } r_{au} > 5.2$$

$$T = 107 r_{au}^{-0.57} \; ; \; \rho = 1.31 \, 10^{-9} \, r_{au}^{-1.79} \text{ for } r_{au} \leq 5.2$$
$$(\beta = 0.1)$$

$$T = 239 r_{au}^{-1.01} \; ; \; \rho = 2.38 \, 10^{-8} \, r_{au}^{-2.5} \text{ for } r_{au} > 5.2$$

$$T = 56 r_{au}^{-0.75} \; ; \; \rho = 1.82 \, 10^{-9} \, r_{au}^{-1.88} \text{ for } r_{au} \leq 5.2$$
$$(\beta \to 0)$$

$$T = 56 r_{au}^{-0.75} \; ; \; \rho = 4.92 \, 10^{-8} \, r_{au}^{-2.37} \text{ for } r_{au} > 5.2$$

where $\beta = H/r$ is a free parameter indicative of the flattening of the dust disk with scale height H.

To conclude, we want to show how a reasonable surface distribution can be obtained by the simple assumption of conservation of angular momentum in the primordial matter of the disk.

a) At the starting of the collapse, a cloud of about 1.01 $M_\odot$, radius R and uniform density ρ_o rotates with angular velocity Ω_o as a rigid body.

b) A core of about 1 $M_\odot$ grows at the center of the cloud and for the residual part of the nebula Ω_o and R remain unaltered, the density becoming

$$\rho_o' = \rho_o M_D / (M_D + M_\odot)$$

The infinitesimal cylindrical shells with equal angular momentum $\delta \Psi_N$ correspond to infinitesimal flat ring with the same angular momentum $\delta \Psi_D$ in the disk. Imposing angular momentum conservation in the infinitesimal elements, it results

$$\delta \Psi_N = -4\pi \rho_o' \Omega_o R^5 \cos^3\theta \, sen^2\theta \, d\theta \quad ; \quad \delta \Psi_D = 2\pi r (GM_\odot/r^3)^{\frac{1}{2}} \sigma r^2 dr$$

$$\sigma = -2\rho_o' \Omega_o R^5 (GM_\odot)^{-\frac{1}{2}} r^{-3/2} \cos^3\theta \, sen^2\theta \, (d\theta/dr) \tag{2}$$

where we have supposed that the disk rotates with Keplerian velocity. Angular momentum conservation of a single particle at the boundary of the cloud gives

$$\Omega_o R^2 \cos^2\theta = (GM_\odot/r^3)^{\frac{1}{2}} r^2 \tag{3}$$

and differentiating equation (3) we obtain $d\theta/dr$. Substituting in equation (2), we have

$$\sigma = R\rho_o' (GM_\odot)^{\frac{1}{2}} \{1 - (GM_\odot r)^{\frac{1}{2}} / (\Omega_o R^2)\}^{\frac{1}{2}} / (2\Omega_o r^{3/2})$$

If we consider a disk of 40 AU (the Pluto's orbital radius), putting $\cos\theta = 1$ in equation (3) and $\rho_o = 10^{-19} g/cm^3$ (10), we obtain

$$R = \{(M_D + M_\odot)/(4/3\pi\rho_o)\}^{1/3} = 1.69 \ 10^{17} \ cm$$

$$\Omega_o = (GM_\odot r_D)^{\frac{1}{2}} / R^2 = 9.9 \ 10^{-15} \ s^{-1}$$

and then

$$\sigma = 1700 r_{au}^{-3/2} (1 - 0.158 r_{au}^{\frac{1}{2}})$$

It is to be remarked that this behaviour is not much different from that of the standard model.

3. TURBULENCE IN THE NEBULA AND IN THE DISK

The initial phases of the formation of the Solar System were probably influenced by turbulence:in fact the high value of the Reynolds number and a sufficient energy input indicate that a well developed turbulence can set on.The Reynolds number is defined as

$$Re = u_L L / \nu$$

where ν is the kinematic viscosity of the gas,resulting

$$\nu = (2.38 \ 10^{-6} T^{2/3}) / \rho$$

and u_L,L are characteristic velocity and dimension of the turbulent process.We suppose that turbulence is iso-tropic and follows a Kolmogorov spectrum,a plausible assumption if Re is sufficiently high (11).The energy dissipated per gram and per second by turbulence is (12)

$$\dot{\varepsilon} = u_L^3 / L \qquad (4)$$

Let us now examine some fundamental steps in the formation of the Solar System.
a) Collapse of the protosolar nebula and formation of the accretion disk.If the collapse is spherically symmetric and homologous,the collapse velocity is

$$v_{ff} = \alpha (2GM/R)^{\frac{1}{2}}$$

where M,R are mass and radius of the nebula and $\alpha < 1$ is a factor taking into account the retarding effects,ma-gnetic field and rotation.The release of gravitational energy per gram and per second is then

$$\dot{\varepsilon}_g = (1/M) d/dt (-3GM^2/5R) = 4.8 \ 10^{-11} \alpha M^{2/3} \rho^{5/6}$$

If we assume supersonic turbulence and $L \simeq ct_{ff}$, where c is the sound velocity in the gas and $t_{ff} = R/v_{ff}$ is the collapse timescale, we have

$$L = 1.5 \ 10^7 (T/\rho)^{\frac{1}{2}} \alpha^{-1}$$

$$\dot{\varepsilon} = c^3/L = 3.2 \ 10^4 T \rho^{\frac{1}{2}} \alpha$$

$$\dot{\varepsilon}_g / \dot{\varepsilon} = 1.5 \ 10^{-15} M^{2/3} \rho^{1/3} T^{-1} \qquad (5)$$

$$Re = 4.9 \ 10^{16} T^{1/3} \rho^{\frac{1}{2}} \alpha^{-1}$$

It is possible to identify two phases during the colla-

pse:a')initially density and opacity are low enough to
mantain a nearly isothermal collapse.Following (10,13)
for $10^{-18} < \rho < 2.85 \ 10^{-12}$ g/cm^3,it results T≈10 K and we
obtain

$$\dot{\varepsilon}_g/\dot{\varepsilon} > 1.1 \qquad\qquad Re > 3.3 \ 10^7 \alpha^{-1}$$

this assures an early development of turbulence during
the collapse.a")When the density becomes sufficiently
high, an adiabatic phase starts,with high opacity and
a steep increase in temperature (10):

$$T = 4.97 \ 10^8 \rho^{2/3} \quad for \quad \rho > 2.85 \ 10^{-12} \ g/cm^3$$

Introducing the density dependence on temperature in
equations $(5_3),(5_4)$ it results

$$\dot{\varepsilon}_g/\dot{\varepsilon} = 3 \ 10^{-24} M^{2/3} \rho^{-1/3}$$
$$Re = 3.8 \ 10^{19} \rho^{0.722} \alpha^{-1}_{>>1} \quad \begin{matrix} for \\ \end{matrix} \quad \rho < 4 \ 10^{-5} \ g/cm^3$$

So,turbulence is probably present during the phases of
the collapse up to the time of the formation of the accretion disk.
     b) Evolution of the accretion disk.The accretion
disk formed at the end of the contraction of the protosolar nebula is probably turbulent and is therefore characterized by a complete mixing of gas and grains.If we
assume,following (14),that

$$u_L/L \approx |dv/dr| \approx \Omega \qquad\qquad (6)$$

where v and Ω are the nearly Keplerian linear and angular velocities and $u_L \approx c$,the Reynolds number becomes

$$Re \approx c^2/\Omega\nu$$

Using the standard disk model(1),it results Re≈1.5 10^{12}
at 1AU and Re≈4 10^{10} at 10AU.The energy input can be
given by the gravitational energy release,associated to
the turbulent angular momentum transport.According to
(7)

$$\dot{\varepsilon}_g = \nu_t \ r^2 (d\Omega/dr)^2$$

where $\nu_t = L \ u_L/Re$ is the turbulent kinematic viscosity.
From equations (4,6) it follows

$$\dot{\varepsilon}_g/\dot{\varepsilon} = 18/(4 \ Re) \ <<1$$

and turbulence cannot be sustained anymore.

c) Turbulence in the dust disk.Another type of turbulence can set on when the dust has settled toward the midplane in a disk with a scale height H much lower than that of the gas H_g(15,16).If the properties of the turbulence within a solid disk rotating in the gas are extrapolated to the dust disk,it can be assumed that turbulence developes in the gas-dust boundary layer.In the standard disk model(1),using the same assumptions as in (16),the Reynolds number,as a function of the mass fraction of the solid component ε, results

$$Re = 2 \ 10^8 \ \varepsilon^{-1} \ r_{au}^{-1.55}$$

As ε is of the order of 100 before the onset of instabilities(17),a value of Re = 1000 necessary for turbulence is guaranteed.The energy input for turbulence in this stage is given by the amount of gravitational energy released by the dust disk during its radial contraction(16).If the mean free path of molecules for mutual collisions is smaller than that relative to grain-molecule collisions,the molecules do not freely diffuse through the particles and the interaction between the dust disk and the surrounding gas acts at the boundary of the disk(15).Thus a torque acts between the surrounding gas and the dust disk because of their different velocities, causing an angular momentum transport outwards and an inward radial decay of the dust.In this conditions,following again (16),we have

$$\dot{\varepsilon}_g/\dot{\varepsilon} \simeq 5 \ 10^4 \varepsilon^{-1} \ r_{au}^2 >>1$$

d) Collisional mechanisms between grains.Grains embedded in a turbulent medium are subjected to forces fluctuating in intensity and direction.As far as the mass of a grain is much greater than the mass of a gas molecule,grains follow the velocity variations of the gas with some delay,expressed by the viscous stopping time τ_v=m/$6\pi\eta_p$s,m and s are mass and radius of a particle,and η_p is the drag viscosity;two drag regimes are possible(see next section).The delay causes a non zero limit for the two points velocity correlation function of grains when the relative distance between the two grains goes to zero,and collisions between grains become possible.In general the collision velocity has a a maximum and drops to zero when the mass of the grains goes to zero(because the coupling between grains and gas is strong)or when this mass diverges(because grains and gas are not coupled anymore).

Völk et al.(18,19)have calculated in detail the

rms velocities and mutual collisional velocities of grains, embedded in the gas. An approximate expression of these velocities can be simply given in the form

$$<\delta v^2>^{\frac{1}{2}} = <u_g^2>^{\frac{1}{2}}; \quad <\delta v_i^2>^{\frac{1}{2}} = <u_g^2>^{\frac{1}{2}} (\tau_v/\tau_{ko})^{\frac{1}{2}}$$

for $\tau_v/\tau_{ko} \leq 1$, and

$$<\delta v^2>^{\frac{1}{2}} = <u_g^2>^{\frac{1}{2}} (\tau_v/\tau_{ko})^{-1}; \quad <\delta v_i^2>^{\frac{1}{2}} = <u_g^2>^{\frac{1}{2}} (\tau_v/\tau_{ko})^{-\frac{1}{2}}$$

for $\tau_v/\tau_{ko} > 1$

Here $<\delta v^2>^{\frac{1}{2}}$ and $<u_g^2>^{\frac{1}{2}}$ are the rms turbulent velocities of grains and gas respectively; $<\delta v_i^2>^{\frac{1}{2}}$ is the rms impact velocity between two grains equal in mass; τ_{ko} is the dissipation time of the largest eddy. If the collision occurs between two grains of substantially different masses, then the impact velocity can be approximated by the turbulent velocity of the smaller grain.

4.   COAGULATION AND SETTLING OF THE PARTICLES

An accumulation theory must account for the growth from very small particles to objects of planetary size. The details of such a growth mechanism are not completely known, however in the last years the coagulation of small particles has been studied both theoretically and experimentally. Various sticking mechanisms between two colliding particles have been assumed, such as accretion by electrostatic forces, accretion by Van der Waals forces, accretion by sticky coating, accretion by coalescence of molten or semimolten droplets. Some or all of these sticking processes may be efficient in aggregating the smallest particles up to sizes where collisional accretion based on gravity takes place. In fact, if the relative velocities are the thermal velocities, bodies smaller than 1 cm will not accrete by gravitational attraction since the thermal Brownian velocity exceeds the escape velocity (20).

Probably none of the above mentioned processes is sufficiently effective and it has been generally assuthat local, patch instabilities, as described by (7,15), are necessary to generate kilometer sized objects. However, a certain degree of coagulation of particles may be important also in set on of gravitational instabilities (17).

Non gravitational mechanisms of adhesion depend on parameters that are poorly known, such as chemical composition, roughness of the surfaces, temperature and

electric charge.Nevertheless many meteorites are agglo-
merates of individual grains and rock fragments and therefore are themselves the result of accretion events
(21);moreover the micrometeorites collected by rockets
are,with few exceptions,aggregates of smaller grains
(22).So we can conclude that there is an indication for
effective sticking of micron sized particles under different physical conditions.What is not yet clear is the
mass limit that it is possible to reach through non gravitational sticking mechanisms.

The collision between two particles can be conveniently characterized in terms of the energy of the two
particle system.To determine when a constructive impact
happens,it is necessary to compare the rebound kinetic
energy with the particle-particle potential well.Thus
the efficiency of adhesion,rebound or destruction of
particles depends on their relative velocity.

In the primordial turbulent phases grains experience high collision frequencies.In some previous works
(23,24,25) we have studied adhesion efficiencies of impacting particles embedded in a turbulent collapsing
cloud.The main results obtained are that 1)the adhesion
efficiency is larger for conducting grains,2)the sticking efficiency decreases increasing the mass of the
grains.So we found that, at the end of the collapse,par-
ticles reach a size limit of the order of $5 \ 10^{-4}$cm for
iron and $3 \ 10^{-2}$cm for graphite.On the contrary silicate
particles do not show a significant growth,remaining of
the order of $4 \ 10^{-5}$cm.

At the end of this phase,a thermally steady and
hydrostatic equilibrium state is reached,generally referred to as the "disk phase".Dust grains floating in
the Solar nebula begin to sink toward the equatorial
plane under the action of the vertical component of the
solar gravity.The equation of motion along the z-axis
is

$$dv_z/dt = -6\pi\eta_p s v_z/m - \Omega^2 z \qquad\qquad (8)$$

The first term in the right hand side of equation (8)
represents the drag forces acting on a particle while
the second term is the z-component of the solar gravity.
This equation implicitly contains two timescales:the viscous stopping time $\tau_v = m/6\pi\eta_p s$ and the Keplerian timescale $\tau_K = 1/\Omega$.In standard disk η_p follows the Epstein
relationship everywhere, except in the Mercury zone,whe-
re the Stokes law holds (26).

In the standard disk $\tau_v \ll \tau_K$ in both the viscous regimes,so that the acceleration term can be neglected
and v_z results

$$v_z = - m\Omega^2 z/6\pi\eta_p s \tag{9}$$

and the timescale of sedimentation is

$$\tau_z \simeq z/|v_z| = 6\pi\eta_p s/m\ \Omega^2 \tag{10}$$

Neglecting coagulation as a first approximation, the settling timescale depends on the grain size and it results to be $4\ 10^3 y$ in the Earth zone and $2\ 10^3 y$ in the Jupiter zone for $s = 3\ 10^{-2}$ cm.

In a detailed calculation of settling without coagulation we have obtained that the sedimentation is faster in the outer regions of the disk for the standard model. In a time interval of $2\ 10^4 y$ for $s = 2\ 10^{-2}$ cm the particles reach the height $z = 10^{-4} H_g$. For a more massive disk where the gas density is higher and flattened, the settling time increases outward: particles with $s = 2\ 10^{-2}$ cm reach the same height in about $7\ 10^4 y$. Thus the settling depends in some extent on the disk model (9).

The situation described above is idealized because we have assumed that the particles are all of the same mass; on the contrary it has been shown(25) that at the end of the collapse a mass distribution not very peaked is obtained. Then from equation (9), it is clear that the larger particles settle to the central plane faster than the smaller ones. Larger particles can thus collect a large fraction of the smaller ones during the settling. Numerical simulations recently performed on settling with coagulation show that, if the sticking efficiency is assumed to be equal to 1, the larger particles increase by many orders of magnitude and reach the central plane at 1AU in a time of about $10^3 y$ (16,27). The two quoted models are in substantial agreement in spite of the different numerical procedures and disk model adopted ($s_{max} \simeq 10^4$ cm).

A different model of coagulation during turbulent phases and settling in quiescent disk has been developed by Makalkin (28). The key point of his model is the study of the adhesion efficiency by Van der Waals forces of particles with different chemical composition and size. The result is that, during turbulent phases, the metallic particles reach larger sizes than the silicate ones: the further growth during sedimentation produces closely packed chain-like structures. Collisions between aggregates of this kind yield probably fragmentation instead of accretion. Weidenschilling (16), in contrast, modifying the sticking criterion developed by Dahneke (29) in order to take into account the surface roughness of the particles, gives an evaluation of the adhesion

critical velocities.He finds that Van der Waals forces
can allow the production of centimeter sized particles
independently of their composition.

To simply evaluate the sticking coefficient for
particles with mass m_1 and m_2 during their settling
toward the central plane we must compare their impact
velocity with the critical adhesion velocity v_a.From
equation (9) it results

$$<\delta v_i^2>^{\frac{1}{2}} \simeq v_i \simeq |v_{z1}-v_{z2}| = \delta\Omega^2 z\,|s_1-s_2|/\rho v_T$$

δ is the bulk density of the particles and v_T the molecule thermal velocity.Considering only the contribution
of the Van der Waals forces,the adhesion velocity v_a
can be expressed as

$$v_a = \{(40W/m_r)(s_r/10^{-5})^\gamma\}^{\frac{1}{2}}$$

where W is the Hamaker constant,m_r and s_r are the reduced mass and radius ,and γ is a parameter that depends
on the chemical composition of the particles (28).So the
sticking condition is

$$e<\delta v_i^2>^{\frac{1}{2}}\leq v_a \qquad\qquad\qquad\qquad (11)$$

where e is the restitution coefficient.Recalling that
the largest particles obtained at the end of the turbulent phases are of the order of 10^{-4},it is possible
to evaluate the point below which these particles can
start to collect smaller particles.Following (7,27),the
mass increment of a particle due to coalescence in a
displacement dz is given by

$$dm = -\xi\pi s^2\epsilon_0\rho dz \qquad\qquad\qquad\qquad (12)$$

where ξ is the sticking coefficient,that results to be
1 when condition (11) is verified and $\epsilon_0\simeq10^{-2}$ is assumed constant and equal to the initial mass fraction of
solid particles.Equation (12) is easily integrated to
give the particle radius at the height z.The results
obtained for different compositions of the growing grains
are reported in Table I. Table I shows that the accretion rate for metallic particles is faster.The sedimentation time in this case is shorter than the sedimentation time in absence of coagulation and the disk flattens more rapidly.When the volume density of particles
reaches values such that gas molecule cannot freely
pass through the dust layer,the motion of the grains
becomes Keplerian.Thus,as pointed out by (15),there is
a velocity difference between the dust layer and the

Tab. I

	W (ergs)	γ	e	z_i/H_g	z_f/H_g	m_f (g)	t_f (y)
Fe	$3\ 10^{-12}$	2	0.01	1	10^{-3}	8.7	$2\ 10^3$
Si	10^{-12}	1	0.5	$6\ 10^{-2}$	10^{-3}	$1.7\ 10^{-2}$	$1.7\ 10^4$

The assumed values of the Hamaker constant W
and of the parameters γ and e are here reported. z_i is the initial point below which particles can collect other particles, m_f is the
mass reached by a particle that has fallen
from z_i to z_f in a time interval t_f.

gas disk and, consequently, a drag in the boundary dust-
gas, that can cause the orbital decay of the disk. Weidenschilling (16) noted that, if the particles are small
(s<1 cm), turbulence will set on again preventing the
dust disk from becoming dense enough for gravitational
instabilities. In this phase of autoinduced turbulence,
however, large particles (s>1 cm) continue to settle toward the central plane, while small particles can continue their collisional evolution if a relationship similar to (11) is fulfilled. If a simple treatment of the
turbulence is used (30,23,16) the relative velocity between two impacting grains is

$$<\delta v_i^2>^{\frac{1}{2}} \simeq (s_1+s_2)(2\varepsilon/9\nu)^{\frac{1}{2}} \simeq 1.3\ 10^{-3}(s_1+s_2)$$

In this case, condition (11) is generally fulfilled in
the standard model and accretion could occur. However
the relative velocity is so low that the hierarchical
coagulation timescale, both for metallic and silicate
particles, is very long ($\simeq$1000 y). On the contrary, using
the model developed by (19), the impact velocity is larger, resulting

$$<\delta v_i^2>^{\frac{1}{2}} \simeq (\tau_v/\tau_{ko})^{\frac{1}{2}}<u_g^2>^{\frac{1}{2}} \simeq 4.9\ 10^{-3}\delta s \qquad (13)$$

The numerical value in equation (13) has been obtained
for the standard disk, assuming as characteristic scale
length of the largest eddy the thickness of the disk
($\simeq 10^9$ cm at 1AU) and as characteristic velocity the value given in equation (7). Using equation (13), condition
(11) is fulfilled up to centimeter-size for metallic
particles; subsequently, the smaller particles increase

furtherly.Following (13) it is possible to evaluate the
growth time of particles during this late turbulent pha-
se.We can write

$$s = s_o N^{1/3}$$

where N is the number of original particles constitu-
ting a grain of radius s.The rate of increase of the
number of original particles in any assembly is given
by

$$dN/dt = N\rho\varepsilon\Sigma<\delta v_1^2>^{\frac{1}{2}}/m$$

Σ is the geometrical collision section.For metallic particles in the standard disk model,it results

$$dN/dt \simeq 2 \ 10^{-11}\varepsilon N$$

Considering that $\varepsilon\simeq100$ when $H\simeq10^9$cm and $s_o\simeq2 \ 10^{-2}$ cm,
it is possible to see that the time necessary to reach
centimeter-sizes is of about 100y.This last increase in
size allows the particles to detach from the gas and to
settle toward the central plane.

5.   GRAVITATIONAL INSTABILITIES

     Gravitational instabilities set on when the densi-
ty of the assembly of gas and dust reaches such large
values that a positive perturbation in density can in-
definitely grow in time.This means that selfgravitation
overcomes pressure and rotation effects and,if present,
magnetic field effects.It is generally assumed that,in a
massive protoplanetary disk,gravitational instabilities
are due essentially to the gas,while in a low mass disk,
as the one till now considered,instabilities can set on
in the solid component.In fact in a low mass disk,the
gas component is stable and a large degree of flattening
is necessary in the solid component to reach instability conditions.Safronov,Goldreich and Ward (7,15) asses-
sed that instabilities formed in the particle disk can
give origin to dust condensations.Goldreich and Ward
(15),considering the balance between selfgravitation and
rotation,select among all the possible collapsing objects those able to reach solid density (3 g/cm^3).So
they obtain a "first generation" of solid planetesimals
with masses of the order of $2 \ 10^{14}$ g in the Earth zo-
ne.In a subsequent stage the planetesimals will be grouped in "rotating disk-like associations"containing about
10^4initial members.Each cluster is initially stable

against collapse;subsequently the relative velocity of
the constituents is reduced by the gas drag,the internal energy of the cluster decreses and a slow contraction starts.The coalescence of the original planetesimals in the cluster gives origin to solid bodies of
the order of $2 \cdot 10^{18}$g in a time scale of about 100 y.
    Safronov (7),taking into account also the finite
height of the disk,found that in the Earth zone dust
condensations have masses of the order of $5 \cdot 10^{16}$g,whi-
le in the Jupiter zone they reach masses of about 10^{22}
g.The further evolution of these condensations is slow
and due mainly to their mutual collisions.The aggrega-
tion process led to the formation of "secondary condensations" with masses 10^4-10^6 times larger than the
masses of the primary ones.Safronov also stresses that
the growth process is continuous and there is therefore
no reason to introduce the concept of primary and secondary aggregations:in fact it is sufficient that the
mass of a condensation increases by an order of magnitude to contract to the state of solid body with $\delta = 1$
g/cm^3.Although the two discussed models are quite different,the results are in substantial agreement,indi-
cating in both cases that in a timescale of the order
of 10^3y the formation of kilometer-size bodies in the
Earth zone occurs.In both models there is the assumption that the gas component,if present,does not affect the set up of instabilities.It is generally assumed
that a strong T-Tauri wind can remove the gas from the
protoplanetary disk,but such a violent mechanism,if present at this stage of the life of the disk,could have
drastically modified physical conditions in the disk,
forbidding the set up of instabilities.We are then left
with the two possibilities that either the T-Tauri phase was subsequent to the formation of planetesimals,or
the gas dissipation was due to a more "gentle" mechanism such as thermal escape from the exosphere of the
disk.In the latter case it has been shown that the dissipation time of the gas from a low mass disk ($M_D \simeq 10^{-2}$
$M_\oplus$) is of about 10^8 y in the Earth zone (31).In both
situations there is no reason to assume that the protoplanetary disk was gas free.
    We have studied,according to (32),the onset of gra-
vitational instabilities and the formation of planetesimals in the sedimentating dust component of the protoplanetary disk,taking into account the interactions
between gas and grains.The situation has been described
as follows: during the sedimentation of the grains,the
density increases until the selfgravitation of
the dust overcomes the thermal and the rotational energy,and gravitational instabilities can set on.Local

gravitational instabilities can occur in small
regions where a fluctuation may allow the density to
be higher than the critical value.This unstable region
can detach itself from the evolving and differentially
rotating disk only if its contraction rate is faster
than the rate of evolution of the surrounding medium.
This condition is equivalent to the requirement that
the collapse velocity be larger than:a) the sedimentation velocity;b) the dispersion velocity across a fragment due to the differential rotation.

    At an early stage of the process condition b) is
not fulfilled and local instabilities are inhibited,
but an annular region (ring) can become unstable and
gravitational instabilities can be studied by means of
axisymmetric perturbations.The fulfilment of condition
a) implies that:1) the physical parameters can be considered "frozen" during the collapse;2) the development
of the instability in a given region does not perturb
the surrounding regions.We do not report here the details of the above sketched theory, that can be found
in (17).

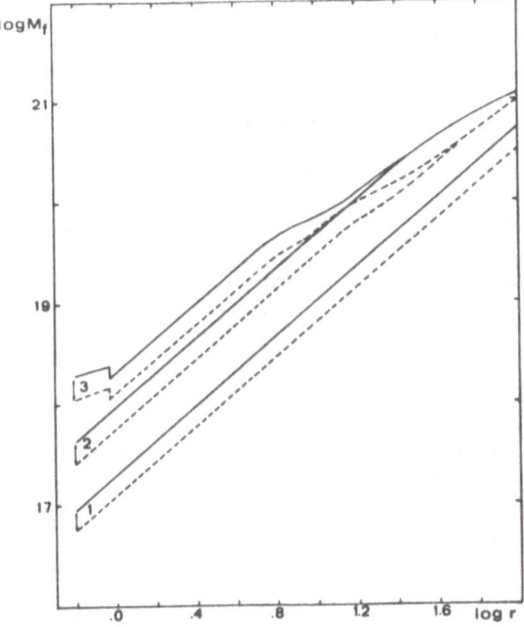

Figure 1. Planetesimal masses
versus radial distance for
three selected values of the mass
of the grains:10^{-4},1 and 10^2 g.

In figure 1 the masses of the planetesimals,resulting from the fragmentation of the ring are shown.We have assumed that fragmentation can take place if a virtual fragment is not disrupted by tangential stresses generated by the differential rotation and by the tidal forces of the bulk of the disk.In the internal region the fragmentation is essentially driven by the differential rotation and the fragments have masses increasing with grain masses and radial distance.In the outer region the masses of the fragments become nearly independent of the mass of the grains.

We can conclude that gas-grain coupling in the protoplanetary disk does not inhibit the onset of gravitational instabilities.However the unstable component is mainly the dust and this is in good agreement with the assumptions of Safronov and Goldreich and Ward (7,15). The main difference is that ring contraction times results longer in the inner zones,and for small grains, comparable with the dissipation time of the disk.Consequently,the mass of the grains is a critical parameter and we want to point out that the growth process by binary collisions and the onset of gravitational instabilities are strictly connected.The strong gas-grain viscous coupling smooths out the collisional random motion of the grains,and the pressure is not large enough to influence the evolution of the disk structure.The influence of the gas on axisymmetric instabilities increases with the distance,so that in the outer regions of the disk the two-fluid model strongly deviates from the one fluid model.

6.   ACCUMULATION OF THE TERRESTRIAL PLANETS

In the previous section we have seen that the formation of planetesimals is possible considering gas-grain interaction.Now we want to investigate the transition from kilometer-size bodies to Earth size objects. In this stage gravitational interactions are dominant; in fact,as shown by (26) for planetesimals larger than 10^{10}g,the gas drag effect is in a first approximation negligible and the planetesimals move on nearly Keplerian orbits.For this reason planetary growth simulations have generally been performed in a gas free medium.The most complete analytical treatment of this theory is given in the book of Safronov (7).A detailed comparison of the results of the analytical and numerical models of terrestrial planet growth is given by Wetherill(33) and Greenberg (34).

In this intermediate stage of the formation of the

terrestrial planets,the scenario is that of a swarm of
planetesimals with mass M $\simeq 10^{18}$g (R$\simeq$4 km,δ= 3g/cm^3)
spread out in a disk width $\leq$ 2AU.The process of accumulation via collisions depends on the coupled distribution of mass and relative velocity.In fact high velocity of impact among the preplanetary bodies will more
easily produce catastrophic destruction rather than ac-
cretion.According to Fujiwara (35) when E/M< 4 10^6erg/g
there is cratering and no catastrophic destruction,
whereas if E/M> 10^8erg/g complete destruction is the
outcome of a collision.As a consequence low impact velocities (v_i<28 m/s) are needed to have accumulation.
      Safronov (7) in his well posed analytical theory,
has demonstrated under various assumptions that decoupling the mass distribution from the velocity distribution in a system of planetesimals differentially rotating around the Sun,the relative velocity,when the equilibrium is reached,is of the order of the escape velocity v_e from the largest body,i.e.

$$v = v_e/(2\theta)^{\frac{1}{2}}$$

The velocities of the bodies tend to an equilibrium value when the gain of energy in close encounters is equal
to the loss of energy in impacts.The non dimensional
parameter θ depends on the properties of the system of
colliding bodies.For an inverse power law of distribution of the masses of the planetesimals

$$n(m)dm = C\ m^{-q}dm$$

with 1.67<q<2,θ is in the range 2 - 7.If q$\simeq$1.67 most
of the mass is in the largest bodies which in turn are
responsible for most of the perturbations.If q$\simeq$2 most
of the mass is in the small bodies and the dominant mode
for energy loss is mutual collisions among these small
bodies.
      Numerical simulations (20,34) and analytical so-
lutions of the coagulation equation (7,36),referred to
this stage of the planetary evolution,show that an "embryo" with radius of 200 - 500 km can be easily formed.
While Safronov and his coworkers have obtained that
most of the mass still resides in the large bodies,Gre-
enberg finds out an opposite outcome from his simulations,i.e.most of the mass is in the small bodies,so
that embryos become to some extent "isolated" (runaway
growth).It must be noted that the numerical algorithm
"particles in a box" of (20) takes also into account
fragmentation and comminution. However, as it has
been recently suggested by Greenberg himself(34),an

improvement in the physical model of collision is necessary in order to definitively assess in which way
the mass distribution evolves.
 In the model of Safronov (7) and Vityazev et al.(37)
the rate of growth of a planetary embryo can be written as

$$dM/dt = 4\pi R_E^2 (1 + 2\theta)\sigma_F/P$$

where $4\pi R_E^2 (1+2\theta)$ is the enhanced cross section due to
gravitational focusing on the embryo with radius R_E,
σ_F is the surface density of solid matter in the"feeding zone" and P the orbital period of revolution.Being
$\theta \approx 2-7$, the cross section can never get much larger than
the geometrical one and a runaway growth is somehow
inhibited.The width of the feeding zone enlarges pro-
portionally to the growing radius of the embryo,being

$$\Delta r \propto r^{3/2} R_E \qquad\qquad\qquad (14)$$

Moreover σ_F decreases as the embryo grows and depletes
its feeding zone,thus

$$\sigma_F = \sigma_p (1-M/M_F) \qquad\qquad\qquad (15)$$

where σ_p is the original surface density of solid matter and M_F is the present mass of the planet.Taking into account the enlargement of the feeding zone (eq. 14),
the depletion of the feeding zone (eq. 15),Vityazev et
al. (37) have shown that,at the first order in the eccentricity,the rate of growth of the embryo results to
be

$$dR_E/dt = (1-A^2 R_E^2)/B \qquad\qquad\qquad (16)$$

where $A^2=(M_0\theta\delta/6\pi)^{\frac{1}{2}}(2r^{5/2}\sigma_F)^{-1}$ and $B=(\delta/\sigma_F)P/(1+2\theta)$.
From equation (16) it follows that the growth time of
the embryo to reach 97% of the present mass of the corresponding planet is given by

$$\tau = 5.3 \ B/2A$$

with reasonable choices of θ and σ_p the obtained masses
of the terrestrial planets are in agreement with the
present ones.τ results of the order of 10^8y for the
terrestrial planets except for Mars ($\tau=1.9 \ 10^9$y).In
this theory the feeding zones are "closed" because they
enlarge up to touch at the end of the process.Weiden-
schilling (38), in his model of accretion,has removed
this constraint,obtaining again accumulation times of

the same order of magnitude.The extremely long time
for the formation of the Red planet is one of the reasons to believe that Jupiter formed before the terrestrial planets and largely depleted in mass the asteroid and the Mars region.It is to be noted also that in
the Weidenschilling model the initial width of the feeding zone is not zero,because of an "initial random velocity" of planetesimals;originally adjacent zones touch themselves and finally largely overlap.Moreover the
original feeding zone has a mass equal to the present
mass of the planet.Weidenschilling (38) claims that
this assumption removes the inconsistency of using
equation (15) when matter is continually added to an
enlarging feeding zone.

We have already seen that analytical theories and
numerical simulations give contrasting size distribution functions during the growth of the embryo.On the
other hand numerical calculations (39) of a non accumulating swarm of 100 planetesimals with R=830 km show
that the average velocity approaches an equilibrium value in agreement with the results of Safronov (7).

This equilibrium value increases linearly with R,and
the values of θ are in the range predicted by the theo-
ry.As Wetherill (39) notes,mutual perturbations cause
a radial diffusion of the planetesimals, and this effect has an analytical counterpart only in a recent
work of Stewart and Kaula (40).

Numerical simulations both bi- and tridimensional
have been performed on simultaneous accumulation of
the terrestrial planets.The timescale for their formation seems to be 10^8,but depending on the value of the
maximum eccentricity e_{max} originally chosen,very few
or too many planets are formed.A recent bidimensional
simulation of Cox and Lewis (41) with e_{max}=0.15 ends
with a state similar to that of the present terrestrial planets,while the tridimensional simulation of
Wetherill (39) leads to a small number of planets when
e_{max}=0.05.It is probable that in a bidimensional simulation a large value of e_{max} is necessary,because there
is some underestimation of the effect of perturbations
caused by encounters with respect to that caused by collisions (39).Therefore we can again argue that the
exact balance between the energy in close encounters
and the energy loss in collisions is not
yet clearly understood.In fact a more recent analytical
developed for the velocity evolution of non accreting
planetesimals,based on Boltzmann and Fokker-Planck
equations (40),shows that the equilibrium random velocities are reached for values of θ larger than those
obtained by Safronov (7) and not in agreement with the

numerical values obtained by Wetherill (39).This effect
is probably due to neglecting the Sun gravitational
field in the Fokker-Planck equation.However the importance of this and a similar work (42) resides in the
attempt to improve the theory for statistical mechanics
of planetesimals in a Keplerian swarm.Cross examinati-
ons of the outcomes of the analytical and numerical
studies can yield a new insight in the intriguing problem of the coupled evolution of the mass and velocity
distribution of a planetesimal population.

REFERENCES

1) Mc Crea,W.H. and Williams,I.P.:1965,Proc.Roy.Soc.
   A. 287,pp.143-164.

2) Cameron,A.G.W.:1978,in "Protostars and Planets",T.
   Gehrels,Ed.,pp.453-478.The Univ. of Arizona Press,
   Tucson.

3) Black,D.C. and Bodenheimer,P.:1976,Ap.J. 206,pp.138-
   149.

4) Bodenheimer,P. and Black,D.C.:1978,in "Protostars
   and Planets",T.Gehrels,Ed.,pp.288-322. The Univ. of
   Arizona Press,Tucson.

5) Weidenschilling,S.J.:1977,Astrophys.Space Sci. 51,
   pp.153-158.

6) Cameron,A.G.W. and Pine,M.R.:1973,Icarus 18,pp.377-
   406.

7) Safronov,V.S.:1969,Evolution of the Protoplanetary
   Cloud and the Formation of the Earth and Planets,Nau-
   ka,Moscow,Trasl.1972,NASA TTF-677.

8) Lewis,J.S.:1974,Science 186,pp.440-443.

9) Coradini,A.,Magni,G. and Federico,C.:1980,The Moon
   and the Planets 22,pp.47-61.

10)Larson,R.B.:1972,in"On the Origin of the Solar System"
   H.Reeves,Ed.,pp.142-150,CNRS,Paris.

11)Batchelor,G.K.:1960,The Theory of Homogeneous Turbulence,Cambridge at the Univ.Press.

12)Landau,L.D. and Lifschitz,E.M.:1959,Fluid Mechanics,

Addison-Wesley Publ.Co.,Reading,Mass.

13) Cameron,A.G.W.:1973,Icarus 18,pp.407-450.

14) Fleck,R.C.Jr.:1980,Ap.J. 242,pp.1019-1022.

15) Goldreich,P. and Ward,W.R.:1973,Ap.J. 183,pp.1051-1061.

16) Weidenschilling,S.J.:1980,Icarus 44,pp.172-189.

17) Coradini,A.,Federico,C. and Magni,G.:1981,Astron.
    Astrophys.98,pp.173-185.

18) Völk,H.J.,Jones,F.C.,Morfill,G. and Rosér,S.:1978,
    The Moon and the Planets 19,pp.221-227.

19) Völk,H.J.,Jones,F.C.,Morfill,G. and Rosér,S.:1980,
    Astron.Astrophys.85,pp.316-325.

20) Greenberg,R.,Hartmann,W.K.,Chapman,C.R. and Wacker,
    J.F.:1978 in"Protostars and Planets",T.Geherels,Ed.,
    pp.599-622.The Univ.of Arizona Press,Tucson.

21) Herndon,J.M. and Wilkening,L.L.:1978,in"Protostars
    and Planets",T.Geherels,Ed.,pp.502-515.The Univ.of
    Arizona Press,Tucson.

22) Brownlee,D.E.:1978 in"Protostars and Planets",T.Ge-
    herels,Ed.,pp.134-150.The Univ.of Arizona Press,Tucson.

23) Carusi,A.,Coradini,A.,Federico,C.,Fulchignoni,M. and
    Magni,G.:1974,in"Exploration of the Planetary System",
    K.A.Woszczyk and C.Iwaniszewska,Eds.,IAU Symp.65,pp.
    21-35.

24) Carusi,A.,Coradini,A.,Federico,C.,Fulchignoni,M. and
    Magni,G.:1975,Astrophys.and Space Sci.33,pp.369-384.

25) Coradini,A.,Magni,G.and Federico,C.:1977,Astrophys.
    and Space Sci.48,pp.79-87.

26) Adachi,I.,Hayashi,C.and Nakazawa,K.:1976,Prog.Theor.
    Phys.56,pp.1756-1771.

27) Nagakawa,Y.,Nakazawa,K.and Hayashi,C.:1981,Icarus
    45,pp.517-528.

28) Makalkin,A.B.:1980,Phys.of the Earth and Planet.Int.
    22,pp.302-312.

29) Dahneke,B.:1971,J.Colloid.Interface Sci.37,pp.342-353.

30) Saffman,P.G.and Turner,J.S.:1956,J.of Fluid Mech.1, pp.16-30.

31) Pechernikova,G.V.and Vityazev,A.V.:1981,Adv.Space Res.1,pp.55-60.

32) Spiegel,E.A.:1972,in"On the Origin of the Solar System",H.Reeves,Ed.,pp.165-178,CNRS,Paris.

33) Wetherill,G.W.:1980,Ann.Rev.Astron.Astrophys.18,pp. 77-113.

34) Greenberg,R.:1981,in"Formation of Planetary System", A.Brahic,Ed.,Sum.School of Space Phys.,Grasse(in press)

35) Fujiwara,A.:1980,Icarus 41,pp.356-364.

36) Pechernikova,G.V.:1974,Sov.Astron.18,pp.1305-1318.

37) Vityazev,A.V.,Pechernikova,G.V.and Safronov,V.S.:1978, Sov.Astron.22,pp.60-63.

38) Weidenschilling,S.J.:1976,Icarus 16,pp.161-170.

39) Wetherill,G.W.:1980,in"The Continental Crust and its Mineral Deposits",D.W.Strangway,Ed.,Geol.Assoc.Canada Spec.Paper 20,pp.3-24.

40) Stewart,G.R.and Kaula,W.M.:1980,Icarus 44,pp.154-171.

41) Cox,L.P.and Lewis,J.S.:1980,Icarus 44,pp.706-721.

42) Nagakawa,Y.:1978,Prog.Theor.Phys.59,pp.1834-1851.

ORIGIN AND EVOLUTION OF THE GIANT PLANETS

Peter Bodenheimer
Lick Observatory, Board of Studies in
Astronomy and Astrophysics, University
of California, Santa Cruz, and Space
Science Division, NASA-Ames Research Center,
Moffett Field, California

ABSTRACT

Two major hypotheses concerning the origin of the giant
planets are discussed: (A) a protoplanet forms in the solar nebula as a gravitationally unstable gaseous subcondensation and
evolves as a chemically homogeneous object until a later stage
when a solid core may form; (B) a solid core forms first by accumulation of planetesimals, after which solar-composition gas
accretes onto the core and eventually becomes unstable to collapse. In general, under either of these scenarios, the evolution falls into three phases: (1) an early cool phase in hydrostatic equilibrium, (2) a hydrodynamic collapse, and (3) a
final phase of hydrostatic contraction and cooling to the present
state. At the final stage the theoretical calculations may be
fitted to present observed properties of the giant planets. The
physical processes that are important in determining the evolutionary characteristics are discussed.

1. INTRODUCTION

One of the critical periods of the formation of the solar
system occurs when the giant planets condense out of the solar
nebula. The processes that occur strongly influence the further
evolution of the nebula itself; also the properties of the nebula strongly influence the evolution of the protoplanets. This
paper is concerned with two major hypotheses that have been proposed for the early history of the giant planets. An important
question that may be asked is whether it is possible to distinguish between these two hypotheses on the basis of our current
knowledge of the solar system. The discussion is based primarily

A. Coradini and M. Fulchignoni (eds.), The Comparative Study of the Planets, 25–48.
Copyright © 1982 by D. Reidel Publishing Company.

on Jupiter and Saturn, which are the best studied of the outer
planets and which are thought to be composed mainly of matter in
the gaseous or liquid state. A brief discussion is also given
regarding the evolution of Uranus and Neptune.

Under the first hypothesis, often referred to under the name
of the "giant gaseous protoplanet" hypothesis, a protoplanet forms
by a process of instability at about the same time as the formation of the Sun. The gravitational instability results in a chemically homogeneous sub-condensation in the solar nebula which
evolves as a unit, subsequently contracting and collapsing under
its self-gravity. Detailed calculations have been carried out
under the assumptions that the object remains spherically symmetric and chemically homogeneous. In fact, it is likely that a
solid core will form during the evolution, by precipitation of
grains toward the center and by gas-drag capture of small stray
objects (1) during the course of the evolution. Three main phases
of evolution occur: (1) an early cool phase, starting with a
radius several thousand times that of the present planet, during
which quasi-static contraction takes place on a time scale of
10^5-10^6 years and hydrogen is in molecular form at temperatures
of less than 2000 K; (2) a hydrodynamic collapse induced by the
dissociation of molecular hydrogen, ending when core temperatures
reach 2×10^4 K, and (3) a final slow contraction and cooling
phase, starting at a radius a few times that of the present
planet, proceeding on a time scale of 10^9 years or more, and
characterized by convective energy transport in the interior.
The details of the physical processes and the evolution under this
scenario are discussed in section 2.

Under the second hypothesis, a solid (probably rocky) core
forms first by the accumulation of small particles in the solar
nebula. The solar nebula model appropriate to this hypothesis is
of relatively low mass (about 0.04 solar mass in the model of
Kusaka et al. (2)) so that it is gravitationally stable. On the
other hand, the production of giant gaseous protoplanets under the
first hypothesis requires a more massive nebula, on the order of
1 solar mass (3). In the low-mass nebula the small dust particles
settle gradually to the mid-plane of the nebula, increasing in
average size as they do so. The thin dust layer then becomes
gravitationally unstable and is able to break up into planetesimals with sizes on the order of a few km (4, 5). By collision
and accretion these small objects then build up into planetary-
size objects of a few earth masses. The time scale for the completion of this process at the position in the solar nebula where the
giant planets formed is long and uncertain. One must ask whether
appreciable gas remained in the nebula after that time so that
the formation of the giant planets could proceed. It is of course
possible that the build-up of the rocky core and the gaseous
envelope proceeded simultaneously once a certain minimum core
mass had been obtained. In any case the hypothesis suggests that

nebular gas accretes onto existing cores of a few earth masses.

The evolution of the gas then again falls into three distinct phases. While the core mass is below a critical value (to be discussed in section 3) the envelope can exist in hydrostatic equilibrium and an analog to the contraction phase (1) will occur. When the core mass exceeds the critical value, the envelope cannot be in hydrostatic equilibrium and it collapses onto the core (phase 2). When collapse and further accretion of gas are complete, phase (3) sets in. Here the evolution must be similar to that obtained under the first hypothesis, involving contraction and cooling on a long time scale with a convective structure. Although calculations to determine the critical core mass have been carried out (section 3), detailed hydrodynamic calculations of the accretion of gas onto the rocky core do not exist.

It is difficult to distinguish on the basis of observational data between the two major hypotheses just discussed, since the only observations available are those for a late stage of planetary evolution (the present). A standard procedure has been to match (non-evolving) theoretical models of Jupiter and Saturn with the observations which have been recently refined by the measurements made by the Pioneer and Voyager spacecraft. A recent summary by Grossman et al. (6) lists the following observed properties that can be used in connection with the theoretical calculations. For Jupiter, the mass is 1.9×10^{30}g (about 320 earth masses), the mean radius is 7.04×10^4 km, the internal luminosity (exclusive of solar absorption) is 8×10^{-10} solar luminosities ($L_\odot$), the surface temperature (including the solar contribution) is 123 K, and the temperature at 1 bar pressure in the atmosphere is 160 K. For Saturn the corresponding quantities are 5.69×10^{29}g (95 earth masses), 5.78×10^4 km, 3.6×10^{-10} $L_\odot$, 97 K, and 140 K. These quantities are compared with models that are assumed to be in hydrostatic equilibrium and to have an adiabatic temperature gradient in the interior due to energy transport by convection. The static models are generally assumed to be in uniform rotation, in which case further comparisons are possible with the observed rotation period and the observed gravitational moments. The calculations are carried out with a wide variety of assumptions regarding the equation of state of the interior and the chemical composition of the gas (typical values are hydrogen mass fraction $X = .74$, helium mass fraction $Y = .24$).

It is generally found that dense cores composed of materials such as Fe, SiO_2, MgO, and Ni are required so that the models will fit the observed gravitational moments and radius. The derived core masses for Saturn are 20-25 earth masses (7) or 15-17 earth masses (8) while for Jupiter the values are 16-18 earth masses (38) or 14-16 earth masses (8). It is still not clear which of the two fundamental hypotheses can best account for the existence of the cores. Evolutionary calculations for the planets can, however, provide an additional constraint on the fit to the observations at the present time, namely, the models must reach the present planetary state at an age of about 4.5×10^9 years, the inferred age of the giant planets. A more detailed account of the evolution under the two basic hypotheses and of the fit to observations will be provided in the following sections.

2. EVOLUTION OF GIANT GASEOUS PROTOPLANETS

2.1 Initial Conditions

The starting point for the development of a protoplanet is the primitive solar nebula, at a stage when the Sun is just beginning to form at the center, so that its mass is much smaller than the nebular mass. Cameron (3) suggests that the nebula is unstable to the formation of rings and that it breaks up early in its history; the rings subsequently evolve to form protoplanets. Other modes of formation are also possible; in any case the requirement for the existence of a self-gravitating subcondensation in the nebula is that its gravitational energy ($-GM^2/R$, where M is the mass and R the radius) be larger in absolute value than the thermal energy (1.5 Rg MT$/\mu$, where T is the mean temperature, Rg is the gas constant, and μ is the mean atomic weight per free particle). Rotation and magnetic fields are neglected in this estimate, but they are not expected to be of great importance. In the outer regions of the solar nebula the temperature is thought to have been about 50 K; thus if we take a Jovian-mass protoplanet ($M_J = 2 \times 10^{30}$g) of solar composition ($\mu = 2$ since the hydrogen is in molecular form), we obtain a maximum radius of about 5000 R_J, where R_J is Jupiter's present radius, and a corresponding mean density ρ of about 10^{-11} g cm^{-3}. If the entire nebula had this mean density and a radius of 50 AU, its mass would be approximately one solar mass.

The nebula will exert a tidal force on the forming protoplanet; however, if the nebula were of roughly uniform density, the tidal radius outside of which the tidal effects would be important is a factor of 10 larger than the maximum radius given above. Tidal effects would therefore not influence the protoplanet. If the Sun were already present as a condensed object, the tidal radius would be a factor of 8 smaller than the maximum radius, and the developing condensation would be disrupted unless its density were a factor of 500 larger than the value given above (in this limit $R_{tidal} = D(m/3M_\odot)^{1/3}$ where D is the distance to the Sun and m is the planetary mass). If we assume, however, that the protoplanets formed at the time when the nebula was still relatively homogeneous and that the nebula evolved in time according to the model of Cameron (3), forming a central Sun in the process, then calculations show that the contracting protoplanet always remains within or near its tidal radius (9). This result is a consequence of the fact, to be further discussed below, that the evolution time of the nebula, and therefore the time scale for shrinking of a protoplanet's tidal radius, is comparable to the contraction time of the protoplanet itself.

2.2 Phase (1): Physics

Most calculations of the evolution of protoplanets have been carried out under the assumption of spherical symmetry, although a rough estimate of the effects of rotation was made by Bodenheimer (10). The protoplanet is assumed to be a homogeneous mixture of gas (X = .74, Y = .24) with dust (presumably interstellar grains with mass fraction .01). During the early contraction phase (1) the physics is relatively simple since the ideal-gas equation of state applies and the object comes into hydrostatic equilibrium if its internal temperatures are above 70 K. The object is quite optically thick and local thermodynamic equilibrium can be assumed, so that radiative energy transport can be calculated according to the simple diffusion approximation (11), which is based on the assumption that the mean free path of a photon is short compared to the distance over which there is a significant temperature change. If the protoplanet is unstable to convection, it is adequate to assume that the temperature gradient is adiabatic.

The magnitude of the radiative opacity, which determines where convection occurs, is the most uncertain part of the physics during phase (1). At the low temperatures and densities expected, the principal source of opacity is the dust grains. At the lowest temperatures water ice condensed onto the grains is the major contributor; the opacity rises approximately as T^2 up to a maximum of about 5 $cm^2 g^{-1}$ at T = 160 K, at which point the ice evaporates. Above that point the following grains have been included (12): magnetite, metallic iron, and hydrated or non-hydrated silicates with or without iron. The opacity displays a complicated behavior but remains between 1 and 10 cm^2/g up to T = 1700 K, at which point the last constituent evaporates and the opacity drops sharply to about 10^{-2} cm^2/g. The principal opacity source at T = 2000 K is water vapor with small contributions from other molecules.

2.3 Phase (1): Evolution in the Isolated Case

The first set of calculations to be described assumes that the protoplanet is isolated – that is, there is no influence on it from the surrounding solar nebula. The surface pressure P can be assumed to be zero, and the object can be assumed to radiate as a black body with $L = 4 \pi R^2 \sigma T_e^4$ where L is the luminosity (erg s^{-1}), σ the Stefan-Boltzman constant, and T_e the temperature of the radiating surface. Starting from the density

$\rho = 10^{-11} g \ cm^{-3}$ where the protoplanet becomes gravitationally bound, the object contracts in quasi-hydrostatic equilibrium on the Kelvin-Helmholtz timescale $t_k = GM^2/(RL)$. Recent calculations have been carried out for four different masses (1.5 M_J, 1.0 M_J, 0.42 M_J, and 0.285 M_J) by Bodenheimer et al.(12). The evolution in the (ρ_c, T_c) plane, where the subscript refers to central values, is shown in Figure 1.

Since the evolution during phase (1) for all four masses is similar, we describe here the case of 1 M_J. The time variation of significant physical quantities is shown in Figure 2. The effective temperature increases slowly but stays on the order of 30 K, which means that the object radiates as a very cool infrared source. The luminosity decreases rapidly at first, then levels off at a value slightly less than 10^{-6} solar luminosities, while the radius contracts from 4000 to 150 R_J. The structure is initially convective except for a thin (< 0.005 M_J) surface radiative zone. When the interior temperatures exceed 170 K the opacity drops due to the evaporation of water ice; however, the drop is insufficient for convective stability to be regained. When T_c reaches 1700 K the opacity drops again, this time by a factor of more than 100, because of the evaporation of the mineral grains. A radiative core then forms which grows in mass as the central regions heat. By the time T_c has increased to 2100 K this core includes about one-third of the mass. The evolution time to this point, which represents the end of phase (1), is about 4×10^5 yr. By way of comparison, a model of Saturn's mass (0.285 M_J), starting at $R = 1640 \ R_s$ (where R_s is Saturn's present radius), contracts in 4.6×10^6 years to $R = 57 \ R_s$ at which point $T_{eff} = 32$ K and log $L/L_\odot = -7.7$. As in the case of Jupiter the structure is convective until grain evaporation results in a radiative core; however, the luminosity is much lower and the evolution time much longer.

A further important point that was discussed by DeCampli and Cameron (9) is connected with the path of evolution of the central regions of the protoplanets in the (ρ, T) diagram (see Figure 1). When the radiative core forms at temperatures above 1400 K, conditions there are such that the evaporated grains can in fact liquify, coalesce, and precipitate toward the center to form a rocky core, as long as the mass of the protoplanet is 1 M_J or less. Since the mass of the radiative core is only a fraction of the total mass of the protoplanet, and since the mass fracture in precipitable grains is less than .01, one would not expect much more than one Earth mass in the core of a Jovian-mass protoplanet. However, DeCampli and Cameron (9) showed that the convective motions in the outer layers could circulate material from the entire protoplanet into the liquid zone and that there would be plenty of time for the circulation to bring all the grains in the protoplanet into the liquid zone during its lifetime and to

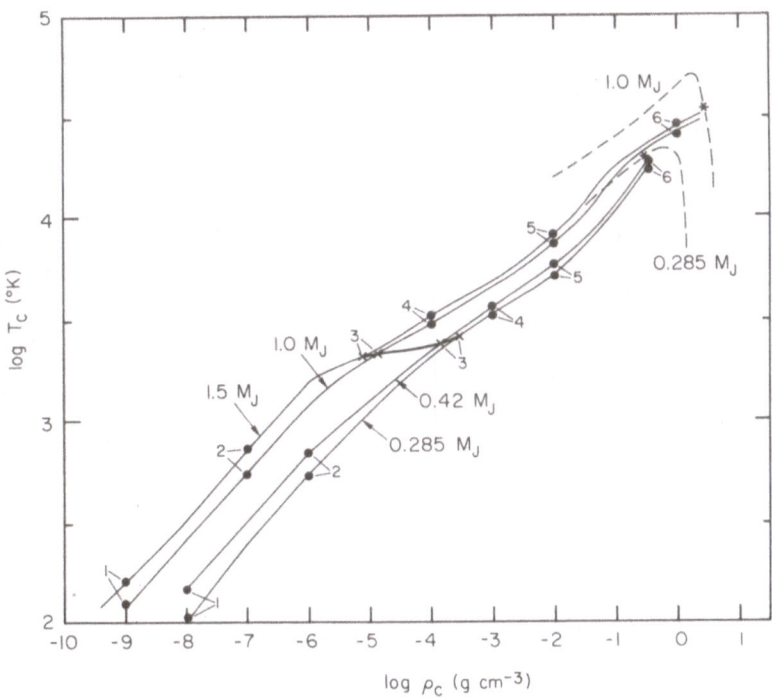

Figure 1. The evolution of the centers of four protoplanets (lightly solid lines) with masses as marked, in the temperature-density diagram. The heavy solid line indicates the onset of dissociation of H_2. The dashed lines show the evolution during the final contraction phase for Jupiter (13) and for Saturn (14). The asterisks indicate the initial models for the final phase as derived from evolutionary calculations of the prior phases (12). The numbered points correspond to the following evolutionary times. From Bodenheimer et al. (12).

POINT	$1.5 \; M_J$	$1.0 \; M_J$	$0.42 \; M_J$	$0.285 \; M_J$
	TIME FROM	INITIAL MODEL (YR)		
1	8.1 (2)	8.6 (2)	5.9 (3)	6.4 (3)
2	2.5 (4)	3.0 (4)	2.0 (5)	2.3 (5)
3	2.1 (5)	4.3 (5)	2.0 (6)	4.6 (6)
	TIME FROM	DISSOCIATION POINT (YR)		
4	0.370	0.210	0.110	0.046
5	0.397	0.242	0.118	0.063
6	0.404	0.253	0.124	0.074

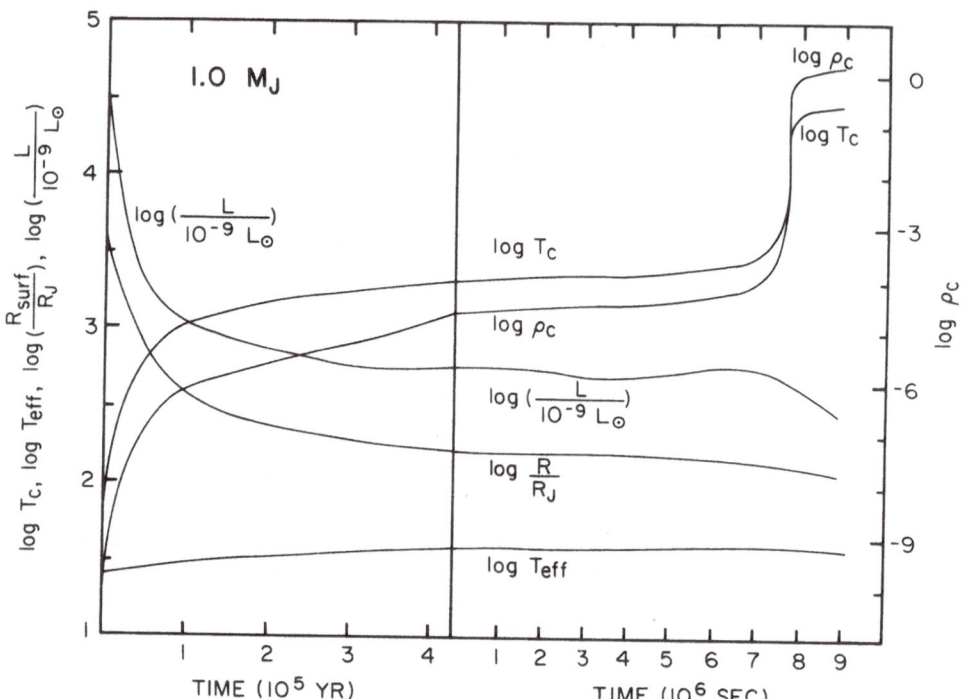

Figure 2. The evolution of central temperature T_c, central density ρ_c, total radius R_{surf}, surface luminosity L, and surface temperature T_{eff} for a gaseous protoplanet of 1 Jovian mass. The left-hand portion represents the contraction phase (1) with time measured from the beginning of the calculation. The right-hand portion represents the hydrodynamic phase (2) with time measured from the onset of collapse. From Bodenheimer et al. (12).

allow them to precipitate. Of course the convection also circulates droplets out of the liquid zone, but if they are carried upward they solidify, retaining their size, and on subsequent passes through the liquid zone they have another chance to precipitate. The details of the core-formation process are complicated, but it is entirely reasonable that a core of the order of an Earth mass could be formed in a giant protoplanet. The size of this core could be supplemented by capture of small stray planetesimals from the solar nebula during the course of the evolution (1), or by mass exchange with the solar nebula which could feed in more grains. Of course during the precipitation process the protoplanet envelope eventually loses all of its grains, and therefore the opacity is sharply reduced. The effect upon the evolution has so far not been calculated, nor has the heat generated by the settling process been included.

2.4 Phase (1): Evolution of Embedded Protoplanets

More recently, calculations have been made that take into account the fact that protoplanets are embedded in a relatively dense solar nebula (15). Two different surface boundary conditions have been employed: (1) a "thermal bath" whose temperature varies with time according to the evolution of the nebula but whose density is set to a small constant value, and (2) a surface temperature and pressure which both vary with time according to the evolutionary calculations of Cameron (3). In all other respects the calculations are carried out with the same physics and assumptions used for isolated protoplanets.

The nebular evolution (3) has two main phases. At the early stages the Sun has a very small mass, and the nebula continues to grow in mass as material rains down upon it from the surrounding protostellar cloud. As a result the temperatures and pressures at the midplane of the nebula increase with time. This phase lasts about 5×10^4 years, after which infall stops, and the evolution is dominated by viscous dissipation in the nebula which results in mass transfer inwards to the Sun and angular momentum transfer outwards. Evaporation of material from the surface of the nebula also occurs. The mass, temperatures, and pressures in the nebula decrease on a time scale of 10^5 years. The initial conditions for the nebula require specification of one parameter known as the nebular adiabat: Cameron has considered two cases known as the "high" adiabat and the "low" adiabat (for details see (15)). For example, the maximum temperature in the nebula at Jupiter's formation point is 166 K for the "high" adiabat and 42 K for the "low" adiabat.

The "thermal bath" boundary condition is appropriate for the case when a "zone of avoidance" is created around the protoplanet. Lin and Papaloizou (16) and Cameron (17) have shown that under certain conditions the protoplanet causes tidal trasnport of angular momentum in the nebular disk and resulting truncation of the disk, so that the density of material in the neighborhood of the protoplanet is reduced. Thus the surface of the protoplanet is influenced by the radiation temperature of the nebula but not by its pressure. In the standard "thermal bath" calculation the surface temperature was increased at the rate of 10^{-3} degrees per year. The protoplanetary radius for the case of 1 M_J decreased to a minimum of 960 R_J after about 3 x 10^4 years, then it began to expand. The central density reached a maximum of 5 x 10^{-8} g cm^{-3}, then began to decrease. After a time of 1.4 x 10^5 years it was clear that the protoplanet would be totally evaporated. After 7 x 10^4 years in fact the luminosity was negative, meaning that energy was being transferred from the warmer nebula into the cooler outer layers of the protoplanet. Since energy is deposited into the protoplanet from the nebula faster than it is being radiated away, the object eventually becomes gravitationally unbound. Basically, this effect occurs because the time scale for significant temperature increase in the nebula is shorter than the protoplanetary contraction time. The schematic time-development of T_c for this case is compared with that for the isolated case in Figure 3.

How is a protoplanet to survive? Two other calculations were carried out in which the temperature variation with time was taken from Cameron's "low adiabat" nebula. Thus, instead of continuously rising, the temperature rose to a maximum, then fell. In the first case the radius decreased to 3600 R_J, then expanded and was dispersed after only a few thousand years. In the second case the protoplanet was placed at Jupiter's present distance. Since the maximum nebular temperature was only 42 K, the protoplanet was able to survive. It contracted continuously and followed a similar evolutionary path to that of the isolated protoplanet, except that the total time for contraction was somewhat longer. Thus it appears that a protoplanet evaporates if it is formed closer to the Sun than the Jupiter position; otherwise, it survives. This conclusion depends, protoplanet evaporates if it is formed closer to the Sun than the Jupiter position, otherwise, it survives. This conclusion depends, of course, on the parameters chosen for the solar nebula, such as its initial adiabat and its evolutionary time scale.

If a zone of avoidance does not appear, the "thermal bath" boundary condition does not apply, and both temperature and pressure variations at the boundary must be considered. Two cases are considered here, the "high" adiabat and the "low" adiabat nebulae, both at the Jupiter position. The evolution is compared with that of the isolated protoplanet in Figure 3. The "high" adiabat nebula corresponds to a situation where the entropy in the nebula is higher than that in the protoplanet, and for a while the flow of energy goes from nebula to protoplanet. The initial contraction results in a radius of 10^3 R_J at about 6 x

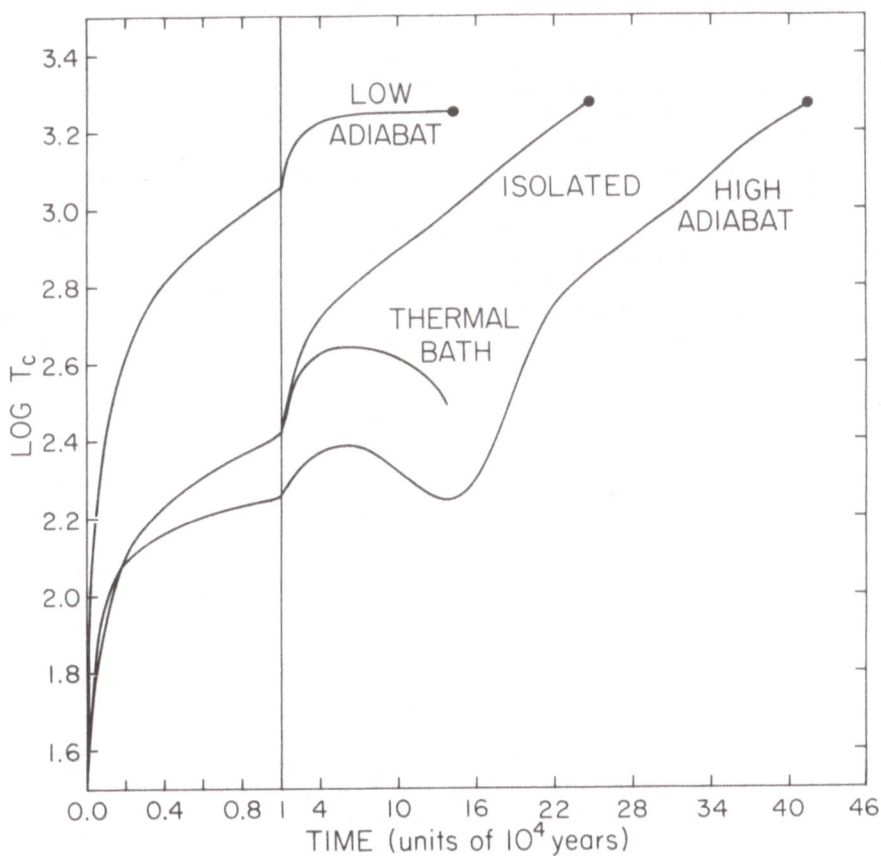

Figure 3. Comparison of the evolution of the central temperatures of protoplanets of 1 Jovian mass with four different boundary conditions: that for an isolated protoplanet, that for a thermal bath rising at 10^{-3} degrees per year, and those for pressure and temperature variations corresponding to the evolution of Cameron's (3) model solar nebulae characterized by a "high" and "low" adiabat, respectively.

10^4 years. At about the same time the maximum boundary temperature of 165 K is attained, corresponding to a maximum central
temperature of 240 K. Expansion takes place until 1.6 x 10^5
years, at which point R = 1660 R_J. During the previous phases,
when the nebular temperature was high, the evaporation was prevented by the high surface pressures, the expansion is caused
primarily by the release of this pressure. When contraction re-
sumes at 1.6 x 10^5 years, the structure and evolution quickly
relax to those found in the isolated case, since the influence of
the nebula has become negligible. The main overall effect of the
boundary condition is to delay the evolution through the supply
of additional thermal energy to the protoplanet, as evidenced by
the short expansion phase. The total time to reach T_c = 1600 K,
just before the radiative core forms, is twice as long as in the
isolated case.

If instead the "low adiabat" boundary condition is taken,
the temperature is lower for a given pressure and the thermal
energy from the nebula never has much influence on the protoplanet. The situation results in a higher entropy in the surface
layers of the protoplanet than in the surrounding nebula. The
dominant effect comes from the high surface pressure, which results in heating of the outer layers and more rapid radiation of
energy from the protoplanet. Thus the contraction is accelerated,
and the protoplanet contracts continuously, reaching T_c = 1500 K
in about half the time taken by the isolated protoplanet (see
Figure 3). The internal structure is very similar to that in the
isolated case. It is clear from these results that the evolution
of protoplanets is closely connected with that of the nebula itself. Further calculations should include additional protoplanet-
ary-nebular interactions, such as accretion of matter onto the
protoplanets and stripping of the outer layers by tidal effects.

2.5 Phase (2): Collapse

When the central temperature of the protoplanet reaches
2000 K an instability sets in. The hydrogen molecules start to
dissociate, and a large fraction of the released gravitational
energy goes into supplying the dissociation energy rather than
into increasing the temperature. As the protostar contracts, the
force of gravity increases more rapidly than the opposing pressure gradient, and hydrostatic equilibrium is no longer possible.
The evolutionary time scale is suddenly reduced to the order of
days as gravitational collapse begins in the central regions and
follows shortly in the outer layers. For 1 M_J the collapse starts
when ρ_c = 2 x 10^{-5} g cm^{-3} and proceeds on a time scale of 0.25
years, at which point the increasing stiffness in the equation of
state, caused by completion of dissociation and non-ideal-gas effects, causes a halt to the collapse. The collapse is practically
adiabatic, since the hydrodynamical collapse time becomes short

compared to the time for radiation to diffuse outward from the
collapsing region. The long diffusion time is due primarily to
the increase in opacity above 2000 K, caused by the increasing
importance of molecules, the negative hydrogen ion, and hydrogen
itself. The collapse stops when $T_c = 2 \times 10^4$ K and $\rho_c = 0.2$ g
cm^{-3}, an increase of four orders of magnitude in density. Hydrogen is partially ionized at this point, but the properties of
the non-ideal equation of state give $\Gamma_1 = [d(\ln p)/d(\ln p)]_s >$
4/3 so that the material is stable against further collapse.

The small hydrostatic core that forms at the center quickly
grows in mass. Soon the infalling outer material develops supersonic velocities relative to the near-equilibrium core, and an
accretion shock forms at the core's outer edge when the core includes about 40% of the protoplanet mass. This shock is initially
located at a radius of about 3 R_J. The calculations (12) have
been carried to the point where more than 95% of the mass has
been accreted onto the core, at which time $\rho_c = 1.8$ g cm^{-3} and
$T_c = 2.9 \times 10^4$ K. A hydrostatic model has been constructed using
the density distribution in the core at this time; this model
represents the starting point for the phase (3) evolution. Properties of this model are: $\rho_c = 2.73$ g cm^{-3}, $T_c = 34700$ K, log
$L/L_\odot = -5.65$, $T_{eff} = 600$ K, and R = 1.3 R_J. Thus the collapse
takes the Jovian protoplanet down to a radius comparable to the
present value. A similar evolution occurs for a Saturn-mass protoplanet. The collapse starts at $\rho_c = 3.7 \times 10^{-4}$ g cm^{-3} and $T_c =$
2500 K. The hydrostatic core forms after 0.06 years with an in-
itial $\rho_c = 0.2$ g cm^{-3} and $T_c = 1.5 \times 10^4$ K. The accretion phase
lasts about 0.01 years, and the resulting hydrostatic model has
$T_c = 18,000$ K, $\rho_c = 0.28$ g cm^{-3}, log $(L/L_\odot) = -4.78$, and R =
2×10^{10} cm or 3.4 times the present radius of Saturn. The evolution of the central regions during phase (2) is illustrated in
Figure 1, and the time variation of various important quantities
for the case of 1 M_J is shown in the right-hand portion of Figure 2.

The structure of models of Jupiter and Saturn near the end
of phase (2) is illustrated in Figure 4. The diagram also shows
the various regions of equation-of-state physics that must be
considered in the evolution of the giant planets. It is evident
that complications set in during the transition to phase (3). In
the outer regions of the models, with $\rho < 10^{-3}$ g cm^{-3}, the ideal
gas law applies; however, the excitation of rotational levels of
molecular hydrogen must be considered in the calculation of the
internal energy of the gas above T = 80 K, the vibrational levels
must be considered above T = 1500 K, molecular dissociation must
be considered in the temperature range 2000-7000 K, and hydrogen
ionization must be considered for $T > 10^4$ K. In the non-ideal
regime at higher densities the equation of state becomes extremely complicated and must be determined by detailed numerical cal-

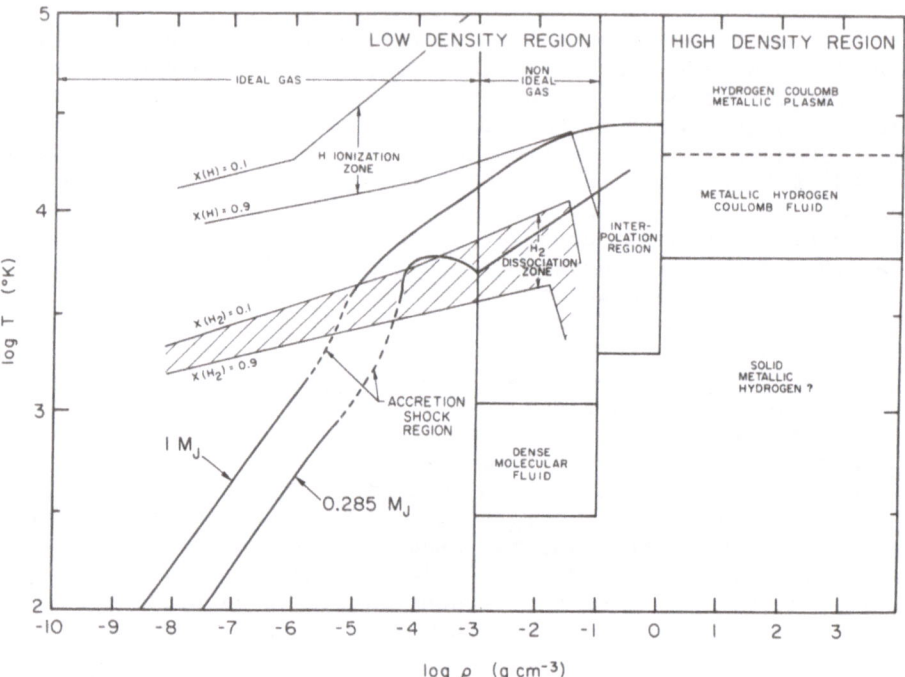

Figure 4. Internal structure (solid lines) of homogeneous
models for Jupiter and Saturn near the end of the hydrodynamic
phase (2). Important equation-of-state regions are indicated
in the ρ,T plane. Interior to the accretion shock region
(dashed lines) the models are approaching hydrostatic
equilibrium. Outside that region a small fraction of the
material is still collapsing. As the models evolve to the
present state, the central regions become cooler and somewhat denser. From Bodenheimer et al. (12).

culations. For the late phases of evolution of the giant planets
the equation of state calculated by Graboske et al. (18) has been
used, as modified by recent improvements summarized by Grossman
et al. (6).

The non-ideal regime may be divided into three density ranges:
the low-density region where $\rho < 10^{-1}$ g cm^{-3}, the high-density region where $\rho > 1$ g cm^{-3}, and the intermediate region. In the low-
density region the equation of state is calculated according to a
Helmholtz free energy minimization technique, taking into account
coulomb corrections, pressure ionization and dissociation, and excluded volume corrections. Three temperature subdivisions can be
considered. From 60 K to 2500 K the hydrogen is in a molecular-
fluid state, from 4000 K to 6000 K the hydrogen is in the dissociation regime, and for temperatures greater than $2 \times 10^4 - 10^5$ K,
depending on density, the hydrogen is ionized. The effects of pressure dissociation and pressure ionization become important for densities in the range 10^{-3} to 10^{-1} g cm^{-3} (see Figure 4). In the
high-density region the equation of state is determined by a modified Thomas-Fermi model (see (18)) with coulomb corrections for
the ions and additive volume effects included. For T < 3000 hydrogen is in a solid metallic phase, while for 3000 < T < 20,000
K it is in a metallic fluid coulomb phase. In the intermediate
region no theoretical model is available, and the equation of
state is determined by interpolation between the high- and low-
density regions. Laboratory data on pressures obtained in shock
waves under similar conditions are used to match the theoretical
determinations. Note that the interior regions of Jupiter and
Saturn models at the end of phase (2) fall within the somewhat
uncertain region of the equation of state.

2.6 Phase (3): Contraction and Cooling

The final evolutionary phase for gaseous giant protoplanets
begins for Jupiter when $R = 1.3$ R_J and $L = 10^{-6}$ $L_\odot$, while for
Saturn the corresponding quantities are $R = 3.4$ R_S and $L = 10^{-5}L_\odot$.
The positions of the centers of the objects at this time in the
(ρ_c, T_c) diagram are shown as asterisks in Figure 1. During
phase (3) the protoplanets are in hydrostatic equilibrium and,
due to the very high opacity to radiation in the interior, they
are convectively unstable. The convective regions may be assumed
to have an adiabatic temperature gradient; thus the main complica-
tion in construction of models, apart from the equation of state,
lies in the specification of the surface boundary condition. The
outer, low-density regions near optical depth unity are radiative,
and thus the energy loss from the planet is determined by the
opacity of these layers. In practice, detailed non-gray model
atmospheres are fitted to the convective interiors so that an accurate boundary condition can be obtained. The principal opacity
source is pressure-induced transitions of molecular hydrogen (19).

In the interior, the equation of state is basically given by the
calculations of Graboske et al. (18), whose main features were
described in the preceding paragraph.

As the protoplanet evolves, the loss of energy from the surface must be supplied by gravitational energy released by contraction. If the object is composed of an ideal gas, it can be
shown by use of the virial theorem that half of the released
gravitational energy goes into internal heating, the remainder
is radiated away. For planetary radii about five times the present
values for Jupiter and Saturn, this situation would hold, and one
would expect an increase of internal temperature. However at
smaller radii the gas quickly becomes more and more non-ideal and
les and less compressible; eventually gravitational contraction
becomes too slow to make up for the loss of energy at the surface.
At that point the interior begins to cool down in order to supply
radiated energy at the expense of internal energy (which of
course was produced by previously liberated gravitational energy).
At the present stage the interiors of the giant planets are practically incompressible, that is, the objects are close to the
minimum radius for their mass. They are in the cooling phase,
and one would expect that there was a maximum in T_c sometime in
the past.

Calculations for the non-rotating, homogeneous evolution of
Jupiter and Saturn were carried out by Graboske et al. (13) and
Pollack et al. (14), respectively. Since they did not consider
the previous evolutionary phases, they started at an arbitrary
planetary radius roughly ten times the present value. The contraction was followed for 4.5×10^9 years. At the outset the
luminosities and effective temperatures were far higher than
present values; both decreased with time. The contraction was
rapid for the first million years. Saturn reached a central temperature maximum of 21,300 K at 1.2×10^6 years; Jupiter reached
51,400 K at 1.2×10^5 years. Thereafter cooling set in and the
evolution time lengthened considerably. These equilibrium calculations for Jupiter and Saturn are illustrated in the (ρ_c, T_c)
diagram in Figure 1. It is clear from that diagram, however,
that the earlier portions of those evolutionary tracks are never
reached if one considers the phase (1) and phase (2) evolution.
The collapse calculations (12) show that Saturn does not come
into hydrostatic equilibrium until near the point of temperature
maximum in the center, while Jupiter comes into equilibrium
when it is already in the cooling portion of its evolution. The
actual maximum temperature in Jupiter was probably only 35,000 K.

The fit to the present observed properties of the planets
is not affected by the slight discrepancy just discussed, since
the evolution of a fully convective structure is practically independent of initial conditions. The results for Jupiter (13)

show that at an age of 4.6×10^9 years (the approximate age of
the solar system) the theoretical luminosity agrees with the
measured excess radiation (solar effects subtracted) to within
the observational error. The calculated and observed radii agree
to within 0.1 percent. The results for Saturn (14), however, did
not agree as well. At an age of 4.6×10^9 years the calculated
luminosity is a factor of 3 less than the observed value, while
the theoretical radius is about 9 percent too large. The calculated age at the point where the calculated and observed luminosities agreed was only 2×10^9 years. In both of these calculations,
a complete fit to all the observations was not made, however,
since the models were not rotating and therefore could not be
compared with the observed gravitational moments. More recently
the evolutionary calculations of phase (3) were repeated with
the inclusion of improvements in the equation of state, rotation,
and dense rocky cores (6) and the agreement with observations
reconsidered. The results of this investigation will be considered in the next section. For the moment let us conclude the
discussion of homogeneously evolving protoplanets by remarking
that the evolutionary calculations are clearly able to account
for the present excess luminosity of Jupiter in terms of gravitational energy. At the present epoch only 25% of the energy
released can be accounted for by the current rate of contraction; the remaining 75% comes from loss of internal energy,
which, however was generated by more rapid contraction in the
past. In the case of Saturn the situation is not so clear,
since a good model fit has so far not been obtained. A possible
additional energy source, which seems to be required, is
discussed in the next section.

3. THE EFFECT OF DENSE CORES ON THE EVOLUTION OF GIANT PLANETS

Extensive calculations have been made for the evolution of
giant planets under the assumption of chemical homogeneity, as
discussed in the previous section. However, it is fairly clear
from comparison of theoretical models with observations of
Jupiter and Saturn that the inclusion of a rocky core with 6% of
the mass for Jupiter or 20% for Saturn brings about the best
agreement. While it is possible that the core could have formed
by precipitation or accretion during the evolution of the planet,
it is also quite reasonable to assume that the core formed first
and the gaseous envelope accumulated later. Here we consider this
hypothesis in more detail. Calculations of phases (1) and (2) of
the evolution under this hypothesis have not been nearly as detailed as those for the homogeneous case, and it is not entirely
clear that the early hydrostatic phase (1) exists at all for the
gas. Detailed evolutionary models have been constructed for the
contraction-cooling phase (3). We therefore discuss the evolution

in two parts, first considering the early evolution up to the point where the entire object comes finally into hydrostatic equilibrium, then discussing the final contraction to the present state. Many of the important physical processes are the same as those already discussed in the previous section.

3.1 Giant Planets with Cores: Early History

Following Hayashi et al. (20) we may divide the formation of a giant planet into four stages. First, a rotating disk-like nebula of about 0.05 $M_\odot$ forms around the Sun. Dust grains grow by collision and weak electrical attraction to sizes of about a centimeter. The dust then settles toward the central plane of the nebula, forming a relatively dense thin layer there. A recent estimate (21) indicates that this process takes 10^4 years at the Jupiter position in the model of the solar nebula obtained by Kusaka et al. (2). Second, when the dust layer becomes dense enough it undergoes a gravitational instability and fragments into a number of planetesimals of approximately kilometer size (5, 4, 22). Third, collisions between these planetesimals result in the building of more massive objects (5), and later the more massive objects grow by capture of some of the remaining planetesimals (20). This accretion process has been considered by a number of authors (for an interesting recent review see (23)) and will not be discussed in detail here. The time scale for build-up of Earth-mass objects is of course of considerable importance. Various estimates have been made for this time scale. Hayashi (24) quotes 10^6 years as the time required to build the earth up to its present mass and 10^7 years as the time to form a rocky-icy core of ten Earth masses for Jupiter. This calculation includes the effects of gas drag, collisions, and gravitational encounters. Safronov and Ruscol (25), using a more massive model of the solar nebula, give the time scale for accreting objects of about 2 Earth masses as 3×10^7 years at Jupiter's position and 2×10^8 years at Saturn's. The fourth stage of evolution then involves the capture of some of the nebular gas by the protoplanetary cores (26, 27, 28) and the eventual dissipation of the remainder of the nebula (e.g., 29).

Once the first three stages have been accomplished and the rocky core has been formed, it has a respectable tidal radius in the solar nebula and thus a reasonable amount of gas is gravitationally attached to it. We now consider the question of what happens to this gas. The most complete treatment is that given by Mizuno (28), who constructs spherically symmetric hydrostatic models of the gaseous envelope surrounding a rocky core of a given mass. The gas is considered to be ideal and both radiative and convective energy transport are considered. The tidal radius is calculated under the assumption that the Sun is already present at the center of the nebula. At that point the temperature and density of the nebula as calculated by Kusaka et. al.

(2) are applied as outer boundary conditions. These conditions vary as a function of distance to the Sun — for example, at Jupiter's distance $\rho = 1.5 \times 10^{-10}$ g cm^{-3} and T = 97 K. The luminosity of the envelope is supplied by the dissipation of energy of planetesimals, which are accreting onto the core at a given rate. Opacity due to molecular gas and grains is included.

Mizuno's results indicate that a typical model contains a nearly isothermal region near the photosphere, a radiative region, and then an inner convection zone. For relatively low values of the core mass, the envelope mass is roughly equal to the core mass. A critical value of the core mass is reached beyond which no envelope structure can exist in strict hydrostatic equilibrium. For standard values of the opacity this critical mass is about 10 Earth masses; for lower values of the opacity the critical mass is reduced. Of particular interest is the fact that the value of the critical mass does not depend on the position in the solar nebula. This finding may be compared with the deduced core masses of the four giant planets, all of which have very similar values of 15–20 Earth masses. For core masses above the critical value, Mizuno (28) concludes that the envelope must collapse onto the core and form a dense, relatively compact structure. The hydrodynamics of this collapse has not been calculated in detail, but the process may be considered to be analogous to the phase (2) collapse of a giant gaseous protoplanet. When the critical core mass is attained, the envelope mass is about the same as the core mass. Therefore, during and after the collapse additional material must be accreted onto the core in order to build the envelope up to the mass of Jupiter or Saturn. The rate at which this material can be accreted and some details of the process have been discussed by Safronov and Ruscol (25). The core can continue to grow by accretion of planetesimals at the same time. Future inclusion of hydrodynamic and accretion effects in time-dependent calculations is necessary to clarify this evolutionary picture, but for the moment we may assume that the end result of these effects is a Jupiter- or Saturn-mass protoplanet with a radius a few times the present value, in hydrostatic equilibrium, with a rocky core having 15 or 20 Earth masses in either case. The radius at this stage is somewhat uncertain, since rotational effects probably leave the outer parts of the object in a surrounding disk.

3.2 Giant Planets with Cores: Final Evolutionary Stages

Once the accretion process stops and the planet maintains a constant mass, the evolution through the phase (3) cooling-contraction is similar to that of a homogeneous gaseous planet. In the case of Jupiter and Saturn the evolution is dominated by the structure of the envelope, which remains fully convective with a rate of energy loss determined by the opacity in the thin

surface radiative layer. The luminosity and surface temperature
decrease on a time scale of 5×10^9 years. In fitting the numerical models to the observations it is assumed that the early
accretion phases contribute only a small fraction of the total
evolutionary time to the present state.

Grossman et al. (6) have presented evolutionary sequences
for Jupiter and Saturn with the inclusion of a slow rotation and
with cores of various masses. The cores consist of SiO_2, MgO,
Fe, and Ni in solar abundance ratios; the envelopes are a homogeneous gas of solar composition. The equation of state is also
somewhat modified over earlier calculations; because of this and
rotational effects the fit to observed parameters discussed earlier (section 2.6) is no longer obtained for models without a
rocky core. For Jupiter, it was found that the best fit to observed values of radius, luminosity, temperature at 1 bar pressure, and the gravitational moments J_2 and J_4 was obtained with
envelope composition X = .77, Y = .21 with a core mass of 6.5%
(19 Earth masses) at an age of 4.9×10^9 years. For Saturn, as
in the homogeneous case discussed above, the results were not as
good. The radius, luminosity, and temperature at 1 bar pressure
can be fitted with the same chemical composition as for Jupiter
and with a core mass of 21% (also 19 Earth masses), but the evolution time to this point is only 2.6×10^9 years, a factor of two
too short. Also, the values deduced for J_2 and J_4 with this core
mass are well below the observed values. The authors suggest that
the short evolutionary time can be corrected if the limited solubility of helium in metallic hydrogen is considered (30). At
temperatures below about 8000 K this effect results in precipitation of a substantial fraction of the helium and the accompanying
release of additional gravitational energy. (In Jupiter the solubility effect is very small because of the higher internal temperatures.) This effect is unlikely to resolve the discrepancy in
the gravitational moments. We note, however, that Hubbard et al.
(31) have obtained better agreement with Saturn's observed moments,
without considering evolutionary effects, using a slightly different equation of state.

To conclude this section we summarize the present state of
Jupiter and Saturn (data from Grossman et al. (6)). Jupiter has
a rock core of approximately 20 Earth masses whose central density is about 34 g cm^{-3} and whose central temperature is 27,000 K.
At the boundary of the core in the solar-mix envelope the density
is about 4.3 g cm^{-3} and the temperature about 16,000 K. The inner
part of the envelope (about 70% of the total mass) is in the metallic hydrogen state, while most of the remainder is in the fluid
molecular hydrogen state. No first-order phase transition occurs
between the observed atmosphere and the molecular-metallic boundary, whose properties are somewhat uncertain. In the case of
Saturn, the core also is deduced to have about 20 Earth masses

with a central density of 28 g cm^{-3} and central temperature of
25,000 K. At the core-envelope interface the gas density is about
2 g cm^{-3} and the temperature about 10,000 K. The metallic hydrogen inner layer of the envelope is smaller in mass than in the
case of Jupiter, including only 20% of the mass; the remainder
is primarily fluid molecular hydrogen. The atmospheric temperatures (about 123 K and 97 K for Jupiter and Saturn, respectively)
are sufficiently different to significantly affect the observed
spectrum. The source of the internal luminosity of Jupiter is apparently mainly the cooling of the interior, while in the case of
Saturn the settling of the helium toward the center due to limited
solubility may provide an additional important contribution. The
fact that the concentration of helium in the surface layers of
Saturn ($Y = 0.11 \pm .04$) appears to be less than that in Jupiter
($Y = 0.19 \pm .05$), according to the infrared observations from
Voyager 1 (32), is consistent with this hypothesis.

4. CONCLUSION: NOTE ON URANUS AND NEPTUNE

Although the two outermost giant planets have not been studied nearly as extensively as the inner two, it is clear that their
observed properties and theoretical models of their structure
could provide additional important clues to the answer to the basic question posed in this paper – the nature of the formation
mechanism for the giant planets. Two unusual observational facts
present themselves. First, the observed masses and radii of Uranus
and Neptune, when combined with theoretical mass-radius relations
of cold objects (33) indicate that a substantial fraction of mass
must be in elements heavier than hydrogen and helium. Second,
Uranus has no measurable excess luminosity (the only giant planet
to lack it) while Neptune radiates about twice the energy received
from the Sun.

These characteristics can in general be accounted for by
theoretical models. Hubbard and MacFarlane (34) have calculated
three-layer models consisting of central rocky cores, mantles of
water, methane, and ammonia (known as "ices"), and helium-hydro-
gen envelopes. In the case of Uranus, a model that matches observations has 11% of the mass in the envelope, 65% in the "ice" layer, and 24% in the core. For Neptune, the corresponding quantities
are 7%, 68%, and 25%. The abundance ratio of rocky material to
"icy" material is therefore similar to that in the Sun. Podolak
and Reynolds (35) consider two-layer models consisting of the
rocky core plus hydrogen and helium mixed with the "ices" in the
envelope. In general, the models that fit the observations have
an "ice"-to-rock ratio that is less than solar. In this connection
it is also suggested from observations (36) that Uranus has a
deficiency of nitrogen in its atmosphere. Turning now to the
luminosities of the two planets, Hubbard (37) has calculated the

evolution of Uranus and Neptune by assuming that the hydrogen-
helium layer and the "ice" layer (probably actually liquid) are
cooling, convective, adiabatic structures. A cooling time, at
constant mass, of 4.6×10^9 years gives surface temperatures consistent with the observations. Although the two planets are in-
trinsically very similar, the apparent discrepancy in their luminosities can be accounted for by the difference in the amounts of
sunlight they receive, with the intensity being 2.4 times less at
Neptune than at Uranus. The evolutionary calculations are consistent with the suggestion (37) that solar heating effects dominate in the atmosphere of Uranus (but not in Neptune) and sup-
press or make unmeasurably small the luminosity due to internal
cooling. The calculations also indicate that the two planets never passed through a high-luminosity stage analogous to that of
Jupiter and Saturn at the onset of phase (3), but that the luminosity in the past was never more than a factor of two higher than
it is now.

What do the characteristics of Uranus and Neptune tell us
about their possible origin? Podolak (36) suggests that the accretion theory is preferred. Thus the rock and that part of the
"ice" that was actually frozen at the temperature in the outer
part of the solar nebula accreted to form a solid core. When the
critical core mass was reached, nebular gas accreted to form the
envelope. One difficulty is that so little gas was accreted by
the outer giant planets as compared with the inner ones. It is
possible that the gas density was lower at the positions of Uranus
and Neptune or that the core accretion times were sufficiently
long that little gas was left by the time the critical mass was
attained. An even greater difficulty is that current theories
predict accretion times for Uranus and Neptune that are far too
long to be of interest. However, the accretion theory can account
for the deduced compositions obtained from the models, and the
apparent lack of nitrogen in Uranus can be accounted for by non-
equilibrium chemical processes in the solar nebula during the accretion phase. The evolutionary calculations are also consistent
with (but do not prove) this mode of origin. On the other hand,
if the gravitational condensation theory applies, one has the difficulty of removing the major portion of the (presumed large) gaseous envelope by a process that would not do the same to Jupiter
and Saturn. No entirely satisfactory process has been proposed
so far. Various compositional constraints (such as the low nitrogen abundance) are discussed by Podolak (36) who shows that it is
possible for these to be satisfied by models based on the giant
gaseous protoplanet theory, but this theory is by no means required to explain them. It is clear that we have taken only the
first steps toward clarification of the origin of the giant planets, but we can still tentatively summarize the situation as follows: (1) the terrestrial planets were formed by the accretion
process, since giant protoplanets formed in the inner solar system

are subject to evaporation and tidal effects. (2) Jupiter and Saturn are likely to have been formed as giant gaseous proto-planets; however, the accretion process followed by gas capture and collapse is also plausible, subject to the difficulty of long core accretion times, particularly for Saturn. The challenge now is to find a significant theoretical or observational test that can distinguish between these two possibilities. (3) Uranus and Neptune were formed by the accretion-gas-capture process; however, in this case both suggested formation processes run into serious difficulties.

ACKNOWLEDGMENT

The author's research on cosmogony was supported in part by NSF grant AST 79-21263.

REFERENCES

(1) Pollack, J.B., Burns, J.A., and Tauber, M.E.: 1979, Icarus 37, pp. 587-611.
(2) Kusaka, T., Nakano, T., and Hayashi, C.: 1970, Progr. Theor. Phys. 44, pp. 1580-1595.
(3) Cameron, A.G.W.: 1978, Moon and Planets 18, pp. 5-40.
(4) Goldreich, P., and Ward, W.R.: 1973, Astrophys. J. 183, pp. 1051-1061.
(5) Safronov, V.S.: 1969, Evolution of Protoplanetary Cloud and Formation of the Earth and Planets, NASA TTF-677.
(6) Grossman, A.S., Pollack, J.B., Reynolds, R.T., Summers, A.S., and Graboske, H.C., Jr.: 1980, Icarus 42, pp. 358-379.
(7) Podolak, M.: 1978, Icarus 33, pp. 342-348.
(8) Slattery, W.L.: 1977, Icarus 32, pp. 58-72.
(9) DeCampli, W.M., and Cameron, A.G.W.: 1979, Icarus 38, pp. 367-391.
(10) Bodenheimer, P.: 1977, Icarus 31, pp. 356-368.
(11) Schwarzschild, M.: 1958, Structure and Evolution of the Stars (Princeton University Press), pp. 37-42.
(12) Bodenheimer, P., Grossman, A.S., DeCampli, W.M., Marcy, G., and Pollack, J.B.: 1980, Icarus 41, pp. 293-308.
(13) Graboske, H.C., Jr., Pollack, J.B., Grossman, A.S., and Olness, R.J.: 1975, Astrophys. J. 199, pp. 265-281.
(14) Pollack, J.B., Grossman, A.S., Moore, R., and Graboske, H.C., Jr.: 1977, Icarus 30, pp. 111-128.
(15) Cameron, A.G.W., DeCampli, W.M., and Bodenheimer, P.: 1981, preprint.
(16) Lin, D.N.C., and Papaloizou, J.: 1979, Mon. Not. R. Astron. Soc. 186, pp. 799-812.
(17) Cameron, A.G.W.: 1979, Moon and Planets 21, pp. 173-183.
(18) Graboske, H.C., Jr., Olness, R.J., and Grossman, A.S.: 1975,

Astrophys. J. 199, pp. 255-264.

(19) Pollack, J.B., and Ohring, G.: 1973, Icarus 19, pp. 34-42.

(20) Hayashi, C., Nakazawa, K., and Adachi, I.: 1977, Publ. Astron. Soc. Japan 29, pp. 163-196.

(21) Nakagawa, Y., Nakazawa, K., and Hayashi, C.: 1981, Icarus 45, pp. 517-528.

(22) Coradini, A., Federico, C., and Magni, G.: 1981, Astron. Astrophys. 98, pp. 173-185.

(23) Wetherill, G.W.: 1981, Scientific American 244, no. 6, pp. 162-174.

(24) Hayashi, C.: 1981, in: Fundamental Problems in the Theory of Stellar Evolution, ed. D. Sugimoto et al. (Dordrecht: Reidel).

(25) Safronov, V.S., and Ruscol, E.L.: 1981, preprint.

(26) Perri, F., and Cameron, A.G.W.: 1974, Icarus 22, pp. 416-425.

(27) Mizuno, H., Nakazawa, K., and Hayashi, C.: 1978, Progr. Theor. Phys. 60, pp. 699-710.

(28) Mizuno, H.: 1980, Progr. Theor. Phys. 64, pp. 544-557.

(29) Elmegreen, B.G.: 1978, Moon and Planets 19, pp. 261-277.

(30) Stevenson, D.J., and Salpeter, E.E.: 1977, Astrophys. J. Suppl. 35, pp. 239-261.

(31) Hubbard, W.B., MacFarlane, J.J., Anderson, J.D., Null, G.W., and Biller, E.D.: 1980, J. Geophys. Res. 85, pp. 5909-5916.

(32) Hanel, R., et al.: 1981, Science 212, pp. 192-200.

(33) Zapolsky, H.S., and Salpeter, E.E.: 1969, Astrophys. J. 158, pp. 809-813.

(34) Hubbard, W.B., and MacFarlane, J.J.: 1980, J. Geophys. Res. 85, pp. 225-234.

(35) Podolak, M., and Reynolds, R.T.: 1981, Icarus 45, in press.

(36) Podolak, M.: 1981, in: IAU Colloquium No. 61, ed. G.E. Hunt (Cambridge University Press), in press.

(37) Hubbard, W.B.: 1978, Icarus 35, pp. 177-181.

(38) Podolak, M.: 1977, Icarus 30, pp. 155-162.

ORIGIN OF REGULAR SATELLITES

Stuart J. Weidenschilling

Planetary Science Institute, Tucson, Arizona, USA

ABSTRACT   The regular satellites of Jupiter and Saturn are generally believed to have accreted within cooling circumplanetary nebulae. Small silicate bodies are lost into the planet by gas drag before ice can condense. Larger silicate protosatellites survive by exerting tidal torques on the gas, clearing low-density "tunnels" around their orbits. The nebula is thus divided into series of gas rings depleted in silicates. Cooling eventually allows ice condensation, yielding another generation of icy bodies. Collisional accretion of these objects accounts for stochastic density variations of Saturn's inner satellites. High dynamic pressure may have prevented accretion in the inner part of the Jovian nebula; J5 may be an ablated remnant of a larger body.

INTRODUCTION

A planet with a retinue of satellites resembles the solar system in miniature. This concept dates from Galileo's discovery of Jupiter's large satellites, which he used as an argument in support of the Copernican system. The analogy persists in modern theories of cosmogony that assume that planets and satellites formed by similar processes. This assumption is plausible if the correspondence is not required to be exact. However, the two types of system differ in size and mass by several orders of magnitude. Scaling of complex processes over such a range may lead to significant differences in behavior. With this caveat, I shall address the formation of satellites within a disk-shaped circumplanetary nebula (CN), assumed to be a small-scale version of the solar nebula.

A. Coradini and M. Fulchignoni (eds.), The Comparative Study of the Planets, 49–59.

To preserve the planet-satellite analogy, I adopt a restrictive definition of regular satellites. They have prograde orbits of low inclination with respect to the primary's equator, and low eccentricity. Mass ratios much greater than Jupiter/Sun (i.e., Earth's and Pluto's satellites) are excluded. The small Martian satellites may be captured, rather than formed in situ, and are not considered here. Three planets then are defined as having regular satellite systems: Jupiter, Saturn, and Uranus. It may be significant that all are gas giants, and also possess rings.

Table I lists relevant properties of the principal regular satellites of Jupiter and Saturn. Density is the principal clue to bulk composition, and an important constraint on theories of their origin. It has long been known that the Galilean satellites have densities that decrease with increasing distance from the planet. This trend is consistent with compositions of rocky cores of roughly equal sizes, with the addition of different amounts of water ice. Saturn's satellite system is dominated by Titan, which contains some 97% of the mass, and appears to have nearly cosmic proportions of rock and ice. The small inner satellites appear to be depleted in silicates relative to ice. They show real differences in density, but no clear trend with orbital radius. Sizes and masses of the Uranian satellites are known too poorly for meaningful density estimates.

TABLE I: Principal Regular Satellites

	a/R	m(g)	m/M	r(km)	ρ(gcm^{-3})
Jupiter					
Rings	1.77				
5 Amalthea	2.55	$\sim 10^{22}$	$\sim 10^{-8}$	75×130	
1 Io	5.95	8.9×10^{25}	4.7×10^{-5}	1815	3.53
2 Europa	9.47	4.9×10^{25}	2.6×10^{-5}	1565	3.03
3 Ganymede	15.1	1.5×10^{26}	7.8×10^{-5}	2640	1.93
4 Callisto	26.6	1.1×10^{26}	5.7×10^{-5}	2410	1.79
Saturn					
Rings	1.2-2.5	$<10^{24}$	$<10^{-6}$		
1 Mimas	3.1	4×10^{22}	6.6×10^{-8}	195	1.2
2 Enceladus	4.0	8×10^{22}	1.3×10^{-7}	250	1.1
3 Tethys	4.9	6.4×10^{23}	1.1×10^{-6}	525	1.0
4 Dione	6.3	1.1×10^{24}	1.9×10^{-6}	560	1.4
5 Rhea	8.8	2.3×10^{24}	9×10^{-6}	765	1.2
6 Titan	20.2	1.4×10^{26}	2.5×10^{-4}	2560	1.9

THE CIRCUMPLANETARY NEBULA

Current theories of the formation of Jupiter and Saturn (1)

involve hydrodynamic collapse of a giant gaseous protoplanet (2)
or accretion of gas onto a massive solid core (3). Either process
could plausibly leave a disk-shaped CN, in which satellites could
form (4,5). Both produce a planet which is initially hot and
distended. Its high luminosity results in a strong radial temperature gradient in the nebula by which Pollack and Reynolds (6)
explained the compositions of the Galilean satellites. They assumed that silicates could condense in the nebula, but that the
initial temperature was too high for ice to condense. As Jupiter
cooled, the limit of ice stability moved inward, barely reaching
Europa's zone when the nebula dissipated. Jovian thermal histories (7) imply this took $\sim 10^6$ years. Pollack et al. (8) performed a similar calculation for Saturn. The smaller planet
cooled faster, but evidently its nebula lasted long enough for
ice to condense near the planet. Again, the implied lifetime of
the nebula is $\sim 10^6$ years.

A crude estimate of the mass and structure of a CN can be
obtained by adding the cosmic complement of H and He to the heavy
element contents of the satellites, and spreading this material
in zones around their present orbits. The assumptions and limitations of such a reconstruction are discussed by Weidenschilling
(9). For the solar nebula, this method provides a firm lower mass
limit. The properties of a CN are more speculative, as it need
not have had cosmic composition. Its material could have been
depleted in heavy elements by prior formation of the protoplanet's
core, or augmented by capturing planetesimals from heliocentric
orbits. Still, this approach provides a zero-order estimate which
may suggest differences between heliocentric and circumplanetary
environments.

Results of such a calculation are shown in Figure 1. The
Jovian system shows a trend of surface density, σ, approximately
as a^{-2}, where a is the distance from the planet, in the region of
the Galilean satellites. It is plausible that the original distribution of σ was monotonic with a. In that case, the sharp drop
inside Io's orbit indicates loss of condensable matter in that
zone. Saturn's inner satellites show a remarkably similar pattern,
with a peak in σ at Tethy's orbit. However, Titan, with most of
the system's mass, is a strong exception. If the initial distribution of σ was monotonic through Titan's zone, either a very
massive nebula ($\sim 1/3$ of Saturn's mass) experienced extreme loss
of condensable matter (both silicates and ice) from its inner region, or that σ increased with a out to Titan's distance. Neither
possibility can be ruled out. Still, we shall proceed with the
assumption that at least at some stage of its evolution Saturn's
inner CN had the surface density shown in the trend from Tethys
to Rhea.

Properties of the model Jovian CN are given in Table II. The

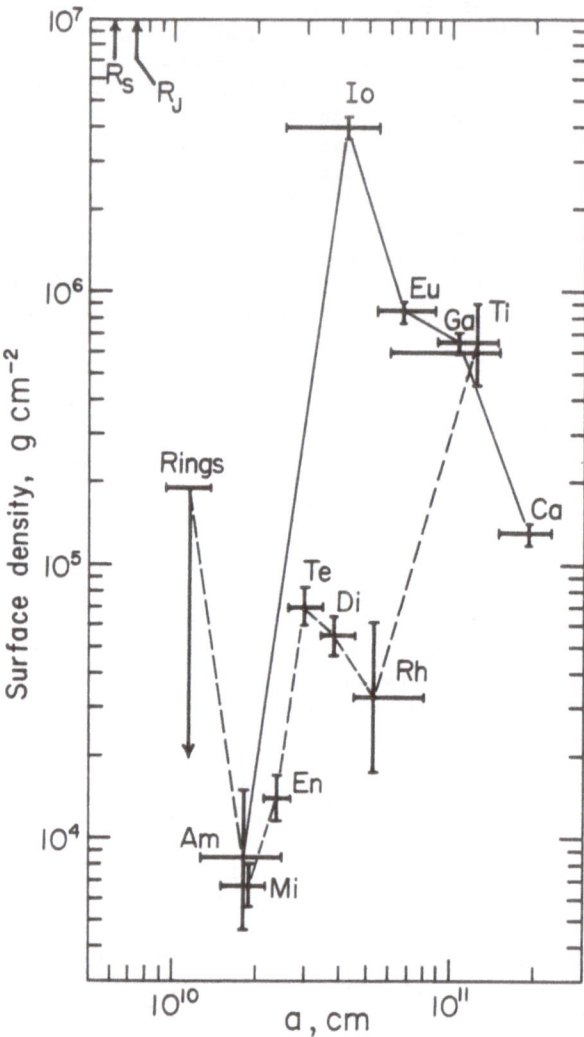

Figure 1. Equivalent surface densities obtained by adding H and He in solar proportion to the heavy element contents of the satellites of Jupiter (solid line) and Saturn (dashed line), and spreading this mass in contiguous zones around their present orbits. Saturn's rings represent an upper limit only.

surface density is ~10^3 times that in the low-mass models of the solar nebula. The pressure in the central plane is simply related to σ by:

$$P_c = \sigma \Omega \bar{v}/4 \ , \tag{1}$$

where Ω is the local Keplerian angular frequency, and $\bar{v}$ is the main thermal velocity in the gas (10). In Io's zone, the pressure is several bars. Plausible temperatures would allow liquid H_2O or H_2O-NH_3 solution to be stable in a substantial region of the nebula.

TABLE II: Jovian Nebula Model

Mass = 0.05 M_J Surface density $\sigma \propto a^{-2}$
Temperature $T \propto a^{-1}$ Pressure $P \propto a^{-4}$

Conditions in Io's Zone (a = 6 R_J)
   σ(gas) = 3×10^6 g cm^{-2} σ_s(solids) = 1×10^4 g cm^{-2}
   T = 530°K P_c = 6 bars $\rho = 3 \times 10^{-4}$ g cm^{-3}
   V_K - V(gas) = 2.3×10^4 cm s^{-1}

GAS-SOLID INTERACTIONS

The high gas density, ρ, in the Jovian CN results in a mean free path of gas molecules ~10^{-3} - 10^{-5} cm. Even small dust particles are in the continuum flow regime, subject to Stokes drag. They settle toward the central plane, reaching it on a timescale of ~$(1/s^2)$ yr, where s is the particle radius in cm. This contrasts with the solar nebula, where the mean free path is typically ~1 - 10^2 cm; the free-molecular regime yields settling in ~$(10^3/s)$ yr. If fine dust is present in the CN, its slow settling allows high opacity and a steep temperature gradient to persist for the lifetime of the nebula; hence, I assume $T \propto 1/a$. The Stokes settling time is independent of gas density, so this may also hold for Saturn's CN, though the longer mean free path makes the choice of appropriate drag law somewhat uncertain.

The $1/s^2$ dependence of settling time means that particles larger than 10^{-1} cm reach the central plane of the CN more rapidly than in the solar nebula. If particles coagulate when they collide, the settling time is extremely short. Large particles grow by sweeping up smaller ones due to different settling velocities. This produces a "runaway" of the largest ones. Weidenschilling (11) has described a numerical model for simultaneous settling and coagulation in the solar nebula. Application to conditions in the Jovian CN shows that if solids are initially present solely as micron-sized dust particles, meter-sized bodies accrete and reach the central plane in only a few days. The settling is non-homologous, i.e., there is a sharp concentration of mass in the central plane, while small particles which escape being swept up remain suspended. The dense layer is gravitationally unstable, and breaks up into bodies of mass

$$m \sim 16\pi^4 G^2 \sigma_s^3 / \Omega^4 \qquad (2)$$

where σ_s is the surface density of solid matter in the layer, and G the gravitational constant (12). In the Jovian CN, $M \sim 10^{18}$ g ($\sigma^3 \propto \Omega^4$, so m does not depend on position in this model).

The subsequent evolution of these bodies is controlled by the nebular gas. From Eq. (1) we see that in general the pressure is not uniform, and the Jovian nebula model has $P \propto a^{-4}$. The pressure gradient gives the gas a radial acceleration equal to $(-1/\rho) \partial P / \partial a$ (13). Since $\partial P / \partial a < 0$ in the model CN, the gas is supported against the planet's gravity, and its rotation is slower than Keplerian. The deviation from the Kepler velocity is

$$\Delta V = V_k - V_{gas} = -(\Delta g / 2g) V_k \qquad (3)$$

where $g = GM/a^2$ is the gravity of the central body. We define

$$\Delta g = (1/\rho)(\partial P / \partial a) = -nRT/\mu a \qquad (4)$$

where $P \propto a^{-n}$, R is the gas constant, T the temperature, and μ the molecular weight of the gas (14). In the model CN, $\Delta V \simeq 10^4$ cm s^{-1}.

Solid bodies are not supported by the pressure gradient. They are perturbed by drag as they move with respect to the gas. The effects of drag are described in detail in refs. (14) and (15). Briefly, small bodies are carried with the gas at its angular velocity, and draft radially as seen in a frame moving with the gas. Large bodies maintain nearly Keplerian orbits which are perturbed by a tangential drag force (a "headwind" when $V_{gas} < V_k$). The maximum radial velocity, equal to $-\Delta V$, is reached at some intermediate size which depends on the densities of the gas and of the solid body. In the Jovian model CN, this critical size is ~1 km. For a centrally condensed nebula, the direction of motion is inward, and solids may be lost into the planet. However, note that the direction of motion depends on the sign of $\partial P / \partial a$. In general, solids move toward regions of higher pressure, and can move outward if the pressure gradient is reversed.

The high density and small size of the CN compared to the solar nebula causes lifetimes against loss by drag to be very short. Above the critical size, larger bodies have lower radial velocities. Protosatellites can slow their orbital decay by accretion, but this alone does not prevent their loss. Table III

lists relevant timescales for the Galilean satellites. Gravitational accretion times are for their silicate cores, using
Safronov's (10) expression. This may be an overestimate, as gas
drag tends to flatten the system of solid bodies, and drag-induced
differential motions increase the collision rate. In any case,
accretion times are shorter than orbital decay times (computed
assuming silicate density of 3 g cm^{-3} and a drag coefficient of
0.4). Ice condensation times are estimated from Jupiter thermal
history models (6), revised downward to account for the higher
pressures in the CN. We see that a simple model does not work.
If a Jovian CN had the structure implied by the mass distribution
of the satellite system, the inner Galilean satellites would have
been lost, or at least had their orbits drastically altered, long
before ice condensed in the outermost zone. Shorter condensation
times are not consistent with Jovian thermal history. If instead
we assume that the gas abundance was low, i.e., the silicate/gas
ratio was much greater than solar, there is no source of H_2O for
the outer satellites. It appears necessary to alter the structure
of the nebula. Fortunately, this occurs naturally as a result of
tidal interactions of protosatellites with the gas.

TIDAL INTERACTIONS AND SATELLITE FORMATION

As a satellite orbits within the nebula, its gravity deflects
the motion of nearby gas. The gas inside its orbit loses angular
momentum, while that on the outside gains (a similar phenomenon
results in confinement of Saturn's F ring by "shepherding" satellites). The tidal torques push the gas away from the satellite's
orbit (16). This is opposed by viscosity, which redistributes
angular momentum (17). If the satellite has enough mass, the
tidal torque dominates, clearing a "tunnel" of low density. This
reduces or eliminates orbital decay due to drag.

Coradini et al. (18) estimate the mass required for tunneling
by equating characteristic timescales for tidal and viscous transport of angular momentum. However, it is possible to equate local
torques near the satellite's orbit. The expression for tidal
torque on a thin disk (16) must be modified when the nebula's
half-thickness H may exceed the tunnel's width. To keep the
tunnel open, the tidal torque at a distance H from the central
plane must equal the viscous torque. This gives:

$$\alpha \nu \Omega \sim 4G^2m^2/9\pi\Omega^2x^2H^2 \tag{5}$$

Here α is a dimensionless constant of order unity which depends
on the nebular structure, ν is the kinematic viscosity, and x the
half-width of the tunnel, assumed to be the Hill radius, $x =
a(m/M)^{1/3}$, where M is the planet's mass. As turbulence would
decay rapidly after the nebula formed (18), the molecular viscosity

TABLE III: Timescales (yr) in Jovian Satellite Zones

Satellite	Accretion	Orbital Decay	H_2O Condensation
1	2×10^2	7×10^2	$1\text{-}3 \times 10^6$
2	1×10^3	3×10^3	$.3\text{-}2 \times 10^6$
3	5×10^3	2×10^4	$1.5\text{-}9 \times 10^5$
4	4×10^4	1×10^5	$.4\text{-}4 \times 10^5$

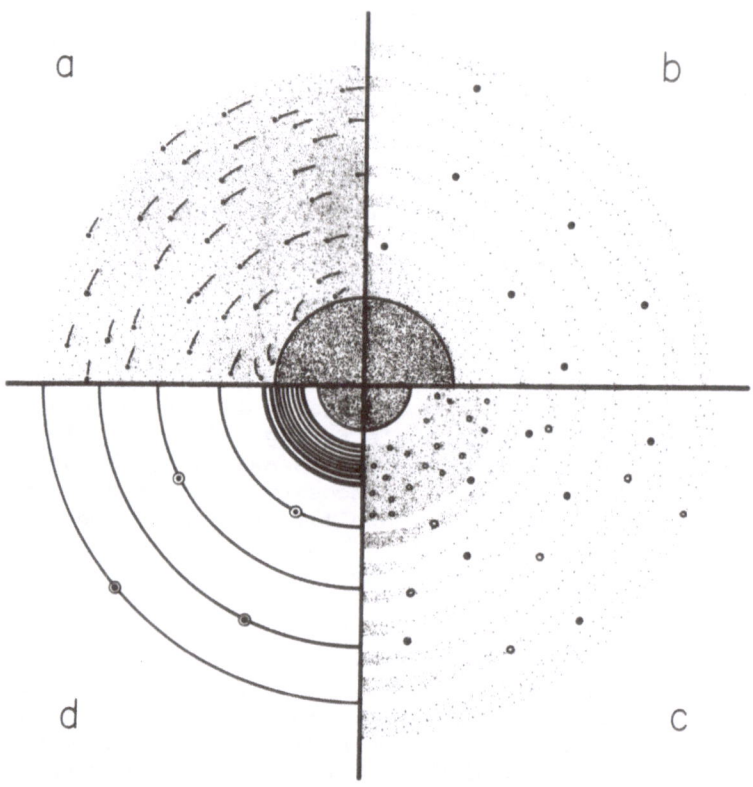

Figure 2. Stages of satellite formation, as described in text.
a) Silicate bodies accrete in warm nebula. b) Silicate protosatellites grow large enough for tidal torques to clear "tunnels"
in the gas. The rings of gas are depleted in silicates. c) The
protoplanet cools, allowing ice to condense and accrete in the
gas rings. d) Stochastic accretion of the two populations yields
satellites with varied compositions.

is appropriate. Using $\nu = 1$ cm^2 s^{-1} and $H = 0.1a$, the critical mass in Io's zone is $\sim 10^{17}$ g. However, bodies in adjacent orbits must be spaced far enough so they do not interfere: $\Delta a \gtrsim 2x$. This implies

$$m \sim (4\pi\sigma_s a^2)^{3/2} M^{-1/2} \tag{6}$$

In the Jovian CN, $m \gtrsim 10^{24}$ g, independent of a. The lower value of σ_s in the Saturnian nebula gives $m \gtrsim 10^{21}$ g.

Once tunneling occurs, growth of the protosatellite slows, or ceases altogether. This happens because gas drag now tends to move smaller solid bodies away from the tunnel, toward regions of higher pressure (the pressure gradient is no longer monotonic). The critical masses to initiate tunneling appear to be one or two orders of magnitude less than those of the present regular satellites. This result implies a multi-stage scenario for their formation, illustrated schematically in Figure 2:

a) The circumplanetary nebula is initially too hot for ice to condense. Silicate bodies form by gravitational instability and/or collisional accretion. Gas drag causes their orbits to decay, and the innermost region is depleted of silicates.

b) The accreting silicate bodies grow large enough for tidal tunneling, and orbital decay ceases. The gas, confined in a series of rings between these protosatellites, is depleted in silicates.

c) The protoplanet cools and contracts. The temperature in the gas rings drops low enough for ice to condense. Ice particles are driven to the centers of the rings by drag forces, where they accrete into another generation of icy protosatellites.

d) The closely spaced orbits of these bodies are unstable, allowing close encounters and collisions. They accrete into larger bodies of mixed composition -- the present satellites. At about this time, the remaining gas dissipates. The innermost silicate-depleted region may yield a ring system composed mostly of water ice, if its temperature is low enough.

This scenario may explain a puzzling feature of Saturn's system. Data from Voyager I showed that the inner satellites, Mimas through Rhea, have densities, and hence silicate/ice ratios, which differ in seemingly random fashion. The Voyager Imaging Team (19) concluded:

"...we speculate that these icy bodies are so small that the stochastic character of accretion itself may be recorded in their bulk properties. If the planetesimals -- the lumps of matter accumulating in the final stages of accretion -- were heterogeneous in their bulk properties, then random factors would control the final makeup of small bodies. By chance,

Dione and Mimas may have gathered a few more rocky planetesimals than did Tethys and Rhea. If so, we have found a
class of objects in the solar system that reflect the
character of accretion itself."
Implicit in such a model is the need to make an earlier generation
of bodies, some rocky and others icy, which are not much smaller
than the final satellites. Tidal "tunneling" appears to do this.
It remains to be shown that the orbits of the first-generation
protosatellites would in fact be so closely spaced as to be unstable. Otherwise, we would expect strict alternation of rocky
and icy satellites. The problem is similar to that in models of
accretion of the terrestrial planets, where orbital linkage of
accreting bodies must be maintained until there are a few planets,
rather than many small ones. Wetherill (20) suggested that Jupiter's perturbations might have maintained linkage in that case.
For satellite systems, it may be necessary to invoke bombardment
by heliocentric planetesimals.

The Galilean satellites do not show obvious random variations
in rock/ice ratio, although Eq. (6) predicts that the protosatellites should have been large enough, compared with the final products, for compositional fluctuations to be significant. Apparently, the Jovian CN was never cool enough for ice to condense
in its inner region. Only two satellites contain large quantities
of ice; the difference in composition between Ganymede and Callisto may actually be stochastic.

The inner parts of the reconstructed CN's in Figure 1 show
a sharp drop in apparent surface density. Jupiter and Saturn
were several times larger than their present sizes immediately
after their formation. As they cooled and contracted, the inner
boundaries of their nebulae moved inward due to viscous redistribution of angular momentum (shown schematically in Figure 2).
The peaks in σ may mark the original planetary radii. However,
another effect may have been important in the Jovian nebula. Recall that a solid body on a Keplerian orbit feels a "headwind" of
velocity ΔV. As the gas density in Io's zone was comparable to
that of Earth's atmosphere, the dynamic pressure $P_d = \rho V^2/2$ could
be large. For bodies smaller than the original gravitational
instabilities, P_d exceeds that central pressure due to self-
compression, and could disrupt them outright. Larger objects
which are still too small for tunneling could have their surfaces
ablated. Amalthea may be an example of such an object. Its shape
is definitely not ellipsoidal, but resembles a wedge or axe head
in one profile (21). This unusual object may have been shaped by
aerodynamic forces.

The combination of aerodynamic drag and tidal torques explains,
at least qualitatively, some of the distinctive features of the
satellite systems of Jupiter and Saturn. The greatest stumbling-

block to this seemingly simple picture is Titan. It is much too
large, relative to Saturn's other satellites, to be either explained or ignored. The enigma of Titan is a humbling reminder
of our present state of ignorance.

ACKNOWLEDGMENTS

This research was supported by NASA Contract NASW-3214.
Attendance at the ASI was made possible by grants from NSF and
Consiglio Nazionale Richerche. The Planetary Science Institute
is a division of Science Applications, Inc. This is PSI Contribution No. 171.

REFERENCES

1) Cameron, A.G.W., and Pollack, J.: 1976, in "Jupiter" (T. Gehrels,
   Ed.), U. of Ariz. Press, pp. 61-84.
2) Bodenheimer, P.: 1974, Icarus 23, pp. 319-325.
3) Perri, F., and Cameron, A.G.W.: 1974, Icarus 22, pp. 416-425.
4) Bodenheimer, P.: 1977, Icarus 31, pp. 356-368.
5) Miki, S.: 1980, preprint.
6) Pollack, J., and Reynolds, R.: 1974, Icarus 21, pp. 248-253.
7) Graboske, H., Pollack, J., Grossman, A., and Olness, R.: 1975,
   Astrophys. J. 199, pp. 265-281.
8) Pollack, J., Grossman, A., Moore, R., and Graboske, H.: 1976,
   Icarus 29, pp. 35-48.
9) Weidenschilling, S.J.: 1977, Astrophys. Space Sci. 51, pp. 153-158.
10) Safronov, V.S.: 1972, NASA TTF-677.
11) Weidenschilling, S.J.: 1980, Icarus 44, pp. 172-189.
12) Goldreich, P., and Ward, W.R.: 1973, Astrophys. J. 183, pp. 1051-
    1061.
13) Whipple, F.L.: 1972, in "From Plasma to Planet" (A. Elvius,
    Ed.), Wiley, pp. 211-232.
14) Adachi, I., Hayashi, C., and Nakazawa, K.: 1976, Prog. Theoret.
    Phys. 56, pp. 1756-1771.
15) Weidenschilling, S.J.: 1977, Mon. Not. Roy. Astron. Soc. 180,
    pp. 57-70.
16) Lin, D.C., and Papaloizou, J.: 1979, Mon. Not. Roy. Astron.
    Soc. 186, pp. 799-812.
17) Lynden-Bell, D., and Pringle, J.: 1974, Mon. Not. Roy. Astron.
    Soc. 168, pp. 603-637.
18) Coradini, A., Federico, C., and Magni, G.: 1981, Astron.
    Astrophys. 99, pp. 255-261.
19) Smith, B., plus 26 others: 1981, Science 212, pp. 163-191.
20) Wetherill, G.W.: 1978, in "Protostars and Planets" (T. Gehrels,
    Ed.), U. of Ariz. Press, pp. 565-598.
21) Veverka, J., Thomas, P., Gradie, J., Morrison, D., and Davies,
    M.: 1980, NASA TM-81776, pp. 343-344.

GANYMEDE AND CALLISTO:ACCUMULATION HEAT CONTENT

A.Coradini,C.Federico (*) and P.Lanciano

I.A.S.Reparto di Planetologia,viale dell'Università 11,00181 Roma
(*)also at Istituto di Geologia dell'Università degli Studi di Perugia

ABSTRACT

Using the accumulation theory for the formation of the two outermost regular satellites of Jupiter,the accumulation heat content has been obtained.The early thermal profiles show the onset of convection just at the end of the formation process,having taken into account the energy irreversibly trapped and the projectile penetration dépth in the impacts.The fraction of the gravitational energy converted into thermal energy results to be larger for Callisto than for Ganymede.

1. INTRODUCTION

Accretion mechanism can be the source able to heat planetary material up to many hundrends of degrees thus justifying the initially hot phase of planetary objects. The initial thermal profiles are necessary also as initial conditions in quantitative studies of the subsequent thermal histories not only for terrestrial planets but also for objects like Ganymede and Callisto, whose initial accretion heat content may have been different.
To obtain the primordial thermal profile of a growing planetary object,we must know with some detail how the accretion process took place.In fact accretion by means of "large" colliding objects deposits most of the heat rather deep inside the planetary object,while,if accretion was mostly infall of centimeter sized objects

A. Coradini and M. Fulchignoni (eds.), The Comparative Study of the Planets, 61–70.
Copyright © 1982 by D. Reidel Publishing Company.

the energy can be deposited only on the surface and rapidly radiated away into space.Safronov(1,2)and Kaula (3,4)have already studied this problem in the framework of an accumulation theory,underlining that the initial thermal profile critically depends on the mass distribution function of the impacting bodies,whereas little depends on the accumulation time,if the heat is deeply buried.More recently Coradini et al.(5) have obtained the initial thermal profiles of Earth and Mars taking into account the energy balance in an impact,the projectile penetration depth and the onset of convection.Moreover they have assumed that the accumulation of the Earth and Mars can be modelled as suggested by(1).

Applying the same numerical simulation,we may now compute the accretional temperature profiles of Ganymede and Callisto.

2. THE MODEL

To determine quantitatively the primordial thermal profile of a growing planetary object,it is necessary to solve numerically the differential equation for heat transport,i.e.

$$\partial T/\partial t = (1/r^2)\partial/\partial r(r^2 k \partial T/\partial r) + E/(c_p\rho) \qquad (1)$$

where T is the temperature,r is the distance from the center of the object,c_p the specific heat at constant pressure,ρ the assumed constant internal density of the growing object,k is the thermal diffusivity and E = E(r,t) is the energy generation rate.Both E and k have different contributions,that depend on the physical model for the formation of the object.

The formation of the satellites is not yet clearly understood and for example Harris and Kaula(6),Harris (7) have shown that satellites may have been formed by accumulation of circumsolar planetesimals,captured in the satellite accretion disk during the final phases of the formation of the planet.On the contrary it has been shown(8) that gravitational instabilities in the satellite disk of Jupiter and Saturn give rise to "local" planetesimals whose subsequent evolution may lead to the formation of the regular satellites.It has also been argued that satellite accumulation cannot be entirely modelled as suggested by(1) for terrestrial planets because the total population of planetesimals is not accessible to the growing embryos at conclusion of their growth.So the neighboring feeding zones are not contiguous at the conclusion of the growth.

   To overcome this difficulty,we have tentatively assumed,following(9),that the adjacent feeding zones initially touch.Moreover each initial feeding zone is a flattened torus whose mass is equal to the present mass of the satellite:during the growth of the satellite embryo,the adjacent zones overlap owing to the increase of the relative velocities of the planetesimals. With these assumptions,the initial feeding zones of Ganymede and Callisto are 3.9 R_J and 6.8 R_J wide,R_J being the radius of Jupiter.Consequently if the final adjacent feeding zones overlap about their centers,we obtain a value of θ equal to 1.7 for both satellites.The initial width of the feeding zones is due to an initial relative velocity v_o among the planetesimals in the swarm.It must be remembered that $\theta = GM/v^2R$ is the Safronov number,v the relative velocity,M and R the mass and the radius of the largest planetesimal-the embryo-for a power law size distribution.

   During its growth,the satellite embryo is struck by planetesimals with an impact velocity given by

$$v_i = (v_e^2 + v^2)^{\frac{1}{2}} \tag{2}$$

where v_e is the escape velocity from the embryo and $v^2 = v_o^2 + GM/\theta R$.The mass distribution can be assumed as

$$n(m) \, dm = C \, m^{-q} \, dm \tag{3}$$

where $q=1.83$ and the value of C is fixed by the total mass in the feeding zone and by the mass of the second largest planetesimals m_{max}.According to(1),we have

$$m_{max} = M/(2\theta)^3 \tag{4}$$

The rate of growth of the embryo is given by

$$\dot{R} = (R_f^3/3R^2) \, \dot{f} \tag{5}$$

where R_f is the final radius of the satellite and $f = M/M_f$. The explicit expression of $\dot{f}$ is given in (9).To obtain $R = R(t)$ the equation (5) must be integrated numerically.Using this model,Ganymede reaches 99% of its present mass in a time interval of 2900 years while Callisto does so in 27000 years.

   In the framework of the above sketched theory,that implies accumulation by means of large bodies,the process of impact is strictly connected with the initial thermal history.Thus E in equation (1) requires to be

$$E = E_w + E_r \tag{6}$$

where E_w is the amount of heat per unit time trapped
in the impact process and E_r is the rate of energy pro
duced by the decay of the long lived (U,Th,K) radioactive isotopes.The concentration and the energy production rate for the four isotopes is assumed chondritic
and the values used are those given in(10).Moreover a
rise in temperature is caused by adiabatic compression
and the temperature profile can never lie under the
one due to this compression.

The diffusivity in equation (1) can be expressed
as

$$k = k_i + k_c + k_v \tag{7}$$

where k_i is the "eddy" diffusivity,connected with the
impact stirring(1,2),k_c is the thermal diffusivity of
materials and k_v is a pseudo diffusivity used to mimic
the heat transfer by convection(3).

The numerical solution of equation (1) has been
obtained using the algorithm given in(11).In our simulation we have taken into account the energy balance
in an impact,the projectile penetration depth and the
convection.In fact we have carefully examined the propagation of shock waves in different media,deducing
the amount of the energy irreversibly trapped E_w' as a
function of the distance r_c from the impact center.The
energy equipartition has been studied following(12);in
this approach the Hugoniot curve of the target material is taken as an approximation to its release adia-
bat.Energy expended in irreversibly heating the target
is subtracted from the total energy available to propagate the shock wave into the target.We have assumed
that the initial shock front is a sphere with its center in D,buried at depth d,proportional to the projectile radius r'.This geometry has been adopted in order to allow for projectile penetration depth during
the initial stage of the event.As an approximation of
the state equation in the rarefaction phase,an adiabat
release equation has been used in the form of the Mur-
nagham equation.Using these approximations, a relation
between the pressure P and the radial distance r_c from
the center D of the crater has been obtained.As a consequence,we have deduced the amount of energy irreversibly trapped in the target at a distance $r_c > r_0$ where
r_0 is the radius of the sphere where the pressure is
uniform and reaches its maximum value.For target and
projectile with the same chemical composition,it follows that $r_0 = r'$.It has been also possible to obtain
an expression for the energy irreversibly buried E_w' in
power law form,i.e.

$$E_W'(r_c) = E_W'(r_o)(r_c/r_o)^{-n} \qquad (8)$$

The value of n depends on the chemical composition of the projectile and target material and on v_i. Taking into account the mean density of Ganymede and Callisto, 1,930 and 1,790 kg/m^3 respectively, we have assumed as type of materials characteristic of the growing embryo and of the infalling planetesimals the ice saturated sand, permafrost. If v_i ranges from 2.5 to 4.5 km/s, n lies in the range 4.75 - 4.86.

The fraction of the impact kinetic energy E_i, irreversibly trapped in the target is

$$h = 1/E_i \{4/3 \ \pi r_o^3 \rho_o E_W'(r_o) + 4\pi\rho_o \int_{r_o}^{r_c^*} E_W'(r_c)r_c^2 dr_c\} \qquad (9)$$

where $r_c^* \approx 5r_o$ and corresponds to a pressure value for which the amount of trapped energy is negligible. From equation (9) it is now possible to obtain h, which depends on v_i and generally ranges from 0.10 to 0.20 using materials and velocity characteristic of the accumulation process of the Jovian satellites Ganymede and Callisto. For chemical composition of target and projectile similar to those used by(13), our h values are in agreement with those obtained by(13), who have used a different physical approach.

The geometry already described with respect to the crater center, can be transferred into radial symmetry with respect to the center of the growing object. Then, taking into account the projectile penetration depth, it is possible to evaluate the profile of the energy irreversibly trapped as a function of r. We must remark that in the Jovian satellite conditions the penetration depth ranges from 1 r' to 2.5 r'. Lastly integrating E' over the impacter mass spectrum, infalling in unit time on the satellite embryo, we obtain E_w. The readers interested in a fully detailed description of the above theory are referred to(5,14,15).

Coming back to equation (6), it may be pointed out that k_c can be assumed constant and equal to 2 10^{-6} m^2/s; in any case its contribution to heat transport is usually negligible. k_i has been evaluated following(1,2) and using the experimental results of(16) to establish the energy diameter relation for impact craters in ice saturated sand. k_v can be expressed through the Nusselt number, using a semiempirical relation between convected heat flux and temperature difference across a convective layer. Thus k_v can be represented as

$$k_v = Nu\ k_c = (Ra/Ra_c)^{1/3} k_c \qquad\qquad (10)$$

where Nu is the Nusselt number,defined as the ratio of
the convection heat flux to that which would be transported for the same ΔT by conduction alone.Ra is the
Rayleigh number which entirely governs the convective
onset and Ra_c is its critical value,which we have assumed equal to 2000.

　　Although there are no reliable data on the rheology
of the materials in the interiors of Ganymede and Callisto,the temperature dependence of the viscosity η
may be estimated by the relation

$$\eta = \eta_0 \exp\{a(T_m-T)/(T_m+T)\} \qquad\qquad (11)$$

where $\eta_0 = 10^{15}$ poise and a = 18 (17).The melting temperature T_m as a function of the radius r was computed
using the phase diagram of water and the formula for
pressure of the adiabatic compression of a homogeneous
body of constant density.

3.　　RESULTS AND CONCLUSIONS

　　The investigations of the photographs of the surface of Ganymede and Callisto,returned by the Voyager
spacecrafts,have shown that very different structures
are typical of these two satellites.In fact,for example
grooved terrains have been found exclusively on Ganyme-
de.Thurber et al.(18) and Schubert et al.(19) have sug
gested that these differences may reflect different de
grees of differentiation in the interiors of these two
objects,even if there are uncertainties on the principal heat sources,on the rheological behaviour of the
homogeneous mixture of ice and silicates and on the i-
nitial thermal state.Schubert et al.(19) in fact have
constructed simple parametric models of accretional
temperature profiles of the two outermost regular satellites of Jupiter using h,the portion of the kinetic
impact energy retained as heat,as a free parameter.
　　On the basis of our theory,summarized in the previous section,we can obtain the thermal profiles of Ganymede and Callisto during their accumulation and at
the end of the formation process,in order to assess if
melting and differentiation occurred also during the
formation stages:figures 1 and 2 show these profiles.
　　In the primordial phases,the energy buried by impacts is not sufficient to melt any fraction of the bo-
dies.The melting of the ice silicate mixture takes pla-

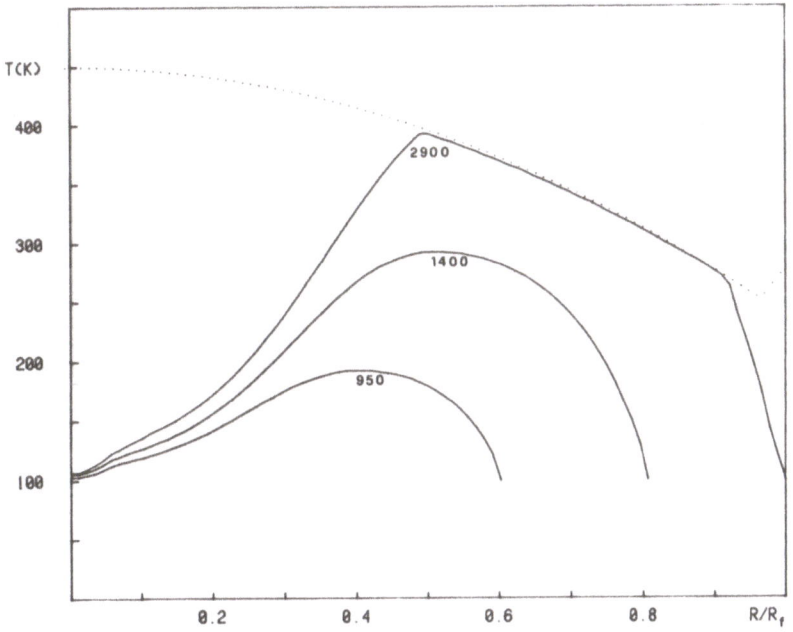

Figure 1. Early thermal profiles for Ganymede.On each curve is labelled the time in years.A large melted and convective shell appears only at the end of the accumulation.The dotted line is the melting curve for ice when the satellite reaches its present mass.

ce just at the end of the accumulation and due to onset of convection the profiles exactly lie on the melting curve.The radioactive energy release in the accumulation time interval ($\tau < 10^4$ years) being negligible, we can calculate along the temperature distribution $T = T(r)$ the conversion of gravitational into thermal energy. This fraction turns out to be 19% of the gravitational energy of Ganymede and 22% of that of Callisto. The convective layer is 960 km wide inside Ganymede and 660 km inside Callisto,being more deeply located into Callisto (depth $\simeq$ 360 km), the melted material amounts to 32% of the present mass in Ganymede and 24% in Callisto.

Comparing our results with those parametrically obtained by Schubert et al.(19), we see that there is some agreement for a value of h in the Schubert et al.(19) model of about 0.4, larger than those obtained by means

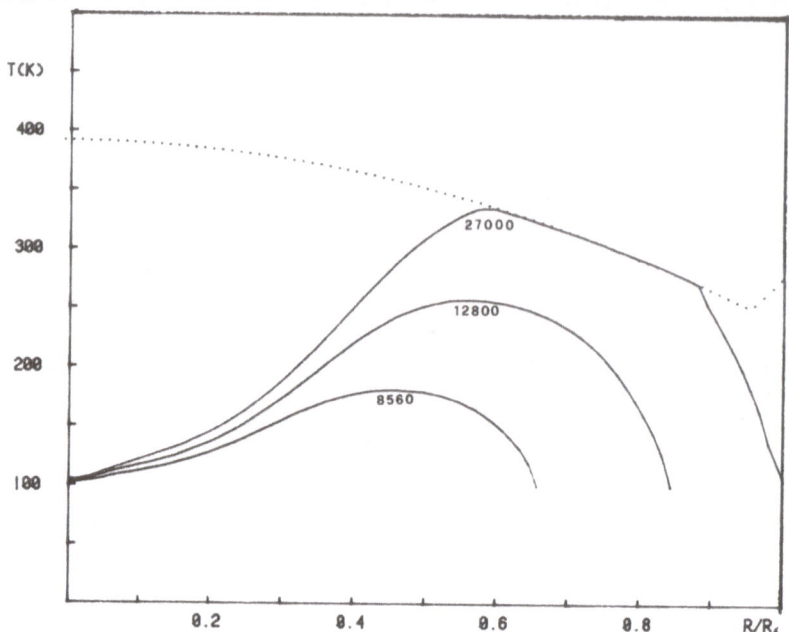

Figure 2. Early thermal profiles for Callisto.In
this case too, melting and convection appear only at the end of the satellite formation.On each
curve is labelled the time in years and the dotted line is the melting curve for ice.

of equation (9).The reasons for this apparent discrepancy is that in our model the impact velocities are
higher and the energy, more deeply buried,is efficiently transferred during the whole process of accumulation. Moreover, taking into account the heat transfer
in the interiors by means of conduction and convection
we conclude that ice melting could not occur in the
outermost layers,but only in the range $0.5\ R_f<r<\xi\ R_f$
where $\xi = 0.92$ for Ganymede and 0.85 for Callisto. As
a consequence at the beginning of their further thermal history , Ganymede was much more prone to internal
differentiation than Callisto, as already suggested by
(19). On the contrary, our initial thermal profiles
drastically differ by those assumed by (18) and this

difference depends on the impacter mass spectrum, on
the impact velocity and thus on the penetration depth
of the projectiles, rather than on how fast the formation process was. Comparing the thermal profiles of

Ganymede and Callisto we would also argue that the crustal expansion, would take place near the end of the heavy bombardment more easily for Ganymede than for Callisto. In fact the crust of the former is thinner ($\xi = 0.92$) favouring the formation on Ganymede of grooved terrains. The presence of a crust on these two objects in our simulation is guaranteed by the fact that the radiative cooling time of the impacted melted zone is shorter than the mean time interval between two subsequent impacts in the same area of planetesimals with $r' > 1$ km. This condition is even more easily verified during the final phases of the accumulation when the rate of impacts, given by (5), rapidly declines. Therefore it is not unrealistic to assume as boundary conditions in solving equation (1) $T(R,t) = T_0 = 100$ K together with $(dT/dr)_{r=0} = 0$.

ACKNOWLEDGEMENT

The authors wish to thank Prof.A.E.Roy for helpful discussions and for his contribution toward improving our English.

REFERENCES

(1) Safronov,V.S.:1969,Evolution of the Protoplanetary Cloud and Formation of the Earth and Planets,Tran. 1972,NASA TT F-67.
(2) Safronov,V.S.:1978,Icarus 33,pp.3-12.
(3) Kaula,W.M.:1979,Jour.of Geophys.Res.84,pp.999-1008.
(4) Kaula,W.M.:1980,in"The Continental Crust and its Mineral Deposits",D.W.Strangway,Ed.,Geol.Assoc.of Canada Special Paper 20,pp.25-34.
(5) Coradini,A.,Federico,C.and Lanciano,P.:1981,Phys. of the Earth Planet.Interiors (submitted).
(6) Harris,A.W.and Kaula,W.M.:1975,Icarus 24,pp.516-524.
(7) Harris,A.W.:1978,Icarus 34,pp.128-145.
(8) Coradini,A.,Federico,C.and Magni,G.:1981,Astron. Astrophys.99,pp.255-261.
(9) Weidenschilling,S.J.:1976,Icarus 27,161-170.
(10)Stacey,F.D.:1977,Physics of the Earth,J.Wiley and Sons,New York.
(11)Tocsŏz,M.N.,Solomon,S.C.,Minear,J.W.and Johnston, D.M.:1972, The Moon 4,pp.190-213.
(12)Gault,D.E.and Heitowit,E.D.:1973,Proc.Hypervelocity Impact Symp.6th,pp.420-456.
(13)O'Keefe,J.D.and Ahrens,T.S.:1977,Proc.Lunar Sci. Conf.8th,pp.3357-3374.

(14)Kieffer,S.W.and Simmonds,C.M.:1980,Rev.of Geophys.
    Space Phys.18,pp.143-181.
(15)Lanciano,P.:1981,Laurea Thesis,Univesity of Roma.
(16)Croft,S.K.,Kieffer,S.W.and Ahrens,T.S.:1979,Jour.
    of Geophys.Res.84,pp.8023-8032.
(17)Reynolds,R.T.and Cassen,P.M.:1979,Geophys.Res.Lett.
    6,pp.121-124.
(18)Thurber,H.C.,Hsui,A.T.and Tocsőz,M.N.:1980,Proc.
    Lunar and Planet.Sci.Conf.11th,1957-1977.
(19)Schubert,G.,Stevenson,D.J.and Ellsworth,K.:1981,
    Icarus (in press).

THE SHAPE OF SMALL SOLAR SYSTEM BODIES:
GRAVITATIONAL EQUILIBRIUM VS. SOLID-STATE INTERACTIONS

P.Farinella°,F.Ferrini°°,A.Milani°°,
A.M.Nobili°°,P.Paolicchi° and V.Zappalà°°°

° Osservatorio Astronomico di Brera
°° Università di Pisa,Ist.Matematico"L.Tonelli"
°°° Osservatorio Astronomico di Torino

ABSTRACT. We discuss the new data on the shape of small planetary satellites and asteroids, comparing them with the theory of equilibrium figures for self-gravitating masses. This comparison allows to obtain interesting informations on the structure, density and strenght of these bodies.

1. INTRODUCTION

Since Newton's investigation on the figure of the Earth (Principia, Book III), the theory of the ellipsoidal shapes of gravitational equilibrium for homogeneous rotating bodies has been developed to a high degree of generality and mathematical refinement (Chandrasekhar, 1969). Of particular interest are the discoveries by Jacobi (who first recognized the possibility of triaxial ellipsoids as equilibrium figures for spinning objects, provided their angular momentum exceeds a critical value) and by Roche (who determined the conditions for equilibrium of a corotating satellite of negligible mass with respect to the primary). Nevertheless, the application of this classical theory to real celestial bodies has been problematic and controversial, for different reasons. As regards massive objects, like stars and planets, the density gradient from the center to the surface is large, providing a substantial deviation from the assumption of homogeneous material; moreover, in most cases the rotation is too slow to produce more than first-order departures from the spherical symmetry. On the other hand, for small bodies, like the majority of asteroids, solid-state interactions are stronger than gravitational ones and we have solid rocky objects which can easily maintain highly irregular shapes substained by the mate-

71

A. Coradini and M. Fulchignoni (eds.), The Comparative Study of the Planets, 71–77.
Copyright © 1982 by D. Reidel Publishing Company.

rial's mechanical strenght.

An order-of-magnitude estimate of the typical size for which gravitational forces begin to dominate the solid-state rigidity can
be obtained in the following way (cf., discussion by Hartmann
following Cook, 1971): we derive the radius r of a spherical body
of density ρ having a central pressure equal to the crushing
strenght s, so that the core loses rigidity and relaxation to e-
quilibrium is possible, and we get

$$r=(3s/2\pi G\rho^2)^{1/2} \ . \tag{1}$$

The strenght of celestial bodies is unknown, and data for terrestrial materials and meteorites allow a wide range of possible values, say from 10^6 to 10^9 dyn cm^{-2}. Using a reasonable value of
the density (~ 2.5 g cm^{-3}), we find that r is in the range from
~ 10 to ~ 300 km. We are inclined to believe that in most cases a
value close to the lower estimate is realistic, because small solar system objects should consist of weakly consolidated aggregates which were accumulated under conditions of weak self-gravitation. Moreover, for km-sized and larger objects the strenght could
be substantially lower than for small laboratory samples of similar materials, owing to the presence of large defects or cracks
which (we quote from Oberbeck and Aoyagi, 1972) "should certainly
be present if meteroids have resulted from breakup of a parent
body caused by collision... Even if the large meteroids formed as
separate bodies, large cracks should be present. Differential
thermal contraction or different rock-forming minerals produce
stresses that commonly fracture new rock formations on the earth".
In the following we shall discuss how the theory of equilibrium
figures fits with the new data obtained in the past few years for
four small satellites imaged by interplanetary probes (Phobos and
Deimos, Amalthea, Mimas) and for the fairly large sample of asteroids studied from earth by photometric techniques. Some indication
on the overall structure, strenght and mean density of these bodies is a simple by-product of the comparison between theory and
observations.

As regards the small satellites, we recall that the appropriate
theoretical treatment is the study of the so-called "Roche ellipsoids" (Chandrasekhar, 1969, ch.8), i.e., the rotationally and
tidally distorted shapes of a corotating satellite (whose spin and
orbital periods are equal) around a primary of much larger mass.
The satellite's material is supposed to be a homogeneous "liquid",
relaxing instantaneously to the equilibrium figure, and the theory
gives a relation between semi-axes ratios and density. As an example, for Amalthea at the present orbital distance, one of these
relationships (the two smaller ellipsoid axes have about the same
lenght) is displayed in Fig.1 .

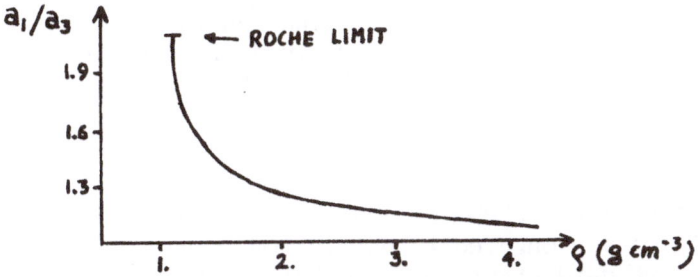

FIGURE 1. Major to minor axis ratio versus density for Amalthea
        at the present orbital distance in the Roche model.

As regards the comparison with real satellites, we must remember
that an important discrepancy could have been caused by tidal
evolution of the orbits (see Nobili, 1978), if these bodies formed and/or solified at different orbital separations than the present ones: provided a significant lag time is present in the relaxation of the figure to equilibrium, the observed shape will
correspond to an earlier orbital distance (which is generally
smaller, with the exception of Phobos).

2. PHOBOS AND DEIMOS

The two small Martian moons (of mean radius 10.5 and 6.5 km, respectively) have been closely imaged by the Viking orbiters, and
seem to be made of a low-density (~ 2 g cm^{-3}) carbonaceous material. Their shapes can be approximated by triaxial ellipsoids
whose axes have ratios of about 1.3 and 1.5 (Veverka and Thomas,
1979). Local deviations from the best-fit ellipsoids reach 1÷2 km.
These data seem to rule out the possibility that the shapes of
Phobos and Deimos correspond to Roche ellipsoids. Indeed, Phobos
is inside or very close (depending on its exact density) to the
classic Roche limit, within which no "liquid" satellite can be in
equilibrium. Its moderate elongation could be ensured only by a
very high mean density (approaching 4 g cm^{-3}; see Soter and
Harris, 1977), which is not consistent with data. As mentioned
before, an interesting possibility could be that Phobos' shape
adjusted to a nearly-equilibrium figure at the time of formation
or at some intermediate epoch, when its orbital distance was 1.2
to 1.3 times the present value; this would imply a reasonable rate
of tidal evolution, according to recent estimates (Smith and
Born, 1976). As regards Deimos, in spite of the fact that its
surface appears to be covered by a significant amount of loose,
downslope-moving material (Veverka and Thomas, 1979), its overall
shape is by far too elongated and irregular to conclude that
self-gravitation plays any important role (the corresponding Roche
figure would deviate from a sphere only by few percents). The
application to Phobos and Deimos of equation (1) yields a minimum

strenght of about 10^6 dyn cm^{-2}, a value consistent with the estimate by Thomas and Veverka (1979) derived from consideration of
the major impact events displayed by the satellites.

3. AMALTHEA

The three-dimensional shape of Amalthea, as shown by the Voyager's
pictures, is very elongated (about 265 km long and 140 km in diameter, with standard errors of $\pm$ 20 km). If this shape corresponds
to a Roche figure, this implies a very low density (~ 1.1 g cm^{-3});
even allowing for tidal evolution, the maximum density (corresponding to formation near the synchronous orbit) would be only ~ 1.7g
cm^{-3} (Ferrini et al.,1980). These low values would be plausible
only if Amalthea were similar in constitution to outer-belt or
Trojan asteroids; indeed, the spectrophotometric properties of
Amalthea (low albedo, reddish spectrum and large phase coefficient)
resemble remarkably those of RD asteroids, whose proportion increases rapidly with heliocentric distance in the outer belt (Tedesco and Gradie, 1981). However, the equilibrium models cannot
explain the fairly large irregularities of Amalthea's limb profile, nor the clear asimmetry between the Jupiter-pointing and
the opposite "hemisphere" (Ferrini et al., 1980). An alternative
possibility is that Amalthea is not in gravitational equilibrium,
but is a solid rocky body of rather high strenght ($s \simeq 10^8$ dyn
cm^{-2}). In this case the density could be higher, perhaps consistent with a metallic constitution. This latter seems needed if
a Joule heating of the satellite's interior is present (Simonelli, 1980).

4. MIMAS

Though heavily craterized (with evidence of a collisional event
close to the limit for catastrophic disruption), this 400-km
sized satellite is at present the smallest body in the solar system for which high-resolution images exist showing a nearly-sphe-
rical regular shape, with small-scale surface reliefs reaching
only $\sim 1\%$ of its size. This means that gravitational forces dominate
against solid-state forces (implying $s \lesssim 4x10^8$ dyn cm^{-2}). If we apply
the Roche model, we find for a density of 1.2 ± 0.1 g cm^{-3} (derived
from Voyager's measurements) an ellipsoidal shape whose deviation
from the spherical symmetry reaches $8.9\pm0.7\%$(Farinella et al.,
1981a). We envisage two possibilities: (1) the flattening predicted by the equilibrium theory is present; this would imply a
nearly-homogeneous interior, presumably of icy composition. However, a test tried with the first Voyager images did not show the
predicted elongation of the satellite's limb profile. If future
analyses will confirm this preliminary result, we could deduce
that (2) Mimas' interior is not homogeneous, but composed by a

denser rocky core covered by an icy mantle. As derived by Dermott
(1979), for a differentiated object the flattening of the equilibrium shape would be reduced by a factor ~0.7. If this reduction
will be observed and measured, it will be possible to obtain
interesting informations on the internal structure and even on
the formation process of Mimas (Consolmagno and Lewis, 1977).
Finally, we think that the previous kind of analysis could be
fruitfully applied also to the other small satellites of Jupiter
and Saturn, for which the Voyager mission has recently provided
high-resolution images.

5. ASTEROIDS

A wide range of sizes and (very likely) bulk compositions is an
outstanding property of asteroids, which in the last decade have
been extensively studied by several techniques. Among these,pho-
tometric lightcurves provide an interesting data set (see Tedesco, 1979) on asteroid spin periods and shapes (the maximum
lightcurve amplitude is a rough indicator of the asteroid's tri-
axiality), regarding bodies of diameter D between ~1 and ~10^3km.
An analysis of these data (Tedesco and Zappalà, 1980; Farinella
et al., 1981c) allows to separate a few classes of different typical sizes and physical properties. The largest asteroids
(D ≳ 300 km) have regular nearly-spheroidal shapes and in one case
(Vesta) a differentiated interior has been inferred from data.
Gravitational interactions seem to dominate their equilibrium
structure, while the collisional history affected weakly their
angular momentum of spin. On the contrary, the smallest objects
(D ≲ 100 km) seem mostly solid rocky bodies, whose shape (frequently very elongated) is not correlated with the rotational period.
These bodies are probably fragments arisen from catastrophic collisional events, which in this size range have a timescale less
than the age of the solar system, while self-gravitation is so
weak that there is little possibility for the breakup's debris to
coalesce again (Davis et al., 1979). In the intermediate size
range, around D ≃ 200 km, we can find a variety of interesting
objects:
(1) several candidates for binary and/or multiple systems, identified by secondary events during stellar occultations, by lightcurve peculiarities and by speckle interferometry. They could have
been originated by catastrophic collisions, whose energy did not
exceed the target's self-gravitational energy, and hence the
breakup did not result into dispersion of the fragments "to infinity". When the angular momentum transferred by the impact is
higher than a critical limit, the recombination in a single body
is not possible, and "fission" into (at least) a couple of bodies
results (Zappalà et al., 1980). The shape of the binary components
can relax to the appropriate equilibrium figures, affected by mutual tidal distortion (Darwin ellipsoids; see Chandrasekhar,1969,

ch.8). This model has been proposed by Weidenschilling (1980)
for the largest Trojan asteroid, 624 Hektor, which would be formed by two nearly-contact equal-sized ellipsoids of density 2.45 g
cm^{-3}, yielding the observed 3 : 1 cross-section ratio. We note
also that binary fission, and the subsequent tidal despinning,
could be responsible of the anomalous percentage of very slow
rotators among small and intermediate asteroids (Farinella et al.,
1981c).
(2) A well-defined class of asteroids with large lightcurve amplitude ($\gtrsim$0.2 mag, implying highly elongated shapes) and short
spin period ($\lesssim$6 hr) can be recognized in the intermediate size
range. Farinella et al. (1981b) have called these objects LASPAs,
and proposed that their shape can correspond to Jacobi ellipsoids,
the triaxial equilibrium figures of self-gravitating masses having
high specific angular momentum of spin (Chandrasekhar,1969,ch.6).
This model implies that objects rotating with periods of 6 hr
must have densities between 1.1 and 1.4 g cm^{-3}, while those rotating in 4 hr would have densities between 2.4 and 3.2 g cm^{-3}.
Shorter periods would be consistent with stability only for higher
densities, and this could explain why asteroids spinning in less
than ~4 hr are rare (Weidenschilling, 1981). LASPAs presumably
formed in the same process as binaries (collisional breakup of
parent bodies and subsequent relaxation of the material to the
equilibrium shape), but with a lower collisional transfer of angular momentum. It is also worthwhile to note that in several
cases the largest bodies ot the asteroid dynamical families are
either LASPAs or slowly-rotating, nearly-spheroidal objects(Farinella et al., 1981d): since families were probably originated
by catastrophic collisions, this fact supports the idea that
these events produce remnant bodies of low strenght, which easily
adjust to the equilibrium shapes associated with their angular
momentum. It is clear that all the models and interpretations
discussed above will undergo a decisive test when the asteroid
exploration by ad hoc probes, now in project, will be realized.

REFERENCES

1) Chandrasekhar,S.:1969, Ellipsoidal Figures of Equilibrium,
   Yale Univ. Press, New Haven (Conn.)/London.
2) Consolmagno,G.J., and Lewis,J.S.:1977, in Planetary Satellites
   (J.A.Burns,Ed.),pp.492-500, Univ.of Arizona Press, Tucson.
3) Cook,A.F.:1971, in Physical Studies of Minor Planets (T.Gehrels,
   Ed.),pp.155-172, NASA SP-267, Washington D.C.
4) Davis,D.R.,Chapman,C.R.,Greenberg,R.,Weidenschilling,S.J.,and
   Harris,A.W.:1979, in Asteroids (T.Gehrels,Ed.),pp.528-557,
   Univ. of Arizona Press, Tucson.
5) Dermott,S.F.:1979, Icarus 37,pp.575-586.
6) Farinella,P.,Ferrini,F.,Milani,A.,and Nobili,A.M.:1981a, Moon
   and Planets 24,pp.465-466.

7) Farinella,P.,Paolicchi,P.,Tedesco,E.F., and Zappalà,V.:1981b, Icarus 46,pp.114-123.
8) Farinella,P.,Paolicchi,P., and Zappalà,V.:1981c, Astron. Astrophys., in press.
9) Farinella,P.,Paolicchi,P., and Zappalà,V.:1981d, Int.Rep. Osservatorio Astronomico di Brera 4/81 and in preparation.
10) Ferrini,F.,Milani,A., and Nobili,A.M.:1981, Adv.Space Res. vol.1,pp.191-197.
11) Hege,E.K.,Cocke,W.J., and Hubbard,E.N.:1980, Bull.A.A.S. 12, p.662.
12) Nobili,A.M.:1978, Moon and Planets 18,pp.203-216.
13) Oberbeck,V.R., and Aoyagi,M.:1972, J.Geophys.Res. 77,pp.2419-2432.
14) Simonelli,D.:1980, in Reports of Planetary Geology Program-1980,pp.72-74, NASA Tech.Mem. 82385, Washington D.C.
15) Smith,J.L., and Born,G.H.:1976, Icarus 27,pp.52-54.
16) Soter,S., and Harris,A.W.:1977, Icarus 30,pp.192-199.
17) Tedesco,E.F.:1979, in Asteroids (T.Gehrels,Ed.),pp.1098-1107, Univ.of Arizona Press, Tucson.
18) Tedesco,E.F., and Gradie,J.C.:1981, Icarus, in press.
19) Tedesco,E.F., and Zappalà,V.:1980, Icarus 43,pp.33-50.
20) Thomas,P., and Veverka,J.:1979, Icarus 40,pp.394-405.
21) Veverka,J., and Thomas,P.:1979, in Asteroids (T.Gehrels,Ed.), pp.628-651, Univ.of Arizona Press, Tucson.
22) Weidenschilling,S.J.:1980, Icarus 44, pp.807-809.
23) Weidenschilling,S.J.:1981, Icarus 46, pp.124-126.
24) Zappalà,V.,Scaltriti,F.,Farinella,P.,and Paolicchi,P.:1980, Moon and Planets 22, pp.153-162.

METEORITES AND COSMOCHEMISTRY

G. Turner

Department of Physics, University of Sheffield,
Sheffield, S3 7RH, U.K.

The study of meteorites is important for a number of reasons but
three aspects are of particular significance.   On the one hand
they represent, along with lunar samples, the only
extraterrestrial material available for detailed laboratory
study.   As such they provide an important window on the
chemistry of the solar system, beyond what is directly available
on the Earth, and they give an indication of the variablity of
that chemistry between individual components of the solar system.
Secondly they represent material which for the most part has been
relatively unaffected by geological processes since the formation
of the Solar System and hence provide the most direct evidence
for events and processes which occurred during and immediately
following the birth of the solar system.   Finally, as a result
of periods of exposure to cosmic rays, solar flare particles,
solar wind and micrometeorites prior to arriving on Earth, they,
and the lunar samples provide the only long term record of the
interplanetary flux of these particles and of their effects.

The summary of meteorite science which follows will begin with a
broad factual description of the major features of meteorite
chemistry, mineralogy and petrology.   This will be followed by a
discussion of the relevance of meteorite observations to some of
the current ideas of the origin of the solar system, leading into
a brief reference to the recent exciting discoveries of isotopic
anomalies.   For a more detailed chronological discussion the
reader is referred to the section on Isotopic Dating of the Solar
System.

Meteorites are fragments from small bodies which are in orbits
capable of intersecting the orbit of the Earth and are of a size

A. Coradini and M. Fulchignoni (eds.), The Comparative Study of the Planets, 79–84.
Copyright © 1982 by D. Reidel Publishing Company.

able to survive entry through the atmosphere.   Accurate orbital
information is only available for three meteorites, all of which
were in highly elliptical orbits extending to the asteroid belt.
The mechanisms for placing these objects into Earth crossing
orbits, (which have capture lifetimes two orders of magnitude
less than the age of the solar system) are still the subject of
much theoretical speculation (1), as is the question of whether
the parent bodies are asteroids or in some cases the cores of
short period extinct comets.

The most obvious distinction between meteorite types is between
the so-called stones and irons.   However a more fundamental
distinction can be made between the undifferentiated and
differentiated meteorites.   The undifferentiated meteorites,
also known as chondrites due to the frequent presence of mm.
sized spherical mineral aggregates called chondrules, are
characterized by an essentially solar composition for the
involatile elements.   They are thought by many to be the most
primitive objects because of this.   Mineralogically they are
composed predominantly of iron magnesium silicates (pyroxene and
olivine) and in some cases grains of iron nickel alloy.
Variations in mineralogy are reflected in overall chemical
composition which forms the basis of classification schemes (2)
(3).   The major chemical variations consist of 1) variations in
oxydation state, 2) variations in the total Fe content, 3) the
presence in some of volatiles (water, hydrocarbons and some trace
elements) and hydrated minerals, and 4) the presence in some of
highly refractory minerals (oxides of Ca, Ti and Al) which occur
characteristically as white inclusions.   These differences are
taken to imply a variety of processes in the early solar system
including;   fractionation on the basis of volatility, a
fractionation of metallic iron relative to silicates, and, the
operation of an oxydation/reduction reaction sequence (4).

In addition to these major chemical distinctions the chondrites
show petrological differences which have been taken by some to
indicate the effects of thermal metamorphism on the parent body
(2).   The observational evidence includes the effects of
recrystallization (larger grain size, chondrules indistinct or
absent, absence of glass), variations in the degree of chemical
equilibration of minerals (variability or otherwise of mineral
composition on a microscopic scale), and the concentration of
volatile species such as the noble gases (5).

The differentiated meteorites, which include the irons and a
group of stones referred to as achondrites are, for the most
part, presumed to be the end product of igneous melting processes
on the parent asteroid or asteroids.   The processes involved are
similar to the processes of partial melting and fractional
crystallization which produce igneous rocks on Earth .   In this

view the differentiated irons represent the asteroidal analogue
of a planetary core.   They are classified on the basis of trace
element groupings (6) which as well as indicating origins in
different bodies also show clearly the systematic effects of
fractionation during solidification.   Cooling rates inferred
from Ni concentration profiles in individual mineral grains (7)
indicate parent bodies of the order of 100 km in radius.

The major distinction between achondrites (3) (8) (9) is between
those which are basaltic, representing partial melts or residual
liquids, and those which represent the residue after the removal
of the basaltic liquids (mafic cumulates).   The basalts are
characteristically rich in calcium, which is reflected by the
presence of plagioclase.

In addition to the major distinctions in meteorite types just
described several minor features are present which for the most
part reflect secondary effects occurring throughout the lifetime
of the meteorite.   One such feature is the common presence of
what may be loosely termed shock effects (10), arising from high
velocity collisions between meteorites and/or parent bodies.
These are shown typically by the frequent presence of glassy
veins (impact melt), brecciation and low gas retention ages (11).
A second related aspect is the presence of features analogous to
those found in some lunar breccias which indicate that the
component parts of the meteorite have once formed part of the
surface regolith of a parent body.   One such characteristic is
the presence of high concentrations of implanted solar wind noble
gases.

Meteorite observations have played a key role in the formulation
of ideas about conditions in the early solar system.   The
chronology of events is discussed elsewhere but the major
observational fact is that the major events, all occured early,
in a narrow time interval of 1 - 100 Ma.   The formation of
differentiated meteorites can be understood in principle by the
operation of well known igneous processes.   The major difficulty
so far has been in accounting for the heat sources necessary.
The discovery of extinct ^{26}Al (12) in the last decade may be
the solution to this problem but if so the short half life of
^{26}Al carries the implication that the differentiation was very
early and essentially contemporaneous with or even preceding the
formation of large bodies.

    The undifferentiated meteorites present a further problem in
that many of the variations in chemistry between classes and
between components of individual meteorites appear to be
controlled essentially by differences in volatility of the
elements concerned.   Over the last two decades this has led to
the idea, supported by a great deal of chemical evidence and

thermodynamic calculation (13) (14), that the undifferentiated
meteorites represent a condensation sequence of material formed
from a cooling gas of near solar composition.   These ideas are
now being actively re-examined in the light of detailed
petrological studies and in view of the unpopularity among
astrophysicists of the hot solar nebula hypothesis.
Nevertheless new ideas which emerge must still face the clearly
observed and in many ways dominant effects of volatility on
meteorite chemistry.   Processes which are possibly capable of
reconciling the geochemical, petrological and astrophysical
constraints include distillation in small objects heated by
^{26}Al decay.

Some of the most exciting observations on meteorites, with
profound implications for the early history of the Solar System,
have come in recent years with the discovery of isotopic
anomalies in a number of elements (15) (16) (17).   For the most
part these anomalies appear to be restricted to rather specific
phases, such as the refractory Ca-Al rich inclusions in
carbonaceous chondrites or in the case of some noble gas
anomalies to an obscure carbon rich phase (5).   For the major
element oxygen however isotopic variability appears to be the
rule rather than the exception but is greatest (up to 5%) in the
Ca-Al rich inclusions.   The variations in isotopic composition of
oxygen are in a direction corresponding to the addition of ^{16}O,
which has led to the view that ^{16}O was introduced in solid dust
grains into the pre solar nebula as a result of a discrete
nucleosynthetic event (explosive carbon burning) shortly before
the formation of the Solar System.   Within a given inclusion
large variations between individual 10 um mineral grains suggest
that partial equilibration with a gaseous phase of different
isotopic composition occurred after the minerals now present were
formed.   Differences between meteorite classes and between
meteorites and the Earth testify to a solar system wide
variability in this nucleosynthetic component.

In a few Ca-Al rich inclusions isotopic anomalies are present, at
levels around one part in 10^4, in Ca, Ti, Si, Ba, Sr, Nd and Sm
and appear also to be the result of discrete nucleosynthetic
events coupled with mass fractionation.   Their importance is
twofold;   on the one hand they provide raw data for the
astrophysicist seeking to understand the details of
nucleosynthesis, while on the other they provide students of the
Solar System with a measure of inhomogeneity within the early
solar nebula and indicate the extent to which it is necessary to
postulate the presence of recently synthesised material.   Time
scales for these pre solar events are provided by shortlived (now
extinct) radionuclides and are discussed elsewhere.   A commonly
held view is that much of the anomalous component was produced in
one or more supernovae events which immediately preceded the

formation of the solar system and may indeed have triggered the collapse of the solar nebula.   The association of supernova remnants with regions of new star formation, as observed for example in the Orion nebula, is readily understood in terms of the short lifetimes of massive O and B stars and so the postulated pre solar supernova is on astrophysical grounds not unreasonable.

Detailed considerations of the anomaly patterns show that no single process is capable of accounting for the observations.   In the case of titanium, for example, which has five stable isotopes the range of variation seen is such as to require at least three clearly resolvable nucleosynthetic components.   Titanium is close to iron, the most stable nuclear species in the periodic table, and the high cosmic abundance of the latter is accounted for by production in quasi thermodynamic equilibrium attained in the high temperatures of exploding stars.   Titanium is therefore pariculary susceptible to variations in the operation of this so called e-process, either on account of kinetic effects due to specific reactions on seed nuclei (i.e. non-equilibrium conditions) or by differences in the actual temperatures involved (higher temperatures resulting in an increased production at masses below the iron peak).

An anomalous component of neon (neon-E) has now been clearly characterized (18) as essentially pure ^{22}Ne and as such can only have been produced indirectly by the decay of ^{22}Na.   Since the half life of ^{22}Na is only 2.6 a, the neon-E must have been incorporated into its present carrier phase close to the site of nucleosynthesis.   Anomalies in the heavy noble gas xenon are the longest established having been recognized for two decades (19). In spite of a steadily increasing characterization of both composition and host phases they are still somewhat enigmatic and explanations range from variations in s- and r-process nucleosynthesis to the decay of an extinct 'superheavy' precursor element (5).

The growth of meteorite science has been particularly great in the past two decades, primarily as a result of the application of important technical advances such as the development of the electron microprobe, improved mass spectrometers and techniques of trace element chemistry.   Surprisingly the growth is still continuing with promise of new discoveries in several areas. The isotope anomaly story is probably just beginning, the nebular condensation hypothesis is undergoing a reappraisal, the details of parent body processes and relationships are still far from clear as are the mechanisms necessary to account for the observed flux of meteorites.   These and other problems will continue to exercise our minds for some time.

REFERENCES

1. Wasson, J.T. and Wetherill, G.W: 1979, in Asteroids (ed. T. Gehrels), pp. 926-974, U. of Arizona Press.
2. Van Schmus, W.R. and Wood, J.A: 1967, Geochim. Cosmochim. Acta, 31, pp. 747-765.
3. Wasson, J.T: 1974, Meteorites, Springer-Verlag.
4. Anders, E: 1971, Ann. Rev. Astron. Astrophys., 9, pp.1-34.
5. Anders, E: 1981, Proc. Roy. Soc. London A, 374, pp. 207-238.
6. Scott, E.R.D. and Wasson, J.T: 1975, Rev. Geophys. Space Phys., 13, pp. 527-546.
7. Goldstein, J.I. and Short, J.M: 1967, Geochim.Cosmochim. Acta., 31, pp.1733-1770.
8. Stolper, E., McSween, H.Y. and Hays, J.F: 1978, Geochim. Cosmochim. Acta, 43, pp. 589-602.
9. Dreibus, G. and Wanke, H: 1980, Z. Naturforsch., 35a, pp. 204-216.
10. Heymann, D: 1967, Icarus, 6, pp. 189-221.
11. Bogard, D.D. and Hirsch, W.C: 1980, Geochim. Cosmochim. Acta., 44, 1667-1682.
12. Lee, T., Papanastassiou, D.A. and Wasserburg, G.J: 1977, Ap. J. (Letters), 211, pp. 107-110.
13. Lord, H.C: 1965, Icarus, 4, 279-288.
14. Grossman, L: 1972, Geochim. Cosmochim. Acta., 36, 597-619.
15. Clayton, R.N: 1978, Ann. Rev. Nucl. Part. Sci., 28, pp. 501-522.
16. Wasserburg, G.J., Papanastassiou, D.A. and Lee, T: 1980, in Early Solar System Processes and the Present Solar System, pp. 144-191, Soc. Italiana di Fisica, Bologna.
17. Begemann, F: 1980, Rep. Prog. Phys., 43, pp. 1309-1356.
18 Eberhardt, P., Jungck, M.H.A., Meier, F.O. and Niederer, F.R: 1981, Geochim. Cosmochim. Acta., 45, (in press).
19. Reynolds, J.H. and Turner, G: 1964, J. Geophys. Res., 69, pp. 3263-3281.

AGES OF THE SOLAR SYSTEM: ISOTOPIC DATING

G. Turner

Department of Physics, University of Sheffield,
Sheffield, S3 7RH, U.K.

The major concern of this section will be to outline the ways in
which measurements of isotope abundances have been used to
determine the chronology of the origin and evolution of the solar
system. In passing it should be remembered that the use of
isotopic information is by no means restricted simply to the
measurement of time scales and, particularly in recent years,
isotope abundances have been used to investigate problems as
diverse as the heat sources in the early solar nebula and the
chemical evolution of the Earth's mantle. The fundamental
property of isotopes which makes them especially useful for
dating and other applications is the fact that, apart from a
limited amount of mass fractionation, the composition of an
isotopic mixture is unaffected by chemical processes. In those
cases where mass fractionation does occur this effect may itself
be useful, particularly as a source of information on
temperatures. Since our main theme is time the events discussed
in this section will be most conveniently presented as a
chronological sequence, progressing from some time before the
solar system existed down to the present day.

The time scale of events prior to the formation of the solar
system, and in particular time scales for the manufacture of the
chemical elements by the various astrophysical processes of
nucleosynthesis, can be inferred from a study of the abundances
of those radioactive (parent) species existing at the time when
the solar system first formed 4.55 Ga ago. The basis for all
such calculations is the competition between production (by
nuclear reactions in stellar interiors) and radioactive decay of
the species concerned. In principle measurements of the
abundances of many radioactive species, with widely differing

85

A. Coradini and M. Fulchignoni (eds.), The Comparative Study of the Planets, 85–94.
Copyright © 1982 by D. Reidel Publishing Company.

half-lives, relative to appropriate stable species, can be used
to infer the production rate of the elements as a function of
time since the galaxy formed. In practice the number of suitable
species is rather small, (^{238}U, ^{235}U, ^{232}Th, ^{187}Re,
^{244}Pu, ^{129}I, ^{107}Pd, ^{26}Al and ^{41}Ca) but has grown
significantly over the last decade.

Although a detailed history of nucleosynthesis cannot be deduced
at the present time nevertheless far reaching conclusions can be
made. The long lived species have been used to deduce an age for
the galaxy of around 13 Ga. The short lived species have been
used very recently to draw conclusions about major (from our
point of view) supernova events which occurred immediately prior
to the formation of the solar system. The discovery of ^{26}Mg
anomalies which correlate directly with the Al/Mg ratio of the
minerals carrying the anomaly indicates that live ^{26}Al was
present in the minerals, at a level of 5 parts in 10^5 relative
to ^{27}Al, when they last cooled to temperatures sufficient to
inhibit the diffusion of Mg (1). Studies of thin sections of
meteorites indicate that the minerals concerned (in a particular
meteorite clast) show the textural relationships characteristic
of minerals crystallizing from a melt. Since most of the clasts
are not unusual in any other respect the implication is that they
were last molten, containing live ^{26}Al, while members of the
solar system. The presence in the embryonic solar system of
^{26}Al indicates that the time interval between the supernova
which produced it and the formation of solid meteoritic minerals
was of the order of 10^6 years or less.

This has been confirmed by the further observation of ^{107}Ag
anomalies (from ^{107}Pd) which, being present in differentiated
iron meteorites, indicates that igneous melting of (small?)
planetesimals also occurred on a short time scale (2). Most
recently evidence of anomalous ^{41}K (from ^{41}Ca) has
constrained the time scale even further, suggesting 'formation
intervals' for some mineral phases of only a few hundred thousand
years (3). In addition to clarifying the picture of our local
stellar environment 4.5 Ga ago the detection of 'extinct' ^{26}Al
provided evidence of a major heat source in the early solar
system.

The presence of ^{244}Pu (half-life 83 Ma.) in the early solar
system, coupled with the apparent absence of ^{247}Cm (half-life
15 Ma.) (4), indicates that the late supernova event did not
manufacture the heavier actinide elements to an appreciable
extent. In order to permit the decay of ^{247}Cm the last major
r-process event which did contribute significant amounts of
^{244}Pu and some of the ^{129}I must in fact have occurred 100 Ma
or so prior to the formation of the solar system. By such
observations and arguments the fine structure of the last stages

of pre-solar nucleosynthesis are being clarified.

In the first hundred million years following the formation of the solar system a number of major events occurred, either in rapid succession or as a series of overlapping events. Solid objects accreted and grew, by a combination of gravitational instabilities and mutual collisions, into planetesimals and finally into the planets. A major heat source, possibly ^{26}Al, caused extensive melting of many of the planetesimals at a very early stage (probably within the first million years and probably before collisional accretion was very far advanced). This led to widespread chemical differentiation which is now evidenced by objects such as the basaltic meteorites and the iron meteorites; it may also have been responsible for many of the features observed in the so called undifferentiated (chondritic) meteorites.

For the asteroidal, or possibly cometary, precursors of the meteorites the long lived heat sources (U, Th, K), which drive the geological processes on the large terrestrial planets, were insufficient to sustain the process of melting and differentiation. These objects therefore cooled rather rapidly, within less than a hundred million years, thereby preserving in their shielded interiors the early record. Some of the chondritic meteorites show evidence of thermal metamorphism and recrystallisation dating from this period, while others, such as the carbonaceous chondrites, appear to have remained essentially unchanged since their formation during or immediately following the formation small bodies in the primitive solar nebula. Based on the presence of isotope anomalies unrelated to radioactive decay some of these objects may contain solid material which predates the solar system.

The above picture relies on the development of radiometric techniques capable of resolving events separated in time by a few million or tens of million years but which happened four and a half thousand million years ago. Currently the methods which best permit this are those based on extinct radioactive species, ^{129}I (5) (6) and ^{26}Al, and the method based on measurements of initial $^{87}Sr/^{86}Sr$ ratio (7) (8). In all cases the ages obtained are relative ages and to some degree model dependent. Absolute ages are provided by the U-Pb, Rb-Sr, Sm-Nd and ^{40}Ar-^{39}Ar radiometric clocks (9) (10) (11) (12). Of these the U-Pb system is currently capable of the greatest precision, $\pm$ 10 Ma in favourable circumstances. Uncertainties associated with the other absolute clocks are of the order of 30 - 50 Ma, again in favourable circumstances. When absolute ages are well defined and internally consistent for different mineral phases (isochronous) they indicate a time when isotopic mixing, for example as a result of thermal diffusion, effectively ceased

within the rock or mineral systems being studied. Since
diffusion of isotopes is dependent on both temperature and
chemistry the different clocks may be reset by, and indicate the
times of, quite different events. In view of the likely
complexity of the first 10 Ma or so of solar system history the
unravelling of a unique and detailed sequence of the earliest
events is currently not possible although a number of significant
chronological statements can be made.

Until very recently most geochemical models of the early solar
system were based on the assumption of an initially hot and
(predominantly) gaseous solar nebula from which the
undifferentiated meteorite minerals condensed. This view is
currently being seriously questioned on the grounds that a hot
solar nebula is astrophysically unrealistic. Wanke (16) has
recently suggested an inversion of the conventional view such
that high temperatures arise in a accretionally cold nebula only
as a result of heating by [26]Al in bodies of sufficient size (a
few km.) to retain the heat for times comparable to the [26]Al
half life. This produces differentiation as in earlier models
but also, in combination with collisions between planetesimals,
leads to vapourization and later condensation of material to form
the 'undifferentiated' meteorites. In that the model overcomes
difficulties associated with the hot solar nebula it is
attractive. It is probably true to say that it cannot be ruled
out on the basis of any present isotopic data. The major
difficulty is likely to be in accounting for the retention of a
broadly solar composition for the chondrites.

Based on U-Pb measurements the absolute ages of the earliest
formed objects are close to 4.55 + 0.01 Ga. Based on [129]I
relative age measurements and initial $^{87}Sr/^{86}Sr$ ratios the
formation of primitive chondritic meteorites and the earliest
differentiated achondrites (strongly depleted in Rb) are barely
resolvable on a 1 to 10 Ma time scale, a feature which is
consistent with the view that the earliest differentiation
occurred rapidly in small bodies heated by the decay of [26]Al.
In detail there are problems associated with the assumption of
[26]Al as the only available heat source for the differentiation
of small bodies (in which gravitational energy release is
negligible incidentally). The eucrites, a major class of
differentiated (basaltic) meteorites are critical in this
respect. Based on Rb-Sr and Sm-Nd absolute ages at least one of
them (Juvinas) appears to have cooled, as a basaltic lava, within
50 Ma of the formation of the solar system (9) (29). Its initial
$^{87}Sr/^{86}Sr$ ratio is significantly higher than the most
primitive value observed in the Allende carbonaceous chondrite,
and requires that the Sr isotopic composition evolved in a high
Rb/Sr source region (e.g. solar or chondritic) for a period of 10
- 20 Ma prior to a depletion of the volatile element Rb and

incorporation in the molten and (very) low Rb/Sr parent asteroid
of the eucrites. This implies that a heat source capable of
volatilising the Rb must have been available long after the decay
of ^{26}Al. At least two possibilities exist: either some
planetesimals were sufficiently large ($\sim$ 30 km), when ^{26}Al was
live, to retain the heat for times of the order of 20 Ma, or an
alternative method of storing the ^{26}Al energy was available.
On Wanke's model of the early solar nebula the energy could have
been stored as chemical energy in the form of reduced volatile
poor (low Rb/Sr) planetesimals and oxydized volatile rich (high
Rb/Sr) planetesimals. Collisions between such objects could
release the stored energy at any time during the accretion
period, causing igneous melting, Rb depletion by volatilization,
and provide a source of 'evolved' Sr isotopic composition.

The time scale for the accretion of large bodies such as the
Earth and Moon is less clear. The depletion of Rb which is a
feature of lunar rocks certainly occurred about the same time as
that in the achondritic meteorites but could well be a feature
which predates the actual accretion of the Moon having been
produced in a similar fashion to that of the eucrites. The
isotopic data is so far quite consistent with the theoretical
accretion times for large bodies which are of the order of 10 -
100 Ma.

The K-Ar clock is particularly susceptible to resetting or
partial resetting at quite modest temperatures, around 2 - 300 C.
The ^{40}Ar-^{39}Ar modification of K-Ar dating is therefore useful
at providing temperature and time information related to the
period of metamorphism and cooling (13) (30) as well as
information on the temperatures attained during asteroid
fragmentation events. ^{40}Ar-^{39}Ar ages of meteorites show a
bimodal distribution with ages clustering around 4.5 Ga and
between 0 - 1.5 Ga. The young ages are for the most part the
result of heating during fragmentation or cratering events which
are involved in the process of generating meteorites as small
bodies from their much larger parent bodies. The lifetimes of
objects in space against destruction by fragmentation are
expected to vary roughly as the sixth root of the mass, so that
collision debris may be expected to survive on average for much
shorter times than the larger object from which it is produced.
The relative absence of intermediate ^{40}Ar-^{39}Ar ages is thus
accounted for by the failure to survive of most of the earlier
collision debris (31). Two methods not involving isotopes also
provide temperature information and cooling rates, the
measurement of Ni concentration profiles in meteoritic iron (14)
and the measurement of fossil radiation damage 'tracks' from
extinct ^{244}Pu (15) and are discussed elsewhere in this volume.

For the planets and several of their satellites the first hundred

million years of solar system history merely set the stage for a
long period of geological evolution which has continued down to
the present day. The major concern of the modern isotope
geologist, in addition to providing an absolute time scale of
events, is to make use of isotopes to follow in detail the
differentiation processes which for the Earth and Moon, and
presumably the other terrestrial planets, have generated
chemically distinct crusts overlying mantles of iron magnesium
silicate.

The only major planetary body apart from the Earth for which
samples are available for laboratory analysis is the Moon. The
application of isotopic methods has produced a rather clear
picture of the way in which that body evolved with time (20)
(21). Four major epochs are discernable. At a very early stage
the Moon underwent extensive melting followed by fractional
crystallisation to produce a low density anorthositic crust
overlying an iron magnesium silicate mantle (22). Parts of this
early crust appear to have undergone partial melting within a few
hundred million years of formation leading to distinct subsurface layers characterised by high concentrations of large
ionic radius elements such as uranium, thorium, potassium and the
rare earth elements (KREEP).

Following, and probably overlapping, this igneous epoch the
process of planetary surface evolution was complicated by the
final stages of an intense bombardment of the inner solar system
by fragments, up to 50 km in diameter, of one or more of the
planetesimals left over from the initial accretion phase (17)
(18) (19), and culminating 3.9 Ga ago. The bombardment epoch
produced many large impact basins and extensive cratering on the
lunar highlands or terrae (21) (23). The impacts fragmented and
in places melted the crust thereby producing the present suite of
highland rocks which display great petrological complexity but
whose chemistry still reflects the early igneous processes. Why
this bombardment should be so intense at such a comparatively
late stage is itself a problem of continuing interest to which
new answers may yet emerge. From the point of view of planetary
chronologies based on crater counting, the so called 'lunar
cataclysm' provides a convenient marker horizon for dating
ancient events in the inner solar system. The ability to
distinguish between pre-3.9 Ga and post-3.9 Ga old surfaces has
been particularly useful in helping to understand the geological
evolution of the Galilean satellites of Jupiter.

The significance of the late bombardment episode for the Earth is
as yet not clear. The oldest crustal rocks on Earth immediately
post-date the bombardment. Whether this is merely a curious
coincidence or a significant observation is not known. The mean
rate of energy influx from impacting objects was certainly small

in comparison with the geothermal flux, however the 'stirring' action of the bombardment may conceivably have been an important factor in the early development of crust and upper mantle. The history of the Earth following the formation of the oldest crustal rocks is of course the concern of a very large community of geologists, geochemists and geophysicists and the detailed study of the Earth's geological evolution is outside the scope of this summary.

Immediately following the bombardment and lasting for a period of some 800 Ma was a further period of igneous activity on the Moon during which iron rich basaltic lavas (Mare basalts), produced by partial melting of the upper mantle, were extruded into the giant basins left by the impacts (20) (21) (24) (25). As on the Earth the heat sources responsible for this late igneous activity were the radioactive isotopes of uranium, thorium and to a lesser degree potassium. All other things being equal it is to be expected that the temperature gradient in the crust of a planet should, at a particular epoch, be proportional to the radius of the planet. For this reason the thickness of the lithosphere is expected to scale inversely as the radius. In the case of the Moon, whose radius is only one quarter that of the Earth, the development of a thick lithosphere has prevented the operation of plate tectonics. In the absence of this effective mechanism for the transport of molten lava to the surface therefore surface vulcanism ceased on the Moon some 3 Ga ago. Following this epoch the Moon has been geologically a dead planet with surface activity being restricted to the effects of meteorite and micrometeorite impacts.

Mars is intermediate in size between the Earth and Moon and for the reasons referred to above is expected to have a lithosphere of intermediate thickness. The observed surface features support this view and indicate a planet with a limited amount of ongoing geological activity and incipient rather than active plate tectonics. Currently the chronology of Martian geology is based on crater counting. This method gives unambiguous answers only for times close to 3.9 Ga due to the existence of the lunar cataclysm 'marker horizon' already referred to. For more recent events uncertainty over the manner in which cratering rate varies with heliocentric distance has led to a number of somewhat divergent Martian chronologies which will probably only be resolved by absolute radiometric age determinatons.

The Moon, Mercury, the parent bodies of the meteorites and other small bodies in the solar system are devoid of atmospheres. For this reason the processes responsible for erosion and transport of eroded material on the surfaces of these objects are primarily exogenic, being the result of particulate bombardment of one sort or another. Over the last two decades these processes have been

extensively studied in meteorites and more recently in lunar
samples.

Many of the methods used are based on the analysis of the
products of nuclear reactions induced by cosmic ray and solar
flare protons and by the lower energy secondary protons and
neutrons released by these primary interactions. In addition to
nuclear reactions, radiation damage tracks are produced by cosmic
ray particles of high atomic number and in the case of the solar
wind, irradiation fluxes are sufficiently high for the more
abundant species to be detected directly.

The basic information deduced from such studies is the nature and
duration of the exposure of the samples analysed to the
particular radiation involved. Because of the attenuation of the
radiation with depth each reaction product or implanted species
is characterized by an attenuation length or depth profile which
may vary from meters to umeters. By determining the abundance of
several species it is in principle possible to deduce the
residence time of a given sample at different depths within the
target. Stable reaction products measure integrated doses while
radioactive species measure fluxes averaged over the mean
lifetime of the nuclide concerned. By comparing measurements for
radioactive species with different half-lives the rates of
various surface processes can be inferred.

Particular problems which have been examined by these methods
include (26) (27) (28); the determination of the ages of specific
lunar craters by the measurement of the surface residence times
of ejecta;   related to this the identification of ejecta from
particular young craters;   the determination of erosion rates and
lifetimes against fragmentation by impact of lunar surface rocks
and meteorites;   determination of the depositional and
'gardening' history of the lunar soil layer by the detailed
analysis of core samples;   studies of the processes involved in
the 'maturation' of the lunar soil layer following the ejection
onto the surface of pristine debris;   the recent history of
meteorites as small bodies;   the ablation of meteorites on entry
through the atmosphere;   the development of regolith (soil)
layers on the meteorite parent bodies.

In parallel with the observational work theoretical modelling is
carried out, making use of the experimental data to constrain the
parameters of the models. Because of the random nature of
meteorite and micrometeorite impacts Monte Carlo methods are
frequently used to simulate the stochastic effects. Physical
methods (SEM etc.) are also used to study directly the effects on
exposed surfaces of micrometeorite impacts.

REFERENCES

1. Lee, T., Papanastassiou, D.A. and Wasserburg, G.J: 1977,
    Ap. J. (Letters), 211, pp. 107-110.
2. Kelly, W.R. and Wasserburg, G.J: 1978, Geophys. Res. Lett.,
    5, pp. 1079-1082.
3. Huneke, J.C., Armstrong, J.T. and Wasserburg, G.J: 1981,
    Lunar Planet. Sci. XII, pp. 482-484.
4. Chen, J.H. and Wasserburg, G.J: 1981, Earth Planet. Sci.
    Lett., 52, pp.1-15.
5. Reynolds, J.H: 1963, J. Geophys. Res., 68, pp. 2939-2956.
6. Jordan, J., Kirsten, T. and Richter, H: 1980, Z.
    Naturforsch., 35a, pp.145-170.
7. Papanastassiou, D.A. and Wasserburg, G.J: 1969, Earth Planet.
    Sci. Lett., 5, pp. 361-376.
8. Gray, C.M., Papanastassiou, D.A. and Wasserburg, G.J: 1973,
    Icarus, 20, pp. 213-239.
9. Allegre, C.J., Birck, J.L., Fourcade, S. and Semet, M.P:
    1975, Science, 187, pp.436-438.
10. Tatsumoto, M., Unruh, M. and Desborough, G.A: 1976, Geochim.
    Cosmochim. Acta, 40, pp. 617-634.
11. Chen, J.H. and Tilton, G.R: 1976, Geochim. Cosmochim. Acta.,
    40, pp. 635-643.
12. Wasserburg, G.J., Terra, F., Papanastassiou, D.A. and Huneke,
    J.C: 1977, Earth Planet. Sci. Lett., 35, pp. 294-316.
13. Turner, G., Enright, M.C. and Cadogan, P.H: 1978, Proc. Lunar
    Planet. Sci. Conf. 9th., pp. 989-1025.
14. Wood, J.A: 1979, in Asteroids (ed. T. Gehrels), pp. 849-891,
    U. of Arizona Press.
15. Pellas, P. and Storzer, D: 1981, Proc. Roy. Soc. London A,
    374, pp. 253-270.
16. Wanke, H: 1981, Phil. Trans. Roy. Soc. London A, in press.
17. Turner, G., Cadogan, P.H. and Yonge, C.J: 1973, Proc. Lunar
    Sci. Conf. 4th., pp.1889-1914.
18. Terra, F., Papanastassiou, D.A. and Wasserburg, G.J: 1974,
    Earth Planet. Sci. Lett., 22, pp. 1-21.
19. Wetherill, G.W: 1975, Proc. Lunar Sci. Conf. 6th,
    pp. 1539-1561.
20. Nyquist, L.E: 1977, Phys. Chem. Earth, 10, pp. 103-142.
21. Turner, G: 1977, Phys. Chem. Earth, 10, pp. 145-195.
22. Oberli, F., McCulloch, M.T., Terra, F., Papanastassiou, D.A.
    and Wasserburg, G.J: 1978, Lunar Planet. Sci. IX, pp.832-834.
23. Maurer, P., Eberhardt, P., Geiss, J., Grogler, N., Stettler,
    A., Brown, G.M., Peckett, A. and Krahenbuhl, U: 1978,
    Geochim. Cosmochim. Acta., 42, pp. 1687-1720.
24. Papanastassiou, D.A., DePaulo, D.J. and Wasserburg, G.J:
    1977, Proc. Lunar Sci. Conf. 8th, pp.1639-1672.
25. Guggisberg, S., Eberhardt, P., Geiss, J., Grogler, N.,
    Stettler, A., Brown, G.M. and Peckett, A: 1979, Proc. Lunar
    Planet. Sci. Conf. 10th, pp. 1-39.

26. Burnett, D.S. and Woolum, D: 1977, Phys. Chem. Earth, 10,
    pp. 63-101.
27. Pillinger, C.T: 1979, Rep. Prog. Phys., 42, pp. 897-961.
28. Curtis, D.B. and Wasserburg, G.J: 1977, Proc. Lunar Sci.
    Conf. 8th, pp. 3575-3593.
29. Lugmair, G.W: 1974, Meteoritics, 9, p. 369.
30. Turner, G: 1981, Proc. Roy. Soc. London A, 374, pp. 281 -
    298.
31. Turner, G: 1979, Proc. Lunar Planet. Sci. Conf. 10th,
    pp. 1917 - 1941.

DO THE AGE DIFFERENCES GIVEN BY RELATIVE OR ABSOLUTE CHRONOLOGIES OF THE MOST ANCIENT METEORITES CORRESPOND TO REAL AGE DIFFERENCES ?

P. Pellas

L.A. 286 du CNRS, Lab. Minéralogie du Muséum,
61 rue Buffon 75005 Paris, France.

ABSTRACT
    Recent results from absolute and relative chronologies of the most ancient meteorites are reviewed in order to analyze if they are significant or not. Use of the various chronometers to analyze the same meteoritic sample is shown to be an interesting approach to retrace the prehistory of meteorites and their environments.

    Since two decades the problem of coupling the data obtained by absolute chronologies (primordial radioactive nuclides) with those provided by relative chronologies (based on extinct nuclides) has been a challenge to cosmochronologists. Until the last years there was no general consensus on the chronology for condensation, accretion and early planetary evolution obtained with the various chronometers (3). Very recently, more precise data were accumulated which appear to indicate that the coupling of the different radioactive clocks becomes a real possibility provided that the uncertainties related to the knowledge of some parameters are taken into account. These parameters are :
A) the values of the decay constants
B) possible isotopic heterogeneities
C) the closed system conditions which could -and should- be different for each parent-daughter system and for each specific material (low or high temperature lattice mineral sites, grain-size, glass, etc...).

    The parent-daughter systems mostly used in meteorite chronologies are shown in table 1. In the table are also shown two extinct nuclides not yet detected in meteorites (^{205}Pb, ^{247}Cm), but important for cosmochronologies. For instance, the limit of ^{247}Cm/^{235}U $\leq 4 \times 10^{-3}$ at the time of meteorite formation implies that the last actinide (major r-process) addition to the material in

95

A. Coradini and M. Fulchignoni (eds.), The Comparative Study of the Planets, 95–100.
Copyright © 1982 by D. Reidel Publishing Company.

Table 1 : COSMOCHRONOMETERS

Parent	Decay-type	Daughter	$T_{1/2}$ (yrs)
		Primordial nuclides	
^{147}Sm	α	^{143}Nd	106×10^9
^{87}Rb	β^-	^{87}Sr	48.8
^{187}Re	β^-	^{187}Os	42.9
^{232}Th	6α	^{208}Pb	13.9
^{238}U	8α	^{206}Pb	4.5
^{235}U	7α	^{207}Pb	0.7
(Th, U)		(^4He)	
^{40}K	β^+	^{40}Ar	1.31
		Extinct nuclides	
^{146}Sm	α	^{142}Nd	103×10^6
^{244}Pu	α, sp. fiss.	$(^{232}Th)(Xe_f, tr.)$	82
^{205}Pb	k-capture	^{205}Tl	20
^{129}I	β^-	^{129}Xe	17
^{247}Cm	3α	^{235}U	15.6
^{107}Pd	β^-	^{107}Ag	6.5
^{26}Al	β^+	^{26}Mg	0.74

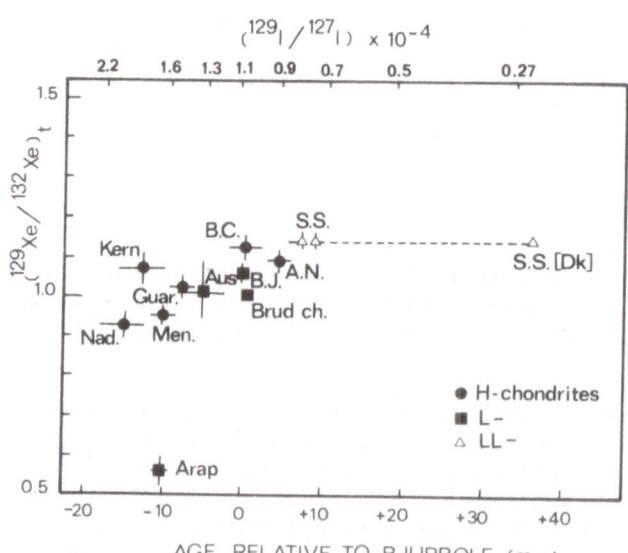

Figure 1 :

A.N.: Ambapur Nagla (H6)
Arap.: Arapohe (L5)
Aus.: Ausson (L5)
B.C.: Beaver Creek (H4)
Brud ch.: Bruderheim
          chondrule (L5)
Guar.: Guarena (H6)
Men.: Menow (H4)
Nad.: Nadiabondi (H5)
S.S.: St. Séverin (LL6)
      (Dk : Dark)

I-Xe ages of ordinary
chondrites relative to
Bjurbole standard (BJ)
(Data from 5, 12, 16, 17,
18, 19).

the protosolar nebula occurred > 70 m.y. prior to the formation of the solar system (1). Some relevant facts are the following ones :

1. The $^{207}Pb/^{206}Pb$ method is the most precise absolute chronology for meteorites (with a decay constant uncertainty of ∿ 5 m.y. for a 207/206 age). Chondrites and some achondrites were apparently formed between 4.559 ± 0.004 b.y. (Allende chondrite) and 4.551 ± 0.004 b.y. (St. Séverin chondrite and Angra dos Reis achondrite), i.e. in a narrow time interval of < 16 m.y. These age differences appear to be significant because they were obtained in one laboratory using the same analytical procedures (1).

2. The values of $^{87}Sr/^{86}Sr$ initial ratios for H, LL and E whole-rock chondrites are indistinguishable from the lowest value obtained for Allende (0.69877 ± 0.00002) (3 ; 8). Assuming a solar Rb/Sr ratio of 0.65, the difference between the $^{87}Sr/^{86}Sr$ initial ratios of chondrites corresponds to a time difference of < 17 m.y. which could correspond to the time elapsed between condensation and the accretion and formation of the chondritic asteroids. In the case of the equilibrated chondrite Guarena (H6), an internal isochron gives an initial ratio $^{87}Sr/^{86}Sr$ of 0.69995 ± 0.00015 which corresponds to a time interval of 100 ± 27 m.y. between the accretion and the blocking temperature of Sr isotopic homogenization in Guarena. This time interval is related to metamorphic processes inside the H asteroid (14 ; 3 ; 8). (Here, one has to note that the whole-rocks isochron of chondrites gives an age of 4.518 ± 0.026 b.y. which is much shorter than the above $^{207}Pb/^{206}Pb$ ages : this difference could be due either to the choice of the ^{87}Rb decay constant or/and to a lower closure temperature for the Rb-Sr than for the U-Pb system).

3. $^{129}I-^{129}Xe$ relative "ages" of meteorites defined by the (high temperature) $^{129}I/^{127}I$ ratio, itself a function of the ^{129}Xe retention in specific lattice mineral sites, cluster within a range of ∿ 50 m.y., corresponding to the spread between Nadiabondi (H5 chondrite) and St. Séverin (LL6 chondrite)(Fig. 1) (5 ; 4 ; 12). The significance of such large relative age differences detected through a short half-life radionuclide are not well understood. It has been suggested that I-Xe ages reflect relaxation of metamorphic conditions at high temperatures a short time after parent-body formation (11). However, materials coming from similar depths inside the H asteroid show very large I-Xe age differences (e.g. ∿ 11 m.y. between Menow and Beaver Creek, both H4 chondrites (5). It could be that the $^{129}I/^{127}I$ ratio for the solar nebula was heterogeneous. If so, the I-Xe ages are meaningless. It could be also that impacts inducing total outgassing occurred during accretion when the already accreted materials were kept at rather high temperatures (∿ 600 - 800°C) by the ^{26}Al decay. In this case, the I-Xe ages could correspond to the resetting of the clock.

Table 2

	ST. SEVERIN (LL6)		Approximate temperature (°C)	GUARENA (H6)	
	Time since t_o (m.y.)	Time (by ago)		Time since t_o (m.y.)	Time (by ago)
^{26}Al	<3	fast condens. and accretion	~200	<3	fast condensation and accretion
^{207}Pb–^{206}Pb	~12	4.551	900 ?	—	—
^{129}I–^{129}Xe	7 – 35	(~4.53)	800	12	(~4.56)
^{136}Xe retention in phosphates	45 (±20)	(~4.52)	700	~20	(~4.55)
^{40}Ar–^{39}Ar	160 (±10)	{ 4.42 ± 0.01 { 4.38 ± 0.01	400 350	80 ± 10	{ (~4.49) 4.44 ± 0.03
(^{87}Sr–^{86}Sr)$_i$	180 (±100)	(~4.40)	300	75 – 100	4.48 ± 0.08
Fission-track retention in phosphates	420	(~4.15)	100	140	(~4.42)
^{4}He retention	420	~4.15	100	—	—

(Radius of LL asteroid : 130 – 150 km)

(Radius of H asteroid : 50 – 70 km)

4. ^{40}Ar–^{39}Ar ages for a number of chondrites fall in the range
$\sim$ 4.37 – 4.52 (13 ; 4 ; 12 ; 2). For silicate inclusions of IA
irons ^{40}Ar–^{39}Ar ages cluster between $\sim$ 4.57 and $\sim$ 4.48 (± 0.03)
b.y. (6 ; 9). The spread in ages is much larger than individual
uncertainties, and here again the age differences could reflect
the different starting times of the clock, related to metamorphic
processes inside the parent bodies (11; 12 ; 2). In particular,
it has been estimated that the closure temperature for argon
ranges between $\sim$ 100° and 400°C in the case of H4-5-6 chondrites
(13). The lowest ^{40}Ar–^{39}Ar plateau age (4.37 ± 0.01 b.y.) corresponds to that of St. Séverin (LL6 chondrite)(12). On the other
hand, by analyzing mineral separates (pyroxene and feldspar) of
five H chondrites, Flohs (2) was able to deduce periods of metamorphism lasting 20 – 50 m.y. which give cooling rates between
pyroxene and feldspar retention temperatures of 2-10°/m.y., in
excellent agreement with fission track cooling rates obtained on
the same objects by Pellas and Storzer (10).

5.   Because fission xenon (from ^{244}Pu) retention in phosphates
took place at a higher temperature ($\sim$ 700 ± 100°C) than fission-
track retention ($\sim$ 100°C) in the same phosphates, it is possible
to define time intervals between the two blocking temperatures.
These Xe-track time intervals are of $\sim$ 120 m.y. and $\sim$ 380 m.y.
for equilibrated H6 and LL6 chondrites giving, for the central
regions of the parent objects cooling rates of $\sim$ 5 and $\sim$ 2°/m.y.
respectively (10). From these values, radii of $\sim$ 60 and $\sim$ 130 km
are estimated for the H and LL asteroids.

6.   With the demonstration of the presence of live ^{26}Al in solar
system material (7), it seems probable that ^{26}Al was the heating
source for the metamorphism of chondritic bodies and the differentiation of achondrites and irons. In this case, the last addition
of freshly synthesized nuclear material to the protosolar nebula,
the condensation and accretion of asteroidal parent bodies of meteorites should have taken place in a time scale extremely short
of $\sim$ 3 m.y. (15).

If all the above results are taken into account and if we try
to put together the various pieces (sometimes not self-consistent !)
of the puzzle, we can derive an approximate scenario relevant to
both Guarena and St. Séverin, the chondrites for which we have the
most abundant data. This attempt is shown in table 2.

REFERENCES

1) Chen J.H. and Wasserburg G.J. (1981), Earth Planet. Sci. Lett.
    52, 1-15.

2) Flohs I. (1981), EOS 62, 203.

3) Gray C.M., Papanastassiou D.A. and Wasserburg G.J. (1973),
    Icarus 20, 213-239

4) Hudson B., Kennedy B.M., Hohenberg C.M. and Podosek F.A. (1979),
    Meteoritics 14, 425-426

5) Jordan J., Kirsten T. and Richter H. (1980), Z. Naturforsch.
    35 a, 145-170

6) Kirsten T. (1973), Meteoritics 8, 400-403

7) Lee T., Papanastassiou D.A. and Wasserburg G.J. (1977), Ap. J.
    Lett. 211, L107-L110

8) Minster J.F. and Allègre C.J. (1979), Earth Planet. Sci. Lett.
    42, 333-347

9) Niemeyer S. (1979), Geochim. Cosmochim. Acta 43, 1829-1840

10) Pellas P. and Storzer D. (1981), Proc. R. Soc. London A 374,
    253-270

11) Podosek F.A. (1979), Meteoritics, 518-520

12) Podosek F.A. (1981), Meteorit. Soc. Meeting, Bern

13) Turner G., Enright M.C. and Cadogan P.H. (1978), Proc. Lun.
    Planet. Sci. Conf. 9th, 989-1025

14) Wasserburg G.J., Papanastassiou D.A. and Sanz H.G. (1969),
    Earth Planet. Sci. Lett. 7, 33-43

15) Wasserburg G.J., Papanastassiou D.A., Lee T. (1980), Early
    Solar System Processes and the Present Solar System, Soc. Ital.
    Fisica, pp. 144-191.

16) Podosek F.A. (1970), Geochim. Cosmochim. Acta 34, 341.

17) Drozd R.J. and Podosek F.A. (1976), Earth Planet. Sci. Lett. 31,
    15.

18) Hohenberg C.M., Hudson B., Kennedy B.M. and Podosek F.A. (1981),
    Geochim. Cosmochim. Acta 45, 535.

19) Jordan J., Kirsten T. and Richter H. (1979), Meteoritics 14,
    434.

SOLAR SYSTEM CRATERING CHRONOLOGY AND DATING OF THE SURFACE STRUCTURES OF THE TERRESTRIAL-TYPE PLANETS

Gerhard Neukum

Institut für Allgemeine und Angewandte Geologie
der Ludwig-Maximilians-Universität
8000 München 2, FRG

ABSTRACT

The ancient impact record of the terrestrial-type planets
Mercury, Mars, earth's moon, and of the satellites of Jupiter
and Saturn is discussed on the basis of data from spacecraft
imagery. The mass-velocity distribution of the impactors seems
to have been the same or very similar in the inner part of the
solar system and probably also at Jupiter and Saturn. Ancient
impact rates appear to have been comparable. The time dependence of the impact rate in the earth-moon system and by analogy
at the other terrestrial-type planets is in accordance with
a smooth rapid decay during the first 1000 million years of
solar system history rather than with a peak in impact rate
(cataclysm) at 4000 million years ago. From the reconstruction
of the martian impact chronology, Mars appears to have been
geologically active in its early times essentially.

1. INTRODUCTION

Cratering has been an important process in the geologic evolution of most terrestrial-type planets visited by spacecraft so
far. Thus, crater studies (abundances and morphologies) are
very useful for obtaining information on the geologic history
of those planets.

Furthermore, measurement of the crater size-frequency distributions on differently old parts of the terrestrial-type planets
give insight into the development of the mass-velocity spectrum
of the meteorites which produced the craters.

A. Coradini and M. Fulchignoni (eds.), The Comparative Study of the Planets, 101–116.
Copyright © 1982 by D. Reidel Publishing Company.

In the following, data on the cratering records and surface
histories - mainly on the basis of cratering studies - of the
terrestrial-type planets Mercury, earth's moon, Mars and the
moons of Jupiter and Saturn will be discussed.

2. METHODOLOGY

The basic method of determining relative ages of different areas
by crater statistics is to compare frequencies of impact craters
super-imposed on these areas. A premise in this procedure is
that one deals with the undisturbed image (i.e., the craters) of
the mass and velocity distribution of the impactors. A common
problem on planetary surfaces, however, is the interaction of
geologic processes with cratering, resulting in size-selective
destruction of craters. An example of the interaction of geologic processes with cratering is given in Fig. 1. Lava flows on
the volcano have destroyed smaller craters; in the surroundings
permafrost melting has caused analogous effects.

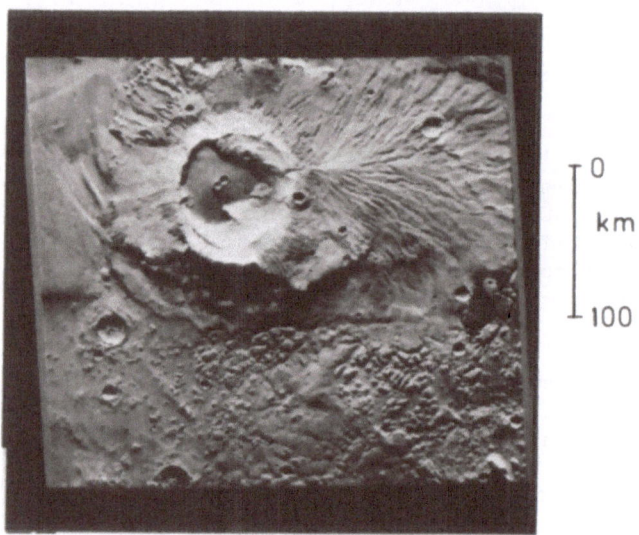

Figure 1. Mars volcano Apollinaris Patera

Therefore the distributions one finds on the terrestrial-type
planets typically do not show the undisturbed production size-
frequency state but have recorded resurfacing effects. Those
effects show up in irregularities in the distributions ('bumps')
or flattening toward smaller sizes. Such irregularities mean
that different sizes of craters had different retention times and

therefore one measures different retention ages at different
sizes, i.e., different ratios of actually determined crater
frequencies to the production size-frequency distribution at
different diameters.

Irregularities in the distributions can also be produced by
different target response to cratering due to factors such as
different lithologies, contrasting strength of the target substrate for craters of different sizes, water content, layering,
etc., as discussed especially for the lunar and/or Martian case
in the literature (References 1-6). Neukum and Hiller (Ref. 7)
discussed the effects for the Martian case. Neukum and Wise
(Ref. 8) also in general discussed the effects of different
impact conditions on a size-frequency distribution curve whose
slope (on a log-log diagram) is different at different crater
sizes. The different effects are shown in Fig. 2.

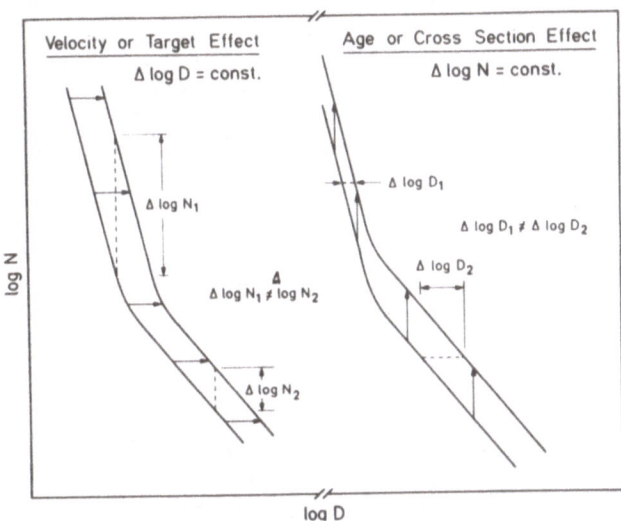

Figure 2. Effects of different impact conditions on a
size distribution curve whose slope is different at
different crater sizes; N is the cumulative number of
craters with diameters greater than a given diameter D.
Velocity or target effects cause the curve to shift
left or right while maintaining a constant change in
log D. The band in the curve produces an inconstant ▲ N
for different diameters. For age or cross-section
effects the curve shifts vertically and the reverse
relations obtain.

Age effects - that is, similar ratios of exposure times of the
surfaces to cratering - will result in similar ratios of crater
frequencies at any diameter (Δlog N = constant for the right
half of Fig. 2). The same is true for different cross-section
effects, in which slower meteorites are more easily deflected
by a planet's gravity. This is equivalent to a difference in
flux. These effects do not change the shape of the curve - that
is, the crater frequency at some diameter in the steep part of
the curve relative to the value at some diameter in the flat
part of the curve. In other words, age differences and cross-
section effects are such that the curves can be shifted vertically and should coincide. The shape of the curve, in terms of
the diameter at which the inflection point occurs in Fig. 2, does
not remain constant if the impact velocity is different on different planets, or if target properties have an effect on the
sizes of the craters produced. Thus, even with the same meteoroid mass distribution, different (average) impact velocities on
different planets or different target properties will result in
different crater sizes for the same projectile mass, and the
shapes or inflection points of the curves for the different
planets will differ. The ratio of crater frequencies at two
fixed diameter values in the steep and flat parts of the curve
will be different, as seen in the left half of Fig. 2 (Δlog N1
$\neq$ Δlog N2 for a diameter shift by a constant factor).

The distributions for the moon and Mars are known rather well
(Ref. 7) as shown in Fig. 3. The difference in the shapes of the
curves can largely be explained by a difference in impact velocity at both planets, being about a factor of 1.5 lower at Mars
than at the moon.

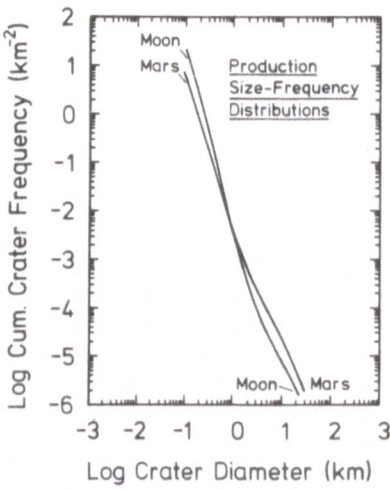

Figure 3. Lunar and Martian
cumulative impact crater
production size-frequency
distributions (standard
curves, reference distributions)

3. ANCIENT LUNAR IMPACT RECORD

An important basis for understanding the cratering time dependance in primarily the inner solar system are results from lunar
studies. Our present knowledge of the lunar cratering chronology
largely stems from km-sized craters. Combination of crater
frequency data and rock ages yield the picture as in Fig. 4.

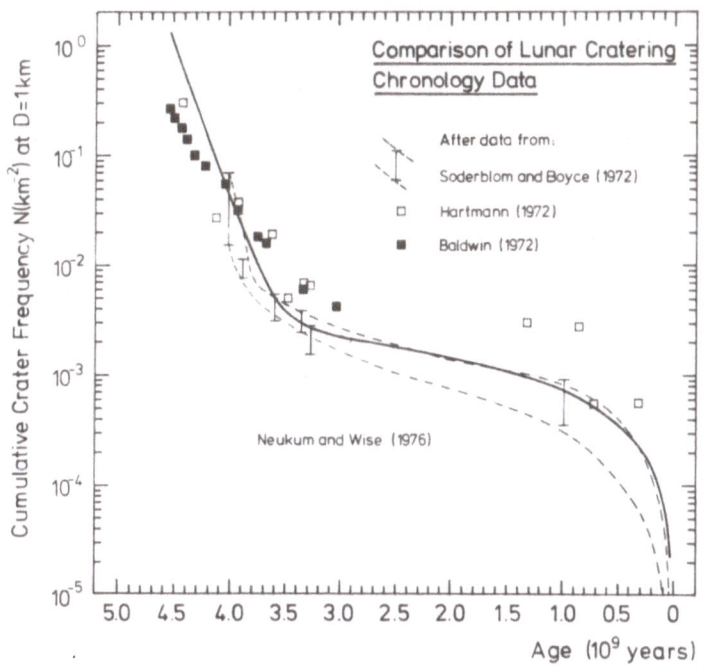

Figure 4. Comparison of lunar cratering chronology data

This is a comparison of lunar cratering chronology data of
different workers. The different data agree within a factor of
2 to 3. Own data are shown in Fig. 5 for details. The radiometric
age data are taken from the literature (the reader is especially
referred to Ref. 9). The rapid decrease of the impact rate in
the first 1.5×10^9 years of the existence of the earth-moon
system is well documented by the data. No data points exist for
the time of between 3 and 1 billion years ago. The comparison
of the data at $\gtrsim 3$ b.y. with those at $\lesssim 1$ b.y. suggests a constant
cratering rate during that time interval and at more recent
times as given by the curve in Fig. 5 for the cumulative crater
frequency (= integrated cratering rate) for that time.

Lunar highland rock age data show a strong peak at ~4 b.y. and
a minor peak at ~4.25 b.y. Those peaks have generally been inter-

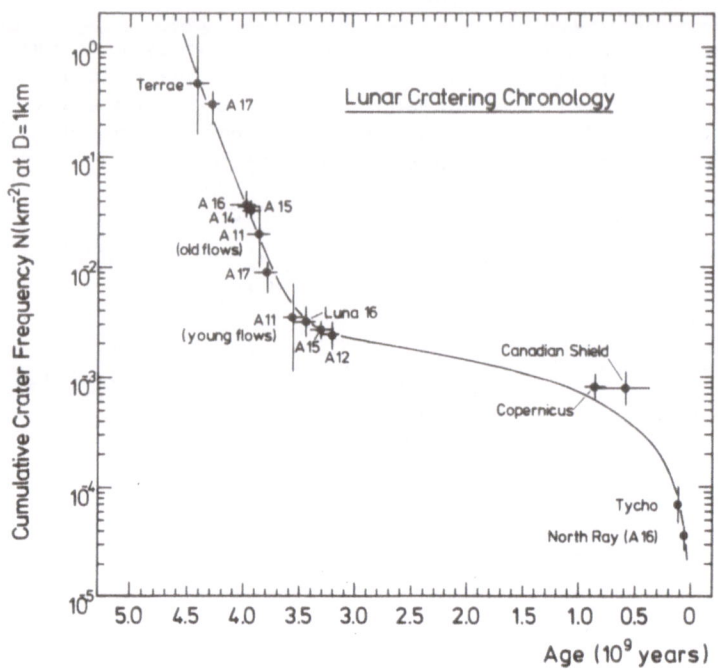

Figure 5. Lunar cratering chronology from own measurement (Ref.)

preted as showing the time of resetting the radometric ages by
basin-forming impacts (e.g. Ref. 10). Much depends on the interpretation of those ages. The clustering of ages around 4 b.y.
has led Tera et al. (Ref. 11) to the interpretation of a "terminal cataclysm" around 4 b.y. ago, i.e. most of the large scale
cratering (basin cratering) at least of the frontside of the
Moon would have happened at that time. The different possibilities
for the cratering rate dependance are shown in Fig. 6.
The cratering rate (= time derivative of the cratering chronology)
on the Moon as a function of time is given with the different poss
ibilities of a smooth decay, a hypothetical "cataclysm" at ca.
4 b.y. ago, and with a theoretical model from Wetherill (Ref. 12).

Recently, we have investigated the cataclysm question in detail.
If there was a terminal cataclysm, then it must show up in the
relative ages of lunar basins and basin production rates as a
function of time. In addition, if there was such a singular activity, one might expect a change in the production size frequency
distribution of lunar craters. I will give arguments for a smooth
decay of the impact rate from crater frequency data superimposed

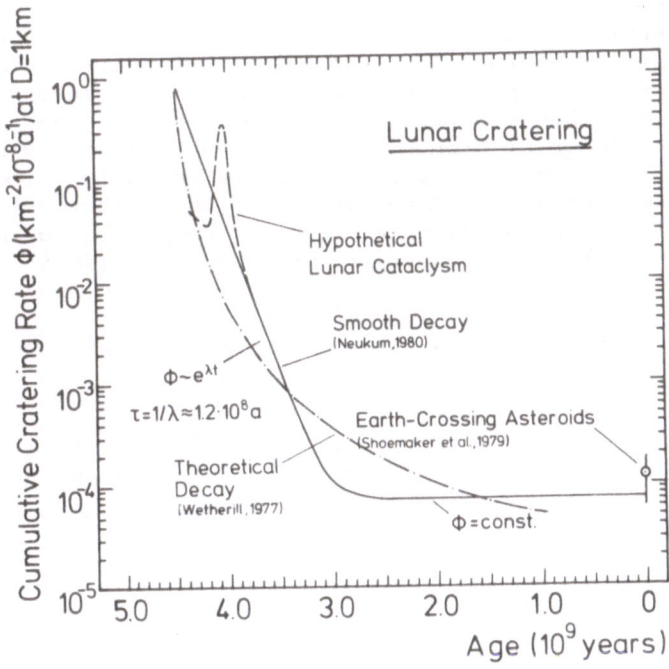

Figure 6. Time dependance of lunar cratering rate

on a number of pre-Nectarian, Nectarian, and Imbrian basins in comparison with the overall large-crater frequency of the oldest parts of the lunar farside. The relative ages of lunar features can be referred to the lunar stratigraphic scheme as given in Fig.7.

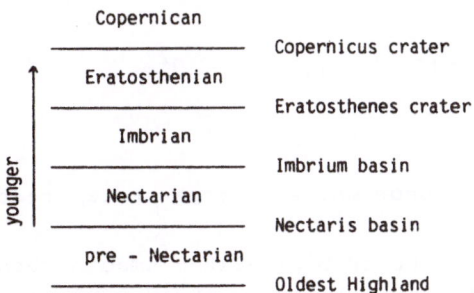

Figure 7. Lunar Stratigraphy

Fig. 8 shows the large-crater frequencies for Imbrian basins,
Nectarian basins, pre-Nectarian basins, and the Lunar Highlands,
and Copernican/Eratosthenian craters (Ref. 13). The data demonstrate constancy of the size-frequency distributions of large craters
over much of the past. The Nectarian crater frequency is a factor
of 3 higher than the Nectarian crater frequency and is slightly
below the lunar highland crater frequency. That means that most
large craters were produced during pre-Nectarian times and not
during Imbrian times. A comparison of the numbers of basins on
the moon (pre-N, N, I) give a similar proportion of production
numbers as crater frequency number relations (Table 1).

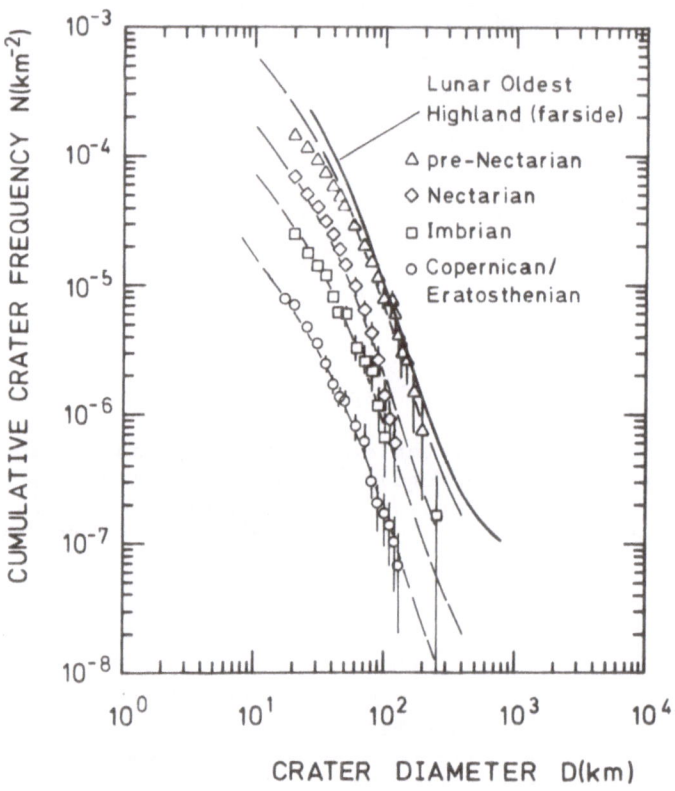

Figure 8. Lunar ancient large-crater record

Consequently there is no problem with a smooth decay history,
but with a terminal cataclysm which would produce most of the
large basins at 4 b.y. ago.

Table 1: Lunar basin occurrence

Total number of basins on the moon:
(D>265 km)

    43
    (including 12 probable or possible ones)

Thereof 30   pre-Nectarian
         (including 12 probable or possible ones)

         10   Nectarian

          3   Imbrian

4. ANCIENT IMPACT RECORD OF THE PLANETS MERCURY, MARS, AND OF THE MOONS OF JUPITER AND SATURN

The other terrestrial planets, Mercury and Mars, and as we have learned recently (Ref. 14,15) the moons of Jupiter and Saturn have also experienced heavy bombardments in their early histories. In Fig. 9 the cratered surface of Ganymede is shown as an

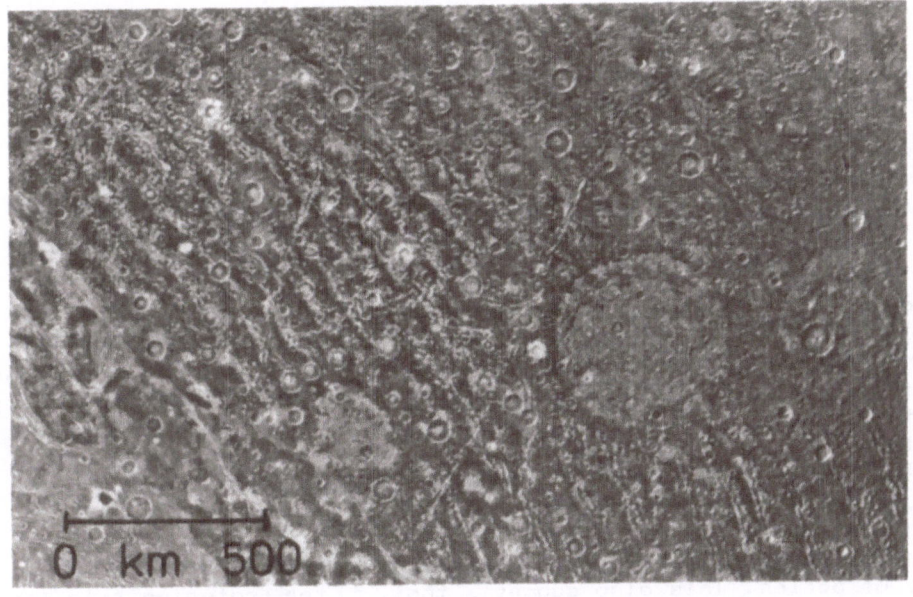

Figure 9. Cratered surface of the oldest (dark) terrain, Galileo Regio, of Jupiter's moon Ganymede

example. Crater frequencies for Mercury, Mars, Ganymede, and for some Saturnian moons have been determined (Fig. 10 and 11).

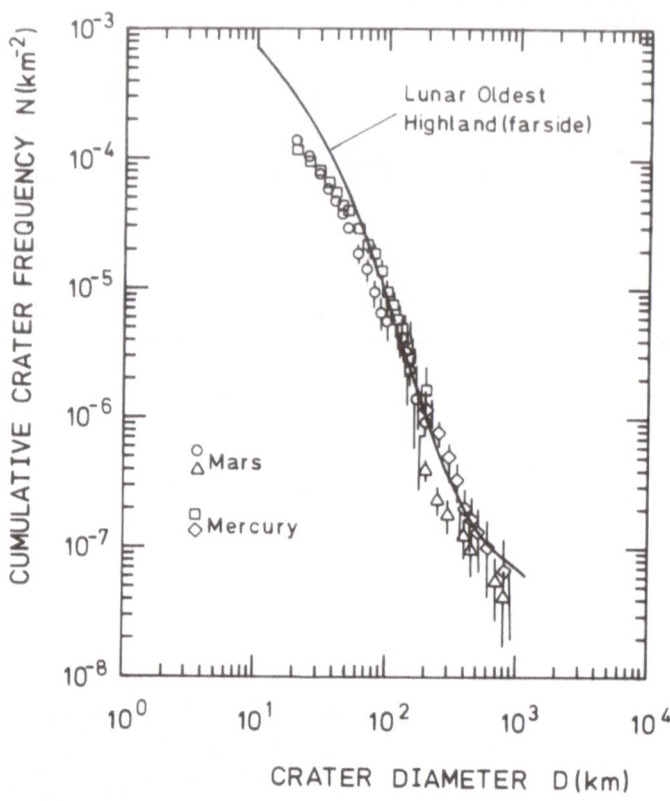

Figure 10. Cratering record of the oldest terrains of Mercury and Mars in comparison with lunar highland cratering

The crater frequencies on the oldest parts of all those terrestrial-type planets look very much alike with respect to the shape of the distributions and with respect to the absolute crater densities. In fact, with the assumption of reasonable impact velocity differences at the different planets which would result in diameter shifts with respect to the lunar curve taken as a basis (cf. previous discussion and Fig.2), the distributions for Mars and Mercury would fall right on the lunar curve. This suggests, that the mass spectrum of the impactors was the same throughout the inner solar system which would probably mean the same family of bodies. This also suggests that, if the cratering rate was comparable, that we in all cases look back into the very early history of the solar system 100 to 200 million years after its

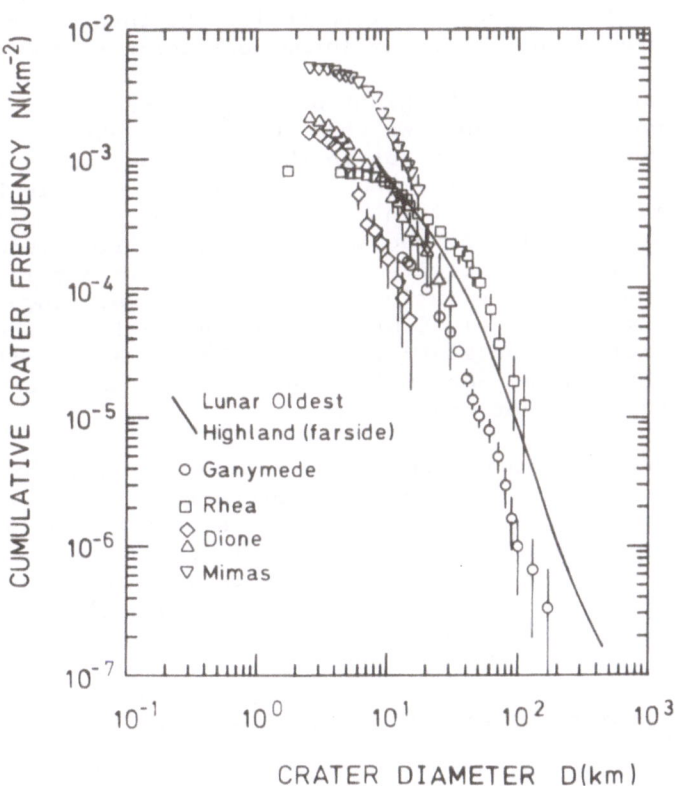

Figure 11. Crater frequencies on Jupiter's satellite Ganymede (Galileo Regio) and on Saturn's moons Rhea, Dione, and Mimas

Table 2: Saturn Satellite Impact Conditions

	Distance from Saturn	Impact velocity from Saturn's gravitational acceleration ($v_\infty = 0$)	Impact velocity from Saturn's grav. acceleration for rel. velocity at infinity of $v_\infty = 5$ km/s
	r (km)	V_{acc} (km/s)	V_{imp} (km/s)
MIMAS	$1.86 \cdot 10^5$	20.2	20.8
DIONE	$3.77 \cdot 10^5$	14.2	15.1
RHEA	$5.27 \cdot 10^5$	12.0	13.0

formation. Recently, it could be shown by various workers (e.g. Ref. 16) that the impact rates for the inner planets were indeed comparable very probably throughout solar system history.

For the moons of the giant planets Jupiter and Saturn, the situation is different: we know very little about impact rates, neither current ones nor in the past. The impact situation is peculiar as shown for the Saturn system in Table 2.
The giant planet Saturn controls the impact conditions on its moons, as similarly Jupiter in its satellite system. The impact velocity is practically determined by its gravitational attraction. But the gravitational impact cross-section, i.e. the cratering rate is largely controlled by the ratio of the velocity from gravitational acceleration to the relative velocity of the encountering meteoroidal bodies at infinity taken 5 km/s here but which is not really known.

The unique composition of the Saturnian moons, however, will give us a means to determine those properties. The moons consist largely of ice with a small silicate component as deduced from their densities and reflectivities. Radioactive heating and resurfacing was probably minor. Thus we look very far back into the past, probably close to 4.5 billion years. Detailed measurements and calculations will for the first time allow us in the near future to narrow down the meteoroid fluxes and relative encounter velocities at infinity for the Saturnian system for the time right after the formation of the solar system.

5. MARS IMPACT CHRONOLOGY AND SURFACE HISTORY

In case of the planet Mars, it has been possible to derive a more detailed knowledge of its cratering history and consequently of its geologic history. Fig. 12 shows the present knowledge of the martian cratering chronology. (The crater frequencies N(4) - N(10) given for 4 to 10 km craters can directly be compared with 1 km values through the application of the Mars production size-frequency distribution as given in Fig. 3.

A detailed discussion is given in Refs. 7,17-19. Though there are still great uncertainties, the model chronologies of Fig. 12 (model I: solid curve; model II: dashed curve) in combination with a variety of crater statistics data (Ref. 7) allow a reconstruction of the gross features of martian geologic history as given in Fig. 13.

Model independent, it is evident, that all major geological events like various types of volcanism, large scale erosion and tectonism occurred simultaneously during the early times of martian history. After this period of activity (model dependent after

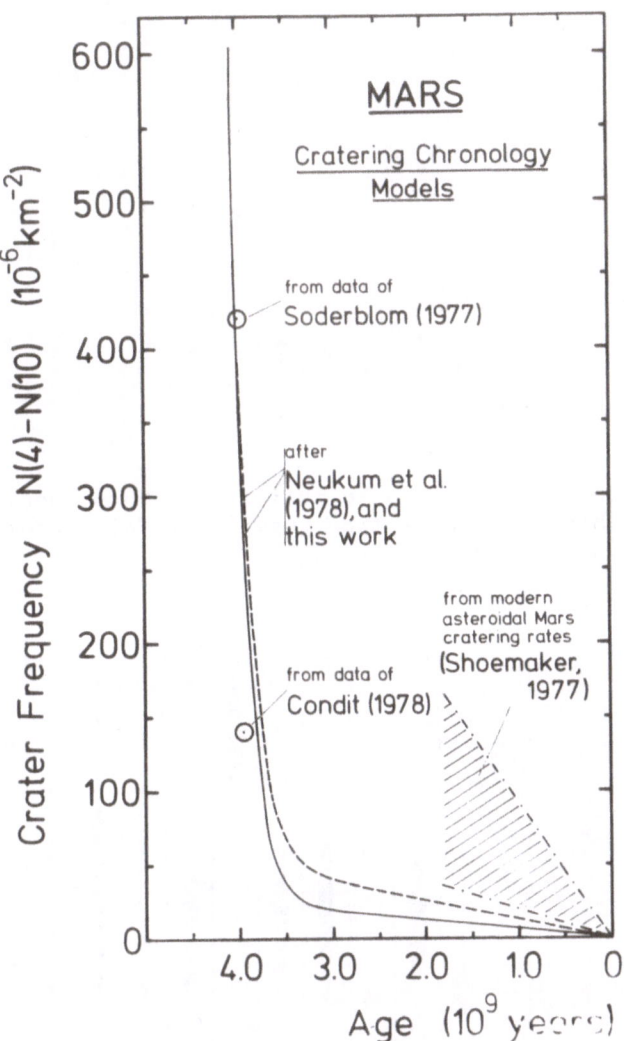

Figure 12. Mars cratering chronology models

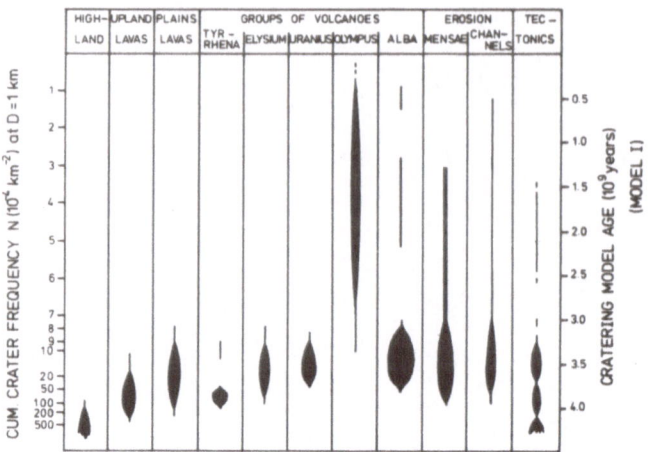

Figure 13 a. Geologic activity of Mars, occurrence of events. Interpretation based on model I cratering chronology.

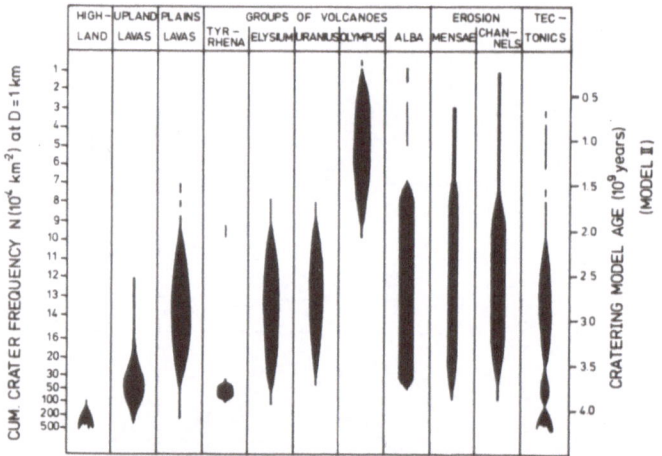

Figure 13 b. Geologic activity of Mars, occurrence of events. Interpretation based on model II cratering chronology.

3.0 or 1.5 aeons ago) shield volcanism in the Tharsis volcanic region, some tectonism and minor wind erosion were the only forces reshaping the surface of Mars probably until several 100 million years ago.

6. REFERENCES

1. Boyce, J. M., and Roddy D.J., 1978, Martian rampart craters: Crater processes that may affect diameter-frequency distribution, Reports of Planetary Geology Program, 1977-1978, NASA Tech. Memo., TM 79729, 162-165.

2. Boyce, J.M., 1979, Diameter enlargement effects on crater populations resulting from impact into wet or icy targets, Reports of Planetary Geology Program, 1978-1979, NASA Tech. Memo., TM 80339, 119-122.

3. Croft, S.K., Kieffer, S.W., and Ahrens, T.J., 1979, Low velocity impact craters in ice and permafrost with implications for Martian cratercount ages, NASA Conf. Publ., 2072, 18.

4. Mouginis-Mark, P., 1979, Martian fluidized crater morphology, Variations with crater size, latitude, altitude, and target material, J. Geophys. Res., 82, 4379-4388.

5. Schultz, P.H., and Spencer, J., 1979, Effects of substrate strength on crater statistics: Implications for surface ages and gravity scaling (abstract), Lunar Planet, Sci. Conf. 10th 1081-1083.

6. Young, R.A., Brennan, W.J., and Nichols, D.J., 1974, Problems in the interpretation of lunar mare stratigraphy and relative ages indicated by ejecta from small impact craters, Proc. Lunar Sci., Conf. 5th, 159-170.

7. Neukum G., and Hiller, K., 1981, Martian Ages, J. Geophy. Res. 86, 3097-3121.

8. Neukum, G., and Wise, D.U., 1976, Mars: A standard crater curve and possible new time scale, Science 194, 1381-1387.

9. Turner, G., 1977, Potassium-argon chronology of the moon, Phys. Chem. Earth 10, 145 ff.

10. Kirsten, T., 1979, Lunar Highland Chronology- In: Origin and Distribution of the Elements (Editor Ahrens, T . J.), Pergamon Press, 91-98.

11. Tera, F., Papanastassiou D., and Wasserburg. G., 1974, Isotopic evidence for a terminal lunar cataclysm. Earth Planet Sci. Lett. 22, 1-21.

12. Wetherill, G.H., 1977, Evolution of the earth's planetesimal swarm subsequent to the formation of the earth and moon, Proc. Lunar Sci. Conf. 8th, 1-16.

13. Wilhelms. E., 1980, personal communication of raw data except for lunar highlands.

14. Smith, B.A., Soderblom, L.A., Beebe, R., Boyce, J., Briggs, G., Carr, M., Collins, S.A., Cook II, A.F., Danielson, G.E., Davies, M.E., Hunt, G.E., Ingersoll, A., Johnson, T.V., Masursky, H., McCauley, J., Morrison, D., Owen, T., Sagan, C., Shoemaker, E.M., Strom, R., Suomi, V.E., and Veverka, J., 1979, The Galilean Satellites and Jupiter: Voyager 2 Imaging Science Results, Science 206, 927-950.

15. Smith, B.A., Soderblom, L., Beebe, R., Boyce, J., Briggs, G., Bunker, A., Collins, S.A., Hansen, C.J., Johnson, T.V., Mitchell, J.L., Terrile, R.J., Carr, M., Cook II, A.F., Cuzzi, J., Pollack, J.B., Danielson, G.E., Ingersoll, A., Davies, M.E., Hunt, G.E., Masursky, H., Shoemaker, E., Morrison, D., Owen, T., Sagan, C., Veverka, J., Strom, R., Suomi, V.E., 1981, Encounter with Saturn: Voyager 1 Imaging Science Results, Science 212, 163-190.

16. Soderblom, L.A., 1977, Historical variations in the density and distribution of impacting debris in the inner solar system: Evidence from planetary imaging. In: Impact and Explosion Cratering (Pergamon Press), 629-633.

17. Hartmann, W.K., 1978, Martian Cratering V: Toward an Empirecal Martian Chronology, and its Implications, Geophys. Res. Lett. 5, 450-452.

18. Shoemaker, E.M., 1977, Astronomically observable crater-forming projectiles. In: Impact and Explosion Cratering (Pergamon Press), 617-628.

19. Condit, D.C., 1978, Distributions and relations of 4- to 10-km-diameter craters to global geologic units of Mars, Icarus 34, 465-478.

THE STABILITY OF THE SOLAR SYSTEM

Archie E. Roy

Department of Astronomy, Glasgow University, Scotland.

The empirical stability criteria approach to the stability of hierarchical dynamical n-body systems is described. Application of this approach is made to the problem of the long-term stability of the solar system's hierarchical systems.

1. INTRODUCTION

The solar system consists of the Sun and its family of planets, the planetary satellite systems, the asteroids, comets and meteors and the three ring systems belonging to Jupiter, Saturn and Uranus. In what follows we will omit from consideration the asteroids, comets and meteors, the ring systems and their attendant newly-discovered satellites.

The planetary system and the satellite systems are obviously examples of many-body hierarchical dynamical systems. A hierarchical dynamical system is one in which it is possible to order the orbits in ascending size, to use conveniently a coordinate system such as Jacobi's where the radius vector of the i-th mass is referred to the centre of mass of the first (i-1) masses, and to expect the original ordering of the orbital sizes to be maintained for a period of time at least as long as the largest orbital period of revolution.

All the systems considered in the solar system are capable of being described in this hierarchical mode. For example, the Earth-Moon-Sun system is a general three-body hierarchical dynamical system (HDS) with the Moon (m_2) in orbit about the Earth (m_1) and the Sun (m_3) considered to be in orbit about the Earth-Moon barycentre. Again, in the Sun and planets case, the

117

A. Coradini and M. Fulchignoni (eds.), The Comparative Study of the Planets, 117–124.
Copyright © 1982 by D. Reidel Publishing Company.

members of this system form a HDS, beginning with the Sun as m_1, Mercury as m_2, Venus as m_3 and so on.

The equations of motion of the n bodies forming a HDS, in Jacobian form, are expressed as follows (5)

$$\frac{m_i M_{i-1}}{M_i} \ddot{\rho}_i = \nabla_i U, \qquad (1)$$

where

$$U = \tfrac{1}{2} G \sum_{k=1}^{n} \sum_{\ell=1}^{n} \frac{m_k m_\ell}{r_{k\ell}} , \quad \ell \neq k \qquad (2)$$

is the force function. In these equations

m_i denotes the ith mass, $i = 0,1,2,\ldots,n$; $(m_0 = 0)$,

$M_i = \sum\limits_{j=0}^{i} m_j$,

$\rho_i = R_i - \bar{R}_{i-1}$, R_i and $\bar{R}_i$ being the position vectors of

m_i and the mass-centre of $(m_1,m_2,\ldots,m_i)$ respectively in an inertial system,

$\nabla_i = i \dfrac{\partial}{\partial x_i} + j \dfrac{\partial}{\partial y_i} + k \dfrac{\partial}{\partial z_i}$, i, j, k being unit vectors,

$r_k = R - R_k$; $r_{k\ell} = |r_{k\ell}|$.

Each body's radius vector is taken from the centre of mass of all the bodies lower down in the hierarchy. Thus Jupiter's radius vector is drawn from the centre of mass of Sun, Mercury, Venus, Earth and Mars.

In particular when n = 3 in equations (1) and (2) we have the general three-body equations of motion in Jacobi coordinates.

The usual integrals of energy and angular momentum may be formed. In essence we have already used the system's centre of mass integrals in forming the equations of motion in a Jacobi coordinate system.

No other integrals have ever been discovered. Nevertheless, in the case of the general three-body problem, a very useful time-invariant statement analogous to the Jacobi integral in the restricted three-body problem may be made in the case of certain three-body HD systems.

2.   THE ZARE CRITERION

Let the three-body system consist of three finite point-masses P_1, P_2 and P_3 of masses m_1, m_2 and m_3, respectively. Suppose P_2 is in orbit about P_1, with P_3 in orbit about the centre of mass C_{12} of P_1 and P_2. Then with n = 3 equations (1) and (2) give the behaviour of P_2 with respect to P_1 and of P_3 with respect to C_{12}.

Such a system is obviously a hierarchical dynamical system, consisting as it does of a binary $(P_1 - P_2)$ about which a third body orbits in a larger orbit.

In recent years a number of authors (1, 3, 8, 9) have shown that it is possible to establish a condition enabling a decision to be made about the permanency or otherwise of the binary. This is analogous to the use of surfaces of zero velocity in the restricted three-body problem to investigate whether or not the massless particle must remain in orbit about one of the massive particles.

For example let the energy and angular momentum integrals be formed from equations (1) and (2). Let the total energy be E and the total angular momentum vector be $\mathcal{C}$. Then, following Zare (8, 9), it may be shown that the stability or otherwise of the binary is controlled by the value of the parameter $S = |\mathcal{C}^2| \, E$. The value of S is, of course, known from the initial values of the masses and the position and velocity components appearing in the energy and angular momentum relations. If S is smaller than or equal to a critical quantity, S_{cr}, which can be computed, then the binary cannot be broken up by the third mass. If, however, $S > S_{cr}$, then break-up may occur. Zare's criterion $S \leqslant S_{cr}$, may therefore be usefully applied to any general three-body problem found in nature of the hierarchical type (binary plus third body). Examples of these are triple stellar systems, Planet-Moon-Sun, Sun-Jupiter-Saturn, though in each case, although the general three-body problem model is a close approximation to the system found in nature, the presence of other perturbing bodies cannot be totally disregarded.

The quantity S_{cr} is computed from the values of the three masses, applying the Lagrange collinear solution of the three-body problem (3, 8, 9).

3.   EMPIRICAL STABILITY CRITERIA

The work by Emslie, Roy and Walker (2, 4, 5, 6, 7) on a statistical empirical approach to the question of the stability of hierarchical dynamical systems (n ≥ 3) makes use of Zare's criterion. A HDS is said to be stable if the ordering of the system's orbital radius vectors in a Jacobian coordinate system does not change in a time interval long in terms of the largest period of revolution present.

The approach involves the establishment of a set of parameters, the epsilons. They are a measure of the disturbance of the elliptic orbit of the i-th mass (i = 2,3,...,n) about the centre of mass of the first i-1) masses. It is found that their sizes may be used to predict in a statistical sense a time interval that must elapse before a change in the HDS's hierarchical order is to be expected.

To obtain the epsilons we proceed as follows. It may easily be shown that

$$\underset{\sim}{r}_{k\ell} = \underset{\sim}{\rho}_\ell - \underset{\sim}{\rho}_k + \sum_{j=k}^{\ell-1} \frac{m_j}{M_j} \underset{\sim}{\rho}_j .$$

Applying this relationship to the expansion of U in Equation (2), the following expression, correct to the second order, may be obtained, viz.

$$\underset{\sim}{\ddot{\rho}}_i = G M_i \underset{\sim}{\nabla}_i \left| \frac{1}{\rho_i} \{ 1 + \sum_{k=1}^{i-1} \varepsilon^{ki} P_2(C_{ki}) + \sum_{\ell=i+1}^{n} \varepsilon_{\ell i} P_2(C_{i\ell}) \} \right| , (3)$$

where

$$\varepsilon^{ki} = \frac{m_k M_{k-1}}{M_k M_{i-1}} \alpha_{ki}^2 \; ; \quad \varepsilon_{\ell i} = \frac{m_\ell}{M_i} \alpha_{i\ell}^3 \; , \quad i=1,2,...,n. \quad (4)$$

In these expressions,

$$\alpha_{ij} = \rho_i/\rho_j < 1; \quad C_{ij} = \frac{\underset{\sim}{\rho}_i \cdot \underset{\sim}{\rho}_j}{\rho_i \rho_j} ; \quad \rho_i = |\underset{\sim}{\rho}_i|$$

while P_2 is the Legendre polynomial of order 2.

On examination it is seen that the first term of the right hand side of equation (3) represents the undisturbed elliptic motion of the i-th mass about the mass-centre of the sub-system of masses $m_1,...,m_{i-1}$, while the ε^{ki}, $\varepsilon_{i\ell}$ provide a measure of the disturbance of the elliptic motion by the remaining masses, ie masses other than the i-th. It may be noted that a superscripted ε denotes the disturbance of a body by an inferior body (smaller orbit) while a subscript denotes the disturbance of a body by a superior body.

If n = 3, equations (3) and (4) reduce to

$$\ddot{\underset{\sim}{\rho}}_2 = GM_2 \; \underset{\sim}{\nabla}_2 \left(\frac{1}{\rho_2} (1 + \epsilon_{32} \; P_2 \; (C_{23}))) \right), \qquad (5)$$

$$\ddot{\underset{\sim}{\rho}}_3 = GM_3 \; \underset{\sim}{\nabla}_3 \left(\frac{1}{\rho_3} (1 + \epsilon^{23} \; P_2 \; (C_{23}))) \right), \qquad (6)$$

with

$$\left. \begin{array}{ll} \epsilon^{23} = \dfrac{m_2 \, M_1}{M_2^2} \; \alpha_{23}^2 = \dfrac{m_1 \, m_2}{(m_1 + m_2)^2} \; \alpha_{23}^2, & \\[3ex] \epsilon_{32} = \dfrac{m_3}{M_2} \; \alpha_{23}^3 = \dfrac{m_3}{m_1 + m_2} \; \alpha_{23}^3 \; . & \end{array} \right\} \qquad (7)$$

Thus ϵ_{32} is a measure of the ratio of the disturbance by P_3 on P_2's orbit about P_1, to the central two-body force between P_2 and P_1. Likewise ϵ^{23} is a measure of the ratio of the disturbance by P_1 and P_2 on the orbit of P_3 about the centre of mass of P_1 and P_2, to the central two-body force between P_3 on the one hand and P_1 and P_2, assumed to lie at their mass-centre.

If we introduce μ and μ_3 by the relations

$$\mu = m_2/(m_1 + m_2) \; ; \; \mu_3 = m_3/(m_1 + m_2) , \qquad (8)$$

then

$$\epsilon^{23} = \mu(1 - \mu) \; \alpha_{23}^2 \; ; \; \epsilon_{32} = \mu_3 \, \alpha_{23}^3 \; . \qquad (9)$$

We now examine this picture in the light of Zare's stability criterion (Section 2) based on the quantity $S = |\underset{\sim}{C}|^2 E$, where C and E are the constants appearing respectively in the angular momentum and energy integrals of the general three-body problem. If the three-body system was a hierarchical one (a binary plus a third body in a large orbit about the binary's mass-centre), and $S \leqslant S_{cr}$, the binary could never be broken-up. The critical stability value S_{cr} was derived from the collinear solution of the general three-body problem. To obtain S_{cr}, a ratio X must be found, where X was the solution of Lagrange's quintic equation (3, 8, 9). In its turn $\alpha_{cr} = (\rho_2/\rho_3)_{cr}$ is related to X.

Now X, and therefore α_{cr}, are functions only of the values of m_1, m_2 and m_3, or μ, μ_3, so that S_{cr} is itself directly related in a known way to μ, μ_3, α_{cr}.

The quantity $\alpha = \rho_2/\rho_3$, however, is independent of μ and μ_3 as is S, both being fixed in value by the initial setting-up of the hierarchical three-body problem. If we assume that the three-body system is set off initially in circular, coplanar orbits (P_2 about P_1; P_2 about the mass-centre of P_1 and P_2), then to the stability criterion of Zare, namely $S \leqslant S_{cr}$, there corresponds the stability criterion $\alpha \leqslant \alpha_{cr}$, for a given $\alpha = \rho_2/\rho_3 = a_2/a_3$ (the radii of the initially circular orbits) and a given μ, μ_3. Note that the value of α_{cr} is dictated solely by μ, μ_3 and the solution of Lagrange's quintic equation in μ, μ_3 and X.

Thus for all pairs of possible values of μ and μ_3, plotted on the $\mu - \mu_3$ plane, a surface of values of α_{cr} exists above it in the third dimension α. For a hierarchical three-body problem with initially circular, coplanar orbits, therefore, α is known, as is μ and μ_3. The point μ, μ_3, α can therefore be plotted. If it lies above or on the point μ, μ_3, α_{cr}, the system is stable in the sense that the binary $P_1 - P_2$ cannot be broken-up.

From relations (9), it is obvious that a system may be expressed not only as a set of values μ, μ_3, α but also as a set of values ε^{23}, ε_{32}, α. Calculating α_{cr} from μ, μ_3 and the Lagrange quintic equation gives, by substitution in (9) values $(\varepsilon^{23})_{cr}$, $(\varepsilon_{32})_{cr}$. It is thus possible to use Zare's criterion in relation to the ε-parameters as well as to the μ-parameters.

The solar system and satellite systems can now be broken into hierarchical three-body subsets. Examples might be Sun-Jupiter-Saturn, Earth-Moon-Sun, Jupiter-Io-Europa, Sun-Earth-Uranus, and so on, the first two in each set forming the binary, the third being looked upon as being in orbit about the mass-centre of the first two. If this is done and the relevant ε parameters are computed so that Zare's criterion of stability may be applied, it is found that, with certain exceptions, the criterion is well satisfied with the real alphas being all much smaller than the α_{cr} values for these systems.

Several comments are necessary.

The exceptions include the retrograde satellites of Jupiter, which is satisfactory since they are possibly captured asteroids and could well escape again.

Eccentricities and inclinations have been neglected. It is probable, however, that if they were included, the solar system results would be essentially unchanged. It is planned to carry out this work.

No triple subset is, of course, totally isolated gravitationally from other members of the solar system. The Sun, Jupiter and Saturn have often been spoken of as essentially making up the solar system, with a little bit of debris left over, such as Earth, Venus and so on. But even the triple subset of Sun-Jupiter-Saturn is to some measure disturbed. The important question from the point of view of stability is therefore: what effect in the long term will these additional perturbations produce?

Although the subset now satisfies Zare's criterion, its alpha value lying a good way below α_{cr}, the system is being disturbed. The alpha height that its point in the $\epsilon_{32} - \epsilon^{23} - \alpha$ space lies at will move in a pseudo-random or pseudo-periodic fashion because of the disturbances by the other bodies. As long as the point lies below α_{cr}, the subset is stable in Zare's sense. The orbits of Jupiter and Saturn will not intersect. But if the point wanders in a sort of random walk so that it ultimately reaches a situation where $\alpha > \alpha_{cr}$, then the subset may become unstable.

The same argument applies to other triple subsets. Equations (3) suggest that the epsilons may well be the crucial parameters in a consideration of the long term stability of the solar system. They are a measure of the disturbances that each body produces on the others' orbits. It is not expected that the computation of the epsilons for a HDS would lead to a determination of the precise time when any two orbits will interchange or a body escape from the system. It should be possible, however, to make a statistical prediction from such data. This statistical, empirical approach to solar system stability is being investigated in a long term programme by Roy and Walker, part of the work involving the numerical integration of general n-body systems (n = 3,4). In the case n = 4, studies have been made to see how different initial sets of starting conditions (the ϵ, α and μ values) govern the time it takes for such four-body systems (which are of course composed of triple subsets) to reach a state where one or more of the Zare stability criteria in the system is violated. From such experiments it is becoming clear that it should be possible from an examination of the 'starting conditions' in any n-body system to provide a statement or statistical estimate of its stability – the dynamical equivalent of the life-time of a planetary atmosphere.

The kinetic theory of gases enables a half-life T (the time it will take half the molecules in the atmosphere to escape into space) to be calculated from x, the ratio of the mean molecular velocity to the velocity of escape from the planet.

For x=1, the value of T is very small indeed. As x decreases,

T grows slowly at first and is measured in minutes, hours, weeks. But quite soon a region of x is reached where T shoots up to durations of astronomical length.

It is possible that the stability of the solar system may have to be treated like this. If we begin with a large number of hierarchical dynamical systems (solar systems) where they all have epsilon and alpha values within certain ranges, we may be able to state that the statistical status quo life-time of these systems is of such and such a duration in the sense that such a life-time will have to elapse before half the systems will have suffered any change in the status quo of their ordered orbits.

If this is so, then with the exception of the 'hard' commen-surabilities in the solar system such as the Neptune-Pluto case, or the Titan-Hyperion case, there would appear to be nothing remarkably esoteric about the distribution of solar system orbits or the values of the elements that describe these orbits. In their distribution, near-circularity and near-coplanarity, they merely reflect the sizes of the epsilons and alphas that have reduced the orbits pseudo-random walks to such small strolls, enabling the solar system's status quo to be maintained over a long time, perhaps an astronomically long time.

REFERENCES

1. Marchal, C. and Saari, D.G.: 1975, Celest. Mech., 12, 155.
2. Roy, A.E.: 1970, Instabilities in Dynamical Systems, ed.
        V. Szebehely, Reidel, Dordrecht.
3. Szebehely, V. and Zare, K.: 1977, Astron. Astrophys., 58, 145.
4. Walker, I.W.: 1982, Celest. Mech., (in press).
5. Walker, I.W., Emslie, A.G. and Roy, A.E.: 1980, Celest. Mech.,
        22, 371.
6. Walker, I.W. and Roy, A.E.: 1981, Celest. Mech., 24, 195.
7. Walker, I.W. and Roy, A.E.: 1982, Celest. Mech., (in press)
8. Zare, K.: 1976, Celest. Mech., 14, 73.
9. Zare, K.: 1977, Celest. Mech., 16, 35.

DYNAMICS OF THE ASTEROIDS

H. Scholl
Astronomisches Rechen-Institut, Heidelberg

INTRODUCTION

The main features of asteroidal dynamics are demonstrated in
Figures 1-4 which show the frequency distributions for perihelion
distances, aphelion distances, simimajor axes and inclinations of
the numbered asteroids. The large majority of the known asteroids
is situated in a belt between the orbits of Mars and Jupiter. Obviously, the boundaries of the belt are due to these two planets.
The dynamics of the belt asteroids is at present mainly determined
by Jupiter. In the past, also collisions among asteroids played
an important role which is indicated by the Hirayama families.
Those asteroids which cross the orbits of a planet might have suffered or will suffer drastic changes in their orbits or will even
collide with that planet unless particular protection mechanisms
prevent such close approaches.

MAIN BELT ASTEROIDS

The main belt covers the region between 2.2 and about 3.5 AU
with respect to semimajor axes. The frequency distribution
(fig.3) shows minima at the Kirkwood gaps which are located at
'commensurabilities' or ratios of small intergers in orbital periods with respect to Jupiter's orbital period. Asteroids close to
commensurabilities are called resonant or in resonance with
Jupiter since the period for a given conjunction configuration
Sun-asteroid-Jupiter, where Jupiter's perturbation are largest,
is a small integer multiple of Jupiter's orbital period.

The dynamical behaviour of a non-resonant asteroid is

125

A. Coradini and M. Fulchignoni (eds.), The Comparative Study of the Planets, 125–130.
Copyright © 1982 by D. Reidel Publishing Company.

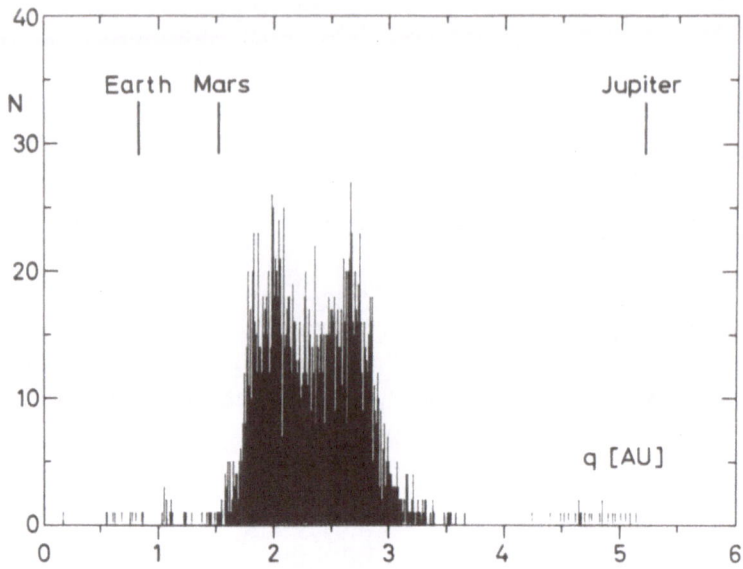

Fig. 1: Frequency distribution perihelion distances

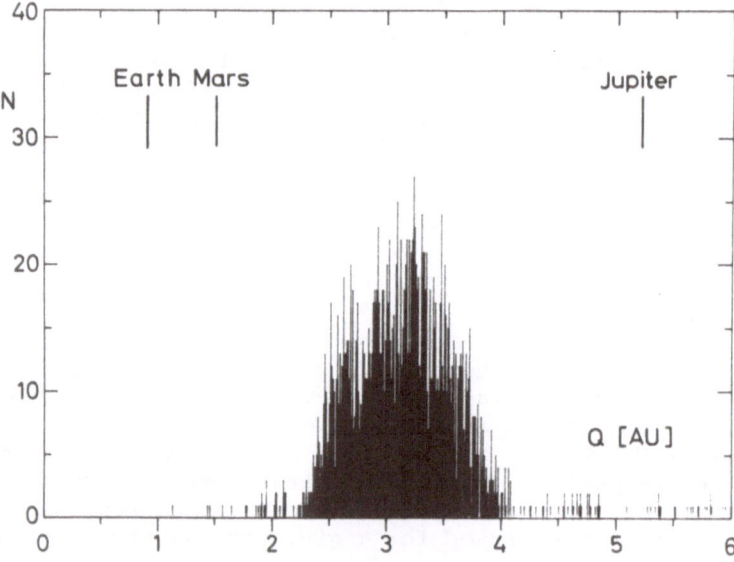

Fig. 2: Frequency distribution of aphelion distances

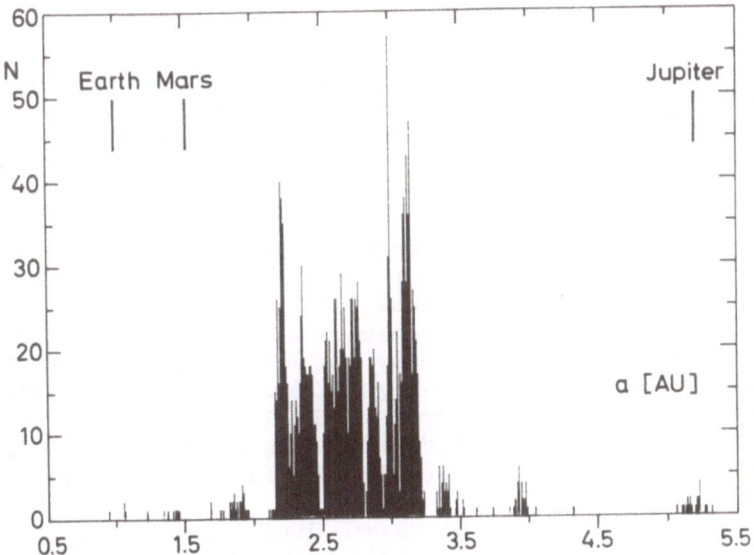

Fig. 3: Frequency distribution of semimajor axes

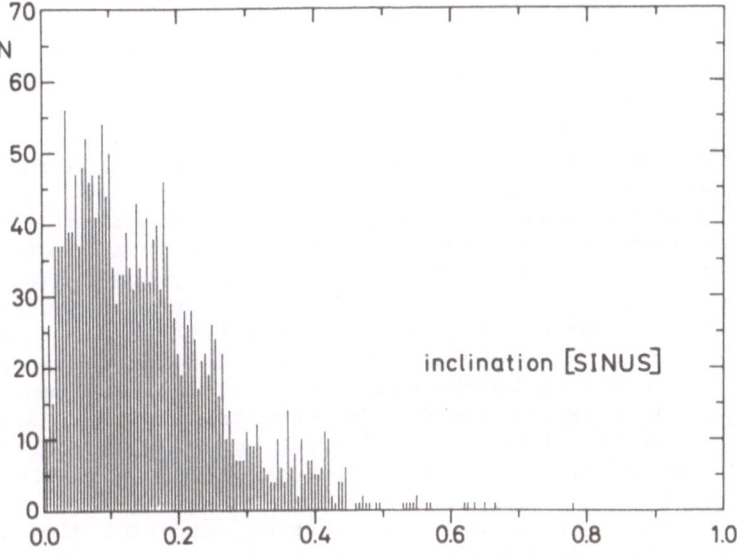

Fig. 4: Frequency distribution of inclinations

described by classical perturbation theories. In most cases it is sufficient to consider the three body problem Sun-asteroid-Jupiter only, since the other planets have either much lower mass or are at much larger distances than Jupiter. Jupiter's perturbations cause a retrograde precession of the asteroidal orbital plane about Jupiter's orbital plane and cause a prograde precession of the orbital ellipse in its plane. The other planets cause normally only slight oscillations in these precessional motions.

The evolution of semimajor axis, inclination and eccentricity of an asteroidal orbit is given in form of first-order differential equations. Since no rigorous solutions exist, solutions are usually given in form of series developments. These series contain short-periodic, long-periodic and secular terms (1). Since the short-periodic terms normally yield only small variations of the orbital elements as compared to the long-periodic terms, they are omitted or averaged in investigations which try to model the dynamical behaviour of asteroids over long time intervals.

The secular terms which are directly time dependent determine the stability of the asteroidal belt: The occurence of a secular term in the series for the semimajor axis for instance means that the asteroid would either leave the solar system or would fall into the Sun. No such secular terms occur for eccentricities and inclinations. There is not yet a definitive answer for the semimajor axis. According to Poisson's theorem (6) on the invariability of the semimajor axes one can say in a simplifying way that if such terms exist, they are very small and of no importance on time scales of 10^7 years. For time scales which are comparable to the age of the solar system, however, the importance of possible secular terms is yet unknown.

The main interest, therefore, concentrates on the long-periodic terms which determine the region an asteroid can cover in the solar system. According to approximating theories, so-called secular theories, the eccentricity and inclination of an asteroidal orbit can be decomposed in a free or synonymously proper and a forced eccentricity or inclination (1). The free eccentricity is the intrinsic originally caused eccentricity. The forced eccentricity is due to the gravitational field generated mainly by Jupiter, to a lesser extent by the other planet and to an even lesser extent by the galactic field. The eccentricity of an asteroidal orbit which is tabulated for a given epoch is the modulus of the vector sum of free and forced eccentricity.

Investigating free eccentricities, Hirayama discovered families of asteroids with similar semimajor axes and free eccentricities (4,7). Members of a family are interpreted to have the same parent body which broke up in a collision. The fragments drifted apart keeping on the average the free eccentricity of the parent body.

RESONANT ASTEROIDS

Since the classical perturbations fail for resonant asteroids, investigations of resonant orbits are based on numerical and semi-analytical methods. Simplifying the results, one can say that the variations in orbital elements are some order of magnitudes larger for resonant than for non-resonant asteroids. It is important to note that apparently the implications of Poisson's theorem concerning the invariability of semimajor axes holds for resonant orbits also (5).

One striking difference between resonant and non-resonant orbits is known: Resonant orbits can change the oscillations in their elements very abruptly. Amplitudes in eccentricity for instance which remained over thousand years nearly constant can multiply instantaneously. Such a behaviour of a non-resonant orbit is known to occur only during close encounter with a planet.

There is a second type of resonance in the motions of asteroids, the so-called secular resonances. If an asteroid's line of apsides or line of nodes revolves with the same speed as that of a planet, the asteroid's eccentricity, inclination and semimajor axis oscillate with large amplitudes. This type of resonance has not yet been investigated as systematically as the orbital resonances described above.

PLANET-CROSSING ASTEROIDS

Asteroids which cross the orbits of planets might suffer strong changes in their orbits (2) unless particular protection mechanisms avoid close encounters. The dynamical lifetime for the Earth-crossing asteroids is about 10^8 years which is significantly shorter than the age of the solar system of 4.5×10^9 years. Subsequently, for these asteroids a source and a delivery mechanism has to be found (10). Besides the main belt, comets have to be considered as a candidate source (8).

Some of the Mars-crossers seem to be able to avoid close encounters with Mars by a phase coupling between variation in eccentricity and motion of the line of apsides: Whenever the line of apsides lies in the orbital plane of Mars, the eccentricity is small enough in order to keep the asteroid's orbit outside of the Martian orbit. The asteroid becomes a non-Mars-crosser.

On the outer edge of the asteroidal belt, there are the Hildas which approach Jupiter's orbit closely, the Trojans and Hidalgo which cross Jupiter's orbit, and the peculiar object Chiron which crosses Saturn's orbit. The Hildas avoid close encounters with Jupiter itself because of their orbital resonances with Jupiter. The Trojans do not approach Jupiter because they are librating around the Lagrangian points L4 and L5. Hidalgo and Chiron on the other hand seem to move on orbits which will be changed drastically by Jupiter or Saturn respectively (3,9).

REFERENCES

(1) Brouwer, D. and Clemence, G.: 1961, Methods of Celestial
    Mechanics, Academic Press New York
(2) Carusi, A. and Valsecchi, G.B.: 1979, Numerical Simulations
    of Close Encounters between Jupiter and Minor Bodies; in
    'Asteroids', Ed. T. Gehrels, The University of Arizona Press,
    Tucson, Arizona, pp. 391-416.
(3) Everhart, E.: 1979, Chaotic Orbits in the Solar System; in
    'Asteroids', Ed. T. Gehrels, The University of Arizona Press,
    Tucson, Arizona, pp. 283-288.
(4) Gradie, J.C., Chapman, C.R. and Williams, J.G.: 1979, Families
    of Minor Planets; in 'Asteroids', Ed. T. Gehrels, The Univer-
    sity of Arizona Press, Tucson, Arizona, pp. 359-390.
(5) Greenberg, R. and Scholl, H.: 1979, Resonances in the Asteroid-
    al Belt; in 'Asteroids', Ed. T. Gehrels, The University of
    Arizona Press, Tucson, Arizona, pp. 310-333.
(6) Hagihara, Y.: 1972, Celestial Mechanics, Vol. II, Part 1, MIT
    Press Cambridge, Mass.
(7) Kozai, Y.: 1979, The Dynamic Evolution of Hirayama Families;
    in 'Asteroids', Ed. T. Gehrels, The University of Arizona
    Press, Tucson, Arizona, pp. 334-358.
(8) Kresák, L.: 1979, Dynamical Interrelations among Comets and
    Asteroids; in 'Asteroids', Ed. T. Gehrels, The University of
    Arizona Press, Tucson, Arizona, pp. 289-309.
(9) Scholl, H.: 1979, History and Evolution of Chiron's Orbit,
    Icarus 40, pp. 345-349.
(10) Shoemaker, E.M., Williams, J.G., Helin, E.F. and Wolfe, R.F.:
    1979, Earth-Crossing Asteroids: Orbital Classes, Collision
    Rates with Earth, and Origin; in 'Asteroids', Ed. T. Gehrels,
    The University of Arizona Press, Tucson, Arizona, pp. 253-282.

ON THE ORBITAL EVOLUTION OF SHORT-PERIOD COMETS HAVING LOW-VELOCITY ENCOUNTERS WITH JUPITER

A. Carusi and G. B. Valsecchi

I.A.S., Rep. Planetologia, Roma, Italy

The idea that both long- and short-period comets come from a large reservoir (the Oort cloud), extending iso-tropically up to a distance of 1 to $2 \cdot 10^5$ AU from the Sun (Weissman, 1981; Fernandez, 1980, 1981)´, is today almost generally accepted.

The Oort cloud hypothesis can explain the appea-rance of "new" long-period comets, via the gravitatio-nal effects of passing stars: then, the "old" long-pe-riod comets are those whose aphelion distances have been reduced by one or more passages through the plane-tary region.

As regards the origin of short-period comets, the situation is more complicated, but it seems improbable that a single close encounter with a giant planet would be sufficient, in a majority of cases, to transform a long-period comet into a short-period one. More pro-bably these comets are those members of a vast popula-tion of bodies revolving in the outer regions of the Solar System in "chaotic orbits" (Everhart, 1979), the aphelia of which have accumulated around the orbit of Jupiter by the maximum possible reduction of energy.

The numerical researches of Kazimirchak-Polonskaya (1972, 1976) demonstrated the relevance of repeated clo-

A. Coradini and M. Fulchignoni (eds.), The Comparative Study of the Planets, 131–148.
Copyright © 1982 by D. Reidel Publishing Company.

se encounters with the giant planets to the dynamical
history of a large fraction of the observed short-period comets.

Everhart (1973a,b) showed that objects in chaotic
orbits can undergo substantial changes in their orbital
elements as a consequence of close encounters, and that
they can enter various types of unstable orbital phases
during their evolution, including temporary satellite
captures by planets and injections into temporary horseshoe and Trojan orbits.

Several numerical studies on close encounters of
fictitious minor bodies with Jupiter (Carusi and Pozzi,
1978; Carusi et al., 1979; Carusi and Valsecchi, 1979,
1980) lead us to the conclusion that the most interesting short period comets among which to search for temporary satellite captures, exchanges of perihelion with
aphelion or vice versa and, more generally, for substantial modifications of orbital parameters at close encounters, are those with Tisserand invariants greater
than 2.9.

The Tisserand invariant, defined by the expression:

$$T = 1/a + 2 \cdot \sqrt{a \cdot (1-e^2)} \cdot \cos(i)$$

where a is in units of Jupiter's semiaxis, is related
to the encounter velocity U with respect to a circular
motion of the planet by the relation (Öpik, 1976):

$$U = \sqrt{3-T}$$

Therefore, T>2.9 implies a low jovicentric velocity at
close encounters.

We have investigated the motion of all comets with
T>2.9 backward in time, starting from the last observation reported in Marsden's Catalogue (1979). The only
comets reported by Marsden and not integrated were P/Enc-
ke, due to the fact that its low aphelion distance pre-
vents it from having close approaches to Jupiter, and P/
Gehrels 3, already integrated (Carusi and Valsecchi,
1979) taking also into account the effects of the Galilean satellites during its 1970 approach to Jupiter.

As it may be seen from figure 1, the orbits of

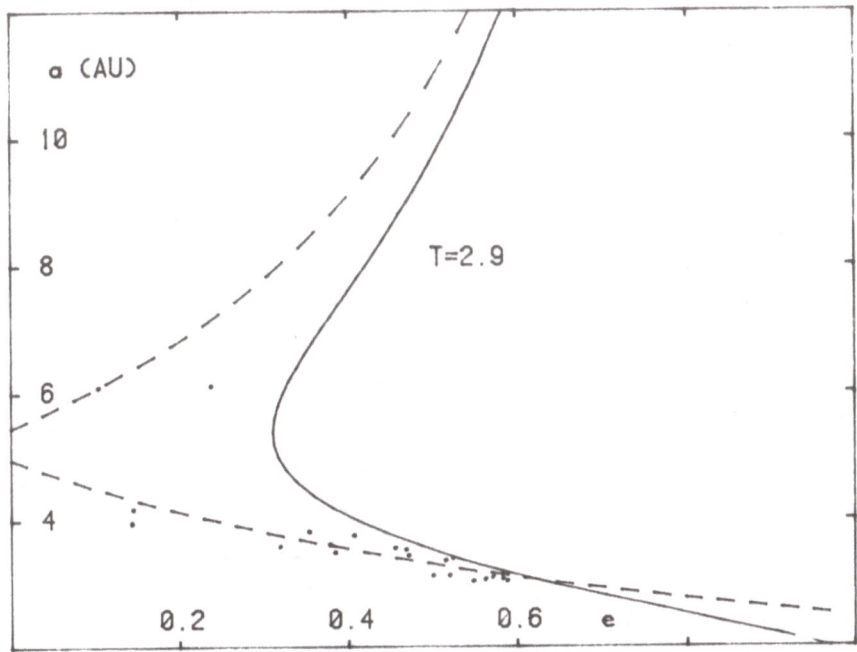

Fig. 1 - Semi-axes and eccentricities of the comets with
T>2.9, reported in Marsden's Catalogue (1979).

many of the comets considered do not even cross the orbit of Jupiter: the points representing them in the a-e
diagram lie outside the curves corresponding to Jupiter's perihelion (lower dashed line) and aphelion (upper
dashed line), thus implying that the orbits of these
comets are completely inside or outside that of Jupiter.
Nevertheless, they can have close encounters with Jupiter, due to the extent of the volume of space in which
the planet's gravity can affect the comet's orbit substantially. In some cases, as that of the 1937 encounter of P/Oterma with Jupiter, examined in detail elsewhere (Carusi et al., 1981), not even a distance of
about 0.6 AU between the two unperturbed orbits can prevent a close approach. Probably the "safe value" for
this distance is close to that of P/Encke: about 0.85 AU.
   The characteristics of the integration procedure
are summarized in Table 1; notice that, with respect to
the integration published by Carusi and Valsecchi (1981),
the initial value of the true anomaly of P/De Vico-Swift

has been corrected and the evolution of the whole sample recomputed.

Table 1. Backwards integration of short-period comets
         with the Tisserand invariant greater than 2.9

- Perturbing bodies: Sun, Jupiter, Saturn, Uranus, Neptune.
- End of backward integration: JD 2400000.5.
- The starting point of each comet is close to the last perihelion passage reported in Marsden's Catalogue: it is given in the list below.

Kowal	2443200.5
Gunn	2442800.5
Smirnova-Chernykh	2442680.5
Schwassmann-Wachmann 1	2442080.5
Clark	2441840.5
Tempel 1	2441400.5
Shajn-Schaldach	2441240.5
Kojima	2440880.5
Schwassmann-Wachmann 2	2439920.5
Reinmuth 2	2439720.5
Tempel 2	2439720.5
De Vico-Swift	2538910.5
Johnson	2438200.5
Whipple	2438160.5
Oterma	2436360.5
Do Toit 2	2431580.5
Du Toit-Neujmin-Delporte	2430200.5
Neujmin 2	2424880.5
Schorr	2421880.5
Spitaler	2411680.5
Brooks 1	2409960.5
Barnard 1	2409320.5

The quality of the initial orbits of the comets

presented here varies widely: some have been observed
for many returns,others just once - the relevant information on this aspect is given in Marsden's Catalogue.

Therefore, the past orbital histories given in Table 2 are not equally reliable.

Table 2. Past orbital histories of short period comets
         with Tisserand quantity greater than 2.9

a(AU)	e	i(°)	ω(°)	Ω(°)	f(°)	Epoch	1/a[+]	T_0[+]	T[+]
P/Kowal									
6.11	0.237	4.4	178.05	28.44	0.37	1977.16	0.85	2.96	2.95
6.12	0.242	4.5	176.68	29.08	356.28	1961.89	0.85	2.95	2.95
6.74	0.292	6.1	173.14	51.12	5.63	1947.43	0.77	2.95	2.94
6.33	0.252	6.2	178.47	52.74	356.06	1931.42	0.82	2.96	2.94
6.32	0.251	6.2	179.41	52.88	4.32	1915.82	0.82	2.96	2.94
6.75	0.290	6.2	183.31	55.23	1.67	1899.95	0.77	2.95	2.94
6.16	0.259	5.6	193.14	67.62	0.15	1885.16	0.84	2.95	2.94
6.14	0.259	5.6	194.03	67.74	3.69	1870.09	0.85	2.95	2.94
P/Gunn									
3.59	0.320	10.4	197.12	68.04	340.98	1969.12	1.45	3.02	3.00
4.00	0.179	10.9	173.88	79.71	3.72	1961.46	1.30	3.03	2.99
4.01	0.176	10.9	173.97	79.73	0.69	1953.39	1.30	3.03	2.99
4.00	0.178	10.9	173.74	79.79	359.31	1945.34	1.30	3.03	3.00
4.15	0.141	10.8	173.56	82.27	2.36	1937.18	1.25	3.02	2.99
4.15	0.138	10.8	174.67	82.28	359.47	1928.72	1.25	3.02	2.99
4.15	0.138	10.8	173.91	82.36	358.83	1920.21	1.25	3.02	2.99
4.15	0.141	10.8	174.84	82.36	3.25	1911.86	1.25	3.02	2.99
4.12	0.151	10.8	176.90	83.06	356.25	1903.40	1.26	3.02	2.99
4.13	0.150	10.8	176.38	83.12	357.03	1895.02	1.26	3.02	2.99
4.11	0.152	10.8	176.40	83.16	355.85	1886.64	1.27	3.02	2.99
4.91	0.052	10.4	200.38	97.79	5.99	1878.52	1.06	3.00	2.97
4.50	0.120	10.5	249.75	104.42	355.92	1869.64	1.16	3.00	2.97
4.51	0.119	10.5	250.20	104.46	357.77	1860.15	1.15	3.00	2.97
P/Smirnova-Chernykh									
4.18	0.146	6.6	90.20	77.10	7.77	1975.73	1.25	3.02	3.01
4.17	0.145	6.6	90.22	77.10	355.23	1975.51	1.25	3.02	3.01
4.16	0.148	6.7	90.56	77.13	356.94	1967.05	1.25	3.02	3.01
5.49	0.044	5.9	83.73	119.79	356.43	1957.78	0.95	3.00	2.99
8.96	0.367	5.7	352.60	139.48	3.49	1926.97	0.58	3.02	3.01
8.90	0.363	5.7	352.65	139.60	358.16	1902.14	0.58	3.02	3.01
8.98	0.367	5.7	352.70	139.68	1.98	1875.53	0.58	3.02	3.01

+ 1/a is in units of Jupiter's semiaxis

Tab. 2 cont.ed

P/Schwassmann-Wachmann 1

6.09	0.105	9.8	14.47	319.64	358.91	1974.09	0.85	3.01	2.97
6.38	0.132	9.5	355.82	321.61	2.34	1957.44	0.82	3.01	2.98
6.39	0.135	9.5	356.24	322.00	357.62	1941.22	0.81	3.01	2.98
6.47	0.153	9.4	359.07	323.06	357.21	1925.26	0.80	3.01	2.98
6.46	0.150	9.4	357.10	323.63	357.28	1908.71	0.80	3.01	2.98
6.50	0.148	9.4	356.64	323.78	0.19	1892.28	0.80	3.01	2.98
6.41	0.136	9.6	3.62	324.79	359.34	1876.40	0.81	3.01	2.98
6.38	0.136	9.6	5.38	324.93	358.68	1860.32	0.82	3.01	2.98

P/Clark

3.12	0.500	9.5	209.13	59.13	8.42	1973.43	1.67	3.01	2.99
3.13	0.499	9.5	209.07	59.15	354.86	1967.85	1.66	3.01	2.99
3.15	0.492	9.5	208.83	59.44	0.53	1962.32	1.65	3.01	2.99
3.15	0.491	9.5	208.69	59.48	358.52	1956.71	1.65	3.01	2.99
3.24	0.456	10.1	204.00	63.19	4.05	1950.96	1.60	3.01	2.99
3.25	0.456	10.1	203.83	63.24	347.30	1945.02	1.60	3.01	2.99
3.39	0.399	11.5	196.11	68.88	10.39	1938.99	1.54	3.02	2.99
3.39	0.399	11.5	195.92	68.93	9.37	1932.74	1.53	3.02	2.99
3.43	0.387	11.4	195.56	69.36	345.53	1926.28	1.52	3.01	2.98
3.43	0.385	11.4	195.43	69.39	17.75	1920.15	1.52	3.01	2.98
3.44	0.384	11.4	195.67	69.46	358.95	1913.65	1.51	3.01	2.98
3.44	0.382	11.4	195.71	69.48	14.16	1907.38	1.51	3.01	2.98
3.44	0.384	11.4	195.91	69.49	355.19	1900.87	1.51	3.01	2.98
3.44	0.382	11.4	196.13	69.49	7.17	1894.59	1.51	3.01	2.98
3.44	0.384	11.4	196.17	69.51	16.36	1888.28	1.51	3.01	2.99
3.43	0.385	11.4	196.42	69.55	10.11	1881.88	1.52	3.01	2.98
3.43	0.386	11.5	196.31	69.59	348.96	1875.36	1.52	3.01	2.98
3.41	0.392	11.5	196.27	69.79	358.20	1869.10	1.52	3.01	2.98
3.41	0.393	11.5	196.06	69.84	10.01	1862.87	1.53	3.01	2.98

P/Tempel 1

3.12	0.520	10.6	179.19	68.35	295.67	1972.23	1.67	2.99	2.97
3.12	0.518	10.5	179.13	68.37	4.70	1967.05	1.67	2.99	2.97
3.13	0.513	10.5	179.06	68.48	348.57	1961.46	1.66	2.99	2.97
3.14	0.513	10.5	178.94	68.51	17.30	1956.02	1.66	2.99	2.97
3.24	0.479	10.6	176.23	69.83	5.03	1950.21	1.61	2.99	2.97
3.24	0.479	10.6	176.09	69.87	351.11	1944.30	1.61	2.99	2.97
3.48	0.405	10.8	168.58	73.10	13.86	1938.08	1.49	2.99	2.96
3.49	0.404	10.8	168.41	73.14	353.84	1931.42	1.49	2.99	2.96
3.50	0.402	10.8	168.56	73.20	356.67	1924.92	1.49	2.99	2.96
3.51	0.400	10.8	168.56	73.23	350.73	1918.32	1.48	2.99	2.96
3.50	0.402	10.8	168.71	73.22	2.67	1911.86	1.49	2.99	2.96
3.50	0.401	10.8	168.96	73.24	351.19	1905.24	1.49	2.99	2.96
3.50	0.402	10.8	168.91	73.27	358.62	1898.75	1.49	2.99	2.96
3.49	0.406	10.8	169.04	73.36	355.85	1892.22	1.49	2.99	2.96
3.48	0.406	10.8	168.87	73.41	356.05	1885.71	1.49	2.99	2.96
3.30	0.462	9.8	159.73	79.61	0.94	1879.36	1.58	2.99	2.97
3.30	0.462	9.8	159.57	79.65	9.04	1873.41	1.58	2.99	2.97
3.17	0.508	6.4	135.08	102.23	19.84	1867.48	1.64	2.99	2.98
3.17	0.508	6.4	134.92	102.28	358.01	1861.72	1.64	2.99	2.98

Tab. 2 cont.ed

P/Shajn-Schaldach

3.75	0.406	6.2	215.12	167.27	5.12	1971.79	1.39	2.94	2.93
3.76	0.405	6.2	215.38	167.33	9.32	1964.56	1.38	2.94	2.93
3.76	0.404	6.2	215.58	167.31	6.50	1957.26	1.38	2.94	2.93
3.75	0.405	6.2	215.32	167.39	10.43	1949.99	1.39	2.94	2.93
4.83	0.111	10.9	173.95	206.33	2.36	1940.65	1.08	2.99	2.96
4.64	0.107	10.8	181.15	209.41	2.58	1930.74	1.12	3.00	2.97
4.66	0.106	10.8	181.03	209.40	359.63	1920.66	1.12	3.00	2.97
4.68	0.106	10.8	180.80	209.50	3.23	1910.63	1.11	3.00	2.96
4.67	0.106	10.8	180.43	209.63	3.32	1900.49	1.11	3.00	2.96
4.64	0.107	10.8	179.85	209.69	357.68	1890.30	1.12	3.00	2.97
4.67	0.110	10.8	181.99	209.72	359.23	1880.39	1.11	3.00	2.96
6.47	0.150	8.2	12.65	227.72	1.65	1859.82	0.80	3.01	2.99

P/Kojima

3.36	0.515	4.1	198.11	291.14	8.40	1970.80	1.55	2.93	2.92
3.36	0.515	4.1	197.94	291.22	1.38	1964.60	1.55	2.93	2.92
3.63	0.443	7.5	173.72	311.47	350.29	1957.78	1.43	2.93	2.92
3.64	0.442	7.5	173.59	311.55	354.84	1950.87	1.43	2.93	2.92
3.64	0.442	7.5	173.79	311.55	8.36	1944.04	1.43	2.93	2.92
3.64	0.441	7.5	173.92	311.55	357.79	1937.03	1.43	2.93	2.92
3.64	0.443	7.5	173.77	311.58	355.58	1930.06	1.43	2.93	2.92
3.51	0.472	7.3	170.07	313.83	356.14	1923.24	1.48	2.93	2.92
3.51	0.472	7.3	169.87	313.90	346.01	1916.59	1.48	2.93	2.92
3.58	0.457	7.3	168.33	314.64	350.86	1909.90	1.45	2.93	2.92
3.59	0.456	7.3	168.21	314.71	356.70	1903.15	1.45	2.93	2.92
3.58	0.456	7.3	168.40	314.72	338.58	1896.26	1.45	2.93	2.92
3.59	0.455	7.3	168.49	314.72	12.06	1889.70	1.45	2.93	2.92
3.58	0.456	7.3	168.43	314.74	12.35	1882.91	1.45	2.93	2.92
3.53	0.467	7.3	167.62	315.26	8.92	1876.16	1.47	2.93	2.92
3.53	0.467	7.3	167.38	315.34	348.58	1869.38	1.47	2.93	2.92
3.68	0.435	7.4	163.03	317.22	354.27	1862.45	1.41	2.93	2.92

P/Schwassmann-Wachmann 2

3.49	0.384	3.7	357.66	125.99	356.77	1968.17	1.49	3.00	3.00
3.50	0.383	3.7	357.70	126.00	356.28	1961.65	1.49	3.00	3.00
3.49	0.384	3.7	357.86	126.01	12.88	1955.25	1.49	3.00	3.00
3.49	0.384	3.7	358.14	126.01	0.75	1948.65	1.49	3.00	3.00
3.49	0.385	3.7	358.05	126.04	358.60	1942.11	1.49	3.00	3.00
3.46	0.394	3.7	358.01	126.29	2.96	1935.68	1.51	3.00	3.00
3.45	0.395	3.7	357.76	126.32	6.77	1929.27	1.51	3.00	3.00
4.41	0.195	0.7	335.56	126.52	358.35	1920.33	1.18	2.99	2.99
4.42	0.193	0.7	335.71	126.50	5.97	1911.20	1.18	2.99	2.99
4.43	0.193	0.7	335.74	126.35	9.40	1901.94	1.17	2.99	2.98
4.41	0.194	0.7	335.61	126.29	3.30	1892.52	1.18	2.99	2.99
4.44	0.196	0.7	337.25	126.21	358.10	1883.20	1.17	2.98	2.98
4.25	0.218	0.7	338.96	126.11	359.62	1874.21	1.22	2.99	2.99
4.28	0.216	0.7	339.03	125.96	1.50	1865.43	1.22	2.99	2.99

Tab. 2 cont.ed

P/Reinmuth 2

3.57	0.455	7.0	45.64	296.09	0.11	1967.63	1.46	2.93	2.92
3.56	0.457	7.0	45.49	296.18	13.64	1960.98	1.46	2.93	2.92
3.52	0.469	7.1	44.25	297.21	347.28	1954.16	1.48	2.93	2.92
3.52	0.469	7.1	43.95	297.37	352.59	1947.59	1.48	2.93	2.92
3.61	0.441	7.4	38.91	302.02	4.66	1940.90	1.44	2.94	2.92
3.62	0.440	7.4	38.76	302.06	346.88	1933.90	1.44	2.94	2.92
3.62	0.440	7.4	38.90	302.11	359.71	1927.11	1.44	2.94	2.92
3.62	0.440	7.4	39.00	302.12	354.91	1920.21	1.44	2.94	2.92
3.61	0.441	7.4	38.96	302.19	349.46	1913.29	1.44	2.94	2.92
3.55	0.457	7.6	36.99	303.61	354.03	1906.52	1.47	2.93	2.92
3.55	0.457	7.6	36.78	303.70	354.60	1899.82	1.47	2.93	2.92
3.60	0.444	7.7	35.65	304.85	7.16	1893.14	1.45	2.94	2.92
3.61	0.442	7.7	35.48	304.91	2.26	1886.26	1.44	2.94	2.92
3.61	0.443	7.7	35.60	304.93	353.01	1879.36	1.44	2.94	2.92
3.60	0.443	7.7	35.82	304.92	358.94	1872.57	1.44	2.94	2.92
3.60	0.444	7.7	35.69	305.00	10.50	1865.81	1.45	2.94	2.92
3.51	0.466	8.0	32.50	307.05	14.31	1859.06	1.48	2.94	2.92

P/Tempel 2

3.03	0.548	12.5	190.96	119.27	3.26	1967.63	1.72	2.99	2.96
3.02	0.549	12.5	191.07	119.28	3.08	1962.37	1.72	2.99	2.96
3.03	0.548	12.5	191.06	119.28	357.67	1957.09	1.72	3.00	2.97
3.04	0.543	12.4	191.03	119.38	14.55	1951.86	1.71	2.99	2.96
3.04	0.542	12.4	190.90	119.41	358.35	1946.49	1.71	2.99	2.96
3.00	0.556	12.7	186.69	121.04	346.15	1941.07	1.74	3.00	2.97
3.00	0.556	12.7	186.59	121.08	340.67	1935.86	1.74	3.00	2.97
2.99	0.558	12.8	186.62	121.14	358.57	1930.74	1.74	3.00	2.97
2.99	0.560	12.8	186.62	121.16	8.65	1925.61	1.74	3.00	2.97
2.99	0.559	12.8	186.70	121.16	17.95	1920.48	1.74	3.00	2.97
2.99	0.559	12.8	186.77	121.17	5.54	1915.28	1.74	3.00	2.97
2.99	0.558	12.8	186.74	121.18	352.98	1910.06	1.74	3.00	2.97
3.03	0.543	12.7	185.94	121.59	10.19	1904.87	1.72	3.00	2.97
3.03	0.543	12.7	185.82	121.62	341.87	1899.49	1.72	3.00	2.97
3.01	0.552	12.8	185.27	121.91	344.19	1894.24	1.73	3.00	2.97
3.00	0.553	12.8	185.19	121.94	352.22	1889.05	1.73	3.00	2.97
3.00	0.553	12.8	185.29	121.96	340.26	1883.81	1.73	3.00	2.97
3.00	0.555	12.8	185.34	121.97	359.17	1878.68	1.73	3.00	2.97
3.00	0.554	12.8	185.37	121.97	22.17	1873.57	1.73	3.00	2.97
3.01	0.552	12.8	185.45	122.00	15.04	1868.34	1.73	3.00	2.97
3.01	0.551	12.7	185.38	122.02	352.55	1863.03	1.73	3.00	2.97

P/De Vico-Swift

3.42	0.525	3.6	325.35	24.42	313.62	1965.41	1.52	2.90	2.90
3.42	0.524	3.6	325.20	24.49	13.99	1959.38	1.52	2.90	2.90
3.47	0.510	3.6	324.21	25.20	353.58	1952.88	1.50	2.90	2.90
3.48	0.509	3.6	324.01	25.29	1.47	1946.44	1.50	2.90	2.90
3.48	0.508	3.6	324.15	25.31	353.31	1939.92	1.49	2.90	2.90
3.49	0.506	3.6	324.12	25.35	18.29	1933.55	1.49	2.90	2.90

Tab. 2 cont.ed

3.48	0.508	3.6	324.22	25.32	8.13	1927.00	1.49 2.90 2.90
3.48	0.508	3.6	324.35	25.33	2.41	1920.48	1.50 2.90 2.90
3.47	0.509	3.6	324.31	25.35	0.81	1913.99	1.50 2.90 2.90
3.45	0.515	3.6	324.23	25.47	12.15	1907.60	1.51 2.90 2.90
3.45	0.516	3.6	324.10	25.52	353.97	1901.10	1.51 2.90 2.90
3.25	0.572	3.0	296.95	49.16	354.22	1894.76	1.60 2.90 2.90
3.25	0.572	3.0	296.78	49.23	354.10	1888.89	1.60 2.90 2.90
3.14	0.607	2.9	280.00	64.05	349.80	1883.06	1.66 2.89 2.89
3.13	0.607	2.9	279.89	64.11	347.39	1877.50	1.66 2.89 2.89
3.11	0.614	2.9	279.33	64.57	3.75	1872.02	1.67 2.89 2.89
3.11	0.615	2.9	279.29	64.60	8.55	1866.54	1.67 2.89 2.89
3.11	0.616	2.9	279.34	64.62	6.72	1861.07	1.68 2.89 2.89

P/Johnson

3.61	0.378	13.9	206.95	118.15	3.88	1963.47	1.44 2.98 2.94
3.62	0.375	13.9	206.02	118.17	351.57	1956.50	1.44 2.98 2.94
3.61	0.377	13.9	206.14	118.19	359.65	1949.70	1.44 2.98 2.94
3.61	0.378	13.9	206.58	118.22	16.83	1943.02	1.44 2.98 2.94
3.60	0.379	13.9	206.33	118.28	356.66	1936.01	1.44 2.98 2.94
3.45	0.416	14.9	196.13	121.51	13.46	1929.27	1.51 2.99 2.94
3.45	0.417	14.9	197.94	121.57	346.86	1922.69	1.51 2.99 2.94
3.50	0.403	14.8	197.40	122.06	9.40	1916.36	1.49 2.99 2.94
3.50	0.401	14.8	197.30	122.11	7.42	1909.80	1.49 2.99 2.94
3.51	0.401	14.8	197.54	122.14	3.24	1903.23	1.48 2.99 2.94
3.51	0.399	14.8	197.66	122.14	1.46	1896.66	1.48 2.99 2.94
3.50	0.401	14.8	197.73	122.16	359.42	1890.09	1.49 2.99 2.94
3.50	0.402	14.8	198.03	122.20	354.34	1883.52	1.49 2.99 2.94
3.49	0.403	14.8	197.89	122.25	359.67	1877.01	1.49 2.99 2.94
3.46	0.414	14.9	197.25	122.70	12.22	1870.61	1.50 2.99 2.94
3.46	0.414	14.9	197.05	122.76	351.27	1864.02	1.50 2.99 2.94

P/Whipple

3.82	0.353	10.2	190.00	188.39	3.76	1963.36	1.36 2.97 2.94
3.80	0.356	10.3	190.45	188.51	351.39	1955.83	1.37 2.97 2.94
3.80	0.356	10.3	190.12	188.60	342.68	1948.32	1.37 2.97 2.94
3.82	0.350	10.2	190.46	188.81	0.91	1941.07	1.36 2.97 2.94
3.83	0.348	10.2	190.57	188.80	356.36	1933.55	1.36 2.97 2.94
3.82	0.350	10.2	190.42	188.86	351.33	1926.01	1.36 2.97 2.94
4.95	0.146	8.7	166.95	206.42	2.99	1916.59	1.05 2.96 2.96
4.64	0.152	8.6	171.46	211.10	6.62	1906.52	1.12 2.99 2.97
4.66	0.152	8.6	171.74	211.09	354.36	1896.26	1.12 2.99 2.97
4.69	0.150	8.6	171.41	211.21	359.64	1886.26	1.11 2.99 2.97
4.68	0.150	8.6	171.00	211.35	2.43	1876.16	1.11 2.99 2.97
4.65	0.151	8.6	170.87	211.45	358.93	1866.00	1.12 2.99 2.97

Tab. 2 cont.ed

P/Oterma

3.96	0.145	4.0	354.87	155.11	359.24	1958.43	1.31	3.04	3.04
3.97	0.143	4.0	354.81	155.13	354.15	1950.44	1.31	3.04	3.04
3.96	0.145	4.0	354.68	155.17	1.30	1942.66	1.31	3.04	3.04
6.89	0.159	3.1	241.62	35.39	358.29	1918.96	0.76	3.03	3.02
7.17	0.199	3.1	238.43	38.00	359.32	1900.87	0.73	3.03	3.02
7.39	0.216	3.1	233.48	38.74	359.99	1880.78	0.70	3.03	3.03
7.36	0.216	3.1	233.00	38.88	357.64	1860.59	0.71	3.03	3.03

P/Du Toit 2

3.03	0.588	6.9	201.53	358.86	14.36	1945.34	1.72	2.95	2.94
3.03	0.587	6.9	201.62	358.86	354.24	1940.01	1.72	2.95	2.94
3.03	0.588	6.9	201.65	358.88	358.75	1934.75	1.72	2.95	2.94
3.03	0.587	6.9	201.67	358.89	346.52	1929.43	1.72	2.95	2.94
3.04	0.585	6.9	201.66	358.95	354.88	1924.17	1.71	2.95	2.94
3.04	0.585	6.9	201.57	358.99	336.58	1918.82	1.71	2.95	2.94
3.14	0.551	8.5	192.50	6.98	20.41	1913.46	1.66	2.95	2.94
3.14	0.551	8.5	192.32	7.04	354.40	1907.78	1.66	2.95	2.94
3.03	0.582	9.1	189.34	8.00	1.12	1902.29	1.72	2.96	2.94
3.03	0.583	9.1	189.26	8.04	340.75	1896.95	1.72	2.96	2.94
3.03	0.584	9.1	189.32	8.05	354.94	1891.73	1.72	2.96	2.94
3.02	0.585	9.1	189.32	8.06	358.42	1886.49	1.72	2.96	2.94
3.03	0.584	9.1	189.38	8.07	357.41	1881.23	1.72	2.96	2.94
3.02	0.584	9.1	189.45	8.08	6.04	1876.00	1.72	2.96	2.94
3.03	0.583	9.1	189.40	8.10	10.21	1870.75	1.72	2.96	2.94
3.05	0.576	9.1	188.90	8.42	4.24	1865.43	1.71	2.96	2.94
3.05	0.575	9.1	188.78	8.46	20.87	1860.15	1.70	2.96	2.94

P/Du Toit-Neujmin-Delporte

3.13	0.583	3.3	69.36	229.59	3.56	1941.56	1.66	2.92	2.92
3.13	0.585	3.3	69.20	229.72	0.91	1936.01	1.66	2.92	2.92
3.13	0.586	3.3	69.16	229.86	346.16	1930.43	1.66	2.92	2.92
3.12	0.587	3.3	69.05	229.96	346.95	1924.92	1.67	2.92	2.92
3.12	0.587	3.3	69.18	229.94	8.07	1919.46	1.67	2.92	2.92
3.12	0.588	3.3	69.05	230.07	13.83	1913.99	1.67	2.92	2.92
3.12	0.587	3.3	69.13	230.05	19.74	1908.50	1.67	2.92	2.92
3.12	0.587	3.3	69.15	230.10	15.15	1902.98	1.67	2.92	2.92
3.13	0.586	3.3	69.12	230.12	1.31	1897.42	1.66	2.92	2.92
3.14	0.582	3.3	68.76	230.50	354.44	1891.85	1.66	2.92	2.92
3.14	0.581	3.3	68.54	230.64	11.56	1886.34	1.66	2.92	2.92
3.21	0.560	3.3	61.09	237.47	352.65	1880.58	1.62	2.92	2.92
3.21	0.560	3.3	60.92	237.57	21.35	1874.93	1.62	2.92	2.92
3.37	0.509	5.0	27.16	269.75	354.57	1868.73	1.54	2.93	2.92
3.37	0.508	5.0	27.01	269.81	19.47	1862.66	1.54	2.93	2.92

Tab. 2 cont.ed

P/Neujmin 2

```
3.09 0.567 10.6 193.72 328.00 347.15    1927.00    1.68 2.95 2.93
3.09 0.568 10.6 193.79 328.02   2.84    1921.63    1.69 2.95 2.93
3.09 0.567 10.6 193.79 328.03   1.33    1916.20    1.68 2.95 2.93
3.09 0.566 10.6 193.86 328.05   7.42    1910.79    1.68 2.95 2.93
3.09 0.565 10.6 193.77 328.09 338.78    1905.24    1.68 2.95 2.93
3.12 0.557 10.6 193.20 328.48 357.73    1899.82    1.67 2.95 2.93
3.12 0.557 10.6 193.07 328.52 339.95    1894.24    1.67 2.95 2.93
3.25 0.500 14.6 165.81 335.48 347.47    1888.51    1.60 2.97 2.93
3.25 0.500 14.6 165.66 335.53 359.13    1882.70    1.60 2.97 2.93
3.49 0.394 18.9 172.86 340.71 359.66    1876.16    1.49 3.00 2.91
3.49 0.393 18.9 172.85 340.77   3.10    1869.64    1.49 3.00 2.92
3.51 0.389 18.9 172.87 340.91 353.77    1863.03    1.48 3.00 2.91
```

P/Schorr

```
3.54 0.468 5.6 279.09 118.34   6.46    1918.78    1.47 2.93 2.92
3.55 0.467 5.6 279.00 118.38  10.29    1912.13    1.47 2.93 2.92
3.55 0.468 5.6 279.16 118.40   1.86    1905.40    1.47 2.93 2.92
3.55 0.466 5.6 279.27 118.41   6.64    1898.75    1.46 2.93 2.92
3.55 0.468 5.6 279.12 118.49 358.01    1892.01    1.47 2.93 2.92
3.52 0.474 5.6 278.85 118.91 357.86    1885.38    1.48 2.93 2.92
3.52 0.475 5.6 278.65 118.99  11.12    1878.84    1.48 2.93 2.92
3.75 0.412 5.5 236.70 158.14   5.67    1871.69    1.39 2.93 2.93
3.76 0.411 5.5 236.53 158.20 358.80    1864.35    1.38 2.93 2.93
```

P/Spitaler

```
3.44 0.471 12.8  13.39 45.89   6.29    1890.86    1.51 2.95 2.91
3.50 0.453 12.9  12.04 46.76   9.47    1884.39    1.49 2.95 2.91
3.50 0.452 12.9  11.90 46.81 353.82    1877.75    1.49 2.95 2.91
3.51 0.450 12.9  12.11 46.86 343.85    1871.14    1.48 2.95 2.91
3.52 0.449 12.9  12.06 46.88   8.73    1864.71    1.48 2.95 2.91
```

P/Brooks 1

```
3.09 0.571 12.7 176.85 54.47 290.42    1886.15    1.68 2.95 2.92
3.09 0.572 12.7 176.90 54.48  22.27    1881.08    1.68 2.95 2.92
3.09 0.574 12.7 176.91 54.49 345.64    1875.53    1.69 2.95 2.92
3.09 0.573 12.7 176.99 54.49 338.19    1870.09    1.68 2.95 2.92
3.09 0.574 12.7 177.06 54.49 351.55    1864.71    1.69 2.95 2.92
3.09 0.572 12.7 177.05 54.51 337.60    1859.24    1.68 2.95 2.92
```

P/Barnard 1

```
3.07 0.583 5.5 301.04 6.07 297.65    1884.39    1.69 2.94 2.94
3.07 0.584 5.5 301.14 6.05 345.82    1879.21    1.70 2.94 2.94
3.07 0.583 5.5 301.07 6.12   4.52    1873.89    1.69 2.94 2.94
3.09 0.579 5.4 300.79 6.44   4.95    1868.49    1.69 2.94 2.94
3.09 0.578 5.4 300.65 6.52 355.77    1863.03    1.68 2.94 2.94
```

For the comets with well determined orbits (e.g., P/Tempel 2) the comparison of the values given in the Table with those reported in Marsden's Catalogue gives an indication of the degree of reliability; for the other comets such a check is impossible. However, we believe that the orbital histories presented here are interesting in themselves, as they represent examples of possible evolutions of objects in chaotic orbits of high Tisserand invariant. This selection characteristic of the comets of our sample makes the occurrence of temporary satellite captures and of exchange of perihelion with aphelion (or vice-versa) rather frequent.

Therefore in Table 2 the orbital elements of each comet are given for the starting point of the integration and for each preceding (in time) perihelion passage, together with the osculating date and the values of $1/a$, $T_o = 1/a + 2\sqrt{a(1-e^2)}$ and $T = 1/a + 2\sqrt{a(1-e^2)} \cdot \cos(i)$, where a is in units of Jupiter's semiaxis.

In Table 3 a list of the encounters with Jupiter within 1 AU is given. Also given in Table 3 are the data concerning temporary satellite captures (Carusi and Valsecchi, 1981), exchanges of perihelion with aphelion and variation of $1/a$ (in units of Jupiter's semiaxis) at each close encounter.

The majority of the comets examined here have rather similar orbital histories, so that we will discuss them first. This group includes P/Clark, P/Tempel 1, P/Kojima, P/Reinmuth 2, P/Tempel 2, P/De Vico-Swift, P/Johnson, P/Du Toit 2, P/Du Toit-Neujmin-Delporte, P/Neujmin 2, P/Schorr, P/Spitaler, P/Brooks 1 and P/Barnard 1.

As shown in Table 2, the semimajor axes of these 14 comets are comprised between 2.99 AU and 3.76 AU during the time span covered by the integration; none of them undergoes a temporary satellite capture and the strongest perturbation at a close encounter (evaluated from the difference of the $1/a$ values at the perihelion passages just preceding and just following the encounter) is 0.12.

The second group includes P/Kowal and P/Schwassmann-Wachmann 1. Their semi-major axes are greater than

that of Jupiter, as they range, over the whole integration, from 6.09 to 6.75 AU.

Table 3. Characteristics of the encounters with Jupiter
within 1 AU of the comets with T greater than
2.9.

Comet	Period spent within 1 AU from Jupiter	$\Delta(1/a)$	Period spent as a temporary satellite
Gunn	1963.9-1966.9	0.15	
Smirnova-Chernykh	1961.0-1964.9	0.30	
Kojima	1961.4-1963.0	0.12	
Smirnova-Chernykh	1954.3-1958.5	0.37	
Clark	1953.6-1954.6	0.05	
Tempel 1	1953.3-1954.1	0.05	
Kowal	1947.4-1949.8	0.03	1944.3-1944.5
Shajn-Schaldach	1944.5-1947.6	0.31	1942.6-1944.0
Tempel 2	1943.0-1944.3	-0.03	
Clark	1941.0-1943.0	0.06	
Tempel 1	1940.9-1942.3	0.12	
Shajn-Schaldach	1940.9-1942.3	(see encounter listed above)	
Oterma	1934.4-1940.5	0.55	1934.5-1939.0
Johnson	1931.1-1932.5	-0.07	
Schwassmann-Wachmann 2	1924.3-1927.5	0.33	1923.0-1924.0
Kojima	1925.2-1925.8	-0.05	
Whipple	1920.7-1923.8	0.31	1919.1-1919.3
Whipple	1916.4-1918.0	(see encounter listed above)	
Du Toit 2	1916.0-1917.1	0.05	
Du Toit 2	1903.7-1904.7	-0.06	
Kowal	1897.3-1900.0	-0.02	1901.9-1902.4
De Vico-Swift	1896.5-1898.3	-0.09	
Neujmin 2	1890.8-1892.3	0.07	
De Vico-Swift	1885.7-1886.2	-0.06	
Gunn	1876.5-1883.6	0.11	1879.0-1880.5

Tab. 3 cont.ed

Tempel 1	1891.1-1882.8	-0.09
Neujmin 2	1878.7-1880.5	0.11
Shajn-Schaldach	1873.9-1877.3	0.31
Schorr	1874.6-1877.0	0.09
Du Toit-Neujmin-Delporte	1871.3-1873.1	0.06
Tempel 1	1869.1-1871.3	-0.06
Kojima	1867.0-1867.7	0.06
Johnson	1859.7-1861.6	0.07
Tempel 2	1859.7-1861.1	0.04
Barnard 1	1860.0-1861.1	0.04

P/Kowal underwent two encounters with Jupiter within 1 AU, without suffering great perturbations ($\Delta(1/a)=0.03$ and $\Delta(1/a)=-0.02$ in the two encounters, if we compute it disregarding the perihelion passages at which the comet actually encountered the planet). In both cases P/Kowal underwent a temporary satellite capture (Carusi and Valsecchi, 1981), but these occurred when the comet was at a distance of more of 1 AU from Jupiter (see Table 3), and were of so short duration that they cannot be considered on the same ground as the satellite capture of P/Oterma mentioned below.

The remaining six comets are the most interesting from the dynamical point of view. Three of them experienced, during the time span examined, a reversal of the apse line at an encounter with Jupiter, exchanging perihelion with aphelion. Among them, P/Oterma undergoes, during that encounter, a rather long satellite capture, whereas both P/Shajn-Schaldach and P/Smirnova-Chernykh, during the encounters that brought them from one to the other side of the orbit of Jupiter, were not captured as temporary satellites. Also, the 1957 perihelion passage of P/Smirnova-Chernykh listed in Table 2 occurred when the comet was still within 1 AU from Jupiter (see Table 3).

Curiously enough, the satellite captures of P/Shajn-Schaldach and P/Whipple reported by Carusi and Valsecchi (1981) took place between two successive encounters of

these comets with Jupiter, i.e. when the distance from
the planet was in excess (probably not too much in excess) of 1 AU.

Also the remaining three comets, P/Gunn, P/Schwas-
smann-Wachmann 2 and P/Whipple, underwent temporary satellite captures (Carusi and Valsecchi, 1981), but only
P/Gunn was within 1 AU from Jupiter at the time of the
event. Also, the orbits of these comets are (not continuously) expanding, as we go backward in time; probably, an extension of the integration for another century
or two would cover also the encounters with Jupiter that
transferred them into the current inner orbits. Kazimir-
ciak-Polonskaya (1967) actually found that such an encounter for the comet P/Whipple occurred between 1840 and
1853, just a few years beyond the end point of our integration.

As it is possible to see in Table 3, the six just
mentioned comets generally suffered stronger perturbations at encounters with Jupiter than the comets belonging to the first two groups that we discussed.

From the above mentioned considerations, we can
summarize the orbital evolution of our sample of comets
with T greater than 2.9 in this way: the three groups
of comets that we found represent three successive stages of the inward evolution of short-period comets.
Starting from orbits like those of P/Kowal and P/Schwas-
smann-Wachmann 1, comets can be transferred inside Jupiter's orbit by a close encounter with that planet.
During the transferring encounter, temporary satellite
captures by Jupiter are possible. After the transition
into the inner orbit, the comets describe extremely unstable patterns like those of P/Oterma, P/Shajn-Schaldach,
P/Smirnova-Chernykh, P/Gunn, P/Whipple, and P/SChwassmann-
Wachmann 2; other encounters follow, during which again
temporary satellite captures or even transitions into
orbits outside that of Jupiter are possible. We know
that both these phenomena in fact happened to P/Oterma
as a consequence of a later encounter with Jupiter which
took place around 1963, _after_ the starting point of P/
Oterma in our backward integration.

The next stage of the evolution of comets with T>

2.9 is represented by the remaining objects examined. When the semiaxis has been reduced enough either by further encounters with Jupiter, or by non-gravitational forces, which become more important for the orbital evolution as the perihelia of the comets get closer to the Sun, the comets begin a gentle wandering in the phase space of orbital elements, as the encounters with Jupiter become less effective. This is due to the geometry of the encounters becoming less favourable for strong perturbations, when the eccentricities increase and the semimajor axes decrease.

The two comets that have not been treated here, P/Encke and P/Gehrels 3, would have been included, in the preceding scheme, in the third and the second group, respectively. In fact, P/Encke is less perturbed by Jupiter than any comet of our sample, as its aphelion is decoupled from Jupiter's orbit. On the other hand, P/Gehrels 3, in the integration made previously (Carusi and Valsecchi, 1979) resulted to have an orbital history closely resembling, for some aspects, that of P/Oterma. Furthermore, again like P/Oterma, it experienced a very long satellite capture by Jupiter during the 1970-1975 encounter (Rickman, 1979; Carusi and Valsecchi, 1979, 1981; Rickman and Malmort, 1981).

The authors wish to thank Dr. Lubor Kresák for helpful discussions, suggestions and for critically reading the manuscript, and Prof. A.E. Roy for reading a part of the manuscript.

References

Carusi, A., and Pozzi, F., 1978. Planetary close encounters between Jupiter and about 3000 fictitious minor bodies. Moon and Planets 19: 71-87.

Carusi, A., Pozzi, F., and Valsecchi, G.B., 1979. Planetary close encounters: an investigation on temporary satellite capture phenomena. In "Dynamics of the Solar System", ed. R.L. Duncombe (Dordrecht: D. Reidel), 185-189.

Carusi, A., Kresák, L., and Valsecchi, G.B., 1981. Perturbations by Jupiter of a chain of objects moving in the orbit of Comet Oterma. Astron. Astrophys. 99: 262-269.

Carusi, A., and Valsecchi, G.B., 1979. Numerical simulations of close encounters between Jupiter and minor bodies. In "Asteroids", ed. T. Gehrels (Tucson: Univ. Arizona Press), 391-416.

Carusi, A., and Valsecchi, G.B., 1980. Planetary close encounters: Importance of nearly tangent orbits. Moon and Planets 22: 113-124.

Carusi, A., and Valsecchi, G.B., 1981. Temporary satellite captures of comets by Jupiter. Astron. Astrophys. 94: 226-228.

Everhart, E., 1973a. Horseshoe and Trojan orbits associated with Jupiter and Saturn. Astron. J. 78: 316-328.

Everhart, E., 1973b. Examination of several ideas of comet origins. Astron. J. 78: 329-337.

Everhart, E., 1979. Chaotic orbits in the Solar System. In "Asteroids", ed. T. Gehrels (Tucson: Univ. Arizona Press), 283-288.

Fernandez, J.A., 1980. Evolution of comet orbits under the perturbing influence of the giant planets and nearby stars. Icarus 42: 406-421.

Fernandez, J.A., 1981. New and evolved comets in the Solar System. Astron. Astrophys. 96: 26-35.

Kazimirchak-Polonskaya, E.I., 1967. Evolution of short period comet orbits from 1660 to 2060, and the role of the outer planets. Sov. Astron.-A.J. 11: 349-365.

Kazimirchak-Polonskaya, E.I., 1972. The major planets
    as powerful transformers of cometary orbits. In
    "The motion, evolution of orbits and origin of
    comets", eds. G.A. Chebotarev, E.I. Kazimirchak-
    Polonskaya and B.G. Marsden (Dordrecht: D. Reidel),
    373-397.

Kazimirchak-Polonskaya, E.I., 1976. Review of investiga-
    tions performed in USSR on close approaches of co-
    mets to Jupiter and the evolutions of cometary or-
    bits. In "The study of comets", eds. B. Donn, M.
    Mumma, W. Jackson, M. A'Hearn and R. Harrington
    (NASA SP-393), 490-536.

Marsden, B.G., 1979. Catalogue of cometary orbits. (Cam-
    bridge: Smithson. Astrophys. Obs.).

Öpik, E.J., 1976. Interplanetary Encounters. (New York:
    Elsevier), 18.

Rickman, H., 1979. Recent dynamical history of the six
    short period comets discovered in 1975. In "Dyna-
    mics of the Solar System", ed. R.L. Duncombe (Dor-
    drecht: D. Reidel), 293-298.

Rickman, H., and Malmort, A.M., 1981. Variations of the
    orbit of comet P/Gehrels 3: temporary satellite
    captures by Jupiter. Submitted to Astron. Astro-
    phys..

Weissman, P.R., 1981. Dynamical history of the Oort
    cloud. Presented at I.A.U. Colloquium 61, Tucson,
    March 1981.

PART   II

PHYSICS OF THE SOLAR SYSTEM

THE INTERNAL STRUCTURE OF THE SUN, ITS PULSATIONS AND THE
NEUTRINO PROBLEM

Lucio Paternò

Osservatorio Astrofisico di Catania,
Istituto di Fisica della Facoltà di Ingegneria
dell'Università di Catania, Italy.

Abstract – The theory of the internal structure of the Sun is described. The standard models of the Sun predict a neutrino flux
which is three time larger than that measured by Davis experiment.
It is not clear whether this dramatic discrepancy depends on the
ideas of stellar structure and evolution theory or on neutrino
physics. A convincing attempt to reconcile theory and observations
has been done constructing a solar model with a very low primordial abundance of heavy elements . This model predicts the correct
neutrino flux, but has a convective zone much thinner than that
of the standard model. The observed spectrum of 5 minute oscillations, which is extremely sensitive to the depth of the convection
zone, can only be reproduced by the standard model indicating
that neutrino flux is high. In recent times, a series of independent experiments have not excluded the possibility that neutrino
is massive and oscillates in three different types.
Should this be the case, assuming and equally probable mixure of
the three neutrinos, the experiment of Davis should actually detect one third of the neutrino flux produced by the Sun at the
Earth's surface.

1. Introduction

No more than ten years ago, the internal structure of the
Sun was thought to be well understood. The theory could more or
less explain the few existing observations, and there was every
confidence that other details would also be explained within the
same theory.
The internal structure of the Sun is determined by evolving
a star of one solar mass and homogeneous initial chemical composi-

A. Coradini and M. Fulchignoni (eds.), The Comparative Study of the Planets, 151–160.

tion for a time equal to the present Sun's age until it reachs the present Sun's luminosity and radius.

This procedure leads to the correct positioning of the Sun on the Hertzprung-Russel (H-R) diagram, if two independent parameters are properly chosen: the initial Helium abundance and the ratio of the mixing-length to the local pressure scale height in the convection theory. The first parameter determines the present Sun's luminosity and the second the present Sun's radius.

The same theory describes in a reasonably good way the evolution of the other stars predicting their position on the H-R diagram for both population I and II stars, and also indicating the instability points, where the stars pulsate.

However the Sun is a star very close to us, which enable us to deeply observe its structure.

The most spectacular experiments, which recently opened a new solar physics, are the measurement of neutrinos emitted from the Sun's core, and the detection of small amplitude solar pulsation. These observations provide a formidable tool for investigating the internal structure of the Sun.

The results of the measurements of the neutrino flux at the Earth's surface, carried out uninterruptly for almost ten years, indicate unambiguously an upper limit that is about three times smaller than the flux predicted by the theory.

The dramatic discrepancy between the predicted and measured neutrino flux open the solar neutrino problem. Does this discrepancy depend on a bad understanding of the physical processes on which the theory of stellar structure and evolution is based, or does it depend on the neutrino physics?

As we shall see the study of solar pulsations and the results of recent experiments on neutrino mass indicate that the discrepancy could be removed and the theory of stellar structure and evolution saved.

2. The standard model of the Sun

The fundamental hypotheses on which the standard theory of the Sun's structure is based are the following: spherical symmetry; hydrostatic balance; thermal equilibrium; energy produced by the thermonuclear conversion of H into He; energy transport by radiation or convection according to the Schwarzschild criterion of stability; convective zones described by the mixing-length theory; mass loss, rotation and magnetic fields negligible.

The differential equations which describe the structure are therefore (Stein 1966):

$$\frac{dp}{dr} = - \frac{GM(r)}{r^2} \rho \qquad \text{hydrostatic balance} \qquad (1)$$

$$\frac{dM\ (r)}{dr} = 4\pi\ r^2 \rho \qquad \text{continuity of mass} \qquad (2)$$

$$\frac{dL\ (r)}{dr} = 4\pi\ r^2 \rho\ \varepsilon \qquad \text{thermal equilibrium} \qquad (3)$$

$$\frac{L\ (r)}{4\pi\ r^2} = F_R + F_C \qquad \text{energy transport} \qquad (4)$$

together with the constitutive equations

$$p = p\ (\rho, T, C_i) \qquad \text{equation of state} \qquad (5)$$

$$\kappa = \kappa\ (\rho, T, C_i) \qquad \text{radiative opacity} \qquad (6)$$

$$\varepsilon = \varepsilon\ (\rho, T, C_i) \qquad \text{energy production} \qquad (7)$$

where r is the radial coordinate, p the total pressure, G the gravitational constant, M (r) the mass enclosed in a sphere of radius
r, ρ the density, L (r) the luminosity at distance r from the
Sun's center, ε the rate of nuclear energy production, F_R and F_C
the radiative and convective flux respectively, T the temperature,
C_i the chemical composition, and κ the radiative opacity.
    The expressions of F_R and F_C into the equation (4) are the
following:

$$F_R = -\frac{16\ \sigma\ T^3}{3\ \kappa\ \rho}\ \frac{dT}{dr} \qquad (8)$$

$$F_C \simeq c_p\ \rho\ (\frac{2g}{T})^{\frac{1}{2}}\ (\frac{dT'}{dr} - \frac{dT}{dr})^{3/2}\ \ell^2 \qquad (9)$$

where σ is the Stefan's constant, c_p the specific heat at constant
pressure, g the gravity, dT'/dr the temperature gradient in the
convective elements, and ℓ the mixing-lenght.
    The mixing-length theory of convection assumes that fluid
elements which have smaller density than the surrounding fluid
rise against gravity through a characteristic mean free path ℓ,
of the order of their size; they then mix abruptly their thermal
properties with the colder surroundings.
    The Schwarzschild's criterion of convective stability esta-
blishes under which circumstances thermal convection is effective
in transporting heat in stellar interiors. The convection takes
places in the regions where the radiative temperature gradient
$(dT/dr)_{rad}$ as defined by equation (8) exceeds its adiabatic counterpart:

$$(\frac{dT}{dr})_{ad} = -\frac{\gamma - 1}{\gamma}\ \frac{\mu g}{R} \qquad (10)$$

where γ is the ratio of the principal specific heats, μ the molecular weight, and R the gas constant.
    The mixing-length theory predicts dT/dr, dT'/dr, and the ef-

ficiency of convection, when the total flux $F_T=F_R+F_C$ is known,
together with $(dT/dr)_{rad}$ and $(dT/dr)_{ad}$. It is then possible to
solve equation (9) determining F_C, and therefore F_R. In the regions where $(dT/dr)_{rad} < (dT/dr)_{ad}$ $F_C=0$, and equation (8) determines F_R (Cox and Giuli 1968).

The radiative gradient is governed by the opacity that is
very sensitive to the temperature. In the surface layers of the
Sun $\kappa \propto T^5$; since T increases inwards, $(dT/dr)_{rad}$ becomes large
causing convective instability and determining the upper boundary
of the convective zone. As soon as we go deeper, the increasing
of T ionizes almost completely the matter, and the opacity depends
only on the electron scattering, following a law of the type
$\kappa \propto T^{-3.5}$. The radiative gradient decreases until it becomes smaller than the adiabatic one, determining the lower boundary of the
convection zone. The mixing length ℓ is taken proportional to the
local pressure scale height $H_p = RT/\mu g$, $\ell=\alpha H_p$. The parameter α
governs drastically the depth of the convection zone.

The main energy production in the Sun is due to the conversion of H into He. This process takes place through two mechanisms:
the proton-proton (p-p) chain (98%) and CNO cycle (2%). The global
reaction links four protons to form a Helium nucleus with the emis
sion of two positrons and two neutrinos, and delivers 26.72 MeV:

$$4p \rightarrow {}^4He + 2\beta^+ + 2\nu + \gamma \tag{11}$$

The rates of energy production due to p-p and CNO processes depend on T and chemical composition (Reeves 1966):

$$\varepsilon_{pp} \simeq 1.9 \ 10^{-29} \ \rho \ X^2 T^4 \qquad \text{erg } g^{-1}s^{-1}$$

$$\varepsilon_{CNO} \simeq 5.7 \ 10^{-130} \ \rho X_{CNO} T^{18} \qquad \text{erg } g^{-1}s^{-1}$$

where X is the mass fraction of H and Z_{CNO} is the mass fraction
of C, N and O.

Reaction (11) consists of many branches in which intermediate products as ${}^3He$ are formed. The most important branch that pro
duces energetic neutrinos detected in the Davis (Davis and Evans
1978) experiment is the following:

$${}^3He + {}^4He \rightarrow {}^7Be + \gamma$$
$${}^7Be + p \rightarrow {}^8B + \gamma$$
$${}^8B \rightarrow {}^8Be^* + \beta^- + \nu$$
$${}^8Be^* \rightarrow 2 \ {}^4He$$

This reaction is extremely sensitive to the temperature. All the
cross sections of the reactions leading to reaction (11) have
been studied at low energies, and then extrapolated to very low
energies of the order of 5-20 keV, as those existing in the core
of stars. The two principal branches of the p-p chain $p+p \rightarrow {}^2D +$
$+ \beta^+ + \nu$, and $p+p+\beta^- \rightarrow {}^2D+\nu$ cannot be reproduced in laboratory,

therefore their cross sections have been determined theoretically
(Fowler et al. 1967, 1975).

The knowledge of the nuclear reactions leads to the determinations of ε that can be described in terms of ρ, T, and C_i.

The opacity can be computed if we know the cross sections of
the absorption processes, and expressed in terms of ρ, T, and C_i
(Cox and Steward, 1970).

The relation (5) describes the equation of state and permits
to determine p as a function of ρ, T, and C_i.

The mixing-length theory permits to determine the gradients
in the convectively unstable regions, otherwise the gradient is
determined by equation (8).

The set of differential equations (1-4) with proper boundary
conditions, and relations (5-7) permit to compute the quantities
p, ρ, T, $M(r)$, $L(r)$, κ, ε as functions of radius, and therefore the
whole solar structure, if the total mass of the Sun and the chemi-
cal composition in each point are known.

Since the Sun's chemical composition is not known a priori,
the only way to determine its structure is to construct evolutionary models. These require two supplementary assumptions that the
initial chemical composition was homogeneous and the age of the
Sun is the same as the age of the Earth, about $4.7 \ 10^9$ years
(Bahcall 1979).

As soon as thermonuclear reactions transform H into He the
chemical composition in the Sun's core varies:

$$(C_i)_{M,t_0+\delta t} = (C_i)_{M,t_0} + \frac{\partial}{\partial t} (C_i)_M \ \delta t \qquad (12)$$

where

$$\frac{\partial}{\partial t} (C_i)_M = f(\rho, T, C_i)$$

Equation (12) describes the variation of C_i as a function of time
in each point of the star, when an initial C_i has been assigned.
Equation (1-4) can then be solved with the new C_i to compute the
new structure at the time $t_0+\delta t$. This procedure is repeated until
the present Sun's age is reached, and therefore its present structure determined.

The problem of which initial homogeneous composition to adopt
is a critical one. The composition is described by the mass fractions of H, He, and all the elements heavier than He, respectively X,Y,Z, with the constraint X+Y+Z = 1, that implies to have only two free parameters.

The generally adopted procedure is to fix the ratio Z/X to
the same value as that observed at the solar surface (Z/X $\simeq$ 0.027)
since the surface Helium abundance is very uncertain, the determinations ranging from Y=0.16 to Y=0.27. Thus a series of models
is constructed with the same Z/X and different Y's, in order to
find that model that at $t=t_\odot$ gives $L=L_\odot$. This model results in a
chemical composition X=0.74, Y=0.24, and Z=0.02. However the lumi-

nosity is not sufficient for placing the Sun in the H-R diagram.
The other necessary ingredient is the radius that strongly depends on α. Thus α is varied until the model gives R=$R_\odot$ at t=$t_\odot$.

Should the same opacity tables, cross sections for nuclear
processes, equation of state, and ratio Z/X be used, the parameter α then determines unambiguously the depth and stratification
of the convection zone.

The standard model predicts a neutrino flux at the Earth's
surface about 5 SNU (1 SNU $\equiv 10^{-36}$ captures per target ^{37}Cl atom per
second).

3. The solar neutrino problem

Solar neutrinos have been detected by the Davis (Davis and
Evans 1978) experiment. This depended on neutrino capture by ^{37}Cl
stored in a 400,000-litre tank of perchloroethylene; the product
is ^{37}Ar which was extracted chemically and subsequently detected
because of its β-decay. The capture or inverse β-decay is described by the following reaction:

$$\nu + {}^{37}Cl \rightarrow {}^{37}Ar + \beta^- \tag{13}$$

The Davis measurements, carried out uninterruptly since 1970,
give an upper limit to neutrino flux 1.6 ± 0.4 SNU. If we compare
this value with that predicted by the standard model of the Sun,
we may conclude that there is something wrong in our ideas of stel
lar structure or nuclear physics.

In recent times a series of exotic solar models producing
low neutrino flux have been proposed to reconcile theory and observations. All these models tend to decrease the central temperature
of the Sun, since the reaction that emits the most neutrinos detected with the Davis experiment is very sensitive to the temperature.

Many unlikely mechanisms have been proposed which invoked a
mixed core, a rapidly rotating core or the presence of very strong
magnetic fields in the core. The same result can be obtained assuming a core very rich in elements heavier than Magnesium on which
H and He later condensed, or very rich in He.

It has also been suggested that a black hole in the center of
the Sun could supply part of its radiated energy, or that the gravity constant decreasing with time produces a decreasing of temperature due to the new conditions of hydrostatic balance (Bahcall
1979).

Even if these models could explain is some way the low neutrino flux, the assumptions were extremely ill defined and not based on observations. On the other hand, the ideas of the standard
stellar evolution could well explain the observations of the other
stars.

The most convincing non-standard model, that does not reject

the standard theory of stellar evolution, is one constructed with
the standard procedure, but with an initial chemical composition
very poor in heavy elements (Christensen-Dalsgaard et.al. 1979).
This model makes use of the not unplausible assumption that the
Sun's surface layers have been enriched in heavy elements during
the life of the Sun until they reached the present composition.
This model with initial $X \simeq 0.84$, $Y \simeq 0.16$, $Z \simeq 0.001$ gives the
observed neutrino flux, but in order to reproduce the correct so-
lar radius needs a convective zone seven times thinner than that
reproduced by the standard model.

   The experiment of solar neutrino detection depends also on
the theory of weak interactions and in particular on the neutrino
properties deduced from this theory. The most important property
is that neutrino does not decay or transform in another particle
with different cross section.

   It has been suggested by Pontecorvo in 1977 that if neutrino
had mass could oscillate between two or more states. If this were
true, neutrino flux measured at the Earth could be reduced by a
factor two or more.

   Experiments carried out in 1979-80 by different groups of
physicists do not exclude the possibility that neutrino is massive and oscillates in three different types (Lubkin 1980).

   In this case the Davis experiment would detect about one
third of the emitted solar neutrinos reconciling the results of
the standard theory with observations.

   At this stage we have two models of the Sun: the standard
model which produces a neutrino flux about 5 SNU, and has a convective zone about 200,000 km deep, and the enriched model which
produces a neutrino flux about 1.6 SNU, and has a convective zone
about 30,000 km deep. These models are undistinguishable between
them from exterior in that they give the same luminosity and solar radius.

4. Solar oscillations as a probe for the Sun's interior

   Oscillations of the solar surface with amplitudes ranging
from 10^{-5} $R_\odot$ to $10^{-6} R_\odot$ have been detected since 1973 (Hill 1978).
These global oscillations can be used to study the internal struc-
ture of the Sun.

   The simplest oscillation modes are radial; also non radial
oscillations have been detected; in this case the Sun expands in
one or more regions and simultaneously contracts in other regions.

   Since the amplitude of oscillation is small with respect the
solar radius and propagation of fluctuations is small with respect
the velocity of sound, it is possible to use the linear theory.

   The problem is described by a set of ordinary differential
equations whose dimensionless form is (Cox 1980):

$$r \frac{dy_1}{dr} = \left[\frac{V}{\Gamma_1} - 3 \right] y_1 + \left[\frac{\ell(\ell+1)}{C_1 \sigma^2} - \frac{V}{\Gamma_1} \right] y_2 + \frac{V}{\Gamma_1} y_3 \qquad (14)$$

$$r \frac{dy_2}{dr} = (C_1 \sigma^2 + Ar) y_1 + (1 - U - Ar) y_2 + Ar y_3 \qquad (15)$$

$$r \frac{dy_3}{dr} = (1 - U) y_3 + y_4 \qquad (16)$$

$$r \frac{dy_4}{dr} = -UAr y_1 + \frac{UV}{\Gamma_1} y_2 + \left[\ell(\ell+1) - \frac{UV}{\Gamma_1} \right] y_3 - U y_4 \qquad (17)$$

where

$$y_1 = \frac{\delta r}{r} = \xi \; , \; y_2 = \frac{1}{gr} \left(\frac{p^1}{\rho} + \psi^1 \right) , y_3 = \frac{1}{gr} \psi^1 , y_4 = \frac{1}{g} \frac{d\psi^1}{dr} \; ;$$

ξ is the radial dispacement, p^1 the pressure perturbation, ψ the gravitational potential, and

$$C_1 = \left(\frac{r}{R_\odot} \right)^3 \frac{M_\odot}{M(r)} \; , \; U = \frac{d \ln M(r)}{d \ln r} , \; V = \frac{g r \rho}{p} \; ,$$

$$A = \frac{1}{\rho} \frac{d\rho}{dr} - \frac{1}{\Gamma_1 p} \frac{dp}{dr} , \Gamma_1 = \left\{ \frac{d \ln p}{d \ln \rho} \right\}_{ad}$$

Each displacement ξ can be expressed in terms of spherical harmonics:

$$\xi_r = \xi_{r,n}^{(\ell)} (r) \; Y_\ell^m (\theta, \phi) \; e^{-i\sigma t} \qquad (18)$$

where n (n = 0, 1, 2, ..) represents the number of nodes in the radial direction, ℓ is the harmonic degree, m the azimuthal number of the harmonic, and σ the angular frequency.

Equations (14-17) must satisfy appropriate boundary conditions at the interior and surface of the Sun. These conditions are simultaneously satisfied only for certain frequencies which are the eingenfrequencies. The corresponding modes of oscillations are the normal modes. If we neglect rotation and magnetic field, normal modes are described only by the numbers n and ℓ.

Modes with $\ell=0$ are radial, and modes with n=0 are called fundamental. Modes with ℓ small penetrate deeply in the Sun, while modes with ℓ large do not.

There are two different physical mechanisms which generate oscillations. The first produces acoustic or pressure waves (p-modes), the second gravity waves (g-modes). In the first case the restoring force is the gas pressure, in the second the gravity. The propagation of g-modes is limited by Brunt-Väisälä frequency so that they decay exponentially in the convective zone. Only g-modes with ℓ very small can be observed at the surface; those

with ℓ large would be observed only if convection zone were extremely thin.

The knowledge of the entire oscillation spectrum would permit in principle to know the internal structure of the Sun, solving the heliological inverse problem, analogously to the geophysical inverse problem.

At the present we can observationally identify with certain-
ty the p-mode spectrum of 5 minute oscillations (Deubner et al.
1979). The ridges in which the power is concentrated define unambiguously the modes with n ranging from 0 to 7, and with ℓ ranging
from about 100 to 1000, in a ℓ,σ diagram.

These modes are confined in the convective zone, and therefore depend strongly on its depth and stratification.

This constraint can be used to test whether the model with a
deep convective zone or the model with a thin convective zone reproduce with a higher degree of accuracy the observed spectrum.

Recent calculations of Berthomieu et al. (1980) have shown
that a standard model with X=0.745 and Z=0.02, and a convection
zone 240,000 deep reproduces with a very high degree of accuracy
the observed spectrum. At the same time calculations of Belvedere
et al. (1981) have clearly shown that the model with low Z abundance and a very thin convective zone does not reproduce at all
the observed spectrum.

5. Discussion

In the light of the results on the reproducibility of the 5
minute oscillation spectrum by the solar standard model, one
would attempt to conclude that the physical processes in the interior of the Sun are correctly described, and that the small
neutrino flux detected by Davis depends on the fact that neutrino
has mass and oscillates in three different types.

On the other hand the observed 5 minute oscillation spectrum
depends only on the depth and stratification of the convective
zone, and not on the details of the internal structure. Even if
the theory suggests that it is possible to construct only one model with a given convective zone depth and stratification, any
conclusion indicating that the reproducibility of the 5 minute
p-mode spectrum implies also a unique solution for the interior
of the Sun must be taken with caution.

Recently all the question has been rediscussed in the framework of a suggestion of Hill (Rosenwald and Hill 1980) who attributed to g-modes with $\ell \simeq 20$ some oscillations observed by himself.
Hill concludes that the convective zone must be thin in agreement
with the non-standard model with low neutrino flux. To overcome
the objection that 5 minute oscillation spectrum cannot be reproduced by a thin convective zone, Hill suggests that the spectrum
can also be reproduced by a thin convective zone if non linear
effects are taken into account at the top boundary. These effects

could actually modify the computed eigenfrequencies.

To test this possibility, Belvedere et al. (1981) have used the method of eigenvalue relaxation inserting the observed eigenfrequencies into the model with low neutrino flux, thus changing the boundary conditions by forcing a mode with a given n computed with non-standard model to coincide with the corresponding observed mode. The results show clearly that the entire observed p-mode spectrum can only be reproduced by the standard model of the Sun with a high neutrino flux.

Does this mean that neutrino is massive? More direct measurements are needed to conclude so, but astrophysics seems to indicate that this conclusion is not unplausible.

References

Bahcall, J.N.: 1979, Space Science Reviews 24, pp. 227-251.
Belvedere, G., Gough, D.O., and Paternò, L.: 1981, preprint.
Berthomieu, G., Cooper, A.J., Gough, D.O., Osaki, Y., Provost, J., and Rocca, A.: 1980, *Nonradial and Nonlinear Stellar Pulsation*, H.A. Hill and W. Dziembowski (eds.), Springer-Verlag, New York, pp. 307-312.
Christensen-Dalsgaard, J., Gough, D.O., and Morgan, J.G.: 1979, Astron. Astrophys. 73, pp. 121-128.
Cox, J.P.: 1980, *Theory of Stellar Pulsation*, Princeton University Press, Princeton, New Jersey.
Cox, J.P., and Giuli, R.T.: 1968, *Principles of Stellar Structure*, Vol. 1, Gordon and Breach, New York.
Cox, A.N., and Steward, J.N.: 1970, Astrophys.J. Suppl. 19, pp. 243-259.
Davis, R., and Evans, J.C.: 1978, *The New Solar Physics*, J.A. Eddy (ed.), Westview Press, Boulder, Colorado, pp. 35-57.
Deubner, F.L., Ulrich, R.K., and Rhodes, E.J.: 1979, Astron. Astrophys. 72, pp. 177-185.
Fowler, W.A., Caughlam, G.R., and Zimmerman, B.A.: 1967, Annual Rev. Astron. Astrophys. 5, pp. 525-570.
Fowler, W.A., Caughlam, G.R., and Zimmerman, B.A.: 1975, Annual Rev. Astron. Astrophys. 13, pp. 69-112.
Hill, H.A.: 1978, *The New Solar Physics*, J.A. Eddy (ed.), Westview Press, Boulder, Colorado, pp. 135-214.
Lubkin, G.B.: 1980, Physics Today, 33,7, pp. 17-19.
Reeves, H.: 1966, *Stellar Evolution*, R.F. Stein and A.G. Cameron (eds.), Plenum Press, New York, pp. 83-122.
Rosenwald, R.D., and Hill, H.A.: 1980, *Nonradial and Nonlinear Stellar Pulsation*, H.A. Hill and W. Dziembowski (eds.), Springer-Verlag, New York, pp. 404-412.
Stein, R.F.: 1966, *Stellar Evolution*, R.F. Stein and A.G. Cameron (eds.), Plenum Press, New York, pp. 3-79.

HEAT TRANSFER AND THE DEVELOPMENT OF INTERNAL STRUCTURE IN THE
TERRESTRIAL PLANETS

D.C.Tozer

School of Physics, The University of Newcastle upon Tyne
Newcastle upon Tyne, NE1 7RU, England.

Abstract
    A theory of planetary differentiation is developed which in-
corporates the tendency of convective heat transfer to maintain
very high average viscosities throughout planetary interiors
under a wide range of conditions. It is concluded that the
separation of a silicate crust from Earth's silicates throughout
that planet's history has been wholly dependent on local and
transient heating events having their origin in strain energy
that is stored throughout the convecting medium. Such slip
events are contingent on cold planetary surface material being
involved in the convective flow. Consideration of the overall
efficiency of conversion of radiogenic to deformational heating
suggests that Earth's silicates would only be partly separated
into a crust and mantle even if there had been no remixing. This
is even more true for smaller planets.
    The separation of core material from silicates is far more
influenced by the gravitational energy this process converts to
heat. In the cases of Earth and Venus it was so important in
catalysing the process that the regulation of a high viscosity
was temporarily broken and the separation of core material
effectively complete at the time one could say planetary
accumulation had finished. Smaller objects may have had a post
accretional phase in which density differences due to core
formation rather than thermal expansion controlled the dynamics.

Introduction
    Many past attempts to discuss the development of a chemically
differentiated planetary structure make only minimal reference to
its connection with the heat transfer process in the planets.
Indeed, it is common to see firm believers in the continuous

161

A. Coradini and M. Fulchignoni (eds.), The Comparative Study of the Planets, 161–180.

development of an internal planetary structure throughout the
last 4.5 billion years discuss the thermal evolution of planets
as though the simultaneous movement of matter a differentiation
process entails would make no difference to conclusions about
the thermal state.   Conversely, the chemical differentiation of
a planet is inevitably associated with the dissipation of some
gravitational potential energy as heat, so that the intimate
connection of the heat transfer and differentiation problems
is obvious.   In fact, were techniques of theoretical analysis
much stronger than they are, both problems would probably be
solved as two aspects of just one problem.

Clearly a common factor controlling the rates of either
process is the rheology of the planetary material, and it is the
different degrees to which this is controlled by internal and
external causes that is perhaps the most fundamental distinction
one can make between the dynamical regimes of the terrestrial and
Jovian planets.   Although the former have much higher surface
temperatures, it is the ratios of these temperatures to the
melting points of the components comprising the planetary
material that are of greater significance.   Judged by such a
rheological criterion, the Jovian planets have always had much
hotter near-surface conditions than the terrestrial planets.
The importance of this difference to the questions of timing and
extent of internal differentiation is that the creep resistance
of virtually all the interiors of the terrestrial planets has
been under the control of the internal heat transfer process for
at least four billion years.   In contrast, the rheology of the
Jovian planets, at least in their outer parts, will, for a very
long time to come, be prevented from getting into a similar
dependence on internal heat transfer by heat from a Sun that will
be radiating at least as much energy as the present one.   The
fact that Jupiter and Saturn still radiate significantly more
energy than they receive from the Sun and in that way raise their
near-surface absolute temperatures by up to ~25%, is an additional factor that makes this rheological comparison with the terrestrial planets even more marked.

Interesting changes in our understanding of terrestrial planet
interiors have come from first attempts to understand this interplay of rheological and thermal questions.   The long standing
tradition of calculating temperatures inside terrestrial planets
on the assumption that it is a heat conducting 'solid' and only
then trying to work out its effective viscosity distribution or
why there might be evidence of large scale deformation in its
surface rocks is now clearly seen to be inconsistent and mis-
leading.   A continuous and very rapid decline with temperature
of the creep resistance of all mineral phases, together with the
newly recognised possibility of disproportionately large
rheological effects arising from minor volatile components in the
planetary material, has made it imperative to examine planetary
heat transfer with a mathematical formalism that allows for

irrecoverable deformability, i.e. a convection theory.

This attempt to see thermal convective movement as the natural response of a planet that has become statically unstable through its thermal expansion and self gravitation should not be confused with the traditional descriptive and kinematic approach to rock deformation now pursued under the title 'The Theory of Plate Tectonics". If there is any 'theory' of Plate Tectonics it is probably that of heat transfer! Unfortunately, this particular attempt to simplify the problems raised by a strongly temperature dependent rheology has rapidly taken on a life of its own and become a new dogma as obstructive to coherent rheological analysis of surface movements[†] as the previous attempt to describe the crust/mantle system as an elastic medium. Heat transfer theorists tend to see the special hypotheses introduced by Plate Tectonics to 'explain' the observed pattern of surface motion or 'plate movements' as simply confusing cause and effect. They would argue that there is no reason to think the observed pattern of movement and tectonic activity should be any more explicable than today's weather would be, given a similar ignorance about the pattern at any previous time. If this lack of information about some 'initial' dynamical state makes it impossible to test a planetary heat transfer theory of tectonic activity against numerous observed phenomena, it should not obscure the fact that this theory does provide a satisfactory account of the spatially and temporally averaged characteristics of this activity - the tectonic climate of the planet, to anyone willing to accept and build upon some unexpected consequences of integrating planetary thermal and rheological problems.

The central result guiding the prediction of a tectonic 'climate' is that the large decrease of creep resistance, or an effective viscosity with temperature, creates such a strong non-linearity in the relationship of the rate of surface heat transfer to the level of temperature reached below quite a shallow depth that one can speak of the horizontally averaged viscosity as a self regulating quantity throughout most of the interior. This has the valuable consequence that it is possible to make usefully precise quantitative predictions about the regime of solutions to the heat transfer problem, despite the large uncertainty that so many olivine obsessed theoreticians seem unable to accept as existing in the relationship of planetary material creep resistance to temperature. This uncertainty about creep resistance shows up in solutions to the heat transfer problem as an uncertainty about the average depth at which the horizontally averaged viscosity becomes a self regulating quantity $\sim 10^{21}$ poise. Leaving aside regions lying beneath the pre Cambrian shields (see below), this depth is no more than about 50 kms in the case of

---

[†] To describe surface material as 'rigid' is not only an extraordinary denial of ubiquitous geological observation, but it would preclude the movements Plate Tectonics seeks to explain.

the Earth and has no direct connection with the depth to the
seismic low velocity layer (Tozer 1981).

The subsidiary, if geologically important question of whether
a planet's superficial rocks are being deformed and translated by
the heat-transfer process is determined by whether an even thinner
layer of material forming the immediate sub surface region, is thin
enough for brittle fracture rather than elasticity to control its
response to the forces that the heat transfer process impresses
upon it.   The condition for surface tectonics can be expressed in
this way because thermally activated creep is utterly negligible
in determining irrecoverable rock deformation at the temperatures
existing near the external surface, while the low pressures make
them relatively very vulnerable to failure by crack propagation.
The predicted thickness of this layer, whose rheology is best
described as a stress limited elasticity, varies from at most a
few kilometres under oceanic ridge crests to perhaps 30 kms under
the oceanic trenches.   The degree of stress concentration this
implies in the near surface rocks has made it readily under-
standable why rocks forming the sea floor move at much the same
relative speeds ($\sim 10^{-7}$cm/sec) as those predicted for the material
in the deep interior where the rheology is both less complicated
and self regulating.

The probability of disrupting such a surface layer with heat
transfer induced stresses would greatly decrease if any change
were to cause an increase in its thickness by even a factor of
two — the stress concentration factor and the possibility of
crack propagation through such a layer is reduced by the higher
pressures associated with a thicker layer.   A relatively greater
average thickness of this sub layer is likely to exist under
continents, produced by a combination of lower surface heat flow,
smaller pore water pressures and a silicate differentiation that
has raised the viscosity of sub continental mantle material vis
à vis the sub oceanic mantle when subjected to the same P, T
conditions.   This difference of thickness and a somewhat lower
density under the same conditions that is also due to the silicate
differentiation, would explain why regions of non zero divergence
in the surface velocity field (spreading and subduction zones) are
mainly confined to the ocean basins.   These rheological differences between sub continental and sub oceanic material, together
with a high probability that the associated chemically induced
density differences exceed the 10^{-4}g /cm^3 thermally induced
density differences that drive convective movements, could easily
be made to explain the elevation of continental surfaces above
those of the sea floor, the closely matching edges of continental
regions after large relative displacement, and why continental
rocks manage to stay near the external surface far longer than
any sea floor material.

It has been said that the rheological contrast of sub continental and sub oceanic material is 'explained' by the 30% lower
heat flow, but this and the more elaborate reasons offered in the

previous paragraph are only a partial answer. They all beg the
question as to why such lateral differences have arisen. If one
accepts that continents are a product of evolution from a more
uniform state of composition for the Earth, one is now immediately
concerned to know how chemical differentiation can have occurred
in a planetary interior in which the regulation of an effective
viscosity at a very large value has precluded any simple picture
of the planetary material being heated by radioactivity to its
melting point - the basis of all previous discussion of this
same problem.

Before we get specifically involved in the separation of
major silicate or core phases, we should briefly consider the
differentiation of perhaps relatively minor phases that can
nevertheless affect these other differentiation processes through
their effect on the overall creep resistance of the planetary
material. This has created new interest in volatile components
like water and carbon dioxide that are known to have disproportionately large creep enhancing effects (Riecker and Rooney 1969,
Murrell and Ismail, 1976) and in particular in those that can avoid
a very prompt and almost complete expulsion from the planetary
material during its accumulation through chemical combination
with the silicate phases. It is known that volatiles are far
more active as creep strength reducers if they are present as
a separate phase, and if that makes these components more prone
to expulsion from the planet, it has also given dehydration and
decarbonation reactions a novel significance to thermal state
calculations - particularly those occurring at such high
temperatures that the radial temperature gradient is convectively
unstable. One is here beginning to see ways in which the heat
transfer process not only regulates the values of a viscosity
function, but also selects the function to be regulated. For
example, using experimental data about the higher temperature
hydration reactions of basalt, I have suggested (Tozer 1973,1981)
that an extremely persistent state of partial differentiation of
an initial silicate mantle water content would arise from a
peculiarity of the amphiboles in dehydrating at lower temperatures
the higher the hydrostatic pressure. In the case of the Earth
this long-lived evolutionary state would involve the existence of
a small quantity of intergranular water trapped at depths
~100 kms and with removal of the possibility of having widespread
melting of silicates at such depths, this now forms the best
basis for interpreting a minimum of S seismic wave velocity and Q
at comparable depths in the sub oceanic mantle. Such hydration
water would be similarly released but trapped at rather greater
depths in the smaller terrestrial planets. One can sum up the
differentiation of volatiles from the silicates in terrestrial
planets by saying that inert gases and chemically reactive
species in excess of the amounts that can combine with the
involatile fraction would be expelled very thoroughly from the
planetary material before it formed objects more than a few

kilometers in radius and then promptly lost to interplanetary
space.   On the other hand, the expulsion of all the chemically
reactive volatile phases is a very protracted process because
their rheological activity promotes the very thermal conditions
that lead to retention.

If we have to attribute the generation and maintenance of a
continental freeboard to differentiation among major silicate
phases, there is growing comparative planetological evidence that
the small water content of terrestrial rock has been decisive in
determining both the tectonic, as well as the erosional evolution
of the Earth's surface.   Even without this recent data (see
below), it has seemed clear that if the creep resistance of in
situ Earth material under particular P, T conditions was essentially the same as that of its most refractory phases, the
resulting sub surface layer of cool and extremely creep
resistant material (effective viscosity $> 10^{27}$ poise) would have
been much too thick to be disrupted by any underlying heat
transfer motions.   The comparative planetological evidence
pointing to the activity of less refractory components in Earth
material controlling its creep resistance, has arisen from the
Pioneer radar examination of the Venusian topography.   As
mentioned by Arvidson at this conference, Venus shows none of the
relief in the low lying parts of its surface that would be
analogous with the oceanic ridge system seen on Earth.   It will
be recalled that this terrestrial relief is now interpreted as a
thermal contraction effect associated with growth in thickness
of a moving boundary whose upper surface actually forms the sea
floor.   One may also notice that there are large areas of the
Venusian surface several kilometres higher than these smooth, low
lying parts.   On the basis of a cratering density, an age of
$> 10^9$ years has been given to the low lying surface rocks – more
than four times the age of Earth's oldest ocean floor.   What is
most striking about this evidence for a relative immunity of the
Venusian surface to disruption is that it exists while the
surface temperature is maintained at ~ 480°C by a massive CO_2
atmosphere.   Since this is 2.7 times the absolute temperature
of Earth's surface, one would expect big differences in
effective creep resistance if the two surfaces were of the same
material.   Clearly, it would be very difficult to explain the
apparent stability of the Venusian surface if the most creep
resistant phases ever likely to be present in substantial
amounts in any terrestrial planet controlled the rheology of a
similar sized Earth having at the same time a cooler but more
tectonically active surface.   Rather than resort to purely ad
hoc assumptions about a difference in composition of the two
planets, the most satisfactory solution to the problem at the
moment is to use the large ratio in amounts of water contained
in Earth's oceans and the Venusian atmosphere ($>10^4$) as an
indicator of the cause of relative creep weakness in terrestrial
rock.   This would suggest that without water, the endogenous

modification of Earth's surface would have ceased long before the
phanerozoic, and the atmosphere would be similar to that of
Venus.   The effect of water in keeping the temperature of the
deep interior of Earth several hundred degrees cooler than that
of Venus may also help to explain why non radiogenic He^3 is still
expelled from the Earth, whereas non radiogenic Ar^{36} seems to have
been expelled in greater quantity to the atmosphere of Venus.   It
is to be hoped that this topographic evidence from Venus, having
such an important bearing on major geotectonic questions, will
soon be greatly refined by the so called VOIR mission to map the
Venusian surface at much higher resolution.

The Differentiation of Major Fractures in Terrestrial Planet Material.   1) The Crust-Mantle Separation .

The facet of observed Earth behaviour that clearly needs reexamination in the light of heat transfer as a regulator of a
very large average viscosity is magmatism, a phenomenon that many
previous writers have connected with this particular separation.
For our purposes, magmatism can be summarised as an ability of
the Earth to produce on average a few cubic kilometers/year of a
material having an effective viscosity more than 10^{15} times
smaller than the values predicted as a horizontal average for the
upper mantle and closely supported by isostatic rebound studies.
Appealing to the fact that the energy associated with active
vulcanism is less than 1% of the total planetary heat flow,
theories of magmatism have traditionally been added to quantitative studies of the internal thermal state as an afterthought.
They normally appeal to the diapiric ascent of material from a
world encircling layer of partially molten silicates. However,
virtually no serious concern seems to have been expressed in the
fact that this model would not explain why magma is ejected in
a series of amounts that are quite trivial compared to the volume
of the presumed semi molten layer and on a time scale quite unlike that we associate with radiogenic heating.   Far from being
the kind of randomly located diapirism we would have to expect
from a world encircling layer, the distribution of active
vulcanism shows obvious correlation with features in the surface
velocity field.

One should notice how different the problem of magmatism now
appears as a result of putting rheological and chemical problems
together.   Although the ability of a seismically 'solid'
mantle to release magma has long puzzled observers, for the
theoretician,the pressing problem has normally been one of
inventing a scheme within heat conduction theory that would
prevent the Earth's upper mantle from now being totally molten.
To this end, the most favoured solution was to locate something
like a half the Earth's total radioactivity
                                        in near surface rocks.
Although, there is plenty of evidence that surface rocks do
contain far more radioactivity than the average planetary rock,

it should not disguise the fact that this solution to the difficulty is inconsistent with the use of heat conduction theory unless one supposes the crustal concentration of radioactivity dates
from the formation of the Earth. The new approach to planetary
heat transfer has certainly removed any necessity to concentrate
radiogenic heat sources near the surface to prevent mantle
melting, and one would now see the crustal concentration of radioactivity simply as one aspect of the whole crust-mantle separation
problem.

The temperature at which any material would be expected to
attain an effective viscosity $\sim 10^{21}$ poise is much less than its
solidus temperature - solid state creep would be unobserved if
that were not true. For example, if water is now present in the
sub oceanic upper mantle at a concentration $\sim 1\%$ that is likely to
persist through the peculiarity of amphibole dehydration and the
subduction of sea floor material (Tozer 1981), horizontally averaged temperatures even at depths of some hundreds of kilometres
might well be little more than $500^{\circ}C$. Whether or not water is
present in amounts that depress the estimates of an upper mantle
temperature, one can always say that the horizontally averaged
temperature is so far below the solidus that the traditional
scheme for making magma by decompression of an initially 'solid'
material on the way to the surface, will no longer work.

Most people's reaction, including my own, to estimates of
horizontally averaged upper mantle temperatures several hundred
degrees less than magma temperatures, is to dismiss them as
absurd. However, I now see the very high average viscosities
and the low temperatures that must accompany them as an essential
part of explaining the salient observational facts about magma
production. As hinted above, the mistake of the traditional
account of magma production was to link it too directly with
radiogenic heating, although there was hardly any alternative so
long as one was committed to heat conduction theory. Convective
heat transfer offers the novel possibility of heating by material
deformation, though one should be careful to note that this
secondary heating does not augment but only redistributes the
pattern of heat flow across the boundaries of a system. A useful image of the convective process is that of a heat engine
driven by radioactivity (or possibly gravitational potential
energy) that has to dissipate its mechanical output internally.
The efficiency of this convective heat engine as a producer of
mechanical energy that has to be dissipated somewhere in the
system grows directly with the linear scale of the convective
motions (see below) and it is this conversion efficiency that
will determine an upper limit to the time average rate of magma
production from a material whose spatially averaged effective
viscosity is being regulated by the heat transfer process at
$\sim 10^{21}$ poise.

If such circumstances do lie far outside any experience we have
or are likely to gain from laboratory convection studies, the

influence of stored energy on the steadiness of shear deformation
has been extensively studied by materials scientists and engineers
in connection with the 'stick-slip' phenomenon observed in the
sliding of materials placed in contact.   Again, the regime of
sliding in these experiments is far removed from that involved in
planetary convection and the task of interpreting the observations
and establishing their relevance to the planetary heat transfer
problem is too complicated to be explained here.   Suffice it to
say that initial conditions have been identified, involving the
temperature of the material, the decrease of its creep resistance
with temperature and the possibility of pore fluid generation,
that in various combinations can give rise to a whole spectrum of
deformation behaviour under a constant applied stress that varies
from a constant and homogeneous shearing of the system, to isolated
slip 'events' in narrow zones.   As a result, I have concluded
that magma generation has been a frequently recurrent phenomenon
of Earth history only because its convection translates and sub-
ducts the coolest superficial rocks and thereby creates the circumstances that lead to its store of shear energy being tapped in
isolated slip events.   Unfortunately, the question of whether
stored energy will be largely converted to heat in a zone of
slippage or first converted to seismic energy and then dissipated
more diffusely depends on several factors that cannot be evaluated
from surface observation, and for similar reasons one cannot
predict the amount of energy likely to be converted to local heat
and/or seismic energy in individual events.   All one can say from
convection theory is that the largest events could involve conversion of the 10^{18} - 10^{19} joules stored in 10^8 - 10^9 km^3 of
convecting material, and that such large events in any particular
region would be separated by intervals $\sim$ a century.   If such an
amount of stored energy were wholly dissipated as heat in a narrow
zone, it is sufficient to create $\sim$ 10 km^3 of material with a magma
like viscosity from the typical $\sim 10^{21}$ poise material.

As far as discussion of a stage reached by crust/mantle separation process is concerned, the fact that creation of the magmatic
like viscosities necessary for relative movement of different
phases in planetary material can now only be understood as an
inherently unsteady and localised heating process is unimportant.
It can be treated as a continuous process whose mean rate is
governed by the mechanical efficiency of the convective heat
engine.   This mean rate of dissipation in a purely thermally
driven convective flow can be represented by a (temporally and
spatially averaged) heat source density $\overline{H}_V$ given by the relation

$$\overline{H}_V \sim 0.1 \frac{g\alpha L}{C_p} (H - H_o) \tag{1}$$

where g is the gravitational acceleration, α the volume expansion
coefficient, C_p a specific heat, L the length scale of the
convective movements, H a uniform primary heat source density
responsible for the convection and H_o the value of H at which

convective heat transfer movements would be initiated in the
planet.   The dimensionless group $\frac{g\alpha L}{C_p}$ can be used as a measure
of the mechanical efficiency of $\frac{}{C_p}$ the convective heat
transfer process.   If a chemical ^{p} separation of the convecting material does occur as a result of this deformational
heating, it is necessarily associated with a reduction in the
planet's self gravitational potential energy and this must be
added to the dissipation occurring throughout the planet.   If we
express this effect by a mean heat source density H_G, we have

$$\bar{H}_V \sim 0.1 \frac{g\alpha L}{C_p} (H - H_o) + H_G \tag{2}$$

Since we have seen that differentiation can only be reasonably
associated with a distribution of H_V that is very strongly peaked
in both space and time, we shall define a time averaged (volume)
rate $\frac{dV}{dt}$, at which planetary material is subject to differentiation
in   shear zones of low viscosity with the equation:

$$\frac{dV}{dt} = \frac{L^3 \bar{H}_V}{\rho C_p \Delta T} \tag{3}$$

where ρ is the density and ΔT the temperature rise necessary for
material normally having a viscosity $\sim 10^{21}$ poise to be made
fluid enough for a differentiation to take place.   I should
emphasise again that (3) is only a meaningful equation for those
planets in which one expects sudden creep events or 'surges' to
occur.   Even then it may well over estimate the rate at which
primary planetary material is being 'processed' in transient low
viscosity zones by its neglect of a possible conversion of some
stored energy into seismic radiation which is then dissipated
more uniformly and therefore insignificantly throughout the planet.
If $\Delta\rho$ is the density difference of the separating materials and
f the volume fraction occupied by one of them in the starting
material, the reduction in gravitational potential energy of the
planet by separating the relevant parts of a volume dV by a
vertical distance L is given by:

$$dU \sim g L \Delta\rho \ f(1 - f)dV \tag{4}$$

and

$$H_G \sim \frac{1}{L^3} \frac{dU}{dt}$$

Combining the above equations we obtain:

$$\bar{H}_V \sim 0.1 \frac{g\alpha L}{C_p} \frac{(H - H_o)}{(1 - \beta)} \tag{5}$$

where

$$\beta = \frac{g L \Delta\rho \ f (1 - f)}{\rho C_p \Delta T} \tag{6}$$

It will be immediately noticed from (5) that the condition
$\beta = 1$ gives a singularity in the rate of dissipation and this
marks a possibility that the gravitational potential energy to be
dissipated as a result of differentiation could equal or exceed
the amount of primary heating that initiated it.   In such circumstances one might envisage the differentiation as a triggered,
self accelerating process that might follow some initial warming
of the system by primary heat sources.   These would loose control
of the thermal development of the system until β returned to some
value < 1 due to the differentiation producing a decrease in the
value of f to be used in equation 6 (see below).

For the case of crustal separation from the Earth's mantle, we
put $g = 10^3$ cm./sec^2 ., $L = 10^8$ cm., $\Delta\rho = 0.3$g/cm^3., $f = 0.2$,
$\rho = 3.4$g/cm^3 and $C_p = 10^7$ ergs/gm^{o}C.   Let us choose a primary
heat source density by the condition that the total heat production balances the present rate of heat loss from the Earth.
With such an amount of heat to be convected away, $\Delta T = 500^o$C is a
representative average figure for the temperature rise necessary
to induce differentiation at the present time.   If we further
suppose that for several billion years this primary heat source
density has been fixed by the radioactivity of K^{40}, U^{238} and
Th^{232} with K/U $\sim 10^4$ and Th/U ~ 4, the value of ΔT would not have
significantly increased throughout most of Earth's history.   The
above values give $\beta = 0.28$ when substituted in (6).   When one
recalls that in formulating eq.(6) we have neglected the unimportant (as far as differentiation is concerned) conversion of
perhaps half the available mechanical energy into seismic waves,
this β value clarifies why there is no evidence for the global
rate of continental crust formation throughout several billion
years of Earth history ever having been very much greater than
its value for recent geological times.

Although it may appear a trivial exercise to demonstrate why
something not observed has not occurred, this result should be
set against the traditional account of vulcanism in terms of a
world encircling, partially molten silicate layer of the upper
mantle.   There is no quantitative indication from those
spherically symmetric models to indicate a mean rate of magma
diapirism from that layer, or indeed why its regional appearances
at the external surface should fluctuate so widely over just a
few decades.   From equation (5) with $\alpha = 2.10^{-5}/^o$C and H =
2.10^{-7} erg/gm. sec. as an average primary heat source density over
the past 4.10^9 years, one may calculate that viscous dissipation
in that period would have been sufficient to differentiate $\sim 7\%$
of the convecting material in the short lived low viscosity zones.
Assuming a perfect separation of the two fractions and with f
for a crustal fraction in the primary planetary material of 0.2,
the lighter fraction would now be enough to form a superficial
shell ~ 15 km. thick.   Given such obvious complications as an
early non radiogenic heating (see below), crustal recycling, an
imperfect separation of the fractions heated in the low viscosity

zones and the ineffective diffuse dissipation of some of the
stored energy in Earth material as seismic waves, it is encouraging that this simple estimation of the stage now reached in
the silicate differentiation of Earth should be in such good
quantitative agreement with the inference of a 30 - 40 km thick
continental crust existing over about a third of the planet's
surface.

From the dependence of β on g and L in eq. (6), it will be
clear that the above remarks about the stability of a radiogenically driven silicate differentiation of Earth are likely to
apply a fortiori to the smaller terrestrial planets.   If we
assume the terrestrial planets formed a set of materially
identical and homogeneous objects at some common time of formation about 4.5 b.y. in the past, a calculation along the lines
indicated above would predict a present degree of silicate
differentiation roughly proportional to the square of the body's
radius above some threshold value $\sim 10^3$ km.   This threshold
value, which might be in error by a factor of two, arises from
the fact that the necessary low viscosity zones would never have
been produced in smaller objects heated by the long lived U, Th,
K isotopes we see active today.   Of course, possible complicating factors for any such calculation are differences in the f
values for a crustal fraction in different objects and differences in the total heat input to the different planetary materials.
Of the latter, the most obvious is the non radiogenic heating
associated with the accumulation of the planetary material into
the familiar objects we observe today.   The total amount of such
'initial' accretional heat energy/unit mass grows approximately
as the square of an object's final radius and even for the Moon
(R $\sim$ 1740 km) it probably exceeds the amount produced by radioactivity in the subsequent 4.5 b.y. of planetary evolution.   If
this total amount of gravitational energy/unit mass to be dissipated is readily calculated, considerable difficulty and controversy surrounds the question as to the fraction of this energy
that would have accumulated as heat in the colliding protoplanetary objects rather than promptly reradiated from their
surfaces.   The distribution of heat energy inside a newly
accreted planet and hence whether it is likely to be convectively
stable or unstable depends on an understanding of this energy
partitioning problem.   Until quite recently, this problem has
been treated almost exclusively by an accretion model due to
Benfield (1950).   He visualised planetary growth as that of a
single 'embryo' body growing at the expense of many much smaller
planetesimals impacting on its surface.   Since the kinetic
energy of planetesimal impacts was assumed to be converted to
heat at the embryo surface, this model maximised the importance
of thermal radiative loss, and always indicated a convectively
stable initial internal temperature distribution having a maximum near the final external surface of the object.   So great
were the radiative losses that when Apollo data indicated a very

early differentiation of the lunar highlands, one needed to
postulate lunar accretion times < 10^4 years to achieve the
necessary low viscosity differentiation conditions.   More
recently, there has been growing interest in the view of planetary accretion as a parallel growth of many proto planetary
objects, with no one object being identifiable as the object we
now refer to as the planet until a few objects were left in non
intersecting heliocentric orbits.   This is statistically much
more likely that the embryo growth picture and quite naturally
makes a case for the newly accreted planetary material being more
uniformly heated because no centre to the accumulation can be
established until the final 'sweeping up' process that did little
but crater the planetary surfaces.   Most of the gravitational
potential energy was dissipated throughout the planetary material
in a small number of large collisions.

   If we now assume a plausible accretion period $\sim 10^8$ years
(Wetherill 1980) and a 100% trapping efficiency of the accretional
energy throughout the planetary material, the average heat source
density H throughout this period when expressed in terms of the
present (radiogenic) value estimated above is about 10^2 for the
Moon and more than 2.10^3 for the Earth.   The significance of
this initial heat input to present observation is rather different
in the two cases.   For Moon, perhaps 30% of the accretional
energy would have been absorbed in warming its material to the
point at which it would start to convect the subsequent heat
input.   The significance of the rate at which the remaining 70%
was dissipated in the lunar interior is that it would induce a
state of convection in which the surface material would be translated and subducted.   This would have promoted the same kind of
shearing instability and intense local heating that radioactivity
has managed to sustain in the much larger Earth right up to the
present day.   The hundred fold larger heat source density in the
Moon predicted from a 10^8 year accretion time is predicted to be
insufficient to create a general liquifaction of the interior,
and this creep instability heating that would have been absent
without accretional heating is quite crucial in explaining the
degree of lunar silicate differentiation now inferred.

   For the Earth and larger terrestrial planets, although their
accretional heating/unit mass is larger, the degree of differentiation it achieves may well be less due to what I shall call the
endogenous and exogenous remixing of their planetary material.
By the former, I mean a remixing of differentiates that is due to
the internal convective heat transport process, while the latter
is a remixing due to the collision of future planetary material
while it still forms separate objects.   Since the amount of
accretional heat/unit mass increases as (planetary radius)^2 the
average radial temperature gradient generated by accretional
heating at the external surface increases as (radius)^3 .   Consequently, not only is convection more vigorous in larger objects,
but the near surface temperature gradient increases so quickly

that any lighter silicate differentiate that does separate is
far more susceptible to shearing, disruption and remixing.
Clearly, an exogenous remixing of the planetary material, though
always present, only becomes a significant process in assessing
the state of chemical differentiation for those cases where the
proto planetary forbears of the existing planets were themselves
large enough to undergo differentiation.   On this point, it is
relevant to notice that an object like Moon, having little more
than 1% the mass of Earth and Venus, shows signs of differentiation very early in its history.   Hence, both the exogenous and
endogenous remixing of materials are likely to have been important processes through a large part of the periods of terrestrial
and Venusian accumulation.   Noticing again that the accretional
heating of Earth material could be twenty times that subsequently
due to radioactive decay, one can only attribute the above
success in calculating the present degree of silicate differentiation from this radiogenic heating acting alone to a near perfect
remixing of Earth material if the mean convection velocity is for
any reason more than perhaps ten times its present value of
$\sim 10^{-7}$cm/sec.

It is also worth noting that the continuing efflux from Earth's
interior into deep ocean water of such a mobile, chemically inert
and non radiogenic isotope as He^3 is an indication that the viscosity regulation provided by convective heat transfer was able
to prevent a thorough degassing of Earth material.   N.B. Heat
transfer theory suggests that horizontally averaged internal
temperatures would not reach silicate liquidus values even if a
heat source density $\sim 10^3$ times the present value were sustained
for longer periods than the accretion time.   With this He^3
evidence, it seems certain that volatiles and particularly
chemically reactive species, would be retained in the less
massive planets.   Although it has proved unpopular among geochemists who have examined the Moon's surface rocks in minute
detail, the presence of a fluid phase-probably water, that
permeates the deep lunar rocks at a temperature at which they
show no tendency to sinter and anneal on a time scale of months,
is needed to explain what is now known about the occurrence of
deep moonquakes (Tozer 1981).

(2)   The Differentiation of Planetary Cores

Although I have indirectly inferred a well-mixed state of the
silicate fraction of Earth material at the end of its accumulation period, a demonstration of that fact could only be said to
follow from a comparison of the rates of fractionation and remixing throughout the accumulation process.   In general, one
must expect the answer to depend on the particular separation
process under consideration.

From a comparison of the fractional uncertainties in the
masses of Earth's present core and crust that result from a lack
of definition in the position of their boundaries one may con-

clude that the endogenous remixing of these differentiates by
convective movement is relatively negligible for core/silicate
differentiations.   Presumably, that reflects the greater
density contrast.   On the other hand, the exogenous remixing
process is quite unspecific in the way it would remix planetary
material.
       One can usefully divide the possible fractionation processes
of planetary material into two classes depending on whether a
particular fraction would have a viscosity much higher or less
than $\sim 10^{21}$ poise under the P,T conditions that would be generated by adiabatic compression in a growing object.   If much
less, like that of the various free volatile phases that could
have been incorporated with silicate in the original terrestrial
planet material, it would be expelled until its concentration
was an extremely small percentage as fast as the planet accumulated (Tozer 1981).   In fact, most of this material would be lost
to interplanetary space, due to a small escape velocity from the
proto planets in which this expulsion would occur.   The crustal/
mantle and core/silicate separations in terrestrial planets are
linked together in the other class of fractionations by the very
high viscosity of the various fractions.   For them, the rate of
separation is not so directly linked to the rate of accumulation,
but is to be viewed as a balance struck between the slowing effect
of the control exerted by the heat transport process over horizontally averaged viscosity values and an accelerating positive
feedback effect of gravitational energy converted to heat by the
particular fractionation.   This latter effect is expressed by
the value of β in equation 6.   A fundamental subdivision of
differentiation processes lying in this second class can be made
on the basis of whether the condition $\beta \geqslant 1$ is likely to be
satisfied during accumulation.

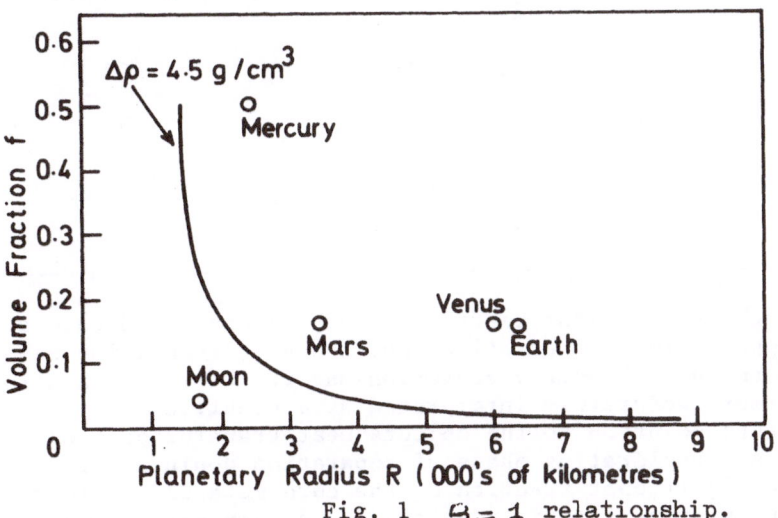

Fig. 1 $\beta = 1$ relationship.

Let us first investigate the circumstances in which the apparent singularity condition $\beta = 1$ (see eq. 5) might be satisfield in the past by the core/silicate composition of the terrestrial planets. From eq. 6 one can express this condition as a function of core/silicate ratio f and planetary radius for a set of well mixed objects (Fig. 1). The value ΔT has, for the moment, been chosen to be $500^{\circ}C$, as if the quasi steady thermal conditions in these objects were being maintained by a heat source density about equal to the radiogenic value we infer for the terrestrial planets. Other parameter values used in eq. 6 apart from $\Delta\rho = 4.5$ g/cm^3, are as used above.

Also shown are the positions well mixed versions of the various terrestrial planets would occupy on such a diagram. Since a position to the right of the $\beta = 1$ relationship indicates separation would be a self accelerating process, we see that core formation in all terrestrial planets with the exception of Moon would have occurred rapidly even if there had been no input of accretional energy. The effect of accretional heating is to advance the time at which such an accelerating phase of core formation commences.

From Table 1 and using a specific heat ~ 1 joule/gm$^{\circ}$C, we see that more than enough energy was dissipated in all these objects to raise their internal temperatures to a value at which core formation would be initiated.

Table 1

	Energy of Accumulation Joules/gm.	Energy dissipated by Core Formation Joules/gm.
Moon	$1.7 \cdot 10^3$	~ 5
Mercury	$5.4 \cdot 10^3$	400
Mars	$7.5 \cdot 10^3$	480
Venus	$3.2 \cdot 10^4$	1500
Earth	$3.8 \cdot 10^4$	1700

Although the simplest reading of eq. (5) would indicate a catastrophic rate of separation once the $\beta = 1$ condition has been attained through planetary accretion and its attendant heating, $\beta \gtrsim 1$ is more accurately interpreted as a condition that no quasi steady state solution to the object's heat transfer problem exists. How long an accelerating phase of separation would continue and whether a significant fraction of the core material would remain mixed with silicates to be separated on a very protracted time

scale, like Earth's silicates, would mainly depend on whether the
heat to be dissipated by a core separation (column 2, Table 1) is
enough to warm the entire planetary material by more than ΔT
degrees, i.e. the temperature rise needed to induce the requisite
low viscosity conditions for differentiation everywhere in the
planet.   This condition allows us to draw a distinction between
core formation in Earth, Venus, and the rest of the terrestrial
planets.   For them, core formation would dissipate at least
three or four times the energy needed to raise the temperature
to the value needed for general differentiation, and core formation would be virtually complete before any slackening of its
rate occurred.   I have estimated a separation time < 10^6 years,
though the fact that this is less than the estimated accretion
time of 10^8 years would imply that cores would have separated in
the proto planets of both Earth and Venus.   The cores we sense
or infer in these bodies at the present time would have been
formed by the agglomeration of these proto cores following
collisions.

Turning to Mercury and Mars, the degree to which their core
material would have separated from a silicate fraction is made
somewhat more uncertain because the energy that has to be dissipated in the course of core formation is only enough to raise the
mean internal temperature by $\sim 500^\circ C$.   This is of the same order
as my estimate of the temperature rise that would be needed to
initiate a general differentiation of the terrestrial planets
ever since a radiogenically driven convection has controlled the
thermal state and horizontally averaged viscosity at $\sim 10^{21}$ poise.
Of course, the regulation of this averaged viscosity by convection
is not perfect and one could expect it to fall to perhaps 10^{15}
poise if the accretion of Mercury and Mars were both spread over
a 10^8 year period.   This would indicate that the appropriate
value of ΔT to use in defining the largest value attained by β
during their accumulation might be only   1/3 of that used in
plotting their position on Fig. 1.   This adjustment makes it
reasonably certain that Mercurian and Martian core formation were
initially self accelerating processes, but there is still some
room for doubt about the completeness of the fractionation by the
end of accretion in view of the decline in β value that occurs as
the separation proceeds, i.e. by reducing the f value in eq.6.
In fact the completeness of core formation in Mars and Mercury at
the end of accretion seems such a delicately balanced question
when the accretion time is chosen to be $\sim 10^8$ years that a
difference in the accumulation time of these objects that one
might predict from the significant difference in mass densities
of Martian and Mercurian material when spread through an
appropriate range of heliocentric orbits, could make a large
difference to the answer.   The following comments seem justified:
(1) Core formation in Mercury and Mars was initiated before their
accumulation was effectively complete, but not so much before as
to make its terminal stages look like an agglomeration of cores

already separated in their respective proto planets.
(2) Particularly in the case of Mars, core formation could well
have been sufficiently incomplete at the end of planetary accumulation for the density differences arising from the continuing
core/mantle separation to have dominated those due to thermal
expansion for some hundreds of millions of years.
(3) Mercury, which unlike Mars, shows no sign as yet of having
water and other volatiles in its constitution, may still contain
a significant percentage of unoxidised core material in its
surface layers.

When one remembers that the surface area of Mercury is only a
half that of Mars, and that the total energy dissipated in these
bodies by accumulation and differentiation is of the same order,
the apparently greater age of the entire Mercurian surface, as
revealed by its cratered appearance, may be further evidence for
the rheological importance of water in weakening the silicates
and hence prolonging the period of disruption of their surfaces
after accretion ends.   In this connection, however, it should
also be noticed that a Martian core is much smaller than a
Mercurian core when compared with the external radii.   Hence,
the length scale of convective movements set up by compositional
and/or thermal density differences in Mars would be much larger
and therefore more likely to have lead to a disruption of the
planet's external surface.   The hemispherical differences in
crater density and mean altitude we now observe in the Martian
surface could well date from the movements of surface material
that accompanied the latest stage of its core separation.

The retention of unoxidised core material near the external
surface of Mercury is of great interest to the question of the
present magnetism of that planet.   Several workers have rather
hastily concluded that a field estimated to be several hundred γs
at the external surface has its origin in an active dynamo in its
core, but this simply overlooks the great difficulty in accounting
for the present liquidity of any significant fraction of a
Mercurian core (Tozer 1978).   When this problem is taken into
account the only tenable explanation of the field sensed by the
Mariner spacecraft is a fossil magnetisation of the near surface
rocks dating from just after the differentiation of the Mercurian
core.   There is certainly enough energy available to liquify an
iron core at that epoch, but convective loss through the surrounding silicate shell would have led to solidification billions of
years ago.   Attractive features of this idea are the fact that
the rapid heat loss would have created vigorous movements in the
Mercurian core, and that the thermal boundary layer near the
planet's external surface would have been rapidly thickening at
just the same time as the core was cooling and solidifying.   The
downward migration of the Curie Point isotherm would facilitate
the thermoremanent magnetisation of the surface rocks.   One could
explain the absence of a similarly intense field on Mars to the
more remote core and therefore smaller magnetising field, and to

the oxidation, tectonic deformation and sedimentation processes that would tend to destroy the magnetisation. It also seems very doubtful that much of a Martian core could still be liquid, since the creep weakening effects of water in its mantle rocks would promote the early freezing of its core even if it were doped with components like sulphur that reduce the melting point of a purely iron core.

Finally, the Moon presents the most interesting case of all to a theory of planetary differentiation. All the arguments used above for early core formation in other terrestrial planets seem insignificant in this case, and yet the magnetisation of its rocks could be indicating the existence of a core of perhaps 500 km radius more than 4 b.y. ago (Runcorn 1981). Perhaps one should turn the argument around by saying that if Moon could differentiate a core so early in its history, all the larger objects discussed would even more surely have extensively differentiated before they finished accumulating.

For an explanation of the lunar problem, appeal has been made to some powerful but shortlived energy source like Al^{26}, but with a half life of 7.10^5 years this looks very ineffective so long as one believes that accretion took $\sim 10^8$ years. Although, one might believe that such a short lived isotope heated proto planetary objects only a few kilometres across to high temperatures, there is plenty of time for these objects to cool off again. Furthermore, this early hot stage occurs when self gravitation is very weak and hence the forces driving any differentiation correspondingly small. There would also be a high probability that differentiates produced at such an early stage would subsequently be remixed by collisions. If one tries to overcome these difficulties by appealing to known or purely hypothetical radioactivities having a half life at least comparable with a 10^8 year accretion time, the chances are high that it should have already been detected experimentally if it had any real importance as a heat source $4.5 \cdot 10^9$ years ago.

In my view, a better explanation of early lunar core separation may well be found in an upward revision of the energy involved in planetary formation, and by giving due attention to the discrete nature of proto planetary collisions. All the energies of planetary accumulation shown in the first column of Table 1 are likely to be underestimates on the grounds that they are based on the assumption that the respective planetary material was at rest when infinitely dispersed. It has been shown by both Safronov (1972) and Wetherill (1980) that when proto planets are moving in heliocentric orbits, mutual perturbations continuously 'pump up' their relative velocities to a value comparable with the velocities of escape from their surfaces. This effect could more than double the amount of energy available for dissipation shown in Table 1. Secondly, since most of the energy of accumulation would be dissipated in relatively few very large impacts that occur towards the end of the accumulation process, the above

attempts to talk of an average rate of energy dissipation during
the period of accretion may give an entirely false impression of
the difficulty of creating the low viscosity conditions necessary
for core/mantle separations.

Conclusions

Previous demonstration that the heat transfer process keeps
the horizontally averaged viscosity in a homogeneous planetary
size object at a value $\sim 10^{21}$ poise carries with it the implication that magmatism and any silicate differentiation now in
progress in the Earth is due to very localised and transient
heating events.   These have their origin in the release of
energy that is normally stored throughout the convecting material
on account of the shear stresses of several bars that are necessarily associated with the movement of such a viscous material.
The inherent unsteadiness of this magma producing process can be
ignored in calculating from the efficiency with which heat energy
can be converted to mechanical energy that less than 10% of Earth's
silicates would have been differentiated due to radiogenic heat
in the last $4 . 10^9$ years.
Study of the energetics of differentiation have shown that
core/silicate separations in several terrestrial planets would
have the character of a triggered, self-accelerating process.
In the case of Venus and Earth, cores would have separated and
coalesced many times in their proto planets before forming the
cores that exist today.   It is suggested that Martian core
formation may have continued long enough after accretion effectively terminated to have been a factor in creating the hemispherical
differences that are now preserved in its external surface.
Vigorous convective cooling of a Mercurian core immediately
following its formation was indirectly responsible for the magnetisation of that planet's surface rocks, a process now recorded in
the external field of that planet.

References

Benfield,A.E.(1950) Trans.Am.Geoph.Union., 31, 53-56.
Murrell, S.A.F. & Ismail, I.A.H.(1976) Tectonophys.,31, 207-258.
Riecker, R.E., Rooney,T.P. (1969) Nature (London) 224, 1299-1301.
Safranov,V.S.(1972) Evolution of the Proto Planetary Cloud and
   formation of the Earth & Planets (Moscow 1969) N.A.S.A.
   translation TTF-677.
Runcorn, S.K. (1981) Phys.Earth and Planet.Ints. 24, 205-217.
Tozer, D.C. (1973) Geofisica Internacional, 13, 363-388.
Tozer, D.C. (1978) The Origin of the Solar System, Ed.S.F.
   Dermott, Wiley, New York 433-462.
Tozer, D.C. (1981) Phys.Earth and Planet.Ints. 25, 280-296.
Tozer, D.C. (1981A) Proceedings of Alpbach Conference of European
   Space Organisation, In the press.
Wetherill, G.W. (1980) Ann.Rev.Astronomy & Astrophysics,18,77-133.

SECULAR TRENDS IN POLAR MOTIONS: A NEW TOOL FOR PROBING THE
VISCOSITY OF THE LOWER MANTLE.

R. Sabadini and E. Boschi

Istituto di Geofisica, Bologna. I.N.F.N., Sez. Bologna
Via Irnerio, 46 - 40126 Bologna, Italy.

D.A. Yuen

Dept. of Geology. Arizona State University.
Tempe, Arizona 85287, U.S.A.

Abstract :   Transient flow in the mantle, induced by glacial
cycles, produce small but discernible variations in the Earth's
rotational wobble and in the length of the day. These geophysical
observables, which arise solely as a consequence of strain fields
endowed with low order spherical harmonics ($\ell = 0$ and $\ell = 2$), can
be usefel in distinguishing the viscosity structure of the lower
mantle, potentially more so than inferences drawn from relative
sea level and gravity anomaly data, whose spectral contents are
dominated by much higher angular orders ($\ell \gtrsim 6$). Results from
dynamical calculations, which are constraint to fit both   the
speed of recent polar wander and the non-tidal acceleration of
the Earth's rotation rate, suggest that some amount of shrinkage
of the Antarctic ice sheet since the late Wisconsin has occurred
and that disintegration of Arctic marine ice sheets in the past
may be a distinct possibility.

Keywords: Mantle viscosity, ice ages, polar wandering.

A. Coradini and M. Fulchignoni (eds.), The Comparative Study of the Planets, 181–194.

The subject of mantle viscosity is both vast and one in which there is at present a great deal of research activity. Indeed it is one in which there have been recently   published   a   number of authoritative works /1-3/ many times   the   length   of   this· article. Any attempt, therefore, at a comprehensive and devoted coverage would be so superficial as to make it valueless   to researchers outside the field.

Hence, we intend to devote this paper to a general discussion of the recent resurgence in the usage of small irregularities in the Earth's rotation to provide new and important insights into the rheological state of the Earth's mantle.   We   begin   by discussing some of the limitations encountered by· the   more traditional means of measuring the viscosity of the lower mantle, an important parameter in the theory of mantle convection.   In the second section of the paper several topics of current interest, concerning rotational dynamics and mantle rheology are selected for discussion in more detail.

Finally, it should be emphasized that no inversion of the geophysical observations to provide a rigorously   defensible rheological structure for the mantle exists.   The   external observations provide a series of tests which must be satisfied by a family of rheological models which are to be seriously entertained. However, much of the uncertainties in the determina- tion of the rheological properties of the mantle may be due to the history of glacial forcings, which must be interpreted in the light of other climatological and geological evidences   as well.

The phenomenon of postglacial rebound was first recognized in the form of rising shorelines around the Baltic. Using a purely viscous, half-space model, Haskell /4/ found the viscosity of the mantle to be about 9×10^{21} P. Since then the relative   sea level data, which measures the time-dependent history of the surface displacement from deglaciation, have been the chief source of information with which theoretical models can be compared. Following this approach, Cathles /1/ and Péltier and Andrews /5/ on the basis of a self-consistent spherical model found that the viscosity of the mantle to be relatively uniform,   again   very close to 10^{22} P. Recently Wu and Peltier /6/ have employed   the free air gravity anomaly data from the Laurentide   region   in conjunction with the relative sea level data to constrain better the lower mantle viscosity. They found that a modest increase of viscosity (less than an order of magnitude) was required to fit the 35-40 mgal negative anomaly above the center of Laurentide.

Yet these two methods are limited in their ability to resolve the viscosity profile of the lower mantle. Our knowledge of the viscosity structure is ultimately determined by the extent to which the displacement fields are excited by the surface load. Eventhough the major spectral content characterizing the Lauren- tide ice sheet lies at a spherical harmonic of order $\ell \approx 6$, quasi -static calculations for an elastic /7/ and a viscoelastic Earth with a uniformly viscous mantle /8/ show that the strain fields, are not significant for $\ell \gtrsim 6$ in the lower mantle. Moreover, the existence of a stiffer lower mantle would tend to drive the displacement fields into the upper mantle. In short, one would expect a trade-off between the resolving length and the amount of rheological stratification in the lower mantle, with the resolu- tion deteriorating with an increasing viscosity gradient.

But it is well known from classical mechanics that temporal changes in the elements of the inertia tensor involve only the $\ell = 0$ and $\ell = 2$ components of the displacement fields, with the $\ell = 0$ contribution restricted to a strictly compressible Earth model. The displacement fields associated with these low angular order modes extend themselves throughout the entire mantle /7/. For this reason fluctuations in the Earth's rotation allow the rheological property of the lower mantle to be sampled more precisely than that obtained from analyses of sea level and gravity anomaly data. The last two years have marked a new period in the investigation of the Earth's viscosity structure from rotational data.

Recent evidence in the form of astronomical data compiled during the last 75 years by the International Latitude Service clearly shows a continuous drift of the pole /9,10/. Theoretical calculations based on a homogenous viscoelastic model /11,12/ indicate that the present secular drift of the pole can be attributed to the Earth's ongoing response to the late Wisconsin deglaciation. For a longer time span, reanalyses of paleomagnetic data, based on a fixed hot spot framework , have pointed out the existence of true polar wandering involving several degrees since the mid-Pliocene /13,14/. Calculations based on a layered visco- elastic Earth /15,16/ have lent support to the hypothesis that there is a cause and effect relationship between the late Cenozoic ice ages and the observed polar wandering. Thus at present there

is gathering a new host of data which clearly addresses the
importance of rotational dynamics in furthering our understanding of the Earth's interior.

In the absence of an external torque, the expression of the
conservation of the system angular momentum associated with the
cryosphere and solid Earth is given in terms of the well known
Liouville equations /17/

$$\frac{d}{dt} (\underset{\approx}{J} \cdot \underset{\sim}{\omega}) + \underset{\sim}{\omega} \times \underset{\approx}{J} \cdot \underset{\sim}{\omega} = 0 \tag{1}$$

which is applicable for a body fixed co-ordinate system rigidly
attached to the Earth as a whole with its origin at the center
of mass and with the axes oriented such that the inertia tensor
$\underset{\approx}{J}$ initially is diagonal. $\underset{\sim}{\omega}$ is the angular velocity vector.

As shown in /12/ the $\underset{\approx}{J}$ tensor may be decomposed as follows

$$\underset{\approx}{J}(t) = I \delta_{ij} + C_{ij}(t) + I_{ij}(t) \tag{2}$$

where I is the inertia of the non-rotating spherical Earth ,
$C_{ij}(t)$ is the component of the inertia tensor due to changes in
the rotation rate, and $I_{ij}(t)$ is the contribution produced by
transient flow in the mantle and by shape deformation due to the
glacial forcing.

The nonlinear version of the Liouville equation must be used
for polar displacements larger than about 20° /12/. But for the
amount of polar drift found in the last 10 Myr /13/ the linearized
Liouville equations may be employed with sufficient accuracy.
The solution has the simple Laplace transform domain form

$$\underset{\sim}{m}(s) = \frac{\psi(s) + \phi(s)}{(1 + \frac{i}{\sigma_r} s)} \tag{3}$$

where $\underset{\sim}{m} = (m_1, im_2)^T$ are the direction cosines of the rotation
axis in the body-fixed coordinate system, s is a complex number
and σ_r is the Chandler wobble frequency for the rigid Earth.
$\psi(s)$ and $\phi(s)$ are respectively the forcing functions arising
from rotational deformation nad the surface loading from ice
masses. They have the explicit forms

$$\psi(s) = \frac{k_2^T(s,a)\, \underset{\sim}{m}(s)}{k_f} \tag{4}$$

$$\phi_1(s) = \frac{I_{13}(s)}{C-A} + \frac{(s\, I_{23}(s) - \bar{I}_{23})}{\Omega(C-A)} \tag{5}$$

$$\phi_2(s) = \frac{I_{23}(s)}{C-A} - \frac{(s\, I_{13}(s) - \bar{I}_{13})}{\Omega(C-A)} \tag{6}$$

where Ω is the diurnal rotation rate, C and A are respectively the axial and equatorial moments of inertia and the overbars denote quantities at time t = 0. $k_2^T(s,a)$ is the tidal love number for $\ell = 2$ evaluated at the surface (r = a) and k_f is the fluid Love number from centrifugal deformation and is about 0.95 /17/.

For a given Earth model, $k_2(s,a)$, $I_{13}(s)$, and $I_{23}(s)$ can be obtained spectrally from the solution of a boundary value problem. By means of a three layer model, which consists of an elastic lithosphere, a viscoelastic mantle, and an inviscid core, Sabadini et al./15/ have obtained analytical expressions for the spectral decomposition of equations (4) to (6), which greatly facilitates the solution of the initial value problem of the polar displacement as a result of glacial forcing.

This three layer model entails all the essential attributes which are necessary to provide a proper description of polar wandering, for the following two reasons:

(1) The lithosphere acts to maintain a non-isostatically compensated top layer which is required to produce a net polar displacement after the completion of a glaciation - deglaciation cycle.

(2) The core serves to confine the transient viscous flow in the mantle and, in so doing, gives rise to another relaxation mode, whose strength of excitation is of the same order as that associated with the mantle mode. This becomes especially important in the course of inferring a mean mantle viscosity from fitting the model to the rotational data.

The importance of the lithosphere as a rheological boundary layer /15/ is illustrated in Fig. 1.

We have considered here the contributions from the two major ice sheets in the Northern Hemisphere, Laurentide and Fennoscandia,

where a total mass of 2.4×10^{19} kg has been used in the unloading
process. The temporal evolution in Fig. 1 does not behave in a
purely exponential manner, owing to the three different relaxation modes in the three layer Earth model.   The dashed   curve
denotes the result from employing the approximation for a non-
compensated crustal layer invoked by Munk and MacDonald /17/. A
20 km thick crust has been assumed as in /11/. The reason for
their using this approximation is that if the Earth behaves as
a fluid as $t \rightarrow \infty$, the perturbed moment of inertia would vanish
altogether, since the final state would then be isostatically
compensated throughout the entire Earth.

   This leads to the feasibility for true polar wandering, as
displayed in Fig. 2.

   In calculating the amount of polar wander which has occurred
since the mid-Pliocene, we have chosen a forcing function, which
has been constructed on the basis of the proxy climatic record,
as revealed by the oxygen isotope data from deep sea cores /18/.
For mathematical convenience a ramp-shaped function has been used
to describe the climatic variability /12/. This time-dependence
can be transformed readily in the Laplace transform domain and
used in equation (3) for calculating the amount of average speed
of polar wandering after many glacial cycles.

   Such a secular motion of the Earth's spin axis relative to
geographical distribution of land masses can have a profound
influence upon the future trend of the current ice age cycles
whose recurrence rate is about 10^5 yr. /19/. We /20/ have, in
fact, proposed that the ultimate termination of ice age cycles
is due to the large polar drift, $O(10°)$, induced by the response
of the planet itself to the continual periodic forcing. Angular
displacements of this magnitude would be sufficient to displace
the present configuration of continents and oceans from the state
of equilibrium it now seems to enjoy. Thenceforth this new set
of surface boundary conditions will not act in concert with
astronomical forcing /21/ to produce large scale glaciation as
formerly.

   As discussed in the introduction, rotational data, such as the
recent secular drift of the pole /9,10/ and the non-tidal
deceleration of the length of the day /22/ contain an abundance
of information about the viscosity structure $\nu(r)$ of the mantle;
some of them are not accessible from other methods. To provide a
quantitative constraint on the deep mantle structure, these
observable quantities must satisfy several basic criteria.

First, it must be cast as a functional D of the rheological model m under consideration /23/

$$D = D(\underset{\sim}{m}(r)) \tag{7}$$

where

$$\underset{\sim}{m}(r) = \underset{\sim}{m} (\ (r),\ldots\ldots\ldots) \tag{8}$$

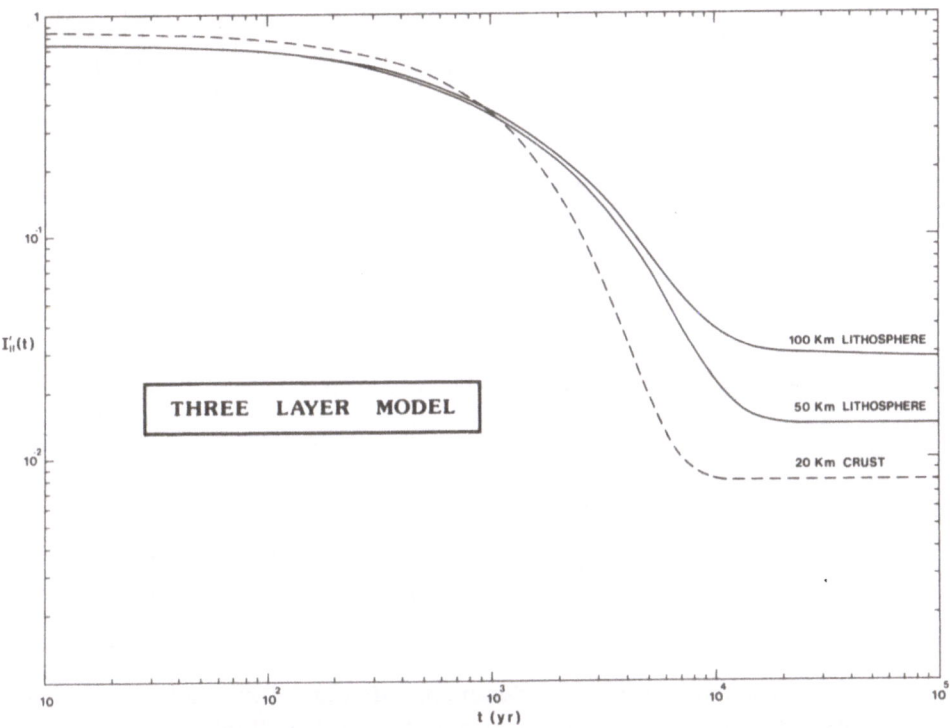

Fig. 1 - Temporal evolution of a perturbed element of the inertia tensor. This has been non-dimensionalized with respect to the increment of the moment of inertia, which is produced by a Heaviside loading function ar t = O.   A mean mantle viscosity of 10^{22} P has been assumed.

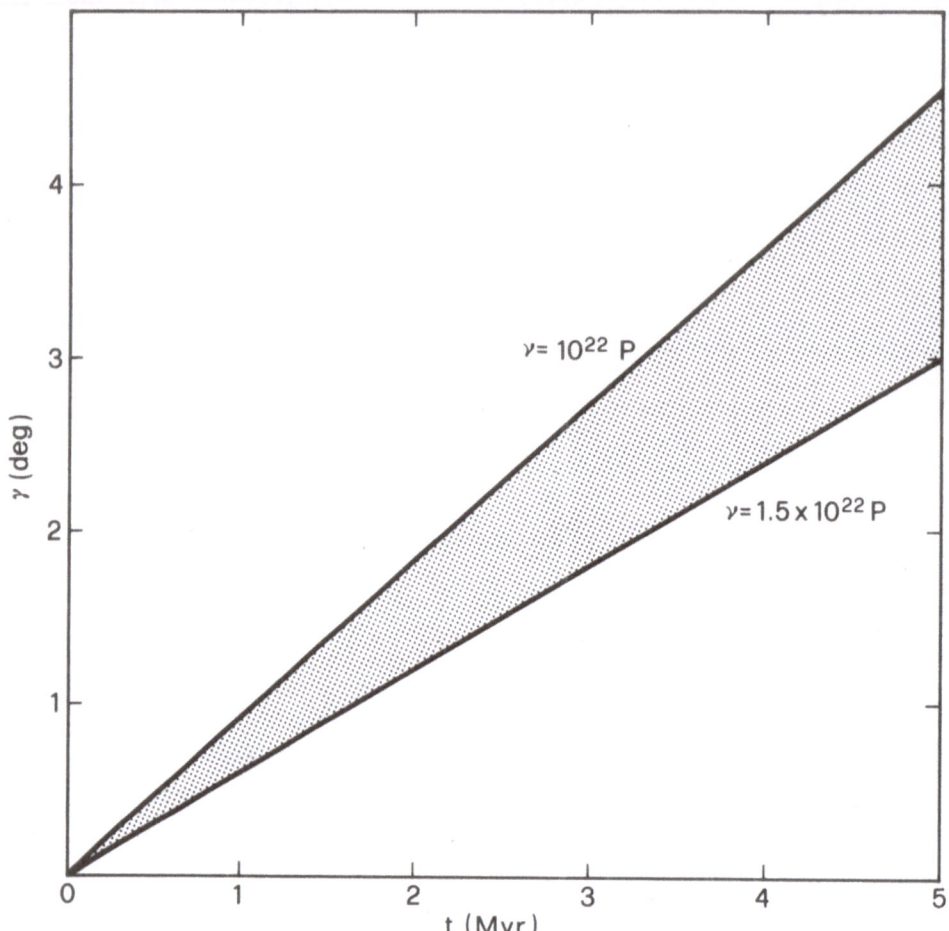

Fig. 2 - Polar wander as a function of time for periodic
forcings from the two major ice sheets in the Northern
Hemisphere. The bounds of mean mantle viscosities, derived
from constraining the 3-layer model to the data, are given
adjacent to the curves.

and r is the radial coordinate. Because of the analytical nature
of our layered viscoelastic model, one can easily construct the
data kernel to deduce the sensitivity of the rotational data to

variations in the viscosity profile /24/.

Second, an efficient algorithm must be available to compute
the data functionals for the various viscosity structures; that
is, the forward problem must be solved in an efficient manner.
Analytical solutions provided by our layered models can again
serve admirably for this purpose. Initial progress on the resolving power of mantle viscosity should be made with a set   of
numerical experiments using synthetic data generated for known
loading functions. They should provide practical information
about such matters as the distribution of the data sites and the
effect of noise in the data upon the viscosity structure.

Finally, these data functionals must correspond to accurately
observable quantities. The estimate of the observational error
of the rotational data /22,9/ can be seen to be comparable or, in
some instances, even smaller than those associated with relative
sea level and gravity anomaly. Hence, rotational data, needed for
the investigation of the rheological state of the lower mantle ,
seem to have the same quality as those which have traditionally
been used for the whole mantle.

Results from our 3-layer model /15/ have demonstrated   the
importance to consider forcings from other ice sheets than the
two, viz. Laurentide and Fennoscandia, which have been   used
exclusively up to now /1,2/. We find it necessary to include some
amount of retreat of the Antarctic ice sheet since the   late
Wisconsin in order to remove the ambiguity present in the two
families of mantle viscosity, which are obtained from fitting the
model to the rotational data set. The lower viscosity branch is
preferred in view of this new forcing function. This view has
been gaining support from some of the glaciologists and Quaternary
geologists /25/.

It is evident from the discussions given by Hughes et al./25/
that the distribution of Northern Hemisphere ice sheets during
the last glaciation is still not completely known in spite of the
ice sheet reconstruction projects during the last decade /26/.
Therefore, it is of importance to consider other possible paleo-
ice sheets in order to force reexamination of the convictions of
the day. From geologic insights afforded by current research in
Antarctica which reveals the importance of the dynamics of ice
streams and ice shelves, Hughes et al./27,25/ have proposed two
reconstructions which emphasize the two extreme possibilities of
minimum and maximum extent of North American ice sheets.   The
greatest difference in areal ice distribution occurs   in   the

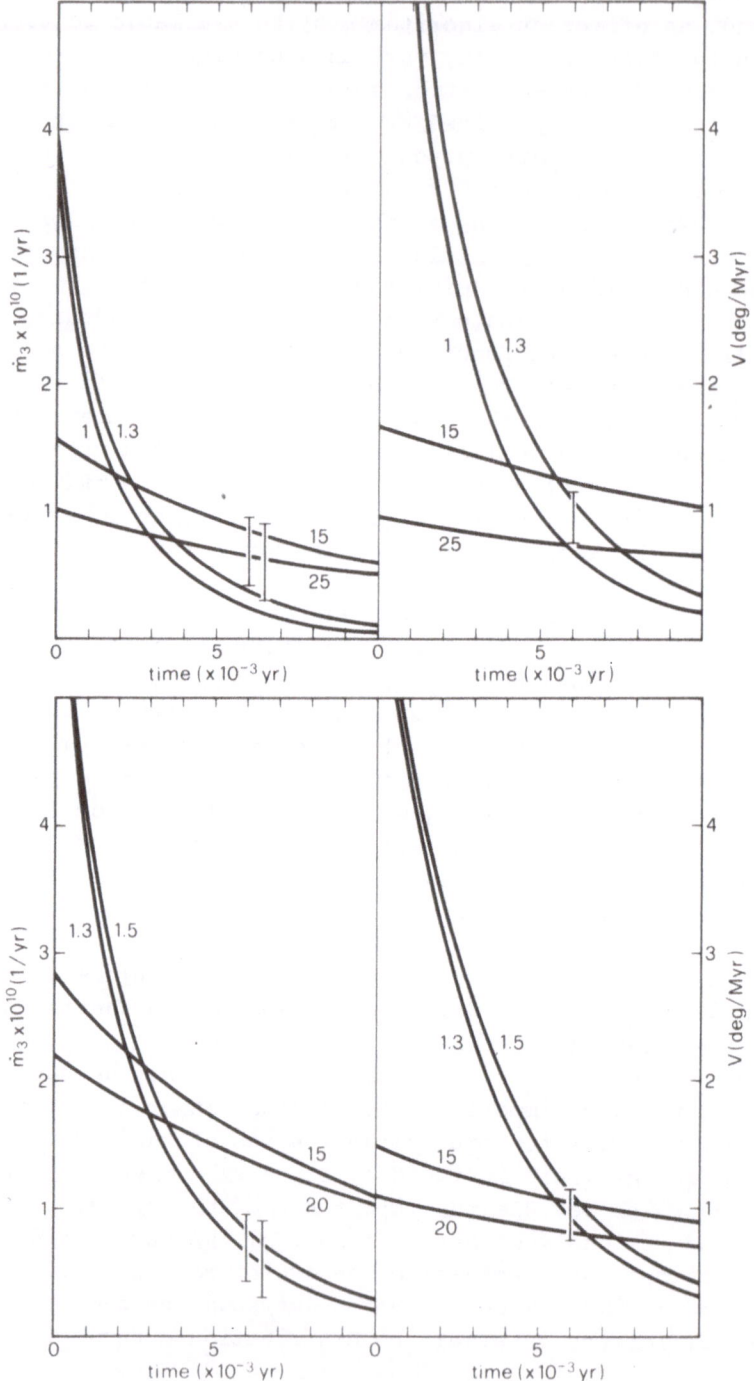

Arctic where it is postulated the large ice sheets actually existed in the East Siberian Sea as an integral portion to the Arctic Ice Sheet. We have included both the Antarctic and Arctic forcings in an effort to see the possible effects this may have on separating the multiple solutions of mantle viscosity which is obtained only when the Laurentide and Fennoscandia ice sheets are considered (see the top half of Fig. 3). The bottom half of Fig. 3 clearly shows that the higher viscosity root may be rejected when additional ice loadings near the polar regions are brought into play.

We have employed the following values for M, the amount of melted ice: for Laurentide, $M = 1.8 \times 10^{19}$ kg; for Fennoscandia, $M = 6 \times 10^{18}$ kg; for Antarctica $M = 1.1 \times 10^{19}$ kg; for the Arctic area $M = 5.1 \times 10^{18}$ kg. As advocated strongly by Hughes et al., /25/, Antarctica could be the second largest contributor to post-glacial sea level rise after Laurentide and the Arctic ice sheet could be comparable in size to Fennoscandia.

To avoid the difficulties in extracting a unique value of the mantle viscosity form rotational data, we have found it necessary to invoke loadings from sites which have been suggested on glaciological grounds /25/. The relative ease of this analytical approach can be further exploited for future models which incorporate both the lower and upper mantles as well. It is hoped that equally important results can be obtained for more sophisticated but still analytically tractable models; for example, what are the bounds of the deep mantle viscosity and what quantitative constraints can we put on the existence of a global low viscosity zone? From the progress made so far, we are confident that rotational data are capable of providing both

---

Fig. 3 - The effects on the fit of the model to the two rotation data from polar forcings. m_3 is the secular variation of the length of the day; V is the instantaneous speed of polar wander. In the top half the forcings only come from Laurentide and Fennoscandia /12/, whereas in the bottom half of the figure excitation from the unstable portions of the ice sheets in both the Antarctic and Arctic regions are taken into account. Geographical parameters for these new forcings are derived from Lingle and Clark /28/ and Hughes et al. /25/. Viscosity solutions, normalized with respect to 10^{22} P, are given next to the curves.

sharp bounds on the variations of mantle viscosity and significant
information concerning the deglaciation history of the last ice
age.

Acknowledgments. This research was sponsored by the Progetto
Finalizzato Geodinamica, the Antarctic Research Division of the
National Science Foundation, and the Research Corporation.

References

/1/   Cathles, L.M. III: 1975, The Viscosity of the Earth's Mantle,
         Princeton Univ. Press.

/2/   Peltier, W.R.: 1980, in Physics of the Earth's Interior, ed.
         by A.M. Dziewonski and E. Boschi, North Holland Publ.
         Co., pp 362-431.

/3/   Peltier, W.R.: 1981, Annu. Rev. Earth & Planet. Sci.9,pp 199-
         225.

/4/   Haskell, N.A.: 1937, Am. J. Sci. 33, pp 22-28.

/5/   Peltier, W.R. and Andrews, J.A.: 1976, Geophys. J.R. astr.
         Soc. 46, pp 605-646.

/6/   Wu, P. and Peltier, W.R.: 1981, Geophys. J.R. astr Soc.,
         submitted.

/7/   Takeuchi, H., Saito, M. and Kobayashi, N.: 1962, J.geophys.
         Res. 67, pp 1141-1154.

/8/ Wu, P. and Peltier, W.R.: 1981, Geophys. J.R. astr. Soc.,
     submitted.

/9/ Dickman, S.R.: 1977, Geophys. J.R. astr. Soc. 51, pp 229-
     244.

/10/ Soler, T. and Mueller, I.: 1978, Bull. Geodes. 52, pp 39-
     57.

/11/ Nakibogly, S.M. and Lambeck, K.: 1980, Geophys. J.R. astr.
     Soc. 62, pp 49-58.

/12/ Sabadini, R. and Peltier, W.R.: 1981, Geophys. J.R. astr.
     Soc. 66, in press.

/13/ Morgan, W.J.: 1981, in The Sea, Vol. 7, ed. by C. Emilia-
     ni, J.C.Wiley, in press.

/14/ Jurdy, D.M.: 1981, Tectonophys, 74, 1-17.

/15/ Sabadini, R., Yuen, D.A. and Boschi, E.: 1981, J. geophys.
     Res., in press.

/16/ Sabadini, R. and Yuen, D.A.: 1982, in The Ninth Internatio-
     nal Symposium on Earth Tides, ed. by J. Kuo and   P.
     Melchior, Schweizerbart-Verlag, Stuttgart, in press.

/17/ Munk, W.H. and MacDonald, G.J.F.: 1960, The Rotation of the
     Earth, Cambridge Univ. P.

/18/ Emiliani, C.: 1955, J. Geol. 63, pp 538-578.

/19/ Hays, J.D.; Imbrie J. and Shackleton, N.J.: 1976, Science
     194, pp 1121-1132.

/20/ Sabadini, R. Yuen, D.A. and Boschi, E.: 1982 Nature, sub-
     mitted.

/21/ Milankovitch, M.: 1930, in Handbuch der Klimatologie, I,
     ed. by W. Koppen and R. Geiger, Gebruder Bornträger,
     pp 1-176.

/22/ Lambeck, K.: 1979, in The Earth: Its Origin, Structure and
        Evolution, ed. by M.W.McElhinny, Academic Press, London,
        pp 59-81.

/23/ Backus, G.E. and Gilbert, J.F.: 1967, Geophys, J.R. astr.
        Soc. 13, pp 247-276.

/24/ Peltier, W.R.: 1976, Geophys. J.R. astr. Soc. 46, pp 669-
        705.

/25/ Hughes, T.J., Denton, G.H., Andersen; B.G., Schilling, D.H.
        Fastook,G.L., and Lingle, C.S.: 1981, in The last Great
        Ice Sheets, ed. by G.H. Denton and T.J. Hughes, J.Wiley
        and Sons, pp 275-308.

/26/ CLIMAP Project Members: 1976, Science 191, pp 1131-1137.

/27/ Hughes, T.J., Denton, G.H. and Grosswald, M.G.: 1977, Nature
        266, pp 596-602.

/28/ Lingle, C.S. and Clark, J.A.: 1979, J.Glaciol. 24, pp 213-
        230.

HIGH PRECISION TRACKING OF SYNCHRONOUS SATELLITES FOR GEOPHYSICAL PURPOSES

L.Anselmo,B.Bertotti*; P.Farinella**;A.Milani,A.M.Nobili,
F.Sacerdote***.
*Istituto di Fisica Teorica,Univ.di Pavia,Italy
**Osservatorio Astronomico di Merate,Italy
***Istituto Matematico"L.Tonelli",Univ.di Pisa,Italy

ABSTRACT

The possibility of tracking very accurately a geosynchronous satellite is very interesting (a) to improve the knowledge of the
resonant geopotential coefficients (hence the knowledge of the
geoid) (b) to determine with higher accuracy the radial departure
of the sea surface from the geoid and eventually its seasonal or
long period variations. We show that both laser (if the spacecraft
has laser retroflector arrays on board) and optical tracking of
many currently used telecommunication satellites, plus a good modelling of non-gravitational perturbations in the orbit propagation, could provide new interesting results.

1. INTRODUCTION

In the usual spherical harmonics expansion of the Earth's gravitational potential V at a given external point (r,φ,λ):

$$V(r,\varphi,\lambda)=GM_e\sum_{\ell=0}^{+\infty}\sum_{m=0}^{\ell}\frac{a_e^\ell}{r^{\ell+1}} P_{\ell m}(\sin\varphi)[C_{\ell m}\cos(m\lambda)+S_{\ell m}\sin(m\lambda)] \quad (1)$$

$(M_e,a_e=$Earth's mass and radius; G=gravitational constant;
$P_{\ell m}(\sin\varphi)=$Legendre associated function of degree ℓ and order m
in the sine of the latitude) the coefficients $C_{\ell m},S_{\ell m}$ with low
ℓ and $(\ell-m)$ even are those which mainly determine long-period
perturbations on the orbit of a synchronous satellite. C_{22} and S_{22}
are the most important terms, representing the ellipticity of the
Earth's equator, i.e., its main departure from a perfect circle
(Allan,1963;Gedeon,1969;Kamel et al.1973). Due to these resonant
coefficients the satellite slowly librates in longitude around

195

one of two stable equilibrium points (at about 75°E and 110°W of longitude) with a libration period of the order of 1000 days. The satellite longitude acceleration, usually referred to as $\ddot{\lambda}$, is of the order of 10^{-3} deg/day^2 (about 10^{-5} cm s^{-2}) and is due for about 80% to the effect of terms containing C_{22} and S_{22}. The important point is that the resonant longitude drift is a long period accumulating perturbation, so that, analysing orbital arcs of long enough duration (compared with the orbital period), it can be easily separated from other perturbations acting on shorter timescales. This allows a precise determination of the involved geopotential coefficients. This method has been widely employed since the sixties by using the tracking data (unfortunately not very accurate) of the first telecommunication satellites (Wagner, 1965;1966;1970) to recover the low-degree resonant geopotential coefficients. The knowledge of these coefficients is very important for many different purposes:

(1) Improvement of the global geopotential models,by reducing the aliasing effects between the uncertainties of low-and high-degree terms. As a matter of fact these models, like the GEMs (Wagner et al.,1977;Lerch et al.,1979), have been obtained by fitting the orbits of many satellites, but mostly low ones: no data from synchronous satellites have been employed,so that it is reasonable to expect that low-degree resonant terms have been rather poorly determined. Indeed, tracking data from satellites not included in the solutions have been already used to test the accuracy of the models (Wagner and Lerch,1978).

(2) Determination of the radial departure of the sea surface from the geoid (the so-called "sea surface topography" or SST), which requires a good knowledge of the geoid. It has been shown (Mather et al.,1978) that the major error source in the low-degree SST determination by satellite altimetry data (a spheric-harmonic expansion of these data has been performed) is, by far, the present uncertainty in the low-degree geopotential harmonics. For this purposes, i.e. to improve the long-wavelength geoid, a wide longitude coverage of the tracking data is necessary and this can be achieved either by organising a world-wide tracking campaign for many satellites positioned at different longitudes or by allowing a given spacecraft to drift in longitude (a geosynchronous satellite with suitable initial conditions can slowly circulate-instead of librating-, covering the whole equator). We will come back to this point in section 3.

(3) Since the SST gives a contribution to the geopotential,likely seasonal variations of the SST can be recovered from seasonal variations in its contribution to gravity (provided that the relative accuracy in the low-degree harmonics is of the order of 10^{-4}!).

(4) Detection of possible secular geopotential's variations of different origin, e.g. due to displacements of mass anomalies near the core-mantle boundary (Wagner,1973). For instance, the well known secular variation of the geomagnetic field ($\sim 0^\delta.5$/yr) could

be correlated with a slow displacement of the mass anomalies causing the C_{22} and S_{22} geopotential terms. No conclusive movement of
the gravity phase angle $\lambda_{22} = \frac{1}{2}$ arc $tg(S_{22}/C_{22})$ was detected by
Wagner (1973), but he concludes that an upper bound on the drift
of $0°.05$/yr is still compatible with possible slow-moving irregularities in the region of the core-mantle boundary. Hence it can
be expected that only a small fraction of the mass anomalies is
connected with the geomagnetic field, yielding a correspondingly
slower drift of the geopotential. As regards points (3) and (4),
we stress that a geosynchronous satellite is ideal for monitoring
changes in the low-order gravitational field. It is in effect a
gravity meter set up at a single longitude for long periods of
time which provides repetitive gravity measurements at the same
location. Moreover, a geosynchronous satellite is sensitive to
only the lowest-degree and lowest-order gravity anomalies and this
benefit is valuable if we aim at measuring Earth-wide time-dependent effects such as global SST seasonal changes or mass displacements in the Earth's interior.
(5) Manoeuvres planning and optimization on telecommunication satellites. In order to reduce fuel consumption an accurate knowledge of the resonant coefficients is essential. In the present
situation, the coefficients relevant for synchronous orbits derived within existing models (Wagner et al.,1977;Balmino et al.,
1976;Gaposchkin,1979) show rather large discrepancies going from
10^{-3} to 10^{-1} of their values.

2. THE PERTURBATIVE MODEL

If we aim at achieving a 10^{-3} relative accuracy in the determination of the resonant coefficients (assuming that accurate enough
tracking methods are available; see section 3), we must be able
to measure the longitude acceleration λ caused by these coefficients with the same accuracy. Since $\lambda \simeq 10^{-5}$ cm s^{-2}, this means
that all the perturbative accelerations greater than 10^{-8} cm s^{-2}
must be taken into account in the orbit propagation model. Table
1 gives a list of the most important dynamical perturbations on
a synchronous satellite of area-to-mass ratio A/M=0.05 cm^2/g.
Moreover, it is necessary to model (1) the kinematical effect of
the lunisolar tides on the Earth-bound tracking stations
($\sim 5.8 \times 10^{-7}$ cm s^{-2}) (2) the apparent perturbations due to polar motions (precession, lunisolar nutation, free nutation) arising when
the equations of motion are integrated in an Earth-fixed frame
(Farinella et al.,1980). If an inertial reference frame is chosen
the effects of polar motions appear in a kinematical way and must
be treated in a different way. The major effort must be devoted
to the modelling of the free nutation, generally referred to as
polar motion, because it is not so well known as precession and
lunisolar nutation, and produces a long-periodic ($\sim$1 year) out-of-
plane a pparent acceleration of the order of 2×10^{-5} cm s^{-2}.

TABLE 1. Dynamical perturbations on a geosynchronous
satellite (A/M=0.05 cm^2/g)

CAUSE	ORDER OF MAGNITUDE IN cm s^{-2}
(Earth's monopole	22.4)
Earth's oblateness	1.76×10^{-3}
Moon's gravity gradient	7.3×10^{-4}
Sun's gravity gradient	3.4×10^{-4}
Ellipticity of the Earth's equator	10^{-5}
*Solar radiation pressure	2.8×10^{-6}
Venus' gravity gradient (at conjunction)	4.3×10^{-8}
Tidal deformations of the solid Earth (dynamical effect)	2.6×10^{-8}
*Earth's albedo radiation pressure	2.3×10^{-8}
*Gas leaks from the thrusters	8.7×10^{-9}
General relativistic correction	7×10^{-9}
Jupiter's gravity gradient (at conjunction)	5×10^{-9}
*Solar wind	5×10^{-9}
Indirect oblation (of the Earth's J_2 on the Moon)	3.1×10^{-9}
Moon's oblateness	3.1×10^{-9}

In Table 1 we have marked by * , among the dynamical perturbations,
the non-gravitational ones which are very difficult to model because depend on unknown or not well known parameters. As shown by
the Table, the solar radiation pressure is the most important one
and much attention must be paid in modelling its long periodic
effects ($\sim$ 1 year) on the satellite orbit. If the spacecraft has
an Earth-pointing antenna on board (telecommunication satellites

do have it) a one-to-one resonance arises between the satellite orbital period and the spin period of the antenna in the inertial frame, such that the interaction between solar radiation and the antenna surface causes large long-periodic effects in longitude (Anselmo et al.,1981). On the contrary, the interaction of solar radiation with the spacecraft body produces longitude effects which are short-periodic, so that they do not accumulate with time and do not need to be very accurately modelled. As an example, for a librating geosynchronous satellite with A/M=0.05 cm^2/g the amplitude of the short-periodic effect in longitude is of the order of 10 m, while the long periodic one, accumulated over about 60 days (assuming no error in the pointing of the antenna) is two orders of magnitude larger, i.e., about 1 km. Moreover, if the antenna is not perfectly pointed towards the Earth but has a little misalignement angle of about 1°(on the average), this effect becomes 10 times larger. If the radiation pressure perturbation is modelled with 10% accuracy, and the free longitude drift of the satellite is long enough to allow the decoupling of the long and short-periodic effects in the data analysis, longitude errors still remain of the order of 100÷1000 m.

3. TRACKING METHODS AND ACHIEVABLE ACCURACY

As we mentioned in the introduction, the tracking data of the telecommunication satellites used by Wagner in the sixties were not very accurate: they allowed a determination of the Earth's equator ellipticity with a relative accuracy not better than few percents. New tracking techniques are now available (LASER, Doppler twin band) of much higher accuracy. As far as we know, no synchronous satellite has -nor is planned to have- the required payload on board to allow Doppler twin-band tracking. On the contrary, the first geosynchronous satellite carrying on board laser retroflector arrays, SIRIO 2, is scheduled for launch by ESA in 1982.The ground laser stations will track SIRIO 2 for a time-transfer experiment on intercontinental baseline (LASSO) and will provide laser ranging data with an uncertainty in the radial direction less than 1 m. Due to the bad geometry of the range measurement, the corresponding uncertainty in longitude will be about 10 times larger. The spacecraft will be positioned in a first stage at the longitude 25°W (± 1°, implying stationkeeping manoeuvres about every 3 months) and subsequently displaced to 20°E; the orbital eccentricity and inclination will be small. Bertotti et al. (1979) suggested to use the laser ranging data to determine the low-degree resonant coefficients of the geopotential. Assuming that the orbit determination of SIRIO 2 could achieve an accuracy comparable to that of the laser measurements (that means neglecting the bad modelling problems for orbital perturbations as radiation pressure), let us evaluate the attainable accuracy in the longitude acceleration $\ddot{\lambda}$ due to the resonant geopotential terms. For a longitude

accuracy of $10^{-5}°$ (corresponding to uncertainties of about 7 m
along track) and assuming an arc between two successive manoeuvres
of 1 month, the $\ddot{\lambda}$ resulting accuracy is of the order of $10^{-8}°$ /day^2,
corresponding to a relative accuracy of 10^{-5} both in $\ddot{\lambda}$ and in C_{22},
S_{22} (it could also increase by allowing longer orbital arcs be-
tween manoeuvres). We remember (Mather et al.,1978) that such a
relative accuracy in C_{22},S_{22} could allow to detect and measure the
seasonal variations of the low-degree geopotential due to SST and
could reveal also geopotential's variations of secular character
which are very interesting from a geophysical point of view. Unfortunately, as we pointed out in section 2,the difficulties in modelling the radiation pressure perturbation on a geosynchronous
satellite with an Earth-pointing antenna on board cause the longitude uncertainties to grow up to 100÷1000 m (i.e.,one or two orders of magnitude worse than the laser data errors). This means
that anyway the resulting relative accuracy in C_{22},S_{22} will not be
better than $10^{-3}÷10^{-4}$. The situation would be entirely different
(making reliable the accuracy estimate based on the errors in the
laser ranging data) if we could get in a geosynchronous orbit a
satellite similar to Starlette or LAGEOS, equipped with a large
number of laser retroflectors and minimizing problems due to non-
gravitational perturbations (small area-to-mass ratio, nearly
spherical shape, no Earth-pointing antenna). Such a space mission,
in our opinion, should be carefully considered by the scientific
community in the next future since it couples to a straightforward
technological feasibility and to a relatively low cost a large
scientific fallout both in geophysics and in oceanography. Until
this mission is realized,we plan anyway to improve, by at least
one order of magnitude, the accuracy in the low-degree resonant
coefficients within the SIRIO 2-LASSO experiment. Moreover we
think that it is worthwhile to analyze again the possibility of
using optical tracking techniques. These techniques do not surely
provide so accurate data as laser techniques, but they have many
advantages: (a) they don't require any ad hoc device on board;
(b) they don't need any particular help from the operative satellite control centers; (c) they are very simple and cheap. On the
other hand, as we have already shown, for currently used telecommunication satellites the accuracy in the tracking data must be
compared with the errors due to bad modelling of radiation pressure perturbations. In the following we describe in some details
the method of optical tracking, since it has never been used in
past to obtain data of geophysical interest for geosynchronous
satellites.
To perform optical tracking of a synchronous satellite (Milani and
Nobili,1980), it is first necessary to estimate its optical magnitude in diffused light. As a matter of fact, only in very peculiar situations (near the equinoxes) it is possible to receive
the sunlight directly reflected from the satellite (in these situations the satellite is several orders of magnitudes brighter),
but we note that for two periods, of about 80 days each,centered

at the equinoxes, no precise orbit determination is possible due
to eclipses. It is not generally easy to predict the satellite
magnitude in diffused light, because the diffusivity coefficients
are not always available and the satellite usually has a complex
shape, different parts having different colors. Anyway it is possible to make an order-of-magnitude calculation which can be subsequently corrected once the first photographic plates are obtained.
As an example, the magnitude of the SIRIO 1 satellite in reflected
light is about 11 (Catalano,1981) but we estimate that its magnitude in diffused light is not less than 17; for SIRIO 2, which
has a larger antenna on board (the white painted antenna is responsible for the major part of sunlight diffusion) we predict a
value of about 15÷16. Since a synchronous satellite is moving
with respect to the stars at a speed of about 15 arcsec per second,
and is generally a faint object, the telescope must track the satellite to allow the impressing of its image on the plate. Then
the stars appear on the same plate as linear trails. As pointed
out by Catalano et al.(1981) it is necessary to have a large e-
nough number of luminous enough catalogue (AGK3,SAO catalogues)
stars, and if the field of the telescope is small a large value
of D^2/F is needed (D= diameter of the main lens or mirror;F=focal
length). On the other hand, a large enough scale is required on
the plate in order to get good angular measurements, and this
means that F cannot be very small. Moreover, unless the star
trails are marked in some way only one angular coordinate, the
N-S one, is determined. The recovery of the E-W coordinate is
possible simply by using a very fast shutter in order to record
the exact initial and final times of the exposure and using the
endpoints of the star trails as reference points,or by moving very
fast the plate in the N-S direction at accurately recorded times.
In this way the star trails would appear stepped and the edges of
the steps could be used as references to determine the E-W coordinate. Finally, many stars and many satellite images, also in
the same plate, are necessary to reduce the errors due both to
diffusion on the plate and to seeing (if the scale is large).
Some trials at the SAO astronomical station of Agassiz and at the
Observatory of Serra la Nave (Catania) have shown that an accuracy of 1÷2 arcsec in the measurement of the satellite angular position is achievable. Since at the distance of a synchronous satellite 1 arcsec corresponds to about 200 m, this means that
optical tracking data could be available with errors of the same
order as the unavoidable uncertainties in the orbit propagation
due to radiation pressure.
Finally we stress that if a world-wide astronomical campaign will
be organised involving small-sized and intermediate-sized tele-
scopes, many synchronous satellites at many different longitudes
could be optically tracked,giving the possibility to achieve a
relative accuracy of $10^{-3}÷10^{-4}$ in the longitude acceleration $\ddot{\lambda}$,
not only at one or two values of λ (as it will happen with
SIRIO 2) but at many different longitudes. As previously noted,

this is an essential requirement to improve the global geoid knowledge.

REFERENCES

1) Allan,R.R.:1963, Planet.Space Sci. 11,pp.1325-1334.
2) Anselmo,L.,Farinella,P.,Milani,A., and Nobili,A.M.:1981, Presented at the International ESA Symposium on "Spacecraft
   Flight Dynamics", Darmstadt,1981. In press.
3) Balmino,G.,Reigber,C.,and Maynot,B.:1976,The GRIM 2 Earth
   Gravity Field Model, Deutsche Geodätische Kommission,München.
4) Bertotti,B.,Bevilacqua,R.,Farinella,P.,Gianni,P.,Milani,A.,
   and Nobili,A.M.:1979, Internal Report, Osservatorio Astronomico di Brera, Merate,n.8/80.
5) Catalano,S.:1981, Private communication .
6) Catalano,S.,Milani,A., and Nobili,A.M.:1981, in preparation.
7) Farinella,P.,Milani,A.,Nobili,A.M., and Sacerdote,F.:1980,in
   Reference Coordinate Systems for Earth Dynamics, vol.86,p.271,
   Reidel Publishing Company,Dordrecht.
8) Gaposchkin,E.M.:1979, Harvard Smithsonian Center for Astrophysics, Preprint n.1092.
9) Gedeon,G.S.:1969, Celestial Mechanics,1, pp.167-189.
10) Kamel,A.,Eckman,D., and Tibbitts,R.:1973, Celestial Mechanics
    8, pp.129-148.
11) Lerch,F.J.,Klosko,S.M.,Laubsher,R.E., and Wagner,C.A.:1979,
    J.Geophys.Res. 84, pp.3897-3916.
12) Mather,R.S.,Lerch,F.J.,Rizos,C.,Masters,E.G., and Hirsch,B.:
    1978, Presented at the International Symposium on "The Use
    of Artificial Satellites for Geodesy and Geodynamics",Lago-
    nissi, Greece 1978.
13) Milani,A., and Nobili,A.M.:1980, Internal Report n.1/80,Grup-
    po di Meccanica Spaziale, Università di Pisa.
14) Wagner,C.A.:1965, J.Geophys.Res. 70, pp.1566-1568.
15) Wagner,C.A.:1966, J.Geophys.Res. 71, pp.1703-1711.
16) Wagner,C.A.:1970, J.Geophys.Res. 75, pp.6662-6674.
17) Wagner,C.A.:1973, J.Geophys.Res. 78, pp.470-475.
18) Wagner,C.A.,Lerch,F.J.,Brownd,J.E., and Richardson,J.A.:1977,
    J.Geophys.Res. 82, pp.901-914.
19) Wagner,C.A., and Lerch,F.J.:1978, Planet.Space Sci.26,
    pp.1081-1140.

ATMOSPHERIC EVOLUTION

Hans-Jürgen Bolle

Institut für Meteorologie und Geophysik
Universität Innsbruck, Austria

ABSTRACT

The abundances of atmospheric gases in planetary atmospheres impose certain boundary conditions upon the planetary formation theories. One of the most important is that the present volatiles did probably not undergo a hot phase during early accretion and must have been added at a later stage if there was such a hot phase.

While in the outer planets, especially Jupiter, a primordial or solar composition of element abundances could survice, the inner planets have suffered major transformations. At Venus and Mars water must have been dissociated and hydrogen escaped. CO_2 was probably chemically bound in the Martian crust while it could not react with the crust material at Venus because of the high temperatures which developed soon due to the greenhouse effect. At Earth, because of its intermediate temperatures, a number of reactions occurred but the most significant change was due to the generation of life which changed the formerly reducing atmosphere into an oxidizing one.

Finally a short overview is given of the present knowledge about the structure of the planetary atmospheres.

A. Coradini and M. Fulchignoni (eds.), The Comparative Study of the Planets, 203–220.
Copyright © 1982 by D. Reidel Publishing Company.

INTRODUCTION

Starting point for all considerations of atmospheric evolutions is the assumption that in an primordial nebular a distribution of stable elements existed which corresponds to the presently observable
relative cosmic abundances. The expectation is, that
it might once be possible to deduce from these primordial abundances the present distribution of elements and molecular compounds in the atmospheres of
planets and moons by physico-chemical reasoning. The
thermal history of the nebular and the planetary bodies, their physical and geophysical properties like
masses, radiation budgets, presense or absence of
magnetospheres, mantle and crust compositions are
essential parameters which enter such deliberations.

Presently no unique theory exists which could explain all features observed in planetary atmospheres.
Even for some of the most obvious differences in
the composition of planetary atmospheres only possible explanations can be offered but not a rigorous
theory. The roots for the uncertainties lie already
in our limited and sometimes incompatible understanding of the first events which took place during the
formation of our solar system. These are discussed in
other contributions of this volume (1 - 5) and elsewhere in the literature (6, 7).

Most of the theories concerned with planetary
formation have to assume that the inner planets went
through a hot phase probably during the first 10^5
years of planet formation when gravitational and radioactive energy, mainly from ^{26}Al, was released. During this time volatile compounds must have been out-
gassed to a large extent. It is very likely that the
released volatiles have been removed from the inner
solar system by high solar activity (T-Tauris phase)
which could have lasted for about 10^7a and that after
this event new atmospheres were accumulated.

Reasons for the assumption that the terrestrial
planets lost their primordial atmospheres are that
the observed distributions of noble gases have relative abundances which are not in accordance with the
solar or cosmic distributions. These gases are inert
against chemical removel from the atmospheres and are
not likely to escape easily because they are too heavy. The solar mass abundance (gram per gram) of neon,
e.g., is about eight orders of magnitude as large as

now observed on Earth. If one considers that the primordial atmospheres consisted mainly of light gases with the elements H, He, O, N, C and S in solar ratios (1000:100:0.6:0.3:0.1:0.02) (8) one can extrapolate backwards to the terrestrial primordial atmosphere containing the same amount of neon as observed today. The result is that this primordial atmosphere can only have had less then 1 % of its present mass and consisted mainly of H and He with a small amount (in the order of 1 %) of O, C and N.

A similar depletion in noble gases has been observed for Venus and Mars so that one can conclude that the inner planets had the same fate in loosing their primordial atmospheres (which has never been replaced in the case of Mercury) (7).

As an example the ratio of nitrogen to ^{36}Ar is much enhanced at the terrestrial planets with respect to the sun (Table 1). An increase from

Table 1: ^{14}N/^{36}Ar ratios in the solar system after Pollack and Black (10).

Sun	Venus	Earth	Mars	Meteorites
37	$0.15.10^4$-1.10^4	20.10^4	5.10^5-50.10^5	$2.5.10^6$

Venus to Mars can be observed and the Martian ratio comes very close to that of meteorites. In this comparison all nitrogen was taken into account which is in the atmosphere as well as in near surface reservoirs (11,12). Also the ^{40}Ar/^{36}Ar ratio shows a trend from about 1 for Venus to about 3000 for Mars which suggests that radioactive potassium, from which ^{40}Ar is derived, is enriched with respect to the noble gases as the distance from the sun increases.

In order to explain such irregularities one has to assume that the material from which the atmosphere was formed, consists of two components: one with primordial composition and one with "planetary" composition. The solution of this puzzle offered by Pollack and Black is, that one component has a C, N and H$_2$O composition which is independent of the position in the nebular, the other is picked up locally in the nebular. The N : C ratio is in fact nearly constant for the planetary objects (0.03 - 0.09; sun: 0.3) and

so are the isotopic ratios for C and N. C, N and H_2O
are fixed in chemical compounds of chondrites. The
rare gases on the other hand are physically attached
to the meteoritic grains in which the chondrites are
imbedded. Solution and adsorption are, however,
strongly temperature dependent and proportional to
the ambient partial pressure. Only if the rare gases
have been imbedded in the meteorites which transported them to the planets at nearly constant temperature
in a planetary disc where a strong pressure gradient
existed, the decrease in rare gas abundancies (Table
2) from Venus to Mars can be explained.

Table 2: Rare gas abundancies in the inner solar
          system after Pollack and Black (10).

Gas	Sun	Venus	Earth	Mars
Ne	21.10^{-3}	$2.10^{-10}-6.10^{-10}$	1.10^{-11}	4.10^{-14}
$^{36}A + ^{38}A$	10^{-4}	$8.10^{-10}-8.10^{-9}$	5.10^{-11}	2.10^{-13}
Kr	10^{-8}	2.10^{-10}	3.10^{-12}	2.10^{-14}
Xe	2.10^{-8}	———	2.10^{-11}	1.10^{-14}

The model requires that the rare gases must have been
accumulated in the meteorites or grains before the
sun developed its radiative power, which would immediately cause the undesired temperature gradient, but
after the gravity field of the sun was strong enough
to produce the necessary density gradient. In order
to sort out the different accretion hypotheses it is
therefore also important to consider the very early
time sequences. A problem which is discussed in another contribution of this publication (13).

     The Pollack and Black model suggests further
that Venus, Earth and Mars have received essentially
the same amount of N_2, CO_2 and H_2O relativ to their
surface, and in fact the N_2 and CO_2 abundances are
about the same for Venus and Earth. Mars is depleted
in N_2, CO_2 and H_2O, and Venus in H_2O. It is, no matter which hypothesis one accepts for the formation
of the planets, very difficult to understand, that
Earth should have received large amounts of water,
and Venus and Mars both orders of magnitudes less.
Let us therefore assume, that the volatiles, except
the rare gases, were offered to Venus, Earth and Mars
in nearly equal quantities and that the present abun-

dancies are the results of further evolution.

Mercury was certainly also in a position to receive volatiles during a phase of a small temperature
gradient throughout the inner planetary system. However, as the sun developed into a star the equilibrium temperature rose to about 500 K compared to about
300 K for Venus. Because of its small mass Mercury
would at this temperatures not be able to retain the
volatiles for considerable times.

The outer planets have obviously not been affected very much by events, which swept out the light
material (H, He) from the inner part of the solar
system since they could retain at least most of their
hydrogen and helium. Jupiter and Saturn may even have
attracted some of the material from the inner part
of the solar system and accumulated in their own atmospheres.

ACCUMMULATION OF ATMOSPHERES AT THE INNER PLANETS

There exist little doubt that the volatile material in the atmospheres of Venus, Earth and Mars
did never went through a hot phase. The D/H-ratio of
water suggests that it has been formed at temperatures between 100 and 220 K, from the oxygen isotop
abundances the temperature should be below 300 K (14).
Candidates for the supply of volatiles are carbonaceous chondrites, mainly of the first type, which
contain up to 20 % H_2O, 5 % carbon, 6 % sulfur and
nitrogen bounded in organic material as well as other
volatiles. Thus one major source can be assumed in
the meteorites which were captured by the planets.
The volatiles incorporated in their grains could have
been released immediately at impact if large objects
are considered. Finer grains would have been deposited without strong heating. The release of their volatiles could have been a long process. Certainly
there will have been occurred also a further outgassing from minerals from which a number of gases can
be produced (N_2, CO, CO_2, NO_x, PO_x, O, Cl_2, F_2, COS,
SO_2 and, by radioactive decay, ^{40}Ar and He). Plutonic
activity contributes to the outgassing of the mantle
(mainly H_2O, CO_2, CO, SO_2, H_2S, H_2, NH_3 and CH_4) (7).

On Earth a substantial fraction of the oceans
was formed within the first billion years. The oldest
sedimentary rocks are more than $3.2.10^9$, perhaps even

$3.7 \cdot 10^9$ years old. The degassing could have been forced by planetary heating due to the release of gravitational contraction or radioactivity. It seems to be possible that the time constant for degassing was of the order of $2 \cdot 10^8$ years. The whole process took probably $2 \cdot 10^9$ years. From that time on no major modification of the earth surface except continental drift is noticable.

As soon as an atmosphere builds up the radiation budget is changed. The surface does not remain in direct radiative equilibrium, the infrared emission layer is lifted, and a greenhouse effect starts at the surface. If we assume water vapor and carbon dioxide as the major constituents, then the infrared transparency of the atmospheres was strongly reduced at nearly all wavelengths.

On Earth the amount of water which can be accumulated in the atmosphere is limited because the Earth is in a temperature range, where all three phases of H_2O can exist. Also the amount of CO_2 is regulated by its solubility in the oceans and by production of carbonate sediments according to the reactions

$$CO_2 + H_2O \longrightarrow H_2CO_3$$

$$Ca^{++} + 2\ HCO_3^- \longrightarrow Ca\ CO_3 + H_2O + CO_2,$$

or by weathering of solid surfaces where reactions like

$$Ca\ SiO_3 + 2\ CO_2 + H_2O \longrightarrow Ca^{++} + 2\ HCO_3^- + SiO_2$$

$$Ca\ CO_3 + SiO_2 \longrightarrow Ca\ Si\ O_3 + CO_2$$

control the equilibrium.

At Venus the temperatures were from the beginning much higher (340 K equilibrium temperature) and the joint action of water vapor and CO_2 drove the greenhouse effect quickly to a stage where the water was completely evaporated and subsequently dissociated, hydrogen escaping to space. The remaining oxygen was probably consumed by oxydation processes at the surface. Carbon dioxide could - because of the

high surface temperatures and the lack of water - not
any more react with the surface materials and remained therefore in the atmosphere.

Mars, on the other side, was from the beginning
so cold that water might have penetrated into deeper
layers which were still heated from the interior,
and froze there after further cooling in a permafrost
layer. The carbon dioxide reacted extensively with
the surface as long as water was present and only a
small amount remained in the atmosphere.

ESCAPE MECHANISMS

If a molecule gains enough kinetic energy to
overcome the gravitational field of a planet, it
might escape into space if the vector of velocity is
directed upwards and if no collisions with other
molecules occur. Light atoms like H and He have a
good chance to escape from upper atmospheres where
the temperatures are high due to absorption or extrem
ultraviolet radiation. Thus some of these molecules
which adopt a Maxwellian velocity distribution gain
high speeds, larger than the critical velocity (Table
3). The free path length must be in the order of the
planetary radius in order to enable the escape. To
achive the maximum possible escape rate it is necessary that there is enough supply of the molecules by
diffusion from below.

This classical thermal escape process is called
Jeans escape. It is rather inefficiant for heavier
species. But there exist other mechanisms which can

Table 3: Critical escape velocities.

Planet	Critical escape velocity in km s^{-1}
Merkur	4.25
Venus	10.4
Earth	11.2
Moon	2.4
Mars	5.05
Jupiter	60.6
Saturn	36.7
Uranus	22.4
Neptune	23.5
Pluto	6.4

provide excess kinetic energy (15) to certain atoms, such as listed in Table 4.

Table 4: Non-thermal acceleration processes after Hunten (15).

Process	Reaction[1]	Possible important for
Charge exchange	$H + H^+ \longrightarrow H^+ + H^*$	
Dissociation recombination	$O_2^+ + e \longrightarrow O^* + O^*$	Mars
	$OH^+ + e \longrightarrow O + H^*$	Venus
Impact or photodissociation	$CO^+ + e \longrightarrow C^* + O^*$	Mars
	$CO_2H^+ + e \longrightarrow CO_2 + H^*$	
	$N_2 + e^* \longrightarrow N^* + N$	Mars
	$O_2 + h\nu \longrightarrow 2 O^*$	
Ion-neutral reactions	$CO^+ + H_2 \longrightarrow CO_2H^+ + H^*$	Venus
	$O^+ + H_2 \longrightarrow OH + H^*$	Venus
Sputtering knock on	$Na + S^{+*} \longrightarrow Na^* + S^{+*}$	Io, most efficient for heavy ions
Magnetospheric wind pick up	$O + h\nu \longrightarrow O^+ + e$ O^+ picked up	
Ion escape	H^+ escape e.g. by open magnetic lines from earth	
Electric fields (B and D)	$X^+ + eV \longrightarrow X^{+*}$	

[1] The star * denotes the accelerated atom.

The mechanisms favour lighter isotopes to escape which can explain certain deviations in the abundances of isotopes.

The present escape fluxes of hydrogen are in the order of 2.10^8 atoms cm^{-2} s^{-1} for Mars and estimated to be 3.10^7 cm^{-2} s^{-1} for Venus. The terrestrial value is 10^8 cm^{-2} s^{-1}.

It must be taken into account that there is also a capture of solar wind particles. The particle
fluxes of H and He is, however, presently 2 - 3 orders of magnitude smaller than the escape flux, and
seems therefore to be unimportant for the H and He
budget.

FURTHER EVOLUTION

Venus

The only processes to remove the water vapor
from the Venusian atmosphere are Jeans escape or
escape due to dissociation-recombination respectively ion-neutral reactions. For the Jeans escape the
ultimate limiting factor is the available ultraviolet radiation energy which is absorbed in the upper
atmosphere of about 10^{-3} W m^{-2}. This is sufficient
to accelerate 10^{10} atoms m^{-2} s^{-1}. The energy available for water vapor dissociation would be sufficient
to produce more H atoms than can thermally be accelerated. If 30 % of the available energy would be used
for escape acceleration, an amount of hydrogen could
be removed in less of 1.10^9 years which is equivalent
to the total water mass on earth (7). The oxygen
which was left behind may have been consumed by rock
oxidation at a smaller rate then the H escape so that
for some time a CO_2 - O_2 - H_2O atmosphere may have
existed.

Mars

Mars remained very probably always colder than
Venus and Earth. Therefore the release of volatiles
was much less efficient and part of it may still be
contained in the solid material. CO_2 can be adsorbed
by the Martian regolith in remarkable quantities,
another part is frozen in the polar caps. A third
part may be in the crust as carbonate minerals. The
Martian crust can also adsorb much more H_2O as the
Earth's crust because of its lower temperature. With
respect to Nitrogen the escape is possible due to the
impact dissociation mechanisms (Table 4), while this
does not work for the Earth and Venus where four times
as high energies would be required.

Mars is object of rather strong mechanical perturbations due to Jupiter. The results are changes
in its orbital parameters, especially the obliquity

of its spin axis, which can vary between 15° and 35° (16,17). This causes a geographical redistribution of the solar energy. Presently the north polar cap never vanishes (water-ice, not CO_2, remains partly fixed) because it is turned towards the sun at aphel (largest distance from the sun). This could easily change at a different configuration e.g. minimum obliquity and closest approach to the sun at equinox when both poles receive solar radiation. In this case it may be that first all frozen CO_2 and then - because of the increasing greenhouse effect - also CO_2 from the regolith reservoir enters the atmosphere where surface pressure and temperatures increase. This would also enable water vapor from a permafrost reservoir to evaporate and similar condition as on earth may develop. In order to explain the features observed at the Martian surface e.g. at Tiu Vallis (Hydapsis Chaos), NE Argyre Planitia or SE Alba Patera it must be concluded that there have been rather rigorous spring like outbreaks in the Martian history. Masursky et al. (18,19) have concluded that these episodes must have occured at different times during the Martian history between 4 and $0.5.10^9$ years before present.

Earth

Very early in the Earth history its atmosphere contained mainly nitrogen, water vapor, carbon dioxide and hydrogen ($\sim 1\,\%$). Chemical processes led to the orign of life via complex organic molecules. It is unlikely that NH_3 (and CH_4) was present in large quantities because this would increase the pH value of the oceans and the carbonate formation would have been increased (7). It is believed that in a reducing atmosphere (20) first metabolic organisms developed which generated their energy from reactions of organic molecules like the alcohol fermentation process

$$C_6H_{12}O_6 \longrightarrow 2\ C_2H_5OH + 2\ CO_2$$

$$\text{neutral} \longrightarrow \text{reduced} + \text{oxidized.}$$

The next step in the evolution may have been the development of bacteria which drew energy from reactions between CO_2 and H_2 (autotrophs):

$$2\ CO_2 + 4\ H_2 \longrightarrow CH_3\ COOH + 2\ H_2O$$
$$CO_2 + 4\ H_2 \longrightarrow CH_4 + 2\ H_2O$$

After the organisms were protected against destruction by UV radiation the solar energy could be used for bacterial photosynthesis by which carbon is obtained from CO_2 due to reduction by means of electron donors like H_2, hydrogen sulfide, thiosulfate or organic molecules. These organisms depended therefore on the supply of H_2. The impact on the atmosphere was probably very small though a decrease of the hydrogen content and a slightly less reducing atmosphere may have resulted accompanied by a rise in the concentration of oxygen due to water photolysis. The last step was the chlorophyll photosynthesis by which water is used to produce the hydrogen, which is necessary for the reduction process, by the bulk reaction

$$6\ CO_2 + 5\ H_2O \longrightarrow C_6H_{10}O_5 + 6\ O_2 - 2.809\ MJ.$$

The first microbes are at least $3.2.10^9$a old (pre-Cambrian). Stromatollites (which are probably still photosynthesis bacteries though they look like algae) have been found, which date back as far as 3.10^9a. Only $2.5.10^9$a ago green plant photosynthesis seem to have started and 1.10^9a back aerobic bacteria have been identified which use oxygen for oxydation processes. Thus free oxygen must have existed at that time. The estimates are that 8.10^8 years back the O_2 concentration was 2 %, and that the present 20 % level was reached $5.8.10^8$ years ago during the Cambrian.

The climatic record of the Earth shows strong variations in its temperature regime. Major changes may have been induced by continental shifts and the respective changes of ocean currents. Others, especially during the last million years, are strongly correlated with the geographical redistribution of solar irradiance due to orbital variations. The temperature variations in the order of a few to 10 degrees have an effect on a number of equilibrium conditions, e.g. the solubility of CO_2 in the oceans. An impact of these climate variations on the composition of the atmosphere is therefore very likely. Indeed, during the last glacial, the CO_2 amount in the atmosphere seems to be decreased by as much as 30 % (21).

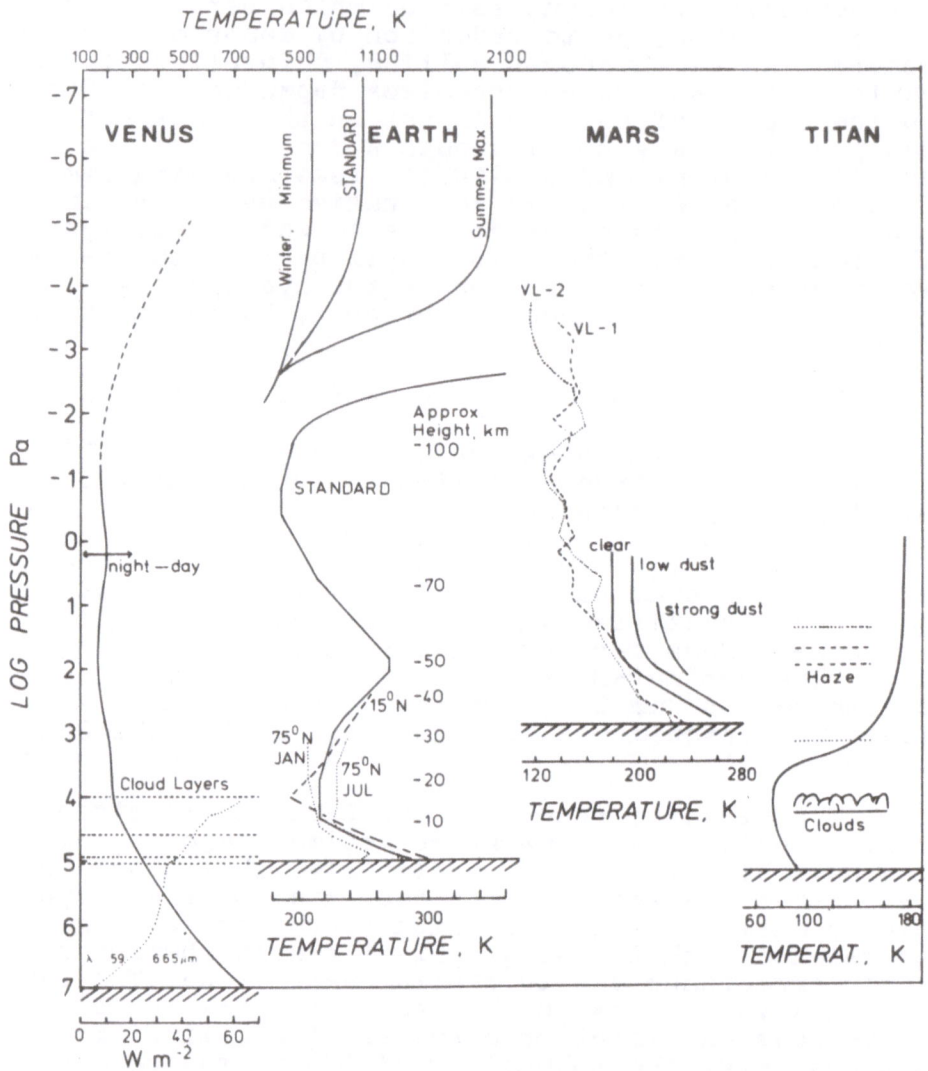

Fig. 1: Vertical atmospheric temperature structure
of the planets Venus, Earth, Mars and of
the Saturnian moon Titan after (26-34).
For Venus also the solar irradiance for the
wavelength range 0.59-0.665 um is given
(38,39).

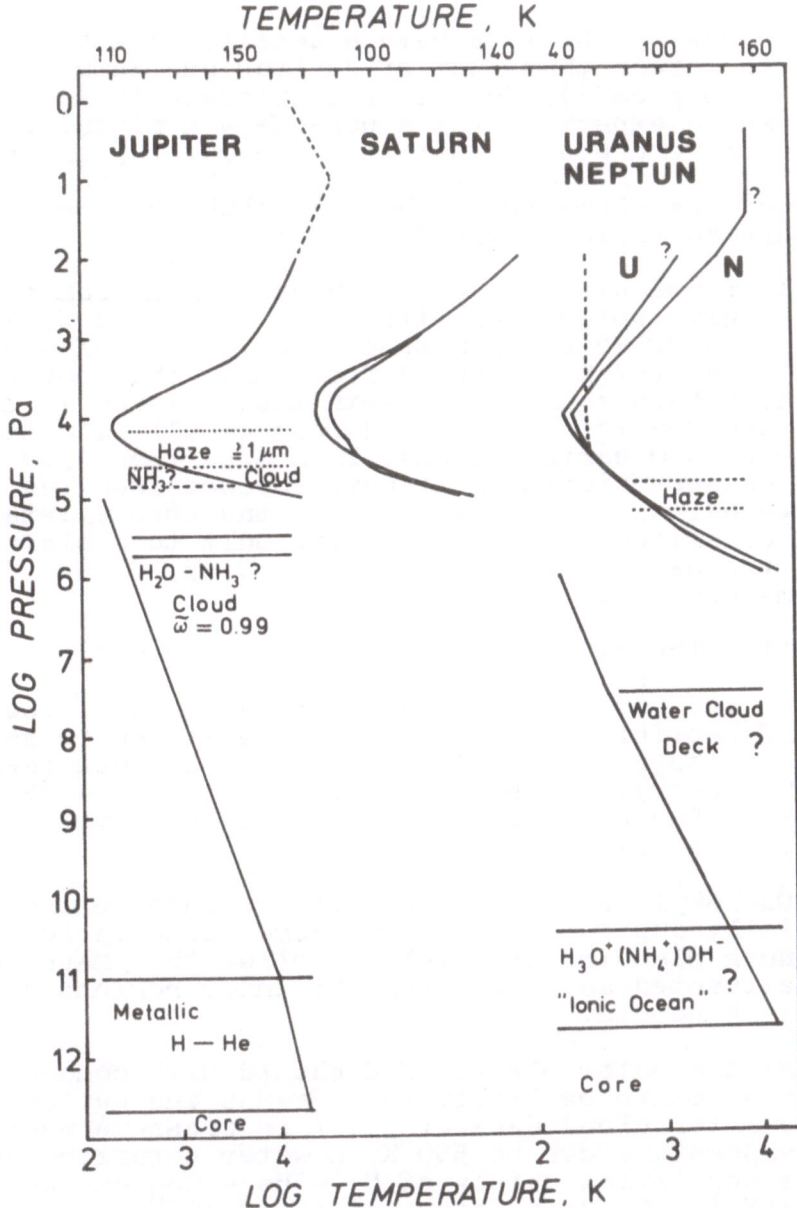

Fig. 2: Vertical atmospheric temperature structure of the planets Jupiter, Saturn, Uranus and Neptune after (35-37).

OUTER PLANETS

Uranus and Neptune have a density of 1.6 g cm^{-3} which is larger than that of Jupiter and Saturn (1.0 ± 0.3 g cm^{-3}). For all four planets this is more than can be expected from a pure He - H mixture. It is therefore probable that also the outer planets have cores of 10 - 20 Earth masses which should be composed of the elements of the next abundant group or of hydrated silicons (22,23).

If these planets are produced from a planetary nebular with solar composition and contain a C, N, O, F, Ne core of 15 earth masses, then the total amount of material (97 % He and H) from which they condensed must have been about 2200 earth masses which is much more than the total condensed mass of all planets together (450 earth masses). Thus more than 1500 earth masses material must have escaped from the solar system. If it is assumed that the core exists of hydrated silicons it remains difficult to explain where the second abundant group of O, N, C, Ne (22 - 5 atoms per 1 atom Si) remained.

Also the different structure of the moons is difficult to explain. Callistro (density 1.8 g cm^{-3}) seems never have reached the melting point of ice, no differentiation due to internal heat can be seen. Ganymede shows a tectonically evolved surface (density 1.9 g cm^{-3}), Titan (1.8 g cm^{-3}) has a N_2 - CH_4 atmosphere, Europe (3.0 g cm^{-3}) a thin ice layer and Io (3.5) is heated by tides.

One explanation may be that the outer planets, especially Jupiter and Saturn, were formed prior to the inner planets, especially earlier then Mars, and have attracted material from the inner regions of the planetary nebula.

At the outer planets H_2O should have condensed and it seems to be likely that Uranus and Neptune have a water cloud deck at 3.10^7 Pa pressure where the temperature can be 350 K, a water saturated envelope consisting mainly of H_2 - He - CH_4 and an H_3O^+ (NH_4^+) OH^- "ion ocean" at p = 3.10^{10} Pa and T = 3000 K.

PRESENT STATE OF ATMOSPHERES AND ATMOSPHERIC EXPLORA-
TION

Recent Earth surface based observations and planetary space missions have returned a wealth of new
informations on planetary atmospheres. These are to
date not all completely evaluated and put into per-
spective with each other. However, it can generally
be stated that the composition and the thermal structure of all planetary atmospheres are basically known
as far as the atmospheres are accessible by presently
available techniques (25). Not measured is the structure of the atmospheres of the outer planets below
the cloud decks. Rather detailed information have
also be gained on the atmosphere of the Saturnian
moon Titan.

All atmospheres consist basically of a troposphere with a negative (and nearly constant) temperature gradient and a Thermosphere/Exosphere with increasing temperatures due to heating by UV radiation.
The Earth atmosphere has also a Stratosphere and a
Mesosphere because of its ozone layer which gives
rise to a warm peak, the statopause, due to an extended absorption spectrum.

Figs. 1 and 2 and Table 5 summarize our present
knowledge about the structure of the atmospheres.

OUTLOOK

Over large time periods Mars and Earth suffer
from climate variations induced by plate tectonics
at the Earth and changes of orbital parameters of
both, Earth and Mars. Such disturbations are smaller
for Venus because of its more stable orbit and much
denser atmosphere. Between surface temperatures and
atmospheric composition exist feedback mechanism because the equilibrium between the gas phase and the
bounded or adsorbed volatiles is temperature dependent. On both planets also the distribution between
gas phase and condensed phase depends critically on
the temperature for water on Earth and carbon dioxide
and water on Mars. It can therefore be predicted that
with future climate variations also the atmospheres
will be affected. These seem primarely to be reversi-
ble processes. However, there are several results of
climate model studies that under certain conditions
a planet may switch from one climate state to another
more stable state (42, 43, 44).

Table [1]: Composition of planetary atmospheres after (38-41)

Gas	Venus Trop.	Venus 150 km	Earth	Mars Trop.	Mars 120 km	Jupiter	Saturn	Uranus	Neptune	Titan	Io
H_2			5.10^{-5}			0.89 ± 0.05	0.94	0.93	0.94	2.10^{-3}	
He		$1.3.10^{7}$	5.10^{-4}			0.11 ± 0.03	0.06				
Ar	3.10^{-3}		1	1.6						~0.12	
Ne	$< 10^{-3}$		2.10^{-3}	$2.5.10^{-4}$		10^{-4}					
Kr				$3 .10^{-5}$							
Xe				$8 .10^{-6}$							
N_2	3.5	5.10^{7}	78	2.7	$1.5\text{-}2.10^{9}$					~ 0.85	
O		$4.4.10^{8}$									
O_2	6.10^{-3}		21	0.13	$1.5.10^{8}$						
O_3	10^{-6}		$10^{-8}\text{-}10^{-5}$	$3 .10^{-6}$							
CO_2	96	8.10^{7}	3.10^{-2}	95.32	10^{11}						
CO	5.10^{-5}	$1.7.10^{7}$	10^{-6}	0.07	$0.5\text{-}1.10^{9}$						
H_2O	$10^{-3}\text{-}10^{-2}$		$10^{-6}\text{-}10^{-2}$	0.03		10^{-3}					10^{-4}
NH_3						10^{-4}	2.10^{-4}				
PH_3						$+$	1.10^{-6}				
CH_4			2.10^{-4}			6.10^{-4}	8.10^{-4}	0.067	0.056	3.10^{-2}	
CH_3D						$+$					
GeH_4						$+$					
C_2H_6						4.10^{-2}	5.10^{-6}			2.10^{-5}	
C_2H_4										8.10^{-7}	
C_2H_2						$+$	2.10^{-8}			5.10^{-6}	
C_3H_4							$+$			6.10^{-8}	
C_3H_8							$+$			1.10^{-5}	
SO_2	6.10^{-2}										0.02
HCN										5.10^{-7}	
NO					5.10^{-6}						
HCl	5.10^{-7}										
HF	10^{-8}										
Unit	vol-%	g/g	vol-%		cm^{-3}	←	— mole-fraction —		→	atm-cm	

At the Earth presently some anthropogenic changes take place in the atmosphere due to the burning of fossil fuels and the injection of nitrous oxides as well as chloro-fluoromethans. These gases cause a slowly increasing greenhouse effect. We do at this time no know whether this will lead to irreversible changes in the atmosphere, something which has to be classified as evolution, or whether this is a small "noise" in the timescales considered here (45).

Nevertheless, there is a certain variability in these atmospheres, they are not "final". At Jupiter, as we have seen, obviously the solar H : He abundance is conserved. Here we are probably dealing with the less evolved planetary atmosphere. Saturn's atmosphere has lost some He due to interaction with lower layers. A modification of the atmospheres of the giant planets is probably an extremly slow process so that we can regard these atmosphers as "quiet" in an evolutionary sense.

REFERENCES

1 Coradini, A.: Primordial phases, this volume.
2 Coradini, A.: Early thermal history of planets. This volume.
3 Wänke, H.: This volume.
4 Bodenheimer, P.: This volume, p. 25.
5 Turner, G.: This volume, p. 79, 85.
6 For references on the orign of planets see also the cited contributions (1-5) of this volume.
7 Walker, J.C.G.: 1977, Evolution of the atmosphere, Macmillan Publ. Co., New York.
8 Unsöld, A.: 1974, The chemical evolution of the galaxies. Proc. First Europ. Astron. Meeting. Vol. III, Springer-Verlag Heidelberg.
9 Windley, B.F. (ed.): 1976, The early history of the Earth. Chichester: Wiley.
10 Pollack, J.B., and Black, D.C.: 1979, Science 205, pp. 56-59.
11 Rubey, W.W.: 1963, Geol. Soc. Amer. Spec. Paper 62, pp. 631-650.
12 Turekian, K.K.: 1972, Chemistry of the Earth. Holt, Rinehart and Winston, N.Y.
13 Pellas, P.: 1982, this volume, p. 95.
14 Reeves, H. (ed.): 1974, on the orign of the Solar System. Conf. Rep.; Nice, 3-7 April 1972. Paris: CNRS.
15 Hunten, D.M.: 1981, Symp. on the origin and evol. of planet. atm., IAMAP, Hamburg.

16 Ward, W.R.: 1974, J. Geophys. Res. 79, pp. 3375.
17 Ward, W.R., Murray, B.C., and Malin, M.C.: 1974,
     J. Geophys. Res. 79, pp. 3387-3395.
18 Masursky, H., et al.: 1977, J. Geoph. Res. 82,
     pp. 4016-4038.
19 Carr, M.H.: 1979, J. Geoph.Res. 84 B, pp. 2995.
20 Abelson, P.H.: 1966, Proc. U.S. Nat. Acad. Sci.
     55, pp. 1365-1372.
21 Oeschger, H., et al.: 1980, Das Klima. Springer-
     Verlag, Berlin, pp. 209-236.
22 Stevenson, J.S.: 1981, The formation of gaseous
     giant planets. Symp. on the orign and evolution
     of planet. atm. IAMAP, Hamburg.
23 Hubbard, W.B.: 1981, Science 214, pp. 145-149.
24 Fahr, H.: 1981, Die Bildung des Sonnensystems:
     Versuch einer Deutung. Phys. Bl. 37, pp. 142.
25 Bolle, H.-J.: 1981, Proc. Summerschool Alpbach,
     1981. ESA, SP, in press.
26 U.S. Standard Atmosphere, 1976. NOAA, NASA, USAF,
     Washington.
27 U.S. Standard Atmosphere Suppl. 1966. Wash., USA.
28 Marov, M.Ya., et al.: 1973, J.Atm.Sci.30,pp. 1210.
29 Seiff, A., et al.: 1979, Scie.205, pp. 46-48.
30 Taylor, F.W., et al.: 1980, Geoph.Res.85 A,pp.7963.
31 Avduevsky, V.S., et al.: 1973, J. Atm. Scie. 30,
     pp. 1215-1217.
32 Tomasko, M.G., et al.: 1979, Scie. 205, pp. 80-82.
33 Seiff, A., et al.: 1977, J. Geoph.Res.82, pp.4364.
34 Hanel, R., et al.: 1972, Icarus 17, pp.423-442;
     1981: Forier spectroscopy on planetary missions
     including Voyager. Proc. Soc. Photo-Opt. Instr.
     Eugrs (SPIE).
35 Trafton, L.: 1981, Rev. Geophys. Space Phys. 19,
     pp. 43-89.
36 Hanel, R., et al.: 1981, Science 212,pp. 192-200.
37 Feston, M.C., et al.: 1981, J.G.R. 86, A7, pp.5715.
38 Nier, A.O., et al.: 1977, J.G.R. 82, pp. 4341.
39 Niemann, H.B., et al.: 1979, Scie. 205, pp. 54.
40 Owen, T., et al.: 1977, J.G.R.82, pp. 4635-4639.
41 Gautier, D., et al.: 1980, NASA Techn. Mem.80722,
     GSFC, Greenbelt, Maryland.
42 Cahalau, R.F., and North, G.R.: 1979, A stability
     theorem for energy-balance climate models.
     J. Atm. Scie. 36, pp. 1178-1188.
43 Fraedrich, K.: 1978, Structural and stochastic
     analysis of a zero-dimensional climatic system.
     Quart. J. Roy. Met. Soc. 104, pp. 461-474.
44 Cordell, B.M.: 1980, Geophys. Res. Letters 7,
     pp. 1065-1068.

CIRCULATION SYSTEMS IN PLANETARY ATMOSPHERES

Hans-Jürgen Bolle

Institut für Meteorologie und Geophysik
Universität Innsbruck

ABSTRACT

The physical processes which generate motions in planetary atmospheres are shortly discussed. The application of the basic principles is illustrated by (i) the relation between the thermal structure of a planetary atmosphere and its wind system and by (ii) the need for CO_2 transports in the Martian atmosphere. The importance of radiative processes for the observed phenomena can be estimated by means of a scale analysis. An overview is given on the circulation systems observed in different planetary atmospheres and their possible mechanisms. Finally the problem of modelling these dynamical systems is shortly stressed.

Key words: Atmospheric Circulation, Planetary Atmospheres

SYMBOLS

a	acceleration	u,v,w	wind components in x,y and z direction
g^*	acceleration due to gravitation	u_g	geostrophic wind
g	acceleration due to gravity	u_*	friction velocity
x	zonal coordinate	r	radius
y	meridional coordinate	Ω	angular velocity
z	vertical coordinate	t	time
z_o	roughness length	φ	geographical latitude
$\vec{v}$	velocity vector	f	Coriolis parameter ($2\Omega \sin\theta$)
		ρ	density

A. Coradini and M. Fulchignoni (eds.), The Comparative Study of the Planets, 221–252.

m	mass	F_{Ce}	centrifugal force
p	pressure	λ	wavelength
T	temperature	Ro	Rossby number (u/L Ω)
Θ	potential temperature	R_i	Richardson number
Γ	temperature gradient	N	Väisälä – Brunt frequency
R	universal gas constant	Q	Energy
R'	individual gas constant (R/M)	Φ	flux, power
		c_p	specific heat capacity
M	molecular mass	H	scale height
K_m	kinematic eddy viscosity	τ	characteristic time
k	von Karman constant	c	speed of light
d	characteristic height of Ekman layer	h	characteristic height
		O	order of magnitude
F_p	force of gravity	τ	stress
F_G	pressure gradient force	ϕ	geopotential
F_C	Coriolis force	s	specific entropy
F_F	friction force	h	diabatic heating rate

INTRODUCTION

Observations of the planets by different techniques (
reveal that in their atmospheres different types of
circulations are manifested and that sometimes very
vigorous transports obviously occur. At Venus, Jupiter
and Saturn the movements in the upper atmospheres can
be observed and studied by optical remote sensing
techniques. The cloud decks obscure the observation
of lower regions which are only accessible by probes
or microwave techniques. Earth, Mars and the moon
Titan are only partly covered by clouds so that in these
cases the whole atmosphere is much more open for detailed studies of its dynamical characteristics.

Motions in atmospheres are the result of *external
forcing*. This forcing can be imposed by differential
heating due to absorption of solar radiation, gravitational attraction by the sun, other planets or moons,
momentum transfer or heating from the lower boundary.
The response of an individual atmosphere by developing
a certain circulation regime depends not only on the
forcing function but also on planetary rotation, the
position of the spin axis relative to the ecliptic, the
structure and roughness of the lower boundary, the types
and concentrations of gases contained in the atmosphere
as well as the presence of clouds and aerosol which
interact with the radiation field.

The range of the planetary parameters that influence atmospheric motions is very broad. We can there-

fore expect a variety of different circulation systems.
The inclination of the spin axis, e.g., varies throughout the solar system between 3 degree for Venus to 82.5
degree for Uranus, the period of rotation between about
10 hours for Jupiter and Saturn to 243 days retrograde
for Venus, and the density at the lower boundary by
several orders of magnitude (2).

Notwithstanding these differences all motions in
planetary atmospheres follow the same physical principles
and *models* of the dynamical behaviour of the planetary
atmospheres can be constructed if the boundary conditions
and input parameters are known with the necessary accu-
racies: one goal for planetary missions. Though numerous
details are already known about the atmospheres of
different planets there is still a lack in the understanding of the generation of some of the observed
basic features. Therefore future planetary missions
have to be designed thoroughly to provide the clues
for the answers to such outstanding questions.

The study of the dynamical behaviour of planetary
atmospheres is an essential tool for the understanding
of atmosphere - surface interactions, it is necessary
for modelling mass transports which affect chemical
reactions and escape probabilities. Furthermore
studies of planetary atmospheres allow us also to understand better some properties of the Earth's atmsophere
where a number of phenomena occur simultanously which
can be observed in a more undisturbed appearance in
the atmospheres of other planets. The study of the
dynamics of planetary atmospheres is furthermore
of great intellectual stimulation in the respect to
assess the physical state of an atmosphere - surface
system, its *climate,* with a minimal number of observations which are then used in physical - mathematical
models, constructed to simulate the time dependence
of atmospheric behaviour.

PHYSICAL PROCESSES THAT LEAD TO MOTIONS

An atmosphere would only be motionless with respect
to the surface of a rotating planet if all molecules
are rotating with the same angular speed as the surface.
As soon as a moleculetravels out of its original latitude
in anequipotential surface it has in its new location
a relative velocity with respect to the planetary surface. Immediately an apparent acceleration due to the
rotation of the planet will be effective which forces
the molecule into a curved orbit relative to the

surface. To conserve mass in a volume element another
molecule has to replace the excursed one and consequently exchange processes are initiated by which
generally momentum is transferred. If an atmosphere is
heated from below, like it is the case on Jupiter,
then *buoyancy* will occur which turns over the gas
vertically.

The static equilibrium is disturbed very efficiently
if *horizontal pressure gradients* build up because of
differential heating due to absorption of solar radiation
If we have a pressure difference of 1 mb = 10^2 Pa =
= 100 N m^{-2} over, say, 1000 km, the acceleration

$$a = \frac{1}{\rho} \frac{\partial p}{\partial x} \tag{1}$$

is 10^{-3} m s^{-2} if the gas density is taken to be 1 kg m^{-3}
(the approximate value near the Earth's surface). If this
gradient is maintained for 1 hour, a gas parcel will
gain a velocity of 3.6 m s^{-1} if no other forces are at
work.

We can consider that a heat source and a sink be
brought into an atmosphere. If the source is placed
some distance above the sink then after a possible
initial temperature exchange the result will be a medium
with less dense masses at higher temperatures on top of
colder and denser masses on the bottom, like in the
oceans. This is a stable situation. If the reverse is
the case, the denser masses at the bottom will perma-
nently be heated and expanded so that buoyancy develops.
The gas which is cooled at higher altitudes and lower
pressures will sink down because its density increases.
From experiments and the consideration of a thermodynamic process in such a situation Sandström (3)
derived the theorem that *a closed steady circulation
can only develop if the heat source is situated at a
higher pressure than the sink* (compare Fig.1)

An important quantity is the geographical distribution of the *radiation budget* of a planet. If the
planet does not produce internal heat and is in
equilibrium with the solar radiation, then there will be
excess absorbed solar energy in equatorial (sub solar)
regions. The sinks due to infrared emissions are more
equally distributed between the poles and the equator
and their variability between day and night, winter and
summer are smaller than the variability of insolation.
In a stationary state therefore heat has to be transported from the regions where energy is deposited in
excess to those where it is lost to space as infrared

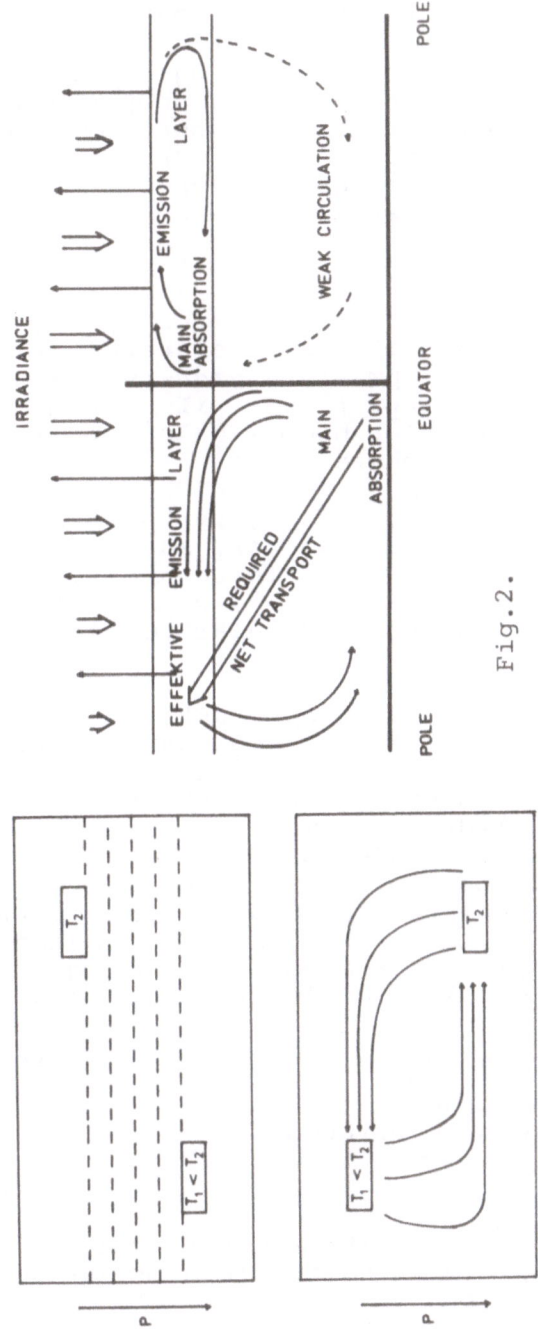

Fig.2.

Fig.1.

Fig.1.: Illustration of the Sandström Principle. A closed steady circulation develops only, if the heat source is placed at higher pressures than the sink. Otherwhise a stable temperature stratification will be established. After Defent (3).

Fig.2.: Energy transport in a transparent (left) and an opaqme (right) atmosphere to maintain a steady state.

radiation. Depending on the transparency of the atmosphere the solar energy will be absorbed near the surface in the case of a less cloudy planet (Mars, Earth) or at higher levels in cases of high lying cloud decks (Venus). This gives rise to more or less deep circulations (Fig.2) since an efficient transport of heat is only possible if it is accompanied by a mass exchange (heat conduction is negligible against large scale transports and turbulent exchange).

ACCELERATING FORCES

The forces which accelerate or decelerate the gas are *gravity* (potential, F_P), *pressure gradient* (F_G), *Coriolis* (F_C), and *friction* (F_F) *forces*, if we neglect electro-magnetical forces which become only important in the upper (ionized) atmosphere.

The *acceleration due to gravity* g is the sum of the acceleration g^* due to the planetary mass (gravitational force) and the centrifugal force due to the rotation of the planet ($\vec{g} = \vec{g}^* + \Omega^2 \vec{r}'$, Ω angular velocity, $\vec{r}$ radius vector perpendicular to planetary spin axis).

The *Coriolis force* (5) is an apparent force which acts on a *moving* mass in a rotating coordinate system. The velocity of a gas parcel is $\vec{v} + \vec{\Omega} \vec{r}$ where $\vec{r}$ is the radius vector from the center of the earth. For the acceleration $d(\vec{v} + \vec{\Omega} \vec{r})/dt$ one component results which is directed at right angle to the velocity vector, the acceleration due to the Coriolis force, and one component normal to the planets spin axis, the *centrifugal acceleration*. The latter can often be neglected at least at the inner planets against other accelerations. The magnitude of the Coriolis acceleration is

$$a_C = 2v\Omega \sin\theta = f v \tag{2}$$

with θ geographical latitude.
The *frictional force* can be written

$$F_F = \frac{\partial}{\partial z} \left(K_m \frac{\partial \vec{v}}{\partial z} \right) \tag{3}$$

K_m is the kinematic eddy viscosity (eddy viscosity divided by the density ρ). In an adiabatic atmosphere it can e.g. be expressed by

$$K_m = \frac{k^2 z(\bar{u} - \bar{u}_1)}{\ln(z/z_1)} \tag{4}$$

where k (= O.4) is a constant (von Karman constant) and
$\bar{u}$, $\bar{u}_1$ the mean wind speed at two levels z and z_1. The
mean wind speeds depend on the structure of the sur-
face which can be parameterized by an empirically to
be determined *roughness length* z_0 and a *friction
velocity* u$_*$

$$u = \frac{u_*}{k} \ln \frac{z}{z_0}$$ (5)

In an atmosphere which is not adiabatically structured
K_m depends also on the temperature profile in the
first km of the atmosphere. The layer through which
surface stress is still effective is called *Planetary
Boundary Layer* (PBL).

     Depending on how these different forces are com-
bined one can distinguish different cases which are
summarized in Fig.3. The *geostrophic* approximation
applies in the free atmosphere where frictional forces
can be neglected and if the pressure system is extended (ridges and troughs). In the vicinity of transient
low or high pressure systems the stream lines are curved
and the centrifugal force F_{Ce} becomes important (d $\vec{v}$/dt =
= $\vec{n}|v|^2/r$, $\vec{n}$ unity vector normal to velocity). This is
the *gradient wind* case. If $F_{Ce} \gg F_C$ (e.g. near the
equator where $\sin\theta \to 0$), F_{Ce} balances F_G and the wind
is *cyclostrophic*. For an *inertial flow* ($F_G \approx 0$) a
balance between Coriolis and centrifugal force must be
established which requires a curved flow with a radius
of curvature of $-|\vec{v}|/f$.

If friction gets into the picture, near the surface, the
wind is necessarely directed into the low pressure
region because now the pressure gradient force has to be
balanced by the sum of the Coriolis and the friction
force. The latter one is always directed opposite to the
velocity vector while $F_C \perp \vec{v}$.

     The wind veering due to the action of surface stress
and subsequent turbulent momentum transport into the PBL
has first been treated theoretically by Ekman (6). In
order to maintain the continuity of mass, the influx
in the PBL into a low pressure system forced by the
action of friction must be compensated by an outflux
at higher levels. The gas is pumped through the low
pressure system across the isobares (Ekman pumping).

     At planets with orographically structured solid
surfaces like Mars and Earth differential heating gene-

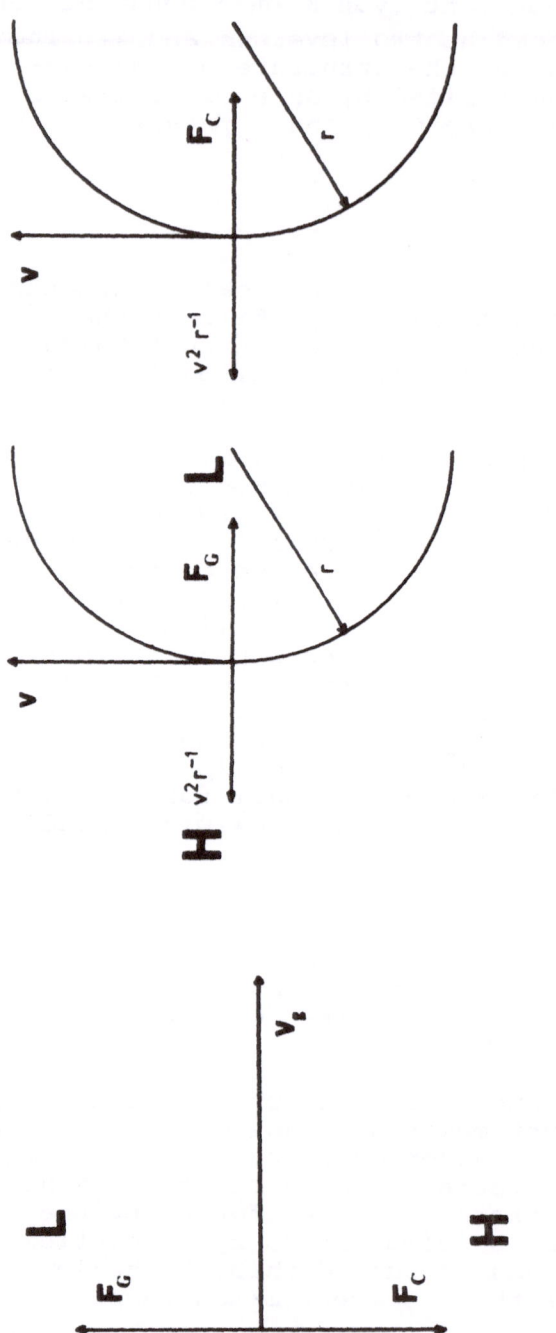

Fig.3. Equilibrium conditions for geostrophic flow (left), cyclostrophic flow (middle) and innertial flow. F_G = gradient force, F_C Coriolis force, r radius of curveture.

rates   small scale circulations in mountainous regions.
Upslope flows occur during maximum solar irradiance
while at night gas cooled on top of higher areas
sinks down into the valleys or the plains due to its
higher density. Such "catabatic" winds are e.g. generated
over high polar plateaus during winter when they act
as cold trap.

If a flow leads over a mountain ridge its velocity
increases because the layer over which the flow ex-
tends is narrowed. Due to this increased velocity also
the Coriolis acceleration increases and the flow is
deflected (to the right in the northern hemisphere of
a prograde rotating planet). Behind the mountain the
curvature is in the opposite direction. Superimposed
can be the effect of the latitudinal variation of the
Coriolis force (β-effect). This can lead to a net
change of the direction of the wind, and moreover to the
generation of waves.

SIMPLE APPLICATIONS OF THE BASIC PRINCIPLES

Temperature - Wind Relation

If we consider the geostrophic wind with equilibrium between Coriolis and pressure gradient forces
for a zonal wind,

$$u_g = -\frac{1}{\rho f}\frac{\partial p}{\partial y},\tag{6}$$

then, with the help of the state equation for an
ideal gas, $\rho = p/R'T$, R' individual gas constant, we
can transform this relation into

$$u_g = -\frac{R'T}{f}\frac{\partial \ln p}{\partial y}.\tag{7}$$

From the hydrostatic equilibrium equation
$\partial p/\partial z = -g\rho$ we derive, using again the state equation,

$$\frac{\partial^2 \ln p}{\partial z \partial y} = \frac{1}{T^2}\frac{g}{R'}\frac{\partial T}{\partial y}\tag{8}$$

If now e.g. (6) is differenciated with respect
to the height z and if for the r.h.s. use is made
of (3),   the following relation can be derived

$$\frac{\partial u_g}{\partial z} = -\frac{g}{fT}\left(\frac{\partial T}{\partial y}\right)_p = \text{const.}\tag{9}$$

which is called the *"thermal wind equation"*. It says:
if there is a meridional temperature gradient with
increasing temperatures towards the equator, then the
wind velocity increases with height proportional to the
temperature gradient scaled by the absolute temperature
and inversely proportional to the Coriolis parameter.
The wind speed is reduced with height if the temperature
gradient is reversed. Assuming that the wind velocity
is zero at the surface one can estimate the geostrophic
wind velocity at level z by integrating equation (9)
from the surface to z. This is only an estimate since
in the PBL relation (9) does not hold anymore. In
regions where large temperature gradients occur therefore
strong winds are present. This can give rise to jet
streams in the upper troposphere which have strong impact
on the generation and movements of transient eddys
(cyclones).

CO_2 - Transport on Mars

In has been observed (7, 8) that the surface
pressure on Mars varies with season and has a maximum
if the extent of the polar caps is smallest. The
pressure variations obviously reflect the CO_2 mass
variations due to deposition and release of CO_2 from the
polar caps. Due to the evaporation of the CO_2 at the summer pole first high pressure is build up at high latitudes.
The CO_2 will subsequently be transported to lower latitudes until it is frozen out again at the winter pole. This
process is obviously fundamental in keeping the energy
budget in balance. The Martian wind system will therefore be adapted to maintain this transport. It is
possible to estimate the required wind velocities
by a simple application of the Ekman layer wind relation
(Fig. 4). The CO_2 mass exported from the polar cap
at latitudes $\theta > \theta'$ causes a rise in pressure at the
rest of the planet (the area of the polar cap is
$2\pi r^2 \left| 1-\sin\theta' \right|$):

$$\frac{dm_{CO_2}}{dt} = \frac{2\pi r^2}{g} (1+\sin\theta)\frac{dp}{dt} . \tag{10}$$

Around the polar cap a zonal wind belt can be expected
like in the Earth's atmosphere with the high pressure
at the pole. The wind vector components in the PBL
can be represented by

$$u(z) = u_g \left\{ 1 - e^{-\frac{\pi z}{d}} \cos(\frac{\pi z}{d}) \right\} \quad \text{(zonal)} \tag{11}$$

$$v(z) = u_g e^{-\frac{\pi z}{d}} \sin(\frac{\pi z}{d}) \quad \text{(meridional)} \tag{12}$$

where

$$d = \pi \sqrt{\frac{K_m}{\Omega \sin \gamma}} \qquad (13)$$

is the characteristic height of Ekman layer, and u_g the geostrophic wind above the PBL. Due to the wind veering in the PBL there exists a cross isobasic meridional transport of the magnitude

$$\frac{dm_{CO_2}}{dt} = 2\pi r \cos\theta \; \bar{v} \; \bar{\rho} \; d \qquad (14)$$

with $\bar{v}$ mean meridional wind velocity and $\bar{\rho}$ mean density in the PBL of depth d. $\bar{v}$ can be computed from e.g. (12) by integration over the height of the boundary layer $0 < z < d$:

$$\bar{v} \cong u_g d / 2\pi \qquad (15)$$

If eq. (14) is set equal eq. (10) it follows a condition for $\bar{v}$ and from eq. (15) a condition for u_g. With $d = 3$ km, $\bar{\rho} = 10^{-2}$ kg m^{-3}, $dp/dt = 2 \cdot 10^{-7}$ mb s^{-1}, $r = 3384$ km follows: $\bar{v} \cong 2.3$ m s^{-1}, $u_g \cong 14$ m s^{-1}. These are very reasonable velocities which are in accordance with wind speeds derived from other observations.

WAVES

Orographic obstacles in laminar flows or shears between flows of e.g. different density can cause disturbations to which the flow reacts by the development of waves. The restoring forces can be gravity or pressure gradients.

Let us recall the remarks made earlier with respect to the overflow of a mountain. In most cases the deviations of the flow due to the action of the Coriolis force do not cancel out so that behind the mountain a net deflection remains and the flow is directed into a latitude where the Coriolis force is not in equilibrium with the gradient force. If the gas masses are exported into a region with a smaller Coriolis parameter its velocity is too large and vice versa. In both cases consequently the gas parcel is turned back towards its equilibrium position. Due to inertia it, however, overshoots this position and starts to oscillate around this equilibrium latitude. A wavelike flow pattern develops, the *Rossby waves*. Because the planet has a limited extension, large scale waves are preferred

with wavelengths, which are an integer fraction of
the length of the zonal circle in which they are
generated ("*planetary waves*").

A characteristic number for the generation of
waves is the ratio of velocity u divided by a characteristic length L to the angular velocity Ω, the
Rossby number $Ro = u/L\Omega$. Here L is the distance over
which major changes of u can be observed.

Typical wavelengths are

$$\lambda = 2\pi \sqrt{u_m} \, h/\beta(h - z_m) \tag{16}$$

where u_m is the eastward directed wind component over the
mountain, z_m the height of the mountain,

$$h = \frac{100 \sin^2\theta}{l^2(l+1)^2} \; , \quad l = 1,2,3,\ldots \tag{17}$$
$$\theta = 90 - \varphi \; \text{(co-latitude)}$$

$$\beta = \frac{2\Omega \sin\theta}{r_p} \tag{18}$$

The waves are stationary or their phases propagate
westward relative to the mean zonal wind speed. Their
speed increases with wavelength. Forced Rossby waves
can only be generated in an eastward flow ($u_m > 0$),
but free Rossby waves, formed by random departures from
geostrophic equilibrium, can also be generated in a
westward flow.

A special case can occur if $R_o < 0.5$ (and $u/\sqrt{2\Omega K_m} \ll 1$)
at an overflow of a mountain. Then the conditions are
given that rotating systems develop which are stagnant
in the flow (*Taylor columns*) (9-10)

Another important type of wave is that of *gravity
waves* which are excited in statically stable atmospheres
in the lee of mountains as lee-waves. These are vertical
oscillations of gas parcels. The restoring force for
their density perturbations is gravity. They are
observed if the wind speed is smaller than the square
root of the acceleration due to gravity times a
characteristic lengths (e.g. the height for 50% density
drop): $u < \sqrt{gL}$. The wavelengths are much smaller than
for Rossby waves, and the periodicity is between a
few minutes to a few hours.

The frequency of the vertical oscillations
depends on the stability of the atmosphere. The charac-

teristic frequency is the *Väisälä - Brunt* (VB) *frequency*

$$N = \frac{g}{\theta} \frac{d\theta}{dz} \tag{19}$$

The horizontal wavelengths of these waves (which are sometimes reflected in the cloud structure) is

$$\lambda_h = 2\pi u_o/N \tag{20}$$

where u_o is the speed over the ridge. Gravity waves can also travel upwards and transport energy from the lower to the upper atmosphere. The vertical wavelength is

$$\lambda_v = \frac{2\pi\bar{u}}{\sqrt{RT/MH}} \ . \tag{21}$$

At high windspeeds pressure and inertial forces are dominant and in these cases ($u > \sqrt{g\ L}$) longitudinal *acoustic waves* occur. In the intermediate case ($u \simeq \sqrt{g\ L}$) a mixture of longitudinal acoustic and transverse gravity waves develop. The waves formed internally by the flow itself or interaction with orography are superimposed by atmospheric *tides* which are generated by gravitational forces due to moon-planet or sun-planet interactions or by the periodic insolation.

Kelvin - Helmholtz Instability

In regions where shear becomes very strong the motion may not remain laminar. This is the case, if the velocity gradient ($\partial u/\partial z$) gets larger than a critical value. This situation can be expressed by the *Richardson number*, defined by the ratio of buoyant to kinetic energy input to the shear:

$$R_i = \frac{g}{\theta_v} \frac{\partial\bar{\theta}_v/\partial z}{(\partial\bar{u}/\partial z)^2} \simeq \frac{g}{T} \frac{\Gamma - \Gamma_{ad}}{(\partial\bar{u}/\partial z)^2} \tag{22}$$

θ_v is the virtual potential temperature, a temperature which the gas takes if it is brought adiabatically to 10^5 Pa pressure and if all condensation heat is transferred to the volume. Γ is the actual, Γ_{ad} the adiabatic temperature gradient.

If a formerly laminar flow starts to get unstable the Richardson number drops below 0.25. The critical Richardson number is smaller than 1 because viscous dissipation of turbulent kinetic energy is not included in the definition of Ri. Then waves develop which

eventually break and form rotating gas parcels, to the
effect that the turbulence spins down and the flow is
forced back into its laminar state (Fig.4).

OBSERVED PLANETARY ATMOSPHERIC CIRCULATION REGIMES

Methods of Assessment

In few cases vertical wind profiles have been
determined by probes which drift with the wind.
These measurements provide local and instant wind
profiles which are not necessarely representative for
the mean circulation. They must therefore be supported
by remote sensing methods which are able to cover
larger areas for sufficiently extended time intervals.
By means of image sequences it is possible to determine
wind vectors from the movement of clouds. The speed
of the clouds reflect, however, not always the wind
speed. It happens frequently that the wind flows
through a system more rapidly than this system is moving.
This is especially true for orographic clouds which are
connected to standing waves.

Another possibility is to derive the wind field
from the temperature field using the thermal wind
equation. The temperature field can be obtained from
infrared or microwave emission measurements in selected
spectral bands. The magnitude of the absorption coefficient determines the layer from which the radiation
emerges, and the radiance depends on the temperature
in this layer. By application of mathematical inversion
methods (11) the temperature profiles can be retrieved
from a set of such measurements. Since the thermal wind
equation allows only to determine vertical changes of
wind speeds it is necessary to tie these measurements
to a reference layer. This can be a cloud layer for
which the wind vectors are known from cloud motions or
it can be the surface for which the wind speed is zero,
if the effects of the PBL are properly accounted for.

Scale Analysis

A first estimate of the dominating processes
responsible for the circulation regime in an atmosphere
can be made by an analysis of the time constants involved to dissipate a disturbance (12-14).

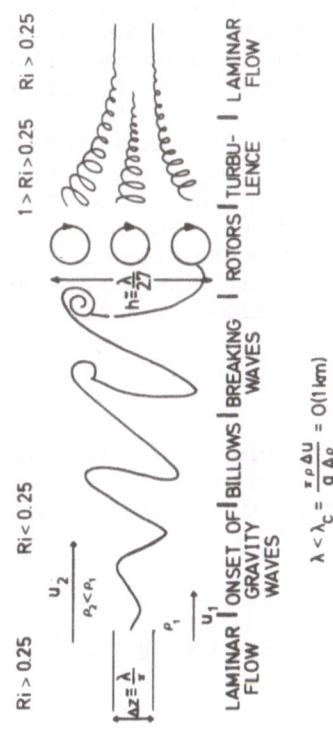

$$\lambda < \lambda_c = \frac{\tau\,\rho\,\Delta u}{g\,\Delta\rho} = O(1\,km)$$

Fig.5.

Fig.4.: Resulting wind vector v(z) in the PBL under the influence of pressure gradient force F_G, Coriolis force F_C and friction force F_F. The wind is directed into the low pressure region (L). V_g is the geostrophic wind outside the PBL where F_F vanishes.

Fig.5.: Generation and decay of Kelvin-Helmholtz-Instability. (After Beer (1)).

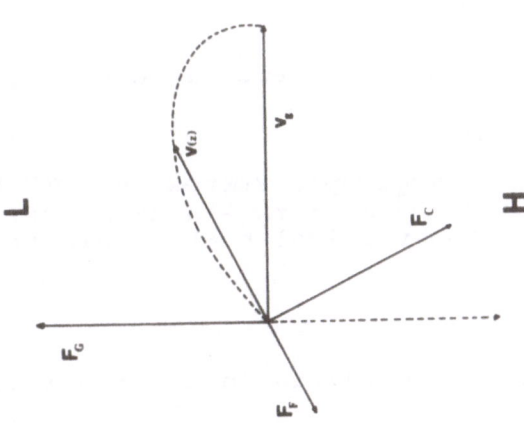

Fig.4.

A characteristic time τ can be defined as the time which elapses until a disturbation is damped to $1/e$ of its initial amplitude or until a reservoir Q of energy is emptied by an outflux $dQ/dt = \phi$ to $1/e$.

For radiation processes one can consider the time τ_{rad} to cool the atmosphere above a pressure level p if the solar heating is suddenly turned off. The heat reservoir is $Q = c_p\, m\, T$ (c_p specific heat, m mass per unit area, T absolute temperature), the flux per unit area $\phi = \sigma\, T^4$ (σ = Stefan Boltzmann constant). As an order of magnitude estimate one can define

$$\tau_{rad} \cong \frac{c_p\, m\, T}{\sigma\, T^4} = \frac{c_p\, H\, p\, M}{R\, \sigma\, T^4} \tag{23}$$

where $H = RT/Mg$ is the scale height, the height of the atmosphere if compressed to the pressure p at its lower boundary (R universal gas constant, M molecular mass, $m = \rho(p) \cdot H = pH/(R/M\ T)$.

A characteristic time for dynamical processes is the propagation of a disturbance (e.g. a temperature anomaly) with the velocity u over planetary dimensions, e.g. the planetary radius r_p:

$$\tau_{dyn} = r_p/u \tag{24}$$

If nothing else is known about the structure of the planetary atmosphere, an appropriate velocity is the velocity of sound.

$$u \cong c = \sqrt{\gamma \frac{RT}{M}} \cong \sqrt{g\ H} \tag{25}$$

Vertical transports can be characterized by

$$\tau_v \approx h\ \sqrt{2/f\ K_m}, \tag{26}$$

the decay time for a barotropic vortex ($h \cong 10$ km, $f = 10^{-4}$ s^{-1}, $K_m = 10^5$ cm^2 s^{-1}; $\tau_v \cong 4$ days) or, if for vertical motion the vertical heat transport is important

$$\tau_v \approx (\frac{\pi R}{h})^2\ \frac{f\theta}{g\cdot\Gamma} \tag{27}$$

Further the length of the day is an important time scale

$$\tau_{day} = q\Omega^{-1} \tag{28}$$

where q is the fraction of rotation which is one day.

The ratio

$$\frac{\tau_{day}}{\tau_{dyn}} \simeq \frac{u}{\Omega} \frac{q}{r_p} \simeq Ro \qquad\qquad (29)$$

is nearly identical with the Rossby number.

Estimates of these parameters show that on Venus below ~40 km

$$\tau_{rad} \gg \tau_{day} \gg \tau_{dyn}$$

and above 56 km

$$\tau_{day} > \tau_{rad} > \tau_{dyn}$$

Further, that $R_O \gg 1$ throughout the Venus atmosphere (with q = 8.44 m s^{-1}, H $\cong$ 15000 m, $u = \sqrt{qH} \cong$ 356 m s^{-1}, Ω = 3 $\cdot$ 10^{-7} s^{-1}, r_p = 6120 km it results Ro $\cong$ 17.

This suggests that radiation processes become essential for the distribution of the temperature adjustment above ~56 km and that below ~40 km dynamical adjustment exists.

For the Earth radiative and turbulent transport relaxation times are of the same order of magnitude (10 days) for disturbations of 1000 km scale length. On Mars with its thin atmosphere the radiative response times are by one order of magnitude shorter. This suggests that radiation has a much stronger influence on the temperature field. In fact the daily temperature wave is very pronounced at Mars. The surface is almost in radiative equilibrium with minimum values at night of 188 K and maximum values at about noon of 244 K (VL). For Mars therefore radiative processes are very important for the temperature distribution.

The geostrophic approximation is valid for large scale phenomena on Earth (O(u) = 10 m s^{-1}, O (f) = 10^{-4} s^{-1}, O(L) = 10^6 m; Ro $\cong$ 0.1) and Mars but not in the atmosphere of Venus (O(f) = 5 $\cdot$ 10^{-7}; Ro $\cong$ 30).

On Jupiter radiative processes are insufficient to maintain the necessary energy transports.

In the following sections the individual regimes will shortly be discussed in more detail.

Earth

At the Earth a large fraction of the solar energy
is absorbed at the surface in the Tropics. This
generates rising air motion in a zone close to the
equator, the Inter-Tropical Convergence Zone (ITCZ).
During ascent the air is cooled adiabatically and
migrates towards both poles in the upper troposphere.
Because of the action of the Coriolis force this meridional flow is bend into a zonal flow near 30° latitude
where the direct polward transport ceases. The cooled
air sinks down and is heated due to the compression.
Part of it flows back to the Tropics in the PBL.

It is evident that a zone of strong temperature
contrast builds up where this warm air borders upon
the cold air which flows out of the polar regions.
Here *jetstreams* develop and also instabilities occur
because air masses of different density and zonal wind
speed confluent. The reaction of the atmosphere to
such shear is the generation of *waves*: cold sectors
protude towards the poles, high pressure ridges
alternate with low pressure troughs. These waves take
such a shape that a net polward energy transport results
In addition travelling *eddies*, cyclones, develop in
this zone, which also add to the polwards transport of
heat by the inclusion of warm air sectors in these
rotating and quickly moving systems. Thus an energy
transport system develops polwards of 30° latitude
which is quite different from that one in the tropics
and which is much more efficient at these high latitudes:
In the tropics we have a circulation around a *horizontal*
axis, the *Hadley Cell*, and in high latitudes there is a
wavelike pattern of standing eddies at which transient
eddies develop with *vertical* axes of rotation. The
influx of warm air into the region polward 30 - 35°
latitude occurs at lower levels while the outflow of
cold air is at higher levels so that in the average
there exists at higher latitudes a net circulation of
opposite direction than in the tropics. The instantanous
picture is, however, much more complicated.

Mars

On Mars the radiative time constant is so short
that the system responses almost imediately to the
daily variation of the solar input. There is furthermore a solid surface with low heat capacity and no
storage in oceans like at the Earth. The winds which
are in geostrophic equilibrium have small influence
on the temperature distribution.

While in the Earth's atmosphere a considerable amount of energy is transported as latent heat (water vapor) and evaporation/condensation processes play an important rôle everywhere (especially in the tropics), the situation is in this respect somewhat different for Mars. Here mainly at the poles the *CO_2 sublimation processes* contribute substantially to the energy budget: cooling to space during winter is partly compensated by release of CO_2 sublimation heat.

A global transport of CO_2 between the summer and winter poles is therefore an essential characteristics of the Martian circulation system, which requires strong westerly winds as we have seen previously. At higher latitudes the atmosphere becomes unstable against (barocline) disturbations. Regionally also mountain induced wind systems develop.

Occasionally these regional winds get so strong that substantial amounts of dust are lifted from the surface. Such local *dust storms* are observed at the edge of the receding south polar cap and in regions where winds generated by tides can be expected to be maximal. The large duststorm of 1977 started in a region (40^oS, 110^oW; Clavitas Fossae) where the topographic influence on the generation of thermally driven winds is believed to be strong (15, 16, 17).

A feedback mechanism between locally risen dust and radiation processes seems to be effective in order to intensify the dust storms. The atmosphere is additionally heated in dusty areas which increases the convective activity similar as the release of latent heat acts in the Earth's atmosphere (especially in the ITCZ). The heating was observed during the Mariner 10 mission when temperature profiles measured by the infrared sounding techniques dropped as the atmosphere got gradually clearer (18).

Venus

Venus rotates very slowly (1 rotation in 243 days, one orbital revolution around the sun in 224.7 days siderial period) around its axis. The area of most intense solar irradiance does therefore change very slowly. Due to the warming at these areas strong winds are generated in the cloud top layer to transport this energy to the antisolar point which is at night for the same long period. Since there are no strong temperature differences between day and night this transport must

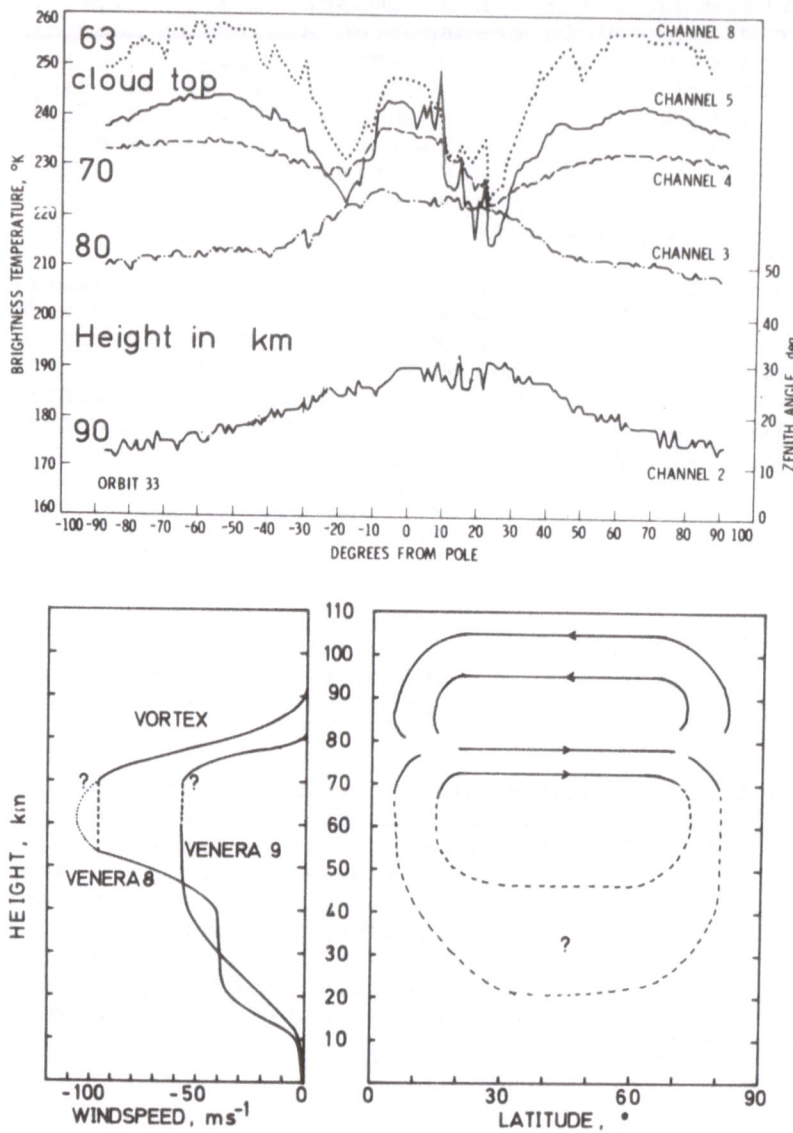

Fig.6.: Effective emission temperatures for infrared channels
tuned to different altitudes, wind profile and probable
circulation regime in the Venus atmosphere after (21,23,24).

be very efficient. It may be suspected for a non-rotating planet that the circulation is radially from the sub-solar point to the anti-solar point. At Venus due to its even very small Coriolis acceleration such a circulation pattern does not occur. Instead there blows a rather steady strong retrograde (east to west) zonal wind of about 100 m s^{-1} velocity in the cloud top layer as seen in ultraviolet light (19). A meridional component, directed poleward, is about one order of magnitude smaller (~10 m s^{-1}). The wind circles the planet in about 4 days (~1.5 radian/day). The observed prominent Y-feature attached to this circulation system seems to be a wave phenomenon rather than a motion system (20).

Near the poles, at about 70° latitude a bright cloud band is often visible which is, as confirmed by infrared measurements, a high reaching cloud system, obviously connected to a circumpolar vortex. It appears and dissappears in time spans of several months (21).

Because of the high wind speeds only a *cyclostrophic* balance can be envisaged at high latitudes. Taylor et al. (21) find, however, departures from cyclostrophic adjustment at lower latitudes and conclude that meridional transport or eddy viscosity must be responsible for this departure.

At the poles probably a sinking motion exists because the emission temperatures increase from about 220 K to 240 K or more. It may even be that the cloud cover is strongly or entirely removed in this area. (fig. 6). The wind speed is maximum in the regions of the cloud tops (60-70 km) and drops quickly on top of this layer (to zero at about 90 km). In the atmosphere above the cloud top an "anti-Hadley" circulation seems to exist with rising air over the poles and sinking air over the equator (22). Below the cloud deck a slow reduction of the velocity (nearly linearly with height) seems to start at 50 km (Venera 8 and 9 measurements) (23). In the lowest few kilometers nearly no wind is present, it drops to 0 - 0.5 m s^{-1}. There is so far no indication of any wind reversal in the lower atmosphere.

The measurements of the Pioneer Venus probes indicate small equatorward wind components in the 10 - 35 km altitude range and a reversal in the cloud layer around 40 km (24).

An interesting distribution has been noted in the

8 - 14 µm wavelength effective emission temperature.
This can be identified with the cloud top temperatures.
It turns out that maximum mean temperatures (252.5 K)
occur at the equator at about 22 hours local time.
There is a secondary afternoon maximum (250 K) around
14:30 hours and minimum (248.5 K) at the terminators
(25, 26). The amplitude of the temperature variations
increases at higher latitudes and also the phase shifts.
At 60° S the maximum (250 K) occurs at about 16:00 hours
and the minimum (236 K) at 04:00 hours local time. This
temperature distribution does not directly reflect
the irradiance distribution and dynamical adjustment
must play an important rôle. The measured temperature
distribution could e.g. be determined by variations
in the cloud top heights. The termination of convection
after sunset can lower the cloud tops into warmer
levels before finally radiative cooling starts from
high latitudes.

Jupiter

     The dark belts and bright zones of Jupiter look
as if they are formed by similar mechanism as roll
vortices which occur in the Earth PBL where sometimes
extended cloud streets (parallel lines of cumulus
clouds) develop at the rising branches of these rolls.
Due to the large Coriolis forces these rolls would have
a helical zonal structure. Thus the Jupiter atmosphere
would suggest to be broken up into a series of Hadley
and anti-Hadley circulations.

     In fact, on Jupiter, similar as in the Earth's
tropical belt, the atmosphere is heated from below (27).
At the Earth by the solar energy absorbed at the surface
and at Jupiter by the release of internal heat. The
radiation budget of Jupiter is not balanced (as it is
for the inner planets). There is an excess of thermal
infrared emission. The energy from the interior of
the planet has to be transported upward into the
main emission layer. Such transports are maintained
by turbulent exchange processes as they occur in the
PBL of the Earth's ITCZ from which more or less
organized convective systems (cloud cluster) develop.
In the Jupiter atmosphere this upward transport must
occur everywhere since the release of internal heat
is not confined to the tropics. In fact at some
Jupiter images at which the polar regions are visible,
irregular cell structures can be recognized at high
latitudes where the zonal structure fades out.

Jupiter's dynamics can therefore basically be looked at as a *turbulent vertical heat exchange system* (28) with less need for meridional transports because radiation does not play a dominating rôle, especially not in the deeper atmosphere. Zonal velocity components develop under the action of the strong Coriolis force which does not allow large excursions from the equilibrium position. The observable latitudinal structure must be connected to vertical motions and associated chemical reactions (e.g. of phosphorus compounds). But there is presently no easy way to demonstrate the generation of this structure. The "roll" - model would require certain regular wind structures related to the boarders of the belts and zones which are not detected in the way such a model would predict it. There are strong prograde jets alternating with narrow retrograde wind systems. Fig. 7 demonstrates their relations to the belts and zones observed at that time (29).

The wind structure shows strong shear zones and the appearance of the clouds suggests that here *Kelvin - Helmholtz instabilities* develop. This gives rise to numerous isolated vortices of which the large Red Spot is the most prominent one. The Red Spot in the southern hemisphere is a high pressure system according to its counter-clockwise rotation. It exposes low temperature at 150 - 200 mb (1.5 - 2 $\cdot$ 10^4 Pa) of e.g. 111 K against 114 K in its surrounding (Voyager 1) but is not apparent as a thermal anomaly at about 500 mb. This suggests sinking motion or convergence at the lower pressures in the haze-top level or a cold cyclonic system on top of a warm anticyclone below the 800 mb (cloud top layer) which is an only dynamically stable system. In fact at about 800 mb a small temperature increase is just discernible at the area of the Red Spot (30, 31). From motion pictures the rotation around the Red Spot does not look like a converging or diverging circulation in the cloud top layer. It rather behaves like a self consistent rotor which sometimes interacts with the surrounding circulation but is not part of the general circulation. Thus it looks like a Tayler column though there is no evidence that it may be generated by an obstacle since the lower boundary of the atmosphere can not be a solid surface.

Other spots detected at Jupiter seem to have opposite thermal structure as the Red Spot and from the Voyager images it must be conlcuded that spots are a common feature in the Jovian circulation regime.

Saturn and Titan

The *Saturnian* atmosphere is similar to the Jovian
one. Also here an internal heat release is an important
energy source for atmospheric circulations, and again
the Coriolis force gives rise to strong zonal winds,
exceeding the velocities measured at Jupiter (Fig. 8)

It is of great interest that Saturn's Moon *Titan*
has an atmosphere which is only a little denser at the
surface than the Earth's atmosphere. By application of
the thermal wind equation to the temperature field
retrieved from spectral infrared measurements (33) a
circulation model for Titan's atmosphere is obtained
which consists in the troposphere of a Hadley Cell
like circulation with rising gas at the equator on the
sinking branches near or at the poles. In the upper
atmosphere the rising tropical gas masses are first
deflected to the summer pole and then turn at pressures
in the order of 1 mb to the north pole where they
decent. Thus rather strong equatorward winds can be
expected in the PBL of the winter hemisphere.

ATMOSPHERIC CIRCULATION MODELS

Basis for modelling atmospheric motions

The motion of a gas parcel in a rotating atmosphere
is determined, as we have seen in an earlier section,
by the acceleration which different forces impose on
the mass. If we exclude electric and magnetic forces –
which become important in the upper ionized atmospheres
– the force per unit mass is given by the equation of
motion (or momentum equation):

$$\frac{\partial \vec{v}}{\partial t} = -\nabla \Phi - \frac{\nabla p}{\rho} - 2\vec{\Omega} \times \vec{v} + \frac{1}{\rho} \frac{\partial \vec{\tau}}{\partial z} \qquad (30)$$

where the terms on the r.h.s. represent the gravitational,
pressure gradient, Coriolis and friction forces. τ is
the stress, or the product of kinematic eddy viscosity
and the vertical wind profile,

$$\vec{\tau} = K_m \, (d \vec{v} / dz) \qquad (31)$$

which represents the momentum transfer between the ground
and the atmosphere as well as between the different
layers of the atmosphere.

In solving this equation it has to be taken care
that mass and energy are conserved in the system. These

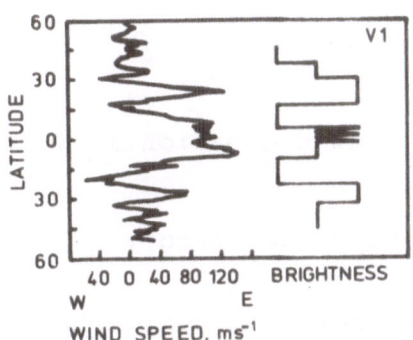

Fig.8. Cloud motion speed for Saturn after Bauer (32)

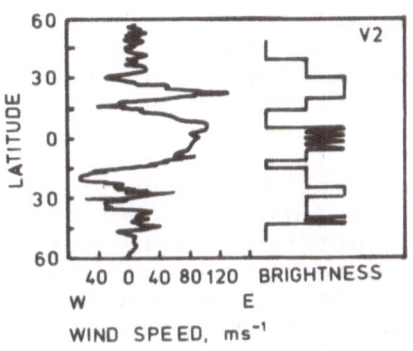

Fig. 7. Cloud motion speeds derived from image sequences from Voyager 1 respectively 2 after (29) compared with a three level brightness scale for the zones in which the speed was obtained.

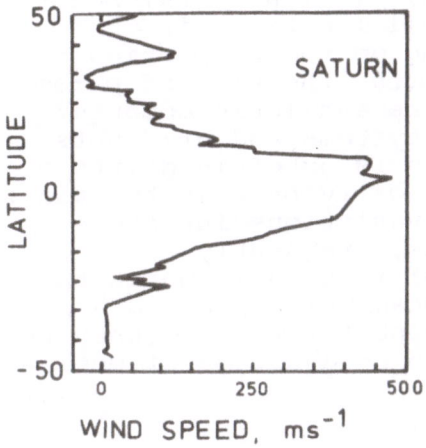

requirements can be expressed by the *continuity equation* (mass conservation):

$$\frac{\partial \rho}{\partial t} = - \text{ div } \rho \vec{v} \tag{32}$$

and the *thermodynamic* equation

$$\frac{ds}{dt} = \frac{h}{T} \tag{33}$$

In addition the hydrostatic equation is valid

$$\frac{\partial z}{\partial p} = - \frac{RT}{gp} \tag{34}$$

In eq. (33) s is the specific entropy

$$ds = c_p \, d \, \ln\Theta \tag{35}$$

and h the heating rate per unit mass. The thermodynamic equation states that the temperature of a gas parcel is changed by pressure changes, which enters through the pressure dependence of Θ, and by external heat sources.

The term h includes the heating rates due to absorption of solar radiation, cooling rates due to infrared emission to space, condensation and evaporation and heating due to the outward transport of heat released from the interior of the planet (as applicable to Jupiter and Saturn). In cases where the solar energy is first absorbed at the ground (Earth, Mars) and then transferred into the atmosphere by secondary processes (heat conduction, latent heat transfer), these processes must carefully be simulated in the models. Often simplified physical models have to be used for the implementation of such processes, which is called "parametrization".

The set of equations (30), (32), (33) and (34) has to be solved, in principle, for all locations in the planetary atmosphere. Depending on the properties of the planets and their atmospheres, the kind of param- terisations involved, and on the available computer capacity different coordinate systems and grid nets are in use (34). The atmosphere is subdivised into a number of vertical layers, typically between two and ten. Since the handling of density variations introduce sometimes an additional problem, frequently the Boussinesq - Approximation (35) is used in which the density ρ is replaced by its mean value $\overline{\rho}$ and $\overline{\rho} - \rho$ is neglected in the equations except for the buoyancy term $\nabla \Phi$ in eq. (30). This results in an equation of motion

$$\overline{\rho} \, \frac{d\vec{v}}{dt} = - \rho \, \nabla \Phi - \nabla p - 2\vec{\Omega} \times \overline{\rho} \vec{v} + \frac{\partial \vec{\tau}}{\partial z} \tag{36}$$

Attempts to model planetary atmospheric circulations

For the Earth very sophisticated numerical simulation models have been developed (36), (37), which are
used for numerical weather forecast and are fed with
thousands of weather observations each day. It is
obvious that such effort leads to a very detailed
description of the general circulation system.
Nevertheless small scale phenomena still sometimes "fall
through the mashes" as the success rate of the daily
weather forecast demonstrates.

It is therefore much more difficult to model planetary atmospheres of which much less observational
details are available. It is especially difficult to
model the deposition of solar energy in the different
layers of the atmosphere, the interaction between atmospheres and the ground in the boundary layer, the eddy
diffusion of heat or the magnitude of condensation and
sublimation processes.

The time scales and basic dynamical and radiation
properties of the Venus atmosphere have been evaluated
by Stone (38). A three dimensional circulation model
for the Venus atmosphere has first been developed by
Kálnay de Rivas (39 a, b) by which it could be demonstrated that even small Coriolis forces break up a
one cell circulation from the sub-solar to the antisolar point into a meridional circulation regime between
the equator and the poles. She could also prove that
the high temperatures at the Venus surface can only be
explained by the greenhouse effect and that the circulation is confined to a top layer if the solar radiation
is absorbed in this layer.

With more refined radiative transfer and heating
rate calculations Pollack and Young (40) could
calculate a vertical temperature profile in agreement
with observations.

The upper stratosphere of a non-rotating planet
Venus where radiative properties become dominant has
been studied by Dickinson (41) and Dickinson and
Ridley (42) who arrived at a direct dayside to night-
side circulation with a return flow at lower levels.
Rossow and Williams (43) examined the region near the
cloud tops. They concluded that the flow resembles that
of a vorticity conserving system with small external
forcing. At a slowly rotating planet the stratospheric
flow is still dominated by the solid body rotation.

A planetary wave with the wavenumber one should be generated which can be recognized in the Y-feature in the cloud structure.

So far the generation of the 4-day retrograde rotation of the Venus atmosphere is not explained satisfactory and Rossow and Williams state:"When the large-scale circulation is properly described, we still may not be able to deduce the nature of the processes driving these motions because the nonlinear balances of the flow are characteristic of the inertial cascade process rather than the forcing. To isolate the forcing processes will require very accurate, highly resolved observation of the smaller, less energetic scales."

For the Mars atmosphere elaborate model computations have been performed by Leovy and Mintz (44) and more recently by Pollack et al.(45). The Martian circulation resembles that one of the terrestrial regime in so far as the generation of low and high pressure systems is concerned, though the exchange between the two hemispheres is much stronger. Detailed budgets of momentum and energy transports have been derived. As for the Earth detailed modelling of the atmosphere-surface interaction is necessary in order to understand quantitatively the observed phenomena (46).

A new approach has been made by Williams (47, 48, 49) to model the Jovian and Saturnian circulation system by means of a quasi-geostrophic model. In his model the circulation system is generated by the same mechanism which is effective at the Earth, baroclinic instability, which involves an inclination of the temperature field with respect to the pressure field. Turbulence is introduced by stochastic forcing functions and the interaction of turbulence and Rossby waves which are generated from the forced zones determine the width and banded structure of the circulation, which is favoured due to the absence of a solid lower boundary.

The general result of these theoretical studies is that the observed phenomena can be simulated if the different parameters which enter the computations like heating rates, eddy diffusion coefficient, temperature field are adjusted to the conditions which can be expected in the individual atmospheres. A number of these values is not known empirically and can not yet be derived from physico-chemical laws. This is esp. true

where chemical processes seem to interact with the
dynamics as in the case of the coloured clouds of
Jupiter and Saturn but also in the case of the H_2SO_4
clouds of Venus where condensation and evaporation
processes can affect the energy budget. Though
it could be demonstrated that e.g. the retrograde
rotation of the Venus atmosphere and the Jupiter
circulation regime or the Red Spot are possible
solutions of the theoretical models, it could not yet
be demonstrated why exactly this system was selected
by nature. One difficulty to explain the reasons for
certain phenomena is, that in some cases small
differences in field parameters, which are difficult
to measure with present techniques, seem to have
strong effects with respect to the large scale
dynamics.

REFERENCES

General introductions into atmospheric physics:

Beer, T.: 1974, Atmospheric Waves, Adam Hilger,
Bristol, G.B.

Fleagle, R.G. and Businger, J.A.: 1980, An Introduction to Atmosüheric Physics, Int.Geophys.
Series 25, Academic Press, New York.

Gossard, E.E. and Hooke, W.H.: 1975, Waves in the
Atmosphere, Elsevier Sciet.Publ.Comp., Amsterdam.

Holton, J.G.: 1972, An Introduction to Dynamic
Meteorology, Int.Geophys.Series 16, Academic Press,
New York.

Houghton J.T.: 1977, The physics of atmospheres,
Cambridge University Press, Cambridge.

Liou, Kuo-Nan: 1980, An Introduction to Atmospheric
Radiation, Int.Geophys.Series 26, Academic Press,
New York.

---

1 Bolle, H.-J.: 1981, Optical and infrared measurements
on planetary atmospheres, Proc.Summer School, The
Solar System and its Exploration, Alpbach, 1981,
ESA SP, in press.

2 Bolle, H.-J.: 1982, Atmospheric evolution, This
volume.

3 Sandström, J.W.: 1908, Dynamische Versuche mit
  Meerwasser, Ann.Hydr.Mar.Met., pp.6 ff.

  Sandström, J.W.: 1908, Deux théoremes fondamentaux
  de la dynamique de la mer, Sven.Hydr.-Biol.Komm.
  Skr.7

4 Defant, A.: 1961, Physical Oceanography, Vol.1,
  Pergamon Press, Oxford, pp. 489 ff

5 Coriolis, G.G.,de: 1835, Mémoire sur les équations
  du mouvement relativ des systèmes de corps,
  J.Ec.Roy.Polyt. 15, pp.142-154.

6 Ekman, V.W.: 1905, On the influence of the Earth's
  rotation on ocean currents, Ark.f.Math.,Astron.
  och Fysik 2, p.52.

7 Taylor, G.I.: 1931, Proc.Roy.Soc. A 132, p.499.

8 Roberts, P.H. and Soward, A.M.:1978,Rotating Fluids in
  Geophysics, Academic Press, London

9 Baines, P.G. and Davies, P.A.: 1980, Laboratory
  studies of topographic effects in rotating and/or
  stratified fluids, In: Orographic effects in
  planetary flows, Joint Scientific Committee,
  GARP Publication Series No.23, ICSU - WMO, Geneve.

10 Leovy, C.B.: 1980, Orographic effects in the atmosphere of other planets, In: Orographic effects in
   planetary flows. Joint Scientific Committee,
   GARP Publication Series No.23, ICSU-WMO, Geneve.

11 Twomey, S.:1977, Introduction to the Mathematics of
   Inversion in Remote Sensing and Indirect Measure-
   ments, Developments in Geomathematics 3, Elsevier,
   Amsterdam.

12 Stone, P.H.: 1975, The dynamics of the atmosphere
   of Venus, J.Atm.Science 32, pp. 1005-1016.

13 Gierasch, P.H. and Goody, R.M.: 1969, Radiative time
   constants in the atmosphere of Jupiter, Journal
   of Applied Science 26, 979-980

14 Goody, R.M. and Belton, M.J.S.:1967, Radiative
   relaxation times for Mars: A discussion of Martian
   atmospheric dynamics, Planet. Span.Scie.15,
   pp. 247-256.

15 Leovy, C.B., Zurek, R.W., and Pollack, J.B.: 1973,
   Mechanisms for Mars dust storms, J.Atm.Scie.30,
   pp.749-762.

16 Iversen,J.D., Greely, R., and Pollack, J.: 1976,
   Windblow Dust on Earth, Mars and Venus, J.Atm.
   Sciences 33, p.2425.

17 Ryan, J.A., et al.: 1981, Geoph.Res.Let.8,pp.899-901

18 Hanel, R., et al.: 1972, Icarus 17, pp. 423-442

19 Limaye, S.S., Suomi, V.E.: 1981, J.Atm.Sci.38,
   pp.1220-1235.

20 Tramb, W.A., Carleton, N.P.: 1975, J.Atm.Scie.32,
   pp.1045-1059.

21 Taylor, F.W., et al.: 1979, Science 205, pp.65-67.

22 Taylor, F.W., et al.: 1980,J.G.R.85, pp.7963-8006.

23 Marov, M.Ya, et al.: 1973, J.Atm.Scie.30,pp.1210-1214

24 Counselman III, C.C. et al.: 1979, Science 205,
   pp.204-206.

25 Ainsworth, J.E. and Herman, J.R.: 1978, J.Geoph.Res.
   83, pp. 3113-3121.

26 Apt, J., Goody, R.M.: 1979, Science 203, pp.785-787.

27 Hanel, R.A. et al.: 1980, Albedo, internal heat, and
   energy balance of Jupiter, preliminary results of
   the Voyager infrared investigation.

28 Williams, G.P., Robinson, J.B.: 1973, J.Atm.Scie.30,
   pp. 684-717.

29 Smith, B.A. et al.: 1979, Science 206, pp. 927-950.

30 Hanel, R., et al.: 1979, Science 204, pp. 972-976.

31 Hanel, R., et al.: 1979, Science 206, pp. 952-956.

32 Bauer, S.: 1981, The atmosphere of Jupiter, Saturn
   and Titan, The solar system and its exploration,
   Proc. Summer School Alpbach 1981, ESA SP in press.

33 Flasar, F.M., et al.: 1981, Global temperature
   distribution and dynamics of Titan's atmosphere,
   Reprint.

34 WMO: 1969, Lectures on numerical short-range weather
   prediction, Hydrometeoizat, Leningrad.

35 Boussinesq, J.: 1877, Mem.Pres.Div.Sav.Acad.Sci.Inst.
   (Sec.Math.phys.) 23, p.380.

36 Mesinger, F., Arakawa, A.: 1976, Vol.I, GARP Publ.
   Series No. 17, Geneva.

37 WMO: 1979, GARP Publ.Ser.No.17,Vol.II

38 Stone,P.H.: 1974, J.Atm.Scie.31, pp.1681-1690

39a K.de Rivas,E.:1973,J.Atm.Scie.31, pp.763-779.

39b Kálnay de Rivas, E.:1975,J.Atm.Scie.,pp.1017-1024.

40 Pollack, J.B. and Young, R.: 1975, Calculations of
   the radiative and dynamical state of the Venus
   atmosphere, J.Atm.Scie.32, pp.1025-1037.

41 Dickinson, R.E.: 1971, Circulation and thermal
   structure of the Venusian thermosphere, J.Atm.
   Scie.28, pp. 885-894.

42 Dickinson, R.E. and Ridley, E.C.: 1972, J.Atm. Scie
   29, pp.1557-1570.

43 Rossow, W.B., and Williams, G.P.: 1979, Large-scale
   motion in the Venus stratosphere, J.Atm.Scie.36,
   pp.377-389.

44 Leovy, C. and Miutz, Y.: 1969, Numerical simulation
   of the atmospheric circulation and climate of
   Mars, J.Atm.Scie.26, pp. 1167-1190.

45 Pollack, J.B., Leovy, C.B. et al.: 1981, A Martian
   general circulation experiment with large topography.

46 Blumsack, S.L. et al.: 1973, An analytical and
   numerical study of the Martian planetary boundary
   layer over slopes, J.Atm.Scie.30, pp.66-82.

47 Williams, G.P.: 1978, Planetary Circulations:
   1. Barotropic representation of Jovian and
   Terrestrial turbulence. J.Atm.Scie.35,pp.1399-1426.

48 Williams, G.P.: 1978, Planetary Circulations: 2. The
   Jovian Quasi-geostrophic regime. J.Atm.Scie.36,
   pp. 932-968.

49 Williams, G.P.: Planetary circulations: 3. The
   Terrestrial quasi-geostrophic regime.

THE CASE FOR A BIMODAL SIZE
DISTRIBUTION IN TITAN'S UPPER HAZE LAYER

Peter H. Smith

Lunar and Planetary Laboratory
University of Arizona, Tucson, AZ 85721

Recent flybys of Titan have provided strong evidence that the
upper haze layer, or layers, on Titan are both highly polarizing
and forward scattering. Spherical particles tend to have either
one quality or the other but not both. A search of prolate
spheroids has been carried out and a class of particles defined
in which large particles (x=3.4) can be highly polarizing. It is
found that even these particles cannot explain the spacecraft
measurements at large phase angles. A second nucleation mode of
tiny aerosols is proposed which can provide the back scattering
necessary to match the geometric albedo. Without this second size
mode the particles must be so bright that the polarization is lost
through multiple scattering.

The Pioneer and Voyager flybys of Titan have given us a
wealth of new data and yet, just as Titan's surface is hidden from
view by a dense cloud cover, the physical properties of the aerosols
are still shrouded in mystery. A preliminary analysis of the
Pioneer polarization data (1) determined that the aerosol mean
size must be near 0.10 μm and quite dark to produce polarizations
in blue light as high as 56.4%. A depolarizing surface at about
2.0 optical depths in blue light (τ_{red}=0.5) was needed to explain
the lower red polarizations ($\sim$48%). This model was found to agree
remarkable well with Danielson's (2) earlier result based on
ground-based measurements. There was even an indication that the
particle size increased with depth in the atmosphere as was predicted by Toon (3) who had developed a cloud physics model for
Titan.

A subsequent analysis of the integrated disk photometry from

A. Coradini and M. Fulchignoni (eds.), The Comparative Study of the Planets, 253–260.
Copyright © 1982 by D. Reidel Publishing Company.

Pioneer out to 96° phase angle and, perhaps more convincingly, the Voyager I high phase angle photometry at 129° and 160° phase angles (4) gave strong evidence that a forward scattering component was present in the upper haze. This new result supported the earlier model of Rages and Pollack (5) based on the variations of the phase coefficient with wavelength from 0° to 6° phase angle and the methane band line strengths. The Rages and Pollack model had been in disrepute because the 0.25 μm particles they needed for the aerosols predicted large negative polarizations in direct conflict with the Pioneer measurements. The agonizing thing about the Rages-Pollack particles is that they are nearly the correct size to explain the forward scattering result.

The conflict as it is presently understood is how can the clouds contain the small, dark particles necessary to explain the extremely large polarizations of the reflected light and at the same time consist of the large, bright forward scattering particles required to explain the Voyager high phase angle data?

Tomasko and Smith (6) explored vertically inhomogeneous models with polarizers above forward scatterers and showed that the polarizing layer needed to extend down to half an optical depth putting the forward scatterers far too deep to be seen at 160° phase. Failing in the attempt to put forward scatterers below polarizers, they tried with some success a model with a thin layer of large forward scattering particles above the smaller polarizing particles. At intermediate phase angles the upper haze has little influence on the high polarization produced by the bottom layer. However, at large phase angles where the slant paths are long, a forward scattering layer of only 0.05 of an optical depth could reproduce the Voyager I observations. This type of situation could occur in an atmosphere if updrafts, say from a large-scale Hadley cell, carried particles upward from the formation altitudes. In fact, this situation is found in the earth's stratosphere over the equatorial zone.

In the present study, I plan to take a different approach and will explore the properties of non-spherical particles to see if forward scattering and large polarization can be produced simultaneously from a homogeneous haze layer. Schaefer (7) has written a computer program capable of finding the scattering phase matrix for spheroids. The region of prolate spheroids out to an aspect ratio of 8:1 appears promising from previous work in this area (8).

The following constraints on the phase matrix were imposed on the search. First, the forward scattering, or diffraction lobe, of the phase function was carefully defined by the Voyager observations. The ratio of the phase function at 20° to 50° scattering angle is approximately 7.2:1. This ratio is necessary to reproduce the observed brightness ratio of 5:1 seen between the brightest areas of the images taken at 160° and 129° phase (4). This reduction in the ratio is due to the dilution effects of multiple scattering. Secondly, the polarization at 90° scattering angle must be greater than or equal to about 80%. The final constraint is that the imaginary index be varied until the observed geometric albedo is obtained. The parameters that can be varied to match these constraints are the aspect ratio, i.e., the ratio of the semimajor axis to the semiminor axis (a/b), the size parameter, and the real index. The calculations are performed for a randomly oriented ensemble of spheroids.

The forward scattering constraint defines the spheroidal size parameter (c) which is fixed in terms of the size parameter of the equivalent volume sphere (x). A value of 3.4 was used in this study. This leaves two variables to match the high polarization. For the particles tried, only the region defined in Fig. 1 satisfied all the constraints. Particles with very low real index work because the scattering efficiencies really depend on the phase lag across the particle ((x(nr-1))). As the real index decreases, the particles begin to act more like Rayleigh scatters which have high polarization, yet they still have the large physical cross-section to maintain the required diffraction peak.

It should be pointed out that solid hydrocarbons, likely to

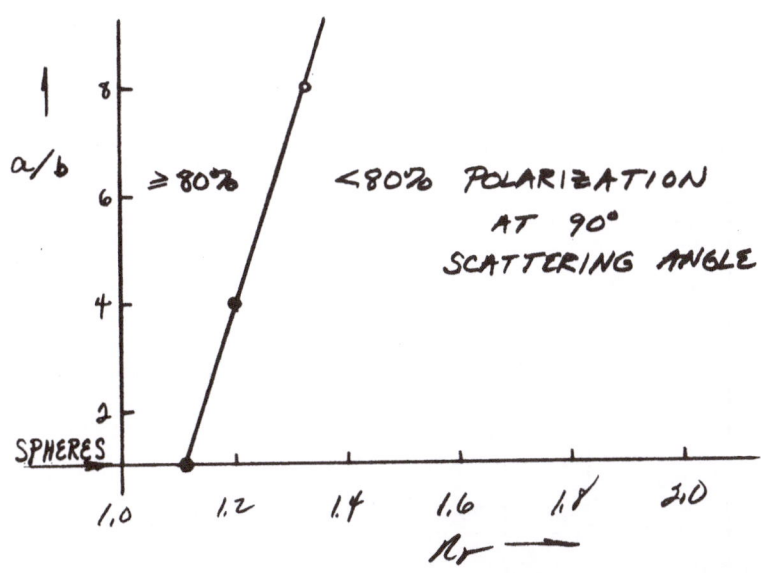

Figure 1. The region for large polarization is shown for prolate spheroids. All particles in the diagram satisfy the forward scattering condition.

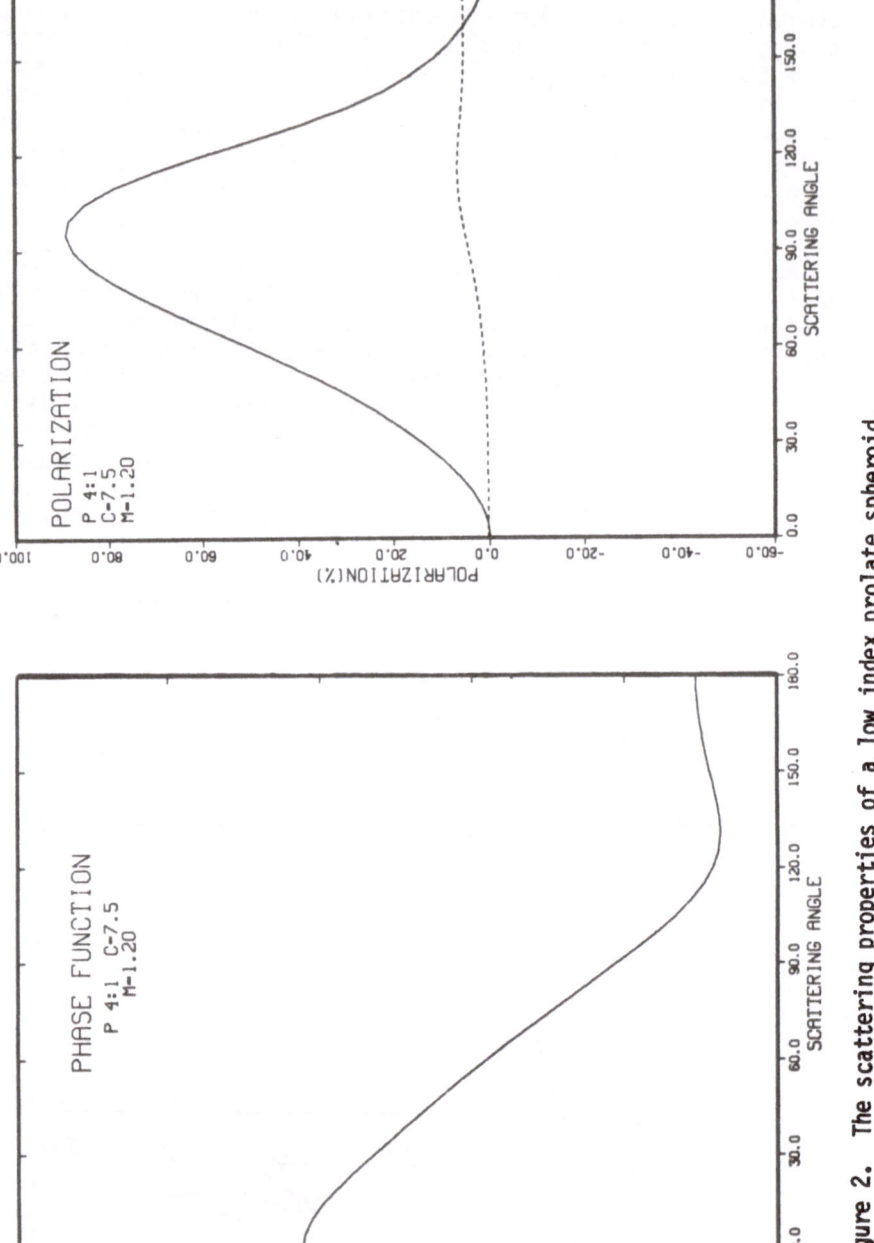

Figure 2.   The scattering properties of a low index prolate spheroid.

be found on Titan, do not have such low indices of refraction. However, it is possible that a low effective index would be measured for loosely packed fluffy particles. The situation is analogous to the radar observations of regoliths, or powdered rocks (9). The dielectric constant measured for powdered substances when compared to the solid form has been shown to obey a Rayleigh mixing formula in many cases:

$$(\varepsilon-1)/(\varepsilon+2)=(\rho/\rho_0)\cdot(\varepsilon_0-1)/(\varepsilon_0+2) \tag{1}$$

where ε_0, ρ_0 are the permittivity and density of the solid material and ε, ρ are those of the powder. In the case of Titan, if a solid material having nr=1.70 were packed into an aerosol with its density reduced by 50% the effective index would decrease to 1.31. Naturally, the "powder" making up the large particle must be much smaller than the wavelength of light for the scattering calculations to be valid.

The phase function and polarization of a fluffy particle with nr=1.20 is shown in Fig. 2. In this example, I have chosen a prolate spheroid with an aspect ratio of 4. The spheroidal size parameter is given by

$$c=kae=7.5 \tag{2}$$

where k is the wavenumber and e is the eccentricity. For blue light (0.44 µm), a=0.54 µm. Note that the linear polarization (solid line) is well above 80% and that the cross polarization (dashed line) is near zero. The forward scattering constraint is satisfied, but almost no light is being back scattered.

A model atmosphere, vertically homogeneous with the phase matrix shown in Fig. 2, was tried and the single scattering albedo varied to match a geometric albedo of 0.11 in the blue. The single scattering albedo necessary was so large (0.92) that multiple scattering diluted the single scattering polarization from greater than 80% to 32%. Since non-spherical particles are known to have reduced back scattering, this situation is very likely the case for any forward scattering, highly polarizing particle. In other words, the phase function must allow the use of single scattering albedos less than about 0.75 to maintain the high polarization. All attempts to add bright underlayers beneath a dark haze failed to solve this problem and, in fact, would be expected to produce a dramatic limb darkening effect contrary to what is observed.

The hypothesis I wish to pursue in this paper involves adding the back scattering component at the same altitude as the forward scattering, polarizing haze. I will do this in the form of a Rayleigh scattering nucleation mode of particles. The additional variable will be the ratio of the scattering optical depth of the

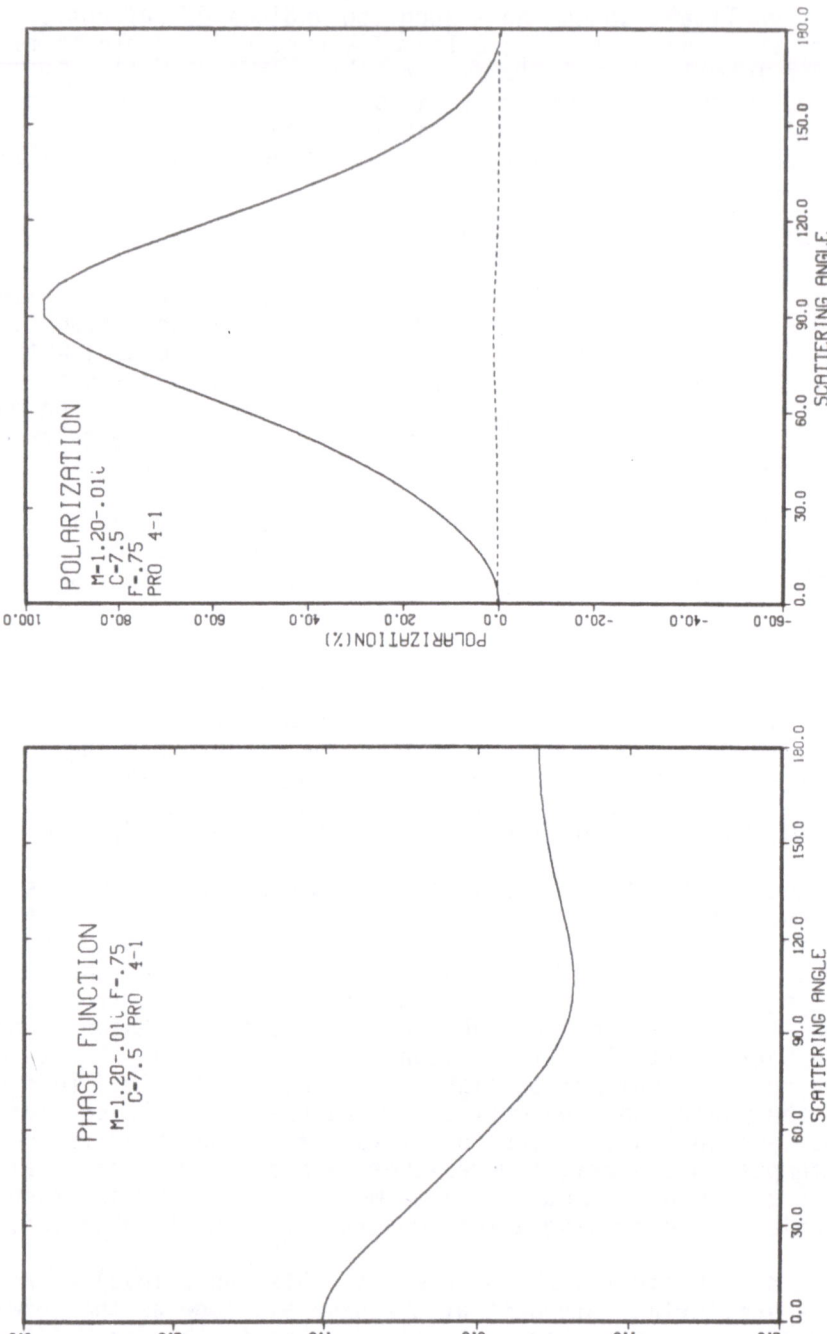

Figure 3. The same as Fig. 2 except that 25% of the total scattering optical depth is contributed by Rayleigh scatterers. A bimodal size distribution.

large particles to the total scattering optical depth of the cloud. This ratio (f) is the fraction that the large particle phase function contributes to the total phase function of the haze.

To determine the plausibility of this type of bimodal size distribution, I have computed a series of multiple-scattering calculations in blue light using the phase function of Fig. 2 with the addition of a Rayleigh phase function. The phase functions are added according to the following formula:

$$p(\theta)=f \cdot P_1(\theta)+(1-f) \cdot P_r(\theta) \tag{3}$$

where 1, r refer to large and Rayleigh. As f is decreased from 1.0 the single scattering albedo rapidly decreases and, as a consequence, so do the number of multiple scatterings. This has the desired effect of preserving the large single scattering polarization. In the present case, when f=0.75 a single scattering albedo of 0.75 gives the right geometric albedo and the polarization is within 3% of the observed value of 56%. The composite phase function for this case is shown in Fig. 3. Note that the forward scattering properties are preserved and that the linear polarization is even larger than in Fig. 2.

It is not my purpose at this time to define the exact parameters necessary to fit all the available data and, indeed, the data are not even completely reduced. Voyager I only gives the forward scattering ratio in their violet filter and we must wait until the Voyager II results are available to know the photometric characteristics at other wavelengths. The parameter space for the bimodal distribution is large enough that it will take a long time to determine if a homogeneous atmosphere of this type will work for all colors. The fraction of large particles will certainly be greater in the red. However, preliminary results are very encouraging and it is felt that a self-consistent model similar to the one shown above can be found to fit over the entire wavelength range.

In conclusion, forward scattering particles, which are also highly polarizing cannot in themselves solve the apparant conflict between the Voyager and Pioneer results. This is because the particles must be too bright in order to match the geometric albedo constraint. A second mode of small particles, perhaps a nucleation mode, with a large back scattering peak is necessary to allow the particles to be dark enough to reduce the polarization dilution caused by multiple scattering. As a cautionary note, I wish to stress that if the requirement for a homogeneous atmosphere is relaxed, layered models (6) with forward scatterers at only the very top of the atmosphere can also work.

ACKNOWLEDGMENTS

This work was performed during a NASA summer fellowship at
Ames Research Center. I wish to thank J.B. Pollack, K.A. Rages
and J. Cuzzi for their support and helpful discussions.

REFERENCES

(1) Tomasko, M.G.: 1980, J. Geophys. Res. 85, pp. 5937-5942.
(2) Podolak, M. and Danielson, R.E.: 1977, Icarus 30, pp. 479-492.
(3) Toon, O.B., Turco, R.P. and Pollack, J.B.: 1981, Icarus (in
     press).
(4) Smith, B.A. et al.: 1981, Science 212, pp. 163-191.
(5) Rages, K. and Pollack, J.B.: 1980, Icarus 41, pp. 119-130.
(6) Tomasko, M.G. and Smith, P.H.: (submitted to Icarus).
(7) Schaefer, R.: 1979, PhD. dissertation, NYU at Syracuse.
(8) Asano, S. and Makoto, S.: 1980, Applied Optics 19, pp. 962-974.
(9) Campbell, M.J. and Ulrichs, J.: 1969, J. Geophys. Res. 74,
     pp. 5867-5881.

ELEMENTS OF COMPARATIVE MAGNETOPLANETOLOGY

F. Mariani

Istituto di Fisica "G. Marconi"
Università di Roma

Abstract – Following the pioneering activities in the last two de
cades, our knowledge of the magnetic and plasma environment around
the planets, the Earth among them, has grown very fast. In this
short review we shall summarize how the observational data are or
ganized, in particular discussing some technical and mathematical
aspects and the known mechanisms capable to generate magnetic
fields outside and inside a planet. We then present a panorama of
results of the observations onboard satellites and space probes.
The interpretation of the results and their implications on the
planetary internal and external structure are also outlined.

1. INTRODUCTION

Planetary and interplanetary magnetic fields play an important ro
le in the physical description of the solar system. The Earth is
the only celestial body for which existence of an intrinsic magne
tic field is known since a long time. The wealth of data collected since the first direct observations in early 1500 by W. Gilbert allows a good insight of the field topology, of its temporal
variations and correlations with other terrestrial and/or solar
phenomena: in addition to a main dipole field, significant quadru
pole and octupole contributions are present, their relative contribution being of the order of 1:0.14:0.10. Most part of the sur
face field is of internal origin, and only the field variations
usually not exceeding 1/1000 of the main field exhibit a substantial external signature. Actually, also the main field becomes ap
preciably and increasingly affected by external sources when observations are made at increasing distances from the Earth, up to
the point where external sources are dominant and intimately cor-

A. Coradini and M. Fulchignoni (eds.), The Comparative Study of the Planets, 261–290.

related with the complex interactions of the solar wind with the
the terrestrial environment. At the other edge of the line, below
the Earth's surface, direct observations are only possible in a
very thin layer, our overall knowledge being based on the interpretation derived by the classical spherical harmonics analysis.
Two important features of the internal geomagnetic field are (i)
the westward drift, at a velocity of the order of tenths of degree per year; (ii) the occurrence during its geological history
of reversals of the dipole orientation. These two features are
to be essential ingredients of any theory of the geomagnetic field.
As regards the external field, the theoretical interpretation of
the observations is very complicate because of the existence of
several ingredients whose relative importance is strongly different in different places and, also, at different times. In particular, while close to the Earth's surface the field is strong
enough to control the distribution of the energetic charged parti
cles, at larger distance (i.e. above a few Earth's radii) charged particle motions can distort the field to the point that its
dipolar topology is drastically changed and unexpected features,
like the so called geomagnetic tail, show up. A complicating factor is represented by the solar wind impinging with variable velocity and carrying its own magnetic field which is rooted in the
inner region of the solar corona.
Information on the intrinsic magnetic field of other bodies in
the solar system, in particular direct in situ determinations,
have only been possible in the last few years (or months). We can
now say that Earth is no longer the only available "magnetic experiment": Earth is one experiment among several others in the
solar system. Other celestial bodies can be used as tests of our
"terrestrial" theories and, viceversa, suggestions and feedbacks
by other "experiments" can prove essential to a better understanding of our planet.

2.   OBSERVATIONAL TECHNIQUES AND DATA INTERPRETATION

2.1. Direct satellite measurements

The vector field can be measured by means of a number of techniques, which are throughly described and discussed in the literature (1). Here only some aspects of the instrumentation which are
specific of on board rockets and space vehicles will be considered. The very accurate, high temporal resolution, determinations
of the field components possible on the Earth's surface are in
principle also possible on rockets and satellites. However
size, power and weight are limited so to lead to a number of con
straints which sometimes can severely reduce our capabilities.
Accuracy may also be a problem: for example a $\pm$ 1 γ ($=10^{-5}$ oersted) uncertainty is well acceptable on the ground where the geomagnetic field is of the order of 0.5 oersted, but it is bad when

the field to measure is the interplanetary field which near Earth
is of the order of a few γ at the Earth's orbit but only a fraction of a γ at Jupiter's orbit. So it is essential to reduce to
an absolute minimum the effect produced by on board spurious sour_
ces like magnetic materials, electric current loops, currents in
the solar panels, modulations, misalignement effects, electronic
noises, etc. Another difficulty stays in the limited capability
of telemetry channels which may not allow the desirable temporal
resolution; this and the additional fact that a space vehicle moves fast on its orbit leads to a spatial resolution which may be
unsatisfactory. An additional consequence is that temporal and
spatial variations may be inextricably mixed, at least when data
by a single spacecraft are available. Aliasing is also a problem,
which although potentially present anytime a discrete sampling
is made, becomes more serious when data from spinning satellites
are processed.

2.2. Indirect observational techniques

Other than in situ measurements, estimates of the magnetic field
in remote or unaccessible regions can be made by indirect techniques. A succesfull example is given by the observation of planetary radioemissions. Assuming a certain model of the emission
mechanism (for example, synchrotron radiation by electrons rotating in a magnetic field) an estimate of the field strength in
the place where the emission takes place can be easily derived.
Faraday rotation and Zeeman effect are other examples of phenomena from which informations on magnetic field can be obtained.
A different approach, complementary to and in some case substitu-
tive of direct field determinations, is based on the study of the
properties of the plasma environment around a planet. There may
be two main sources of charged particles around a planet: low
energy particles produced by photoionization of the atmospheric
components by the U.V. radiation and solar wind particles. However, different physical conditions exist, whether or not an atmosphere and/or an intrinsic magnetic field exist. We shall come
to this point in section 3.

2.3. The mathematical description of the field

Once the vector field, in particular its horizonthal and radial
(i.e. normal) components are known in any point of a spherical
surface at the same time, it can be described by the well known
spherical harmonics expansion, under the generally true additional hypothesis the field is curl-free, i.e. there is no localized current. However, also in this last case a similar approach
can be used with some care at least for some special current
configurations (for example spherical current layers). The sources of the field, i.e. all the coefficients of the spherical harmonics expansion of the field can be uniquely determined so that

the field topology is perfectly known, at the given time, in any place, either inside or outside the spherical surface. The same technique can also be used to study the topology of the field va riations, replacing the field vector variation to the field vector.

The above mathematical analysis is a very powerful tool to separate internal and external sources of the field (or of its varia tions). Let us write the magnetic potential $V(r,\theta,\phi)$ in its more general form:

$$V\left(r,\theta,\phi\right) = a \sum_{n=0}^{\infty} \sum_{m=0}^{\infty} \left[\frac{P_n^m}{r^{n+1}} + q_n^m r^n\right] P_n^m(\theta) \cos\left(m\phi + \epsilon_n^m\right) \qquad [1]$$

where a is the earth's radius; $P_n^m(\mu)$ is the spherical harmonic of degree n and order m; $\mu = \cos\theta$. The quantities $p_n^m, q_n^m, \epsilon_n^m$, are arbitrary constants to be derived by means of the boundary conditions.

Some comment is needed before using $[1]$, since the terms proportional to r^n and to $r^{-(n+1)}$ do not satisfy the necessary require ment that V is finite at infinity and at the origin, respective ly. If the sources of the magnetic field are all internal to the Earth, i.e., are contained in a sphere of radius a, the conditions to be fulfilled by the potential are that $\lim_{r\to\infty} V = 0$ with an asym ptotic behaviour like $1/r^2$ typical of a dipole field: the term $1/r$ cannot exist and all the coefficients q_n^m are zero. If the sources are all external to the sphere of radius a, the field in ternal to it is only expressed by means of r^n powers and all the coefficients p_n^m are zero. The most general case of internal and external sources at finite distances implies contributions from both r^n and $r^{-(n+1)}$. Potentials produced by internal and external sources may then be expressed by the series expansions:

$$V^i\left(r,\theta,\phi\right) = a \sum_{n=1}^{\infty} \left(\frac{a}{r}\right)^{n+1} T_n^i\left(\theta,\phi\right)$$

$$V^e\left(r,\theta,\phi\right) = a \sum_{n=1}^{\infty} \left(\frac{r}{a}\right)^{n} T_n^e\left(\theta,\phi\right) \qquad [2]$$

where

$$T_n^i = \sum_{m=0}^{n} \left(g_n^{m,i} \cos m\phi + h_n^{m,i} \operatorname{sen} m\phi\right) P_n^m(\mu)$$

$$T_n^e = \sum_{m=0}^{n} \left(g_n^{m,e} \cos m\phi + h_n^{m,e} \operatorname{sen} m\phi\right) P_n^m(\mu)$$

The coefficients $g_n^{m,i}, h_n^{m,i}, g_n^{m,e}, h_n^{m,e}$, are the <u>Gauss' coefficients</u>. By means of simple algebra, one gets:

$$V(\tau,\theta,\phi) = a \sum_{n=1}^{\infty} \sum_{m=0}^{n} P_n^m(\mu) \left\{ \left[c_n^m \left(\frac{\tau}{a}\right)^n + \left(1 - c_n^m\right)\left(\frac{a}{\tau}\right)^{n+1} \right] \cdot \right.$$

$$\left. \cdot A_n^m \cos m\phi + \left[s_n^m \left(\frac{\tau}{a}\right)^n + \left(1 - s_n^m\right)\left(\frac{a}{\tau}\right)^{n+1} \right] \cdot B_n^m \sin m\phi \right\} \qquad [3]$$

where the new parameters $c_n^m, s_n^m, A_n^m, B_n^m$ can be easily related to the Gauss' coefficients. The importance of [3] lies in separating explicitly internal and external contributions; $s_n^m = c_n^m = 1$ or 0 respectively means that only external or internal contributions are present. The four sets of parameters in 3 can be determined once the potential $V(a, \theta, \phi)$ and its normal derivative on the sphere of radius a are known. Actually, the field, i.e. the potential derivative, is known everywhere on the ground. Therefore, knowledge of one horizontal field component, for example the north-south component, and of the vertical component is sufficient to completely determine the set of unknown parameters.

A similar technique can in principle be extended to derive the magnetic potential, and then the field configuration, around other planets or satellites. But here the difficulty is that the knowledge of the field vector at <u>any place</u> and at <u>the same time</u> is clearly far from being possible. The best one can hope is to have a continuous survey around the celestial body by an orbiter, so that a large number of determinations can be obtained, although not simultaneous and possibly limited in longitude and/or latitude. In the case of fly-by's the problem is still more complicate since data are collected only during one distance excursion. When these situations occur the most appropriate approach is that of using a "reasonable" model whose free parameters are then best fitted.

3. SOURCES OF PLANETARY MAGNETIC FIELDS

The geomagnetic field at some distance from Earth is significantly affected by contributions attributable to external sources: the implication is that a current density $J \neq 0$ must flow somewhere in the gaseous envelope of the Earth. On the other hand, intrinsic planetary fields are now essentially believed to be produced by electric currents flowing somewhere inside the terrestrial core. The mainly dipolar character clearly indicates it is a planetary phenomenon; the higher order contributions (quadrupole, octupole, etc.) show an increasingly faster radial variations so that they represent the effect from sources closer to the planetary surface: the higher the degree n of a contribution the closer the source region, so the relative importance of the contributions may give important informations on the planetary core. It is quite clear that the intimate interaction of fields and

charged particles in motion plays a fundamental role in the generation and evolution of magnetic fields. So, the magnetohydrodyna
mics represents a basic tool for the physical description of the
phenomena we are talking about.

3.1. General

It is well known that the vector field B can be written as

$$\underline{B} = curl\ \underline{A} = \underline{\nabla} \times T\underline{r} + \underline{\nabla} \times (\underline{\nabla} S \times \underline{r}) = \underline{B}_T + \underline{B}_P \qquad [4]$$

where $\underline{r}$ is the vector distance of any given point from the origin
of the reference system, T and S are two scalar functions of $\underline{r}$.
The vector $\underline{B}_T = \underline{\nabla} \times T\underline{r} = \underline{\nabla} T \times \underline{r}$ is the toroidal field, eve
rywhere normal to $\underline{r}$, i.e. with zero radial component; the vector
$\underline{B}_P = \underline{\nabla} \times (\underline{\nabla} \times S\underline{r}) = curl^2 S\underline{r}$ is the poloidal field, which has
also a radial component. At the surface of the Earth, assuming it
is not-conductive, it can be shown that $\underline{B}_T$ is always zero, so the
only detectable field is the poloidal contribution $\underline{B}_P$.
The mutual actions of particle motions and ambient field B are de
scribed, in addition to the Maxwell's equations by an equation
which relates the bulk motion velocity $\underline{v}$, the gas pressure p, the
field $\underline{B}$, the current density J as follows

$$\frac{d\underline{v}}{dt} = -\frac{1}{\rho} \underline{\nabla}_p - 2\underline{\Omega} \times \underline{v} + \nu \nabla^2 \underline{v} + \underline{F} + \frac{1}{\rho} \underline{J} \times \underline{B} \qquad [5]$$

where ρ is the matter density, ν is the kynematic viscosity of
the fluid, $\underline{\Omega}$ is the angular velocity and $\underline{F}$ is the dynamical force per unit volume. Equation [5] can be simplified whenever one
or more of its terms are negligible or approximations may be used.
Another important equation, which can be derived by the Maxwell's
equations connects the electric conductivity σ and the field $\underline{B}$:

$$\frac{\partial \underline{B}}{\partial t} = \frac{1}{\mu\sigma} \nabla^2 \underline{B} + \underline{\nabla} \times (\underline{v} \times \underline{B}) \qquad [6]$$

where μ is the magnetic permeability; the quantity $\frac{1}{\mu\sigma}$ is also
indicated as ν_m, magnetic viscosity. In principle, use of [5],
[6] and $J = \sigma (\underline{E} + v \times B)$ and Maxwell's equations, allows the so
lution of the problem of the dynamic evolution of the planetary
field distribution. The mathematical difficulties are however so
big that one can only examine very simplified cases. Anyway, some
help can be obtained estimating orders of magnitude. A particularly interesting example is given by equation [6]. Here the ratio of the first to the second term on the right side of the equa
tion can be written as $\nu_m/Lv = 1/R_m$, where L is a characteristic
length of the fluid and R_m is the magnetic Reynolds number. When

R_m<< 1 equation [6] becomes a typical diffusion equation: the
field B diffuses in the ionized gas with a typical time constant
$\tau \approx L^2/\overline{\nu}_m = R_m$ T where T = L/v is a characteristic time of the fluid.
When R_m>>1 (i.e. the conductivity is very large) the diffusion of
the field is inhibited, i.e. only very slow variations of the
field inside the fluid in motion may exist. In other words, a given current loop generating a field can evolve or change its sha
pe still with no appreciable change in the magnetic configuration.
The field is frozen-in. In any intermediate situation, any stable
configuration requires a dynamical equilibrium between diffusion
and generation of new field lines in the source region.

3.2. The interaction of solar wind with celestial bodies.

The study of the solar wind characteristics in proximity of a celestial body is an important complementary tool to get informations on the magnetic field configuration around it. The reason
is that the motion of the charged particles is strongly affected
by and affects the field configuration due to the electric current flow generated by the induced electric field. The interaction is described by the equations given in the previous section.
If the intrinsic planetary field is strong enough (as for Earth,
Jupiter and Saturn) the solar wind will not penetrate deep into
the planetary environment: it will be deflected around a limiting
surface, the so called magnetopause (or more in general magnetoionopause), enclosing the planetary ionized gas. The shape of this
surface is extremely difficult to compute, although some approximations can be used to get reasonable estimates. Basically, the
magnetopause may be originated by two different interactions. The
first case occurs when impinging particles are affected by a planetary intrinsic magnetic field; the second case when only a iono
sphere exists with no intrinsic field. In either case a magnetoionopause is generated, however strongly different in size. The magnetic field lines configuration inside the magnetoionopause is
completely separated by that outside: as a consequence a discontinuity of the field does exist, which also means that the magnetopause becomes a current layer. When no intrinsic field exists
around a planet, the magnetoionopause separates two different plas
mas: the solar wind plasma outside and a relatively stationary
plasma plus an induced magnetic field inside.
In either case, beyond the magnetoionopause the solar wind flow
is strongly perturbed up to another limiting surface which separa
tes the external undisturbed solar wind from that inside. This
surface is a shock-wave, i.e. a surface where the parameters describing the properties of the solar wind and the magnetic field
undergo a pronounced discontinuity, which again can be described
as a complex current layer. The velocity of the impinging solar
wind is much higher than that of the wave propagation in the ma-
gnetoionic gas, so that in analogy with similar hydrodynamical
problems, a shock is generated. It is somewhat surprising that a

good hydrodynamical analogy can be established, although the plasma is rarefied enough to be considered collisionless. As matter
of fact, the ambient or induced magnetic field acts as a powerful
mechanism strongly affecting the plasma flow, although in the absence of collisions in the sense of ordinary elastic fluids. A
sketch of what the field configuration is in a meridian plane,
for the Earth's case is shown in fig. 1, where levels of theoretical description are also indicated in order of increasing difficulty (2).

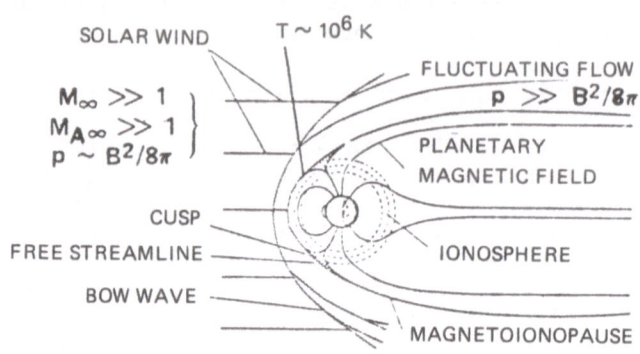

LEVELS OF THEORETICAL DESCRIPTION

1. COLLISIONLESS NONMAGNETIC PLASMA, CHAPMAN-FERRARO
2. HYBRID GAS DYNAMIC AND CHAPMAN-FERRARO
3. DISSIPATIONLESS MAGNETOHYDRODYNAMIC
4. DISSIPATIVE MAGNETOHYDRODYNAMIC
5. ANISOTROPIC PLASMA
6. ANISOTROPIC MULTI-COMPONENT PLASMA

Figure 1. The main features of the solar wind interac-
           tion with a planetary magnetic field (After
           Ref. (2)).

Since the exact mathematical treatment of the problem becomes immediately almost impossible, some approximations are generally ma
de: (i) due to the generally high Alfven Mach number M_A a hydrodynamic approach is taken and field is added as a second step, once
the flow has been computed; (ii) the pressure of the solar wind
on the magnetopause is approximated by $p = k \, n_\infty \, mv_\infty^2 \cos^2 \Psi$ where
k is a constant of the order of 1; n_∞ , v_∞ are the asymptotic
number density and velocity; Ψ is the angle of the flow to the
normal to the magnetopause; (iii) the actual magnetic field is
the dipole field plus an estimated field produced by the currents
flowing on the magnetopause current layer.
A summary of results (2) is shown in fig. 2, where old and new mo
re accurate computed shapes of the magnetopause are shown, for either a meridian and an equatorial cross section.

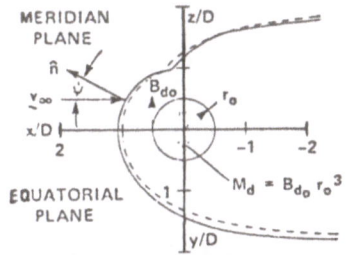

MAGNETOPAUSE COORDINATES, $K \, \rho_\infty/m_p \, n_p = 2$.
--- APPROXIMATE, SPREITER-BRIGGS (1962 a,b)
— EXACT, CHOE et al., (1973)

$$D = \left(\frac{M_d^2}{4\pi \, m_p \, n_p \, v_\infty^2} \right)^{1/6}$$

$M_d = B_{do} \, r_o^3$

COORDINATES OF MAGNETOPAUSE NOSE

COLLISIONLESS MODEL, $p = K \, m_p \, n_p \, V_\infty^2 \cos^2 \psi$, $K = 2$

$$\frac{r_n}{D} = \begin{pmatrix} 1 \\ 1.069 \end{pmatrix} \quad \begin{array}{ll} \text{APPROXIMATE} & B_n = 2 \, Bd_n \\ \text{EXACT} & B_n = 2.443 \, Bd_n \end{array}$$

FLUID MODEL, $p = K(\rho_\infty/m_p \, n_p) \, m_p \, n_p \, v_\infty^2 \cos^2 \psi$
$K = 0.881$ FOR $\gamma = 5/3$, $M_\infty \gg 1$; $\rho_\infty/m_p \, n_p = \Sigma \, m_i \, n_i/m_p \, n_p = 1.16$ FOR 4% He^+

$$\frac{r_n}{D} = \begin{pmatrix} 1.122 \\ 1.118 \\ 1.195 \end{pmatrix}$$
APPROXIMATE, SPREITER et al., (1966), $K \, \rho_\infty/m_p \, n_p = 1$, $B_n = 2B_{d_n}$
IMPROVED PRESSURE, $K = 0.881$, $\rho_\infty/m_p \, n_p = 1.16$, $B_n = 2 \, B_{d_n}$
IMPROVED PRESSURE AND FIELD, $K = 0.881$, $\rho_\infty/m_p \, n_p = 1.16$, $B_n = 2.443 \, B_{d_n}$

NOTE: PRESSURE IS INADEQUATELY REPRESENTED BY $p \sim \cos^2 \psi$ NEAR
NEUTRAL POINT WHERE CUSP DEVELOPS

Figure 2. Computed magnetopause shapes (After Ref.(2)).

Interesting to note that the distance D of the so called stagnation point, i.e. of the nose of the magnetoionopause, is proportional to the 1/3 power of the dipole moment of the planetary field, and only slightly sensistive to the solar wind parameters. As a consequence, the observational value of the distance D, as well as the observed shape of the magnetopause, may be used to estimate the dipole moment. For the Earth's case, with $M_d = 8 \cdot 10^{25}$ gauss·cm³, m = $1.6 \cdot 10^{-24}$ g, $n_p = 5 \cdot cm^{-3}$, $v_\infty = 4 \cdot 10^7$ cm, one gets D ≈ 9 Earth's radii. As fig. 2 shows the distance of magnetopause increases monotonically as the angle Ψ increases. Along the flanks of the magnetopause, where Ψ becomes close to 90°, the approximation on which the theory is based becomes bad. Fortunately, in some special case, one can compute exactly the shock boundaries and the magnetopause (for example when the flow velocity is aligned with the frozen-in magnetic field). Results are shown in fig. 3 where a variety of situations and planets is taken into account (2). A remarkable finding is that the gasdynamic shock description becomes coincident with the magnetohydrodynamic description when the Alfven Mach number is greater than 10. In general the nose gets closer to the planet when M_A becomes smaller; on the flanks of the magnetopause just the opposite occurs, since they move outward.

A somewhat simplier physical situation occurs in the absence of an intrinsic magnetic field, as is the case of the solar wind interaction with the Venusian ionosphere. It is worth to mention that a compression of the frozen-in magnetic field occurs outside the magnetoionopause; in terms of currents this means that a current circulation is required outside the ionopause.

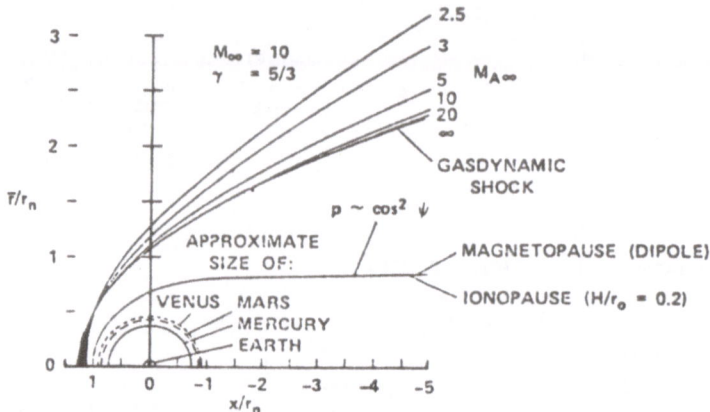

Figure 3. Bow shock shape (After Ref. (2)).

All we have said until now completely neglects any dissipative ef
fect, so that boundary surfaces are considered as geometrical sur
faces. Actually, when dissipative processes (due to fluid viscosi
ty, thermal and electrical conductivity) are taken into account
a finite thickness of such surfaces, is obtained, generally varia
ble from point to point.
A very important feature of the magnetic field topology, in the
antisolar direction is the so called magnetic tail, which is a ma
gnetic flux tube extending in a direction opposite to the Sun whe
re a northern and a southern lobes can be identified with respectively antiparallel field lines: in the case of Earth, field lines are away from Earth in the southern region, below the so called neutral sheet, and earthward north of this plane.
Although not going into details, it is important to remark that
the magnetic tail represents a channel of access of charged parti
cles along the field lines to the innermost part of a planetary
magnetosphere. Because of the instability associated with the clo
seness of antiparallel field lines on the two sides of the neutral sheet, one can expect transient phenomena, like particle acceleration and dumping toward the planet, with associated magnetic field variations. Phenomena of this sort are well observed in
the Earth's polar caps, also in association with auroral activity.

3.3. The magnetic field produced by trapped particle motions

The periodic or pseudoperiodic motions of the trapped particles
close to a magnetized planet are the source of a magnetic field
whose properties strongly depend on the particle energy and angular distribution, chemical composition, etc. In a very general
way the motions can be summarized (3) into three different types,

to which of one an adiabatic invariant can be associated (4):
(i) giration around the field lines with angular frequency eB/mc;
(ii) oscillations along the field lines; (iii) drifts due to cros-
sed electric and magnetic field (in the direction $\underline{E} \times \underline{B})$, to the
gradient of the field intensity (in the direction $\underline{B} \times \underline{\nabla} B^2$) and to
the curvature of the field lines (in the direction of $\underline{B} \times (\underline{B} \cdot \underline{\nabla})\underline{B})$.
In the case of a particle population distributed in space we must
expect that a net current be built up by the individual Larmor
and drift motions, while oscillations along the field lines give
no contribution. The two contributions can be expressed (5) (6)
by means of the field B, its magnetic pressure $P_m = B^2/8\pi$ and
the particle pressure normal P_n and parallels P_s to the field $\underline{B}$
defined as follows:

$$P_n = \frac{1}{2} \int_w \int_\theta m w_n^2 \, n F(w, \theta) \, dw \, d\theta$$

$$P_s = \int_w \int_\theta m w_s^2 \, n F(w, \theta) \, dw \, d\theta \qquad [7]$$

where w_n, w_s are the individual velocity components perpendicular
and parallel to B, $F(w, \theta)$ is the velocity and pitch angle distribution function (w is the total velocity and θ the pitch angle,
i.e. the angle between $\underline{w}$ and $\underline{B}$), m is the mass of each particle.
In general, the total pressures are obtained as the sum of terms
like [7] one per each ionic species. The Larmor current density
$\underline{J}_L$ may be written as follows:

$$\underline{J}_L = \frac{c}{8\pi P_m} \underline{B} \times \left[\underline{\nabla} P_n - \frac{1}{2} \frac{P_n}{P_m} \underline{\nabla} P_m - \frac{P_n}{P_m} (\underline{B} \cdot \underline{\nabla}) \underline{B} / 8\pi \right] \qquad [8]$$

As far as the drifts are concerned, the corresponding velocities
are given by:

$$\underline{v}_1 = c \frac{\underline{E} \times \underline{B}}{B^2}$$

$$\underline{v}_2 = \frac{1}{2} m w_n^2 \frac{c}{e B^4} \underline{B} \times \frac{\underline{\nabla} B^2}{2} \qquad [9]$$

$$\underline{v}_3 = m w_s^2 \frac{c}{e B^2} \underline{B} \times \left[(\underline{B} \cdot \underline{\nabla}) \underline{B} \right]$$

The $\underline{v}_1$ does not depend upon the sign of the individual charge, so
that no contribution to the current density is expected by a neutral mixture of electron and proton gases. On the contrary the ve
locities v_2 and v_3 are opposite for electrons and protons. The
net current density due to drifts can be written:

$$\underline{J}_D = \frac{c}{8\pi\,p_m}\,\underline{B} \times \left[\frac{1}{2}\,\frac{p_n}{p_m}\,\nabla p_m + \frac{p_s}{p_m}\,\left(\underline{B}\cdot\nabla\right)\underline{B}\big/8\pi \right] \qquad [10]$$

A final contribution to the total current density is present as
an effect of the separation of charges, i.e. of the plasma polari
zation, due to inertial forces. This term $\underline{J}_p$ can be expressed by
means of the time derivative of the electric field drift velocity
v_1, as follows

$$\underline{J}_p = \frac{\rho c}{B^2}\,\underline{B} \times \frac{d\underline{v}_1}{dt} \qquad [11]$$

where ρ is the material density of the charged fluid. Once the
particle population is assumed as known, the total current density $\underline{J}$, which is the sum of all the above contributions, depends on
ly upon the total magnetic field $\underline{B}$. The basic procedure to get
the total field $\underline{B}$ is simply that of computing the field $\Delta\,\underline{B}$ produ
ced by the particle current, taking as first approximation $\underline{B} = \underline{B}_o$,
i.e., the field produced by the geomagnetic dipole. The computa-
tion is then iterated by taking a new value $\underline{B}_o' = \underline{B}_o + \Delta\,\underline{B}$ for the
ambient field, and so on until there is no further correction. In
other words, the self-consistent evolution is pursued.
From the experimental view point the effects of electric currents
are clearly identified when satellite observations are compared
with computed fields derived by models based only on near Earth
measurements. The overall configuration of the current circulation in the planetary environment, also because of the superposi-
tion of the effects by the particles trapped in the planetary magnetic field (when existing) and those due to the interaction with
the impinging solar wind, is very complex.

3.4. The origin of the internal planetary magnetism

The generally accepted view is that an internal dynamo is at work
to produce a field (7). Assuming a primordial small field, a rege
nerative action of organized fluid motions is supposed to lead by
a progressive autoexcitation process to a stable field configuration. The mathematical problem is exceedingly complicate also because of the non linearity of the basic equations.
The first question to answer is whether or not generation and amplification of the field connected with electrical fluids symmetric motions is possible. An early general statement (8), i.e.

that stable simple symmetric fields cannot be generated that way,
puts strong constraints on any theory. In a qualitative way we
can say that any symmetric motion around a symmetry axis has the
effect of transporting the axisymmetric field lines without crea
ting any new field line. So, the simultaneous field diffusion due
to the finite conductivity necessarily leads to the progressive
decay of the field itself, which is thus not stable.
The first successful attempt of a non symmetric circulation leading to a "stable" field is due to Parker (9) (10) who showed
that convective fluid motions actually exist, being capable to di
stort the field lines to produce the regeneration mechanism. An
obvious requirement for fluid motions is the existence of a liquid core. This emphasizes that the study of planetary magnetic
field structures can lead to unique informations on the planetary
inner structures.
It has also been convincingly shown that, in addition to stable
magnetic field configurations, also time variable oscillatory solutions exist which might be appropriate to describe the periodic
reversals (like the 22 year reversals of the solar field). However the more or less random temporal distribution of the geomagne
tic field reversals implies that solutions of the basic equations
not simply oscillatory are to be expected when dynamical (rather
than simply kinematical) treatment is used, in particular when a
variability of the fluid velocity is introduced in the theory.
Consideration of the dynamical approach is out of the scope of
this review. We only show in fig. 4 a sketch of a model by Busse
(11) where Coriolis and pressure forces have an essential role to

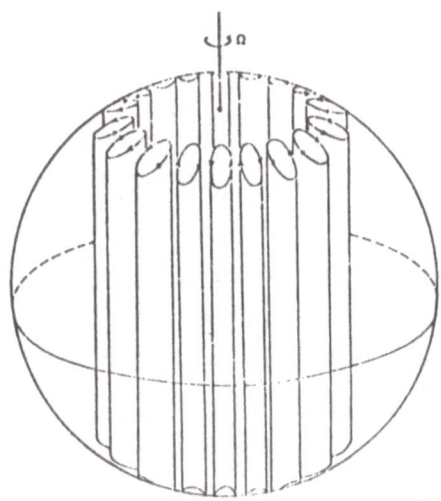

Figure 4. Sketch of the convection motion at the onset
of convection in a rotating internally heated
sphere. (After Ref. (11)).

produce the required non symmetric motion. Although energy considerations show that the growing field tends to an equilibrium amplitude of the field, the details of the equilibrating mechanism
are not known. The driving force of the fluid motion is also unknown: possible candidates are thermal booyancy, chemical separation of lighter constituents, relative motions produced by differential precession rates of the core and the mantle.
The magnetic moment predicted by the Busse model is given by

$$M = B_c R_c^3 = k \left(\mu_0 \rho_0 \right) \Omega R_c^4 \qquad [12]$$

where B_c is the field strength in the core, ρ_c, R_c are its density and radius.
Other ways have been suggested by other authors to estimate magnetic moments on the basis of scaling laws, i.e. more or less empirical relationship between some typical planetary parameters: one
classic example, dating back to Schuster, is the assumption of a
proportionality between magnetic and angular momenta. This relationship is often called magnetic Bode's law (fig. 5). Although
one can bring some physical support to such scaling laws, yet significant discrepancies do exist between the observed values and
those "predicted" by the scaling laws. It is quite clear that the
radius of the liquid core is a critical parameter upon which the
magnetic moments strongly depend. The Earth represents the only

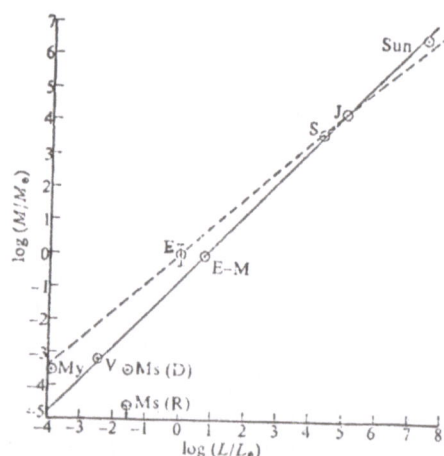

Figure 5. Magnetic dipole moments versus angular momenta
(in units of Earth's values). Dashed and solid
lines when only Earth's or Earth + Moon angular momentum is taken. (After Ref. (12)).

case where the core size is well known by seismic information. In
any other case only estimates are available. Recently some new
ideas have been suggested: Elphic and Russell (13) used as core size the depth at which dipole and higher harmonics give the same
contribution to the RMS magnetic field (which happens to be true
for Earth). Temporal variations of the field as a function of
depth have been used (14) to look for the depth at which the vertical component of the field becomes constant, which means that
the field is frozen-in. In the case of the Earth, such technique
leads to a core size within 2% of that actually known.

4. DIRECT OBSERVATIONS OF PLANETARY MAGNETIC FIELDS

In this section we summarize the results of the recent in situ
measurements of the magnetic field around the planets (and some
of their satellites), also discussing some of their implications.
Physical interpretations are based on spherical harmonics analysis of the field, on scaling of properties from the Earth's magne
tosphere and particle properties or on other indirect observations
(radioemission, optical observations, etc.).

4.1. Mercury

Data taken by Mariner 10 revealed (fig. 6) a small but still surprisingly high field of the order of several hundreds of gammas.
Bow shock and magnetopause similarly to what observed near Earth

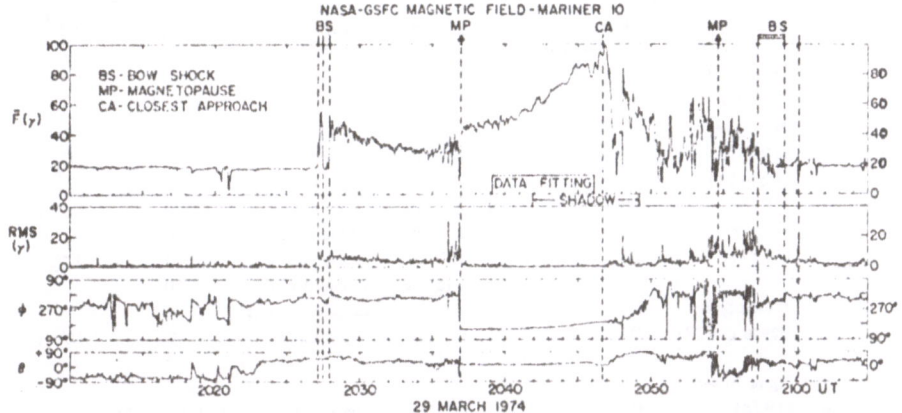

Figure 6 . Magnetic field observations (1.2s averages du
           ring first Mariner 10 encounter with Mercury
           with identification of characteristic discon-
           tinuities of bow shock and magnetopause asso-
           ciated with solar wind interaction. (After Ref.
           (15)).

have been clearly observed. However, due to the lower field inten sity the two surfaces are much closer (in units of planetary ra dii) to the planet and the possibility exists for solar wind par ticles to hit the hermean surface. The external field configura tion strongly suggests existence of a tail-like structure in the antisolar direction. The main field is described by a centered di pole, inclined about 11° to the orbital normal, oriented south ward, just as for the Earth. No radiation belts have been obser ved.

Table 1 gives best fit estimates of the lower multipole coeffi cients as computed by simple use of low order approximations of [1] with the addition in some cases of some simple external cur rent configurations.

As regards the origin of the field (16), two possible mechanisms have been suggested: a dynamo action as inside Earth and/or a rem nant magnetization. A third possibility, induction by the magne tized solar wind magnetic field, has to be disregarded because of the too high observed planetary field, compared with the frozen- -in solar wind field. The fact that the matter density of Mercury is comparable with that of the Earth strongly supports the idea of an internal core essentially made by iron and nickel. If so, it is possible that the magnetic Reynolds number is big enough to allow the internal fluid motions necessary to produce and amplify

Table 1. Mercury's magnetic field parameters according to several approximations or models (After Ref.(16)). where also individual references can be found .

Source	Data	g_1^o	g_2^o	g_3^o	External terms	Offset	Tilt[b]
Ness et al (1974a)	I	227	0	0	0	$0.47 R_M$	30^o_E
Ness et al (1975)	I	350	0	0	$n = 2$	0	10^o_E
Ness et al (1976)	III	342±15	0	0	$n = 1$	0	$11° ±1_O$
Whang (1977)	I& III	266	0	0	2-dimensional	0	
		165	117	0	tail sheet +	0	2.3^o_O
		166		75	48 image dipole	0	
Jackson & Beard (1977)	I & III				scaled terrestrial analogue		
		257	0	0		0	$10°-17^o_O$
		170	114	0		0	
Ng & Beard (1978)	I & III	190	0	0	scaled terrestrial analogue	(0.033, 0.026, 0.189)	1.2^o_O
Ness (1978)	III	330±18	0	0	$n = 2$	0	$14° + 5^o_O$

[a] Using dipole aligned coordinates. Polarity sense is same as at Earth.

[b] Relative to normal to ecliptic or orbital plane, indicated by subscript E or O.

the field at a rate faster than it can be diffused. Some difficulties (16) (17), i.e. that the core is very probably not molten,
can actually be overcome because of the fact that Busse's theory
only requires that part of the core, in particular, an outer shall
be molten. As regards the alternative mechanism, i.e. a remnant
magnetization, it would require a spherical layer several hundreds
 km thick of uniformly magnetized matter at a level comparable
to that of the samples returned from the moon. In this case the
open question is the source of the external contribution. One may
think to a strong polarizing solar field, as big as 10 oersted at
the Mercury orbit; however, since the cooling time is long and
the solar field is known to change polarity periodically one
should not expect a significant effect from the mechanism.
Although also some other arguments have been raised against the
remnant magnetization mechanism, at present it cannot be disregar_
ded. An active dynamo mechanism is more likely. A crucial test
would be to look for a possible secular change, similar to that
on Earth. Actually the few available observations, although some
differences were found between the results from the 1974 and the
1975 flybys, cannot be of much help due to the fact that they are
not out of the experimental uncertainties.
In conclusion, looking at magnetic data and related theoretical
studies, it is possible to think of Mercury interior in terms of
a core whose outer portion is molten so to make possible the required dynamo active circulation. The excitation is to be attribu_
ted to thermal effects, rather than precession induced turbolence
which seems to be weak.

4.2. Venus

Early observations by Mariner 2 (18) and 5 (19) have shown no significant field down to 0.7 planetary radii from the surface, the
upper limit of any possible magnetic momentum being 8.10^{22} $G \cdot cm^3$.
Interpretation of simultaneous data by Venera 4 allowed (20) the
reduction of the upper limit to less than 10^{22} $G \cdot cm^3$. The existen_
ce of a bow shock is convincingly shown (fig. 7) by Venera 9 and‾
10 (21) as well as by the early data from Mariner 10 (22). The
shock is very close to the planetary surface, as a consequence of
the very low magnetic field. Existence of a tail-like is also evident. As a whole, we must say that the effects of a complex exter_
nal current system must be taken out when an intrinsic magnetic‾
moment (if any) is to be extracted from the observational data.
At present, the best available data come from   the Pioneer-Venus orbiter. Existence of very low fields is systematically confir_
med: field strengths up to a few tens of gammas are observed down‾
to the perivenus regions (23). A very peculiar observational feature is the existence of localized ionospheric regions   where
high fields are observed (fig. 8). This feature has been interpre_
ted as an indication of twisted bundles of magnetic flux or flux‾
ropes (24). In principle the root of the observed magnetic flux

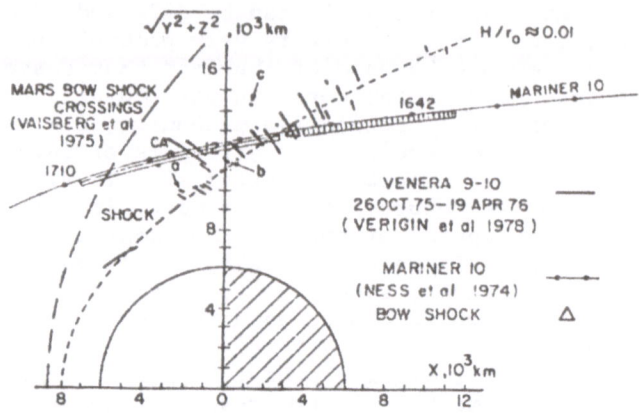

Figure 7. The shape of the bow shocks of Venus and Mars
relative to the planetary surface (After Ref.
(16)).

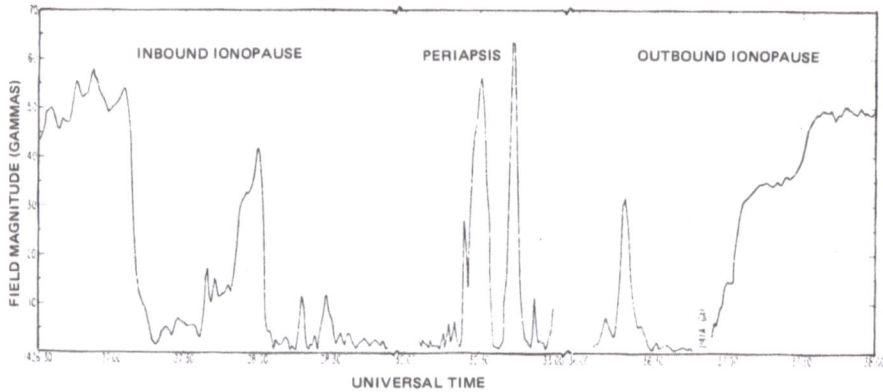

Figure 8. An example of the fine structure venusian
field observed by Pioneer-Venus (After Ref.
(23)).

might be either below the ionosphere or from the magnetosheath;
the authors believe that actually the tubes originate in the magnetosheath, which means that the field is not an intrinsic field
from the planetary body. An estimated upper limit of the planetary dipole moment would then be much lower than 10^{22} G·cm^3. All
above implies that motions in the Venusian core, if at all existing, are extremely slow, because of a less efficient energy sour_
ce and/or a much lower electric conductivity.

4.3. Mars

The most interesting data are in principle those by Mars 2, 3 and
5, since the flyby by Mariner 4 was far enough so that only a gra
zing encounter with the planetary bow shock was possible (24)(25).
The interpretation of the few available magnetic data is very con
troversial, since no general consensus has been reached. In any
case it is clear, that the intrinsic Martian field, if it really
exists, is very low. On one side Dolginov (26) claims that a planetary field exists, corresponding to a magnetic dipole moment of
the order of 2.10^{22} G·cm^3; on the other side, after a data reinter
pretation Russell(27) claims that the observed field might well be
produced by a mechanism similar to that in action around Venus.
Actually, the geometry of the bow shock (28), although only obser
ved on the dusk side of the planet (fig. 7) indicates that a small
internal field may exist. Considerable confusion exists about the
orientation of the dipole. After the initial estimate by Dolginov
(26) who indicated tilt of 72° from the normal to the orbit, the
same author (29) more recently quoted a much different value, i.e.
15°.
The present knowledge of Mars interior is very poor: no doubt that
the lower matter density implies a significantly different consti
tution of the inner core, as compared with that of Earth or Mercu
ry. Size of the core is estimated to be 1500 to 2000 km, seismic
activity appears to be smaller than on Earth. So, although several authors assume   the existence of a fluid core, the possibili
ty it is frozen cannot be disregarded. In the first case, if an
intrinsic field is proved to exist, a dynamo is a real candidate
mechanism. In conclusion, new reliable in situ measurements are
badly needed to prove (or disprove) the very existence of an intrinsic martian field.

4.4. Jupiter

This giant planet is a very special body in the solar system surrounded by a cloud of 15 satellites, one of which, Io,. exhibits
unique features. That Jupiter was highly magnetized was known sin
ce several decades, on the basis of observations of polarized decametric and decimetric radioemissions, interpreted as synchrotron radiation by relativistic electrons in an ambient magnetic
field (30). The first quantitative determinations of the field
gave strengths between some tenths to more than 10 gauss in the
radiation region, which after extrapolation to the planetary surface led to an equatorial field of 3 to 15 gauss, the large mar-
gin of uncertainty being due to those on the theoretical parameters used in the computation. It was not until 1974 that direct
measurements were obtained (fig. 9) onboard Pioneer 10 which led
to estimate magnetic momentum of about $1.5·10^{30}$ G·cm^3 (31). Results from the second flyby, by Pioneer 11 one year later, after
some controversy because of differences between the measurements

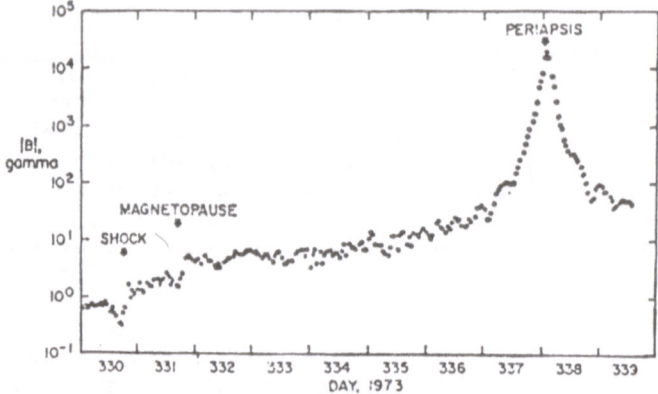

Figure 9. Hourly averages of the field by Pioneer 10
(After Ref. (31)).

by the two onboard magnetometers (a vector helium magnetometer
and a high field fluxgate magnetometer) indicated (32) (33) a sub
stantial agreement on a dipole moment of $1.55 \cdot 10^{30}$ G·cm^3 and a ra
tio of dipole to quadrupole and to octupole moments of 1:0,25:0,20.
By comparison with the Earth's case the non dipole terms are significantly higher in the Jupiter's case. In addition, an inclina
tion of the dipole with respect to the jovian rotation axis of
about 10° was determined   in substantial agreement with the
findings of radioemission observations which were very consistent
with each other although made at different wavelengths (30).
Table 2 gives a list of best fit parameters (moment and dipole
tilt, higher order moments, offset) derived by the Pioneers flybys.

Table 2. A set of parameters for the Jupiter's field. In
the first three lines external contributions up
to quadrupole are assumed; in all four models
internal contributions up to octupole are taken.

Model	Dipole moment (units Gauss·R$_J^3$)	Quadrupole (fraction of dipole)	Octupole (fraction of dipole)	Offset (R_J)	Tilt (degrees)	Reference
P10(3,2)	4.05	0.27	0.29	0.132	11.2	(34)
P11(3,2)	4.17	0.24	0.20	0.128	10.1	(34)
P10,11(3,2)	4.12	0.25	0.20	0.132	10.1	(34)
GSFC O$_4$	4.28	0.24	0.21	0.131	9.6	(33)

The large contributions from higher order moments implies that
the internal source may extend out of the center more than for
Earth. It is worth to enphasize that the dipole points northward,
i.e. opposite to Earth. The interesting question whether or not a
secular variation exists, similarly to Earth, requires several
years of observations. Here we only can say that the dipole moment derived by Pioneer 11 appears to be a few percent higher than
that from Pioneer 10 one year earlier: should this be a real diffe_
rence it might be indicative of a secular variation.
More recently Voyager 1 and 2 visited the jovian environment. Voya_
ger 1 flyby was selected in such a way to get close encounter
with Io, which although implying a higher perijovian distance had
the advantage of allowing the exploration of the dusk side also
including the region where a tail structure was expected, by analo_
gy with the Earth's case. Fig. 10, shows in the upper panel loca-

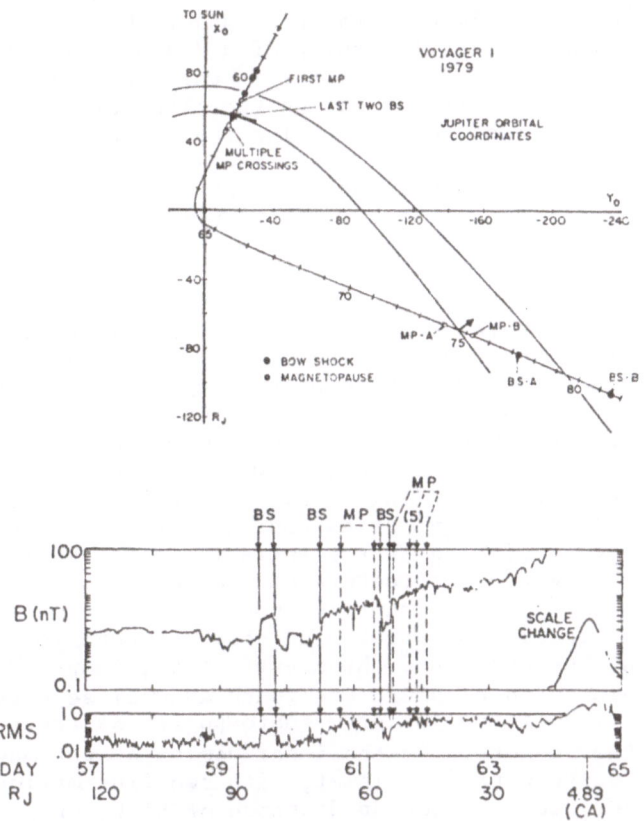

Figure 10. The trajectory of Voyager 1 and the magnetopau_
se and bowshock crossings (upper panel). Details of magnetic field variation up to the
closest approach (CA).(After Ref. (35)).

tion and model shapes of the bow shock and the magnetopause (35),
as derived by the combined magnetic and plasma data. Some detail
of the field strength are shown in the lower panel, where also in
dividual identifications of magnetopause (MP) and bow shock (BS)
crossing are indicated. The magnetopause nose is located about
50 R_j and the bow shock close to 80 R_j, in the sunward direction.
But these distances become several times larger along the dusk si
de of the huge jovian magnetosphere. An important characteristic
of this cavity is its remarkable compressibility as compared with
that of terrestrial magnetosphere.
The rapid rotation of the planet and the several days long residen
ce of the Voyagers close to Jupiter gave a better geographycal di
stribution of sampling, i.e. better informations on the field distribution around the planet. A very interesting result is that
the reversals of field polarity, expected because of the inclination of the magnetic dipole to the orbital plane, are only observed (35) up to a maximum distance of about 80 R_j. This is interpretable as the effect of warping of a diamagnetic plasma sheet
with a thin embedded current sheet in which a field directional
discontinuity is occurring. During the planetary rotation when
the distance exceeds a maximum value the s/c does not reach the
plasma sheet, so that no polarity change is observed. Several explanations have been suggested for such an effect.
One is the centrifugal force distortion in the outer region (36),
which would however lead to a symmetry between the two types of
crossing. Also a spiral shaped distortion has been suggested (37).
In either cases North to South reversals, and viceversa, should
be expected at constant longitudes, and this is only roughly true
looking at the observed crossing longitudes. A significant asymmetry between N→S and S→N reversal longitudes is also observed.
A more plausible idea (35) is that of the bending of a plane current sheet depending upon the angle between the solar wind direction and the dipole direction, in a way very much the same as for
the Earth neutral sheet. This idea is corroborated by the fact
that the field intensity corresponding to the crossing tends to
be smaller and vanishing as the antisolar distance from the planet
increases. The general structure of the jovian magnetosphere as
outlined above by using magnetic observations is well confirmed
(38), by the results of solar wind observations: locations and
shape of the bow shock and the magnetopause, compatible with the
measured solar wind pressure are found as well as a remarkable
compressibility of the jovian magnetosphere. As regards trapped
particles,they corotate in the inner region of the magnetosphere,
i.e. they rotate with the planet, with two flux maxima per rotation. However, above a jovian distance of 85 R_j only one peak remains which concurs with the magnetic data to show the distortion
of the plasma sheet from a planar to a bended configuration.

4.4.1. Jupiter's satellites

An exciting finding of the jovian field (and plasma) observations
stays in the perturbations correlated with the position of jovian
satellites, in particular Io and Ganymede. Voyager 1 was targeted
to cross the field line tube connecting Io to the jovian surface.
The closest approach was 20500 km, i.e. 11 Io radii.
Significant field directional changes, with no intensity changes,
were observed (35) in the few minutes when the flux tube was cros-
sed by the s/c. The field perturbation, superimposed to the ambient jovian field (fig. 11) has been very consistently interpreted as an effect of a twin current system, one flowing toward Jupiter and the other outward, either one aligned with the local
field lines, distributed on a cylindrical tube flux thick just
one Io radius. The existence of no radial variation of the field
strength rules out the possibility of an Io intrinsic field. The
observations give support to the idea of Io's role as a unipolar
dynamo, as originally proposed (39) where the electromotive force
is generated by the Io's motion in the corotating jovian magnetosphere plasma. A power of the order of 10^{12} watts is dissipated
in the current loop. This power, close to that dissipated by tidal
forces, may play an important role inside Io and/or its plasma torus.

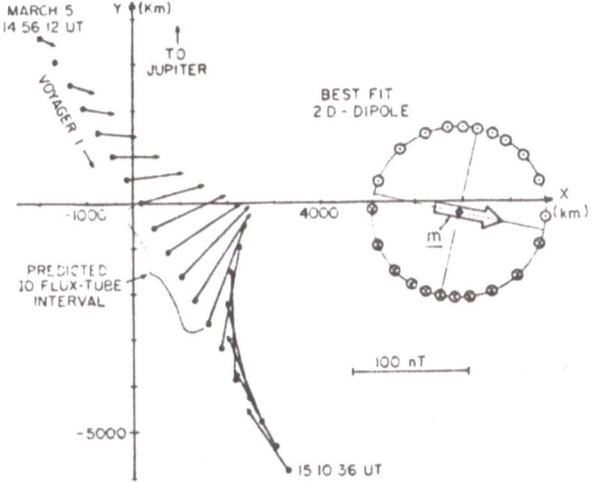

Figure 11. Observed vector field perturbation during the
Io encounter and a cross section of the currents aligned with the field and the resulting dipole moment (After Ref. (35)).

With regard to Ganymede, a distinct perturbation was observed at
distances approximately between −60 and +60 R_G.
No definite statement can be made about the origin of these perturbation. The hypothesis that long wavelength Alfven waves may
be responsible was suggested (40); but also some kind of instability might be the source.

4.5. Saturn

Similarity with Jupiter and the tentative observation of radio
bursts from Saturn at decametric wavelength suggested a magnetic
field of the order of that on Jupiter. But there only was meager
observational evidence whether or not a field actually existed
until the first in situ observations during the Pioneer 11 flyby
(41). Bow shock and magnetopause crossings are clearly identified
(fig. 12). The character of the field inside 10 planetary radii
is of the r^{-3} type, i.e. closely dipolar and symmetric around the
rotation axis of Saturn. Significant discrepancies from the dipole
are found only above 10 R_S. The spherical harmonics analysis shows
that the only internal coefficients (of degree n = 1 and 2) diffe
rent from zero are g_1^o = 0.203 and g_2^o = 0.015. No significant external contribution has been derived inside a few R_S from the ana
lysis. The ratio of quadrupole to dipole moments is about 0.07,
remarkably smaller than the corresponding values for Earth or Jupiter. A small, but significant offset of the dipole along the ro
tation axis (Δz = 0.04 R_S) is sufficient to remove the quadrupole effects. The most striking result of these observations is the

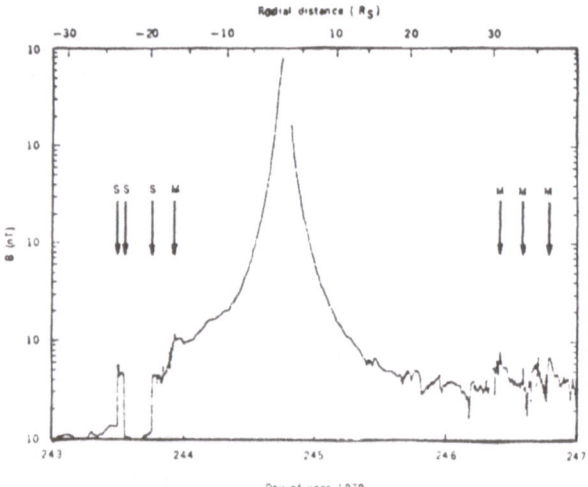

Figure 12. Magnetic field strength from the Pioneer 11
Saturn flyby. Bow shock (S) and magnetopause
(M) crossing are indicated (After Ref. (41)).

near-axial character of the field (i.e. no tilt of the dipole
with respect to the rotational axis). This implies that any plane
tary dynamo theory requiring a tilt or a precession of the dipole
moment must be revised. Another implication is that the high symmetry of the field precludes the possibility of determining the
rotation period by using magnetic field data. The smallness of
the quadrupole moment is interpretable as a consequence of a small
size of the internal conductive core where the dynamo action takes place. In comparison the Jupiter's core is larger. As matter
of fact current models of Saturn and Jupiter internal structure
assume metallic H_2 cores having a radius between 0.2 and 0.5 Saturn radii or between 0.2 and 0.75 Jupiter's radii (42) (43).
About one year after the Pioneer 11 encounter Jupiter's magnetosphere was visited by Voyager 1 on a trajectory which covered a
larger latitudinal extent than Pioneer 11. In addition another
flyby, by Voyager 2, just happened a few weeks ago. Generally
speaking Voyager 1 confirmed (fig. 13) the basical findings by

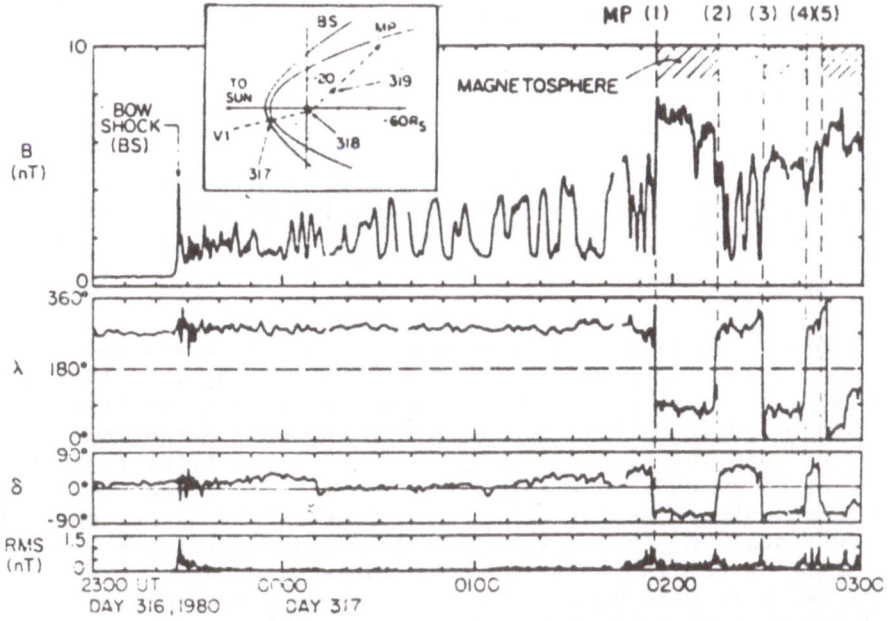

Figure 13. The magnetic field observations (strength F,
            inclination θ, azimuth λ) from Voyager 1. One
            shock and four magnetopause crossings are iden
            tifiable. In the small sketch at the top a
            cross section of these surfaces and the lati-
            tudinal excursion of the s/c trajectory are
            shown (After Ref. (44)).

Pioneer 11. However, the much larger transversal excursion allowed better exploration out of the equatorial plane. (44) (45).
The offset of the dipole was reduced to about 0.02 R_S. Appreciable
contribution by external sources beyond several planetary radii
led to models including external currents (45). A dipole moment of
about 0.2 R_S^3 Gauss · cm^3 is found again; when an equatorial ring
current is added to interpret the external source, there is yet
no change in the RMS residuals, which is interpreted as an effect
of non-potential sources (for example field aligned currents associated with the interaction of the corotating magnetosphere with
a Saturnian satellite or Birkeland currents driven by some asymme
try). The best fit tilt of the dipole is slightly less than 1°.
The orientation of the moment is northward, similarly to the jovian dipole.
The field at larger distances in the antisunward direction shows
a structure interpretable (46) in terms of a magnetotail; on the
sunside a compressed dipole is seen.
Some inference about the field at high latitudes can be made
looking at the radioemission which is definitely proved to be generated by Saturn. This emission shows some asymmetry just interpretable as an effect of axial asymmetry of the field at high latitudes (47).

4.5.1. Titan

A possible interaction of this satellite with the saturnian magne
tosphere has been suggested: the interaction might be explained
as that between a corotating magnetized plasma and a conducting
or magnetized object at a distance of about 145 Titan radii (48).

4.6. The Moon

The lunar field will be dealth with in other lectures so, here we
shall only shortly consider the Moon (which being our closest celestial body is reasonably well known). Following the first measu
rements by Lunik 2 (49) which led to an upper limit of $6 \cdot 10^{21}$
Gauss·cm^3 for the dipole moment based on observations taken as
close as 50 km above the surface, the first extensive measurements
are those onboard Explorer 35 which showed (50)(51) evidence of
no bow shock, i.e. of the fact that solar wind particles were just
impinging on the lunar surface. Occasional increases of field in
the lunar wake were however observed, whose interpretation was
possible during the following Apollo era. This revealed the existence of local fields on the lunar surface as strong as tens up
to more that 100 gammas, with typical sizes of the order of hundreds of kilometers. Continued magnetic measurements on the Moon
surface have been essential to establish the lunar soil conductivity, because of the correlations between field perturbations in
the solar wind and induced perturbations of the field at the sur
face.

The systematic magnetic survey onboard the Apollo subsatellites has been used to compute the lower order coefficients of the spherical harmonics expansion. The dipole term is as low as 0.04 gamma, which corresponds to an upper limit of $1.3 \cdot 10^{18}$ $G \cdot cm^3$ of the magnetic moment.

The obvious question of the origin of the lunar magnetic fields is yet largely unanswered. Runcorn (52) suggested that the magnetization of the returned samples could be due to an internal dynamo active in the past in such a way to not produce sensible fields outside the mantle. Also other less conventional sources have been suggested as possible candidates (for example cometary impacts, thermoelectric currents).

4.7. Uranus and Neptune

Some observations of radioemission onboard IMP 6 have been tentatively attributed to Uranus. This interpretation is however doubtful because at the time of observations Uranus and the Earth were close to each other in the sky so a terrestrial source cannot be excluded (53).

Exploration about Uranus is expecially interesting because of its rotation axis being on the ecliptic. A visit to Uranus and possibly to Neptune is scheduled for Voyager 2, if everything will keep fine until 1986 and beyond.

4.8. Comets

The interaction of the solar wind with the very extended particle environment around is expected to generate magnetic effects, also because of the incoming frozen-in interplanetary magnetic field. The trapping of this field inside the cometary ionosphere may lead to the formation of a magnetotail embedded within the cometary ion tail. Observed rays in the cometary tails might be produced by field aligned ion fluxes. It is also possible that, release of stored magnetic energy by magnetic merging in the tail produce large fluxes of particles. So, although comets are not expected to possess an intrinsic magnetic field, a complex field configuration around them can be predicted. Different authors give different estimates for this field, from a few gammas to several hundred gammas. The first quantitative answers are to be expected from the first cometary flyby of the Halley's comet in early 1986. An all-european mission to this comet, the so called GIOTTO mission, is in preparation.

5. CONCLUSIONS

At present, we can summarize our knowledge on the planetary magnetic fields stating that external sources described by appropriate current systems are always active whenever an atmosphere and/or

an intrinsic planetary magnetic field exists, primarily because
of the interaction of the solar wind with the planetary environment. As regards the internal field, it is to be attributed to the
effect of dynamo currents flowing in a liquid core. The Earth is
the only case for which the internal composition and the core size are known. For any other planet, we can only compare observational data with global results from theories under reasonable as
sumptions for the most relevant parameters. Table 3 summarizes
predicted and observed dipole moments and core sizes for the planets which have been visited by space vehicles. The predicted moments are systematically larger than the observed one, which indi

Table 3. Predicted moment are computed by the Busse's
theory (After Ref. (54)).

Planet	Core Radius, km	Spin Rate, day^{-1}	Dipole Moment. G cm^3		Predicted Core Size, km
			Predicted	Observed	
Mercury	~1800	0.017	9.6×10^{22}	$\sim4 \times 10^{22}$	1450
Venus	~3000	0.0041	1.8×10^{23}	$<1 \times 10^{22}$	<1460
Earth	3486	1.003	...	8×10^{25}	...
Mars	~1700	0.975	3.8×10^{24}	$\lesssim 2.5 \times 10^{22}$	<480
Jupiter	~52.000.	2.44	3.7×10^{30}	1.55×10^{30}	42.000
Saturn	~28.000	2.39	3.1×10^{29}	5×10^{28}	17,700

cates the limits of present theoretical description. The axial sym
metry of the saturnian field is another feature to be fitted in
the theory as compared with the about 10° inclination of the dipo
le to the planetary rotation axes of the other magnetized planets.
From the experimental and observational viewpoint high priority
should be given to orbiters around Mars and the giant external
planets to get extended surveys of their magnetic and particle en
vironment. After all, careful comparative study of all the planets
is necessary step for a quality jump of our level of understanding of the planet we live on.

References

(1) Ness, N.F.: 1977, Spa. Sci. Rev. 11, pp. 459-554.
(2) Spreiter, J.R.: 1976, NASA, Spec. Publ. 397, pp. 135-49.
(3) Alfven, H.: 1950, Cosmical Electrodynamics, Oxford univ.Press.
(4) Northrop, T.G., and Teller, E.: 1960, Phys. Rev. 117, pp.215-
     -25.
(5) Watson, K.M.: 1956, Phys. Rev. 102, pp. 12-19 and 19-27.
(6) Parker, E.N.: 1957, Phys. Rev. 107, pp. 924-933.
(7) Busse, F.N.: 1979, Solar System Plasma Physics, vol. II,
     North Holland, Amsterdam, pp. 295-317.
(8) Cowling, T.G.: 1934, Mon. Notic. Roy. Astron. Soc. 94,
     p. 111-
(9) Parker, E.N.: 1955, Astrophys. J. 122, pp. 293-314.
(10) Parker, E.N.: 1970, Astrophys. J. 162, pp. 665-73.
(11) Busse, F.N.: 1970, J. Fluid. Mech. 44, pp. 441-460.
(12) Russell, C.T.: 1978, Nature 272, pp. 147-8.
(13) Elphic, R.C., and Russell, C.T.: 1978, Geophys. Res. Letters
     5, pp. 211-4.
(14) Hide, R.H.: 1978, Nature 271, pp. 640-1.
(15) Ness, N.F., Behannon, K.W., Lepping, R.P., Whang, Y.C., and
     Schatten, K.H.: 1974, Science 185, pp. 151-60.
(16) Ness, N.F.: 1979, Ann. Rev. Earth. Planet. Sci. 7, pp.249-88.
(17) Stephenson, A.: 1976, Earth. Planet. Sci. Letters 28,
     p. 454.
(18) Smith, E.J., Davis, L. Jr., Coleman, P.J.Jr., and Sonett,
     C.P.: 1963, Science 139, pp. 909-12.
(19) Bridge, H.S., Lazarus, A.J., Snyder, C.W., Smith, E.J.,Davis,
     L. Jr., Coleman, P.J.Jr., and Jones, D.E.: 1967, Science 158,
     pp. 1659-63.
(20) Dolginov, Sh.Sh., Yeroshenko, Ye.G., and Zhuzgov, L.N.: 1968,
     Kosm. Issled. 6, p. 562.
(21) Verigin, M.I., Gringauz, K.I., Gombosi, T., Breus, T.K.,
     Bezrukikh, V.V., Remizov, A.P., and Volkov, C.I.: 1978, J.
     Geophys. Res. 83, pp. 3721-8.
(22) Ness, N.F., Behannon, K.W., Lepping, R.P., Whang, Y.C., and
     Schatten, K.H.: 1974, Science 183, pp. 1301-6.
(23) Russell, C.T., Elphic, R.C., and Slavin, J.A.:1979, Science
     205, pp. 114-6.
(24) Russell, C.T., and Elphic, R.C.: 1979, Nature 279, pp. 616-8.
(25) Smith, E.J., Davis, L. Jr., Coleman, P.J., and Jones, D.E.:
     1965, Science 149, pp. 1241-2.
(26) Dolginov Sh. Sh., Yeroshenko, Ye. G., and Zhuzgov, L.N.: 1975,
     Kosm. Issled. 13, pp. 108-22.
(27) Russell, C.T.: 1978, Geophys. Res. Letters 5, pp. 81-4.
(28) Vaisberg, O.L., Bogdanov, A.V., Smirnov, V.N., and Romanov,
     S.A.: 1976, NASA-Spec. Publ. 397, pp. 21-40.
(29) Dolginov, Sh. Sh.: 1977, Geomagn. Aeron. 17, pp. 569-95.
(30) Smith, E.J., and Gulkis, S.: 1979, Ann. Rev. Earth. Planet.
     Sci. 7, pp. 385-415.

(31) Smith, E.J., Davis, L. Jr., Jones, D.E., Colburn, D.S., Coleman, P.J.Jr., Dyal, P., and Sonett, C.P.: 1974, Science 183, pp. 305-6.

(32) Smith, E.J., Davis, L. Jr., Jones, D.E., Coleman, P.J.Jr., Colburn, D.S., Dyal, P., and Sonett, C.P.: 1975, Science 188, pp. 451-5.

(33) Acuna, M.H., and Ness, N.F.: 1976, J. Geophys. Res. 81, pp. 2917-22.

(34) Davis, L.Jr., Jones, D.E., and Smith, E.J.: 1975, AGU Meeting S. Francisco

(35) Ness, N.F., Acuna, M.H., Lepping, R.P., Burlaga, L.F., Behannon, K.W., and Neubauer, F.M.: 1979, Science 204, pp. 982-7.

(36) Smith, E.J., Davis, L. Jr., Jones, D.E., Coleman, P.J.Jr., Colburn, D.S., Dyal, P., Sonett, C.P., and Frandsen, M.A.: 1974, J. Geophys. Res. 79, pp. 3501-13.

(37) Northrop, C.G., Goertz, C.K., and Thomsen, M.F.: 1974, J. Geophys. Res. 79, pp. 3579-82.

(38) Bridge, H.S., Belcher, J.W., Lazarus, A.J., Sullivan, J.D., McNutt, R.L., Bogenal, F., Scudder, J.D., Sittler, E.C., Siscoe, G.L., Vasyliunas, V.M., Goertz, C.K., and Yeates, C.M.: 1979, Science, pp. 987-91.

(39) Piddington, J.H., and Drake, J.F.: 1968, Nature 217, pp.935-7.

(40) Ness, N.F., Acuna, M.H., Lepping, R.P., Burlaga, L.F., Behannon, K.W., and Neubauer, F.: 1979, Science 206, pp. 966-972.

(41) Smith, E.J., Davis, L. Jr., Jones D.E., Coleman, P.J.Jr., Colburn, D.S., Dyal, P., and Sonett, C.P.: 1980, Science 207, pp. 407-10.

(42) Slattery, W.L.: 1977, Icarus 32, pp. 58-72.

(43) Podolak, M.: 1978, Icarus 33, pp. 342-8.

(44) Acuna, M.H., Connerney, J.E.P., and Ness, F.N.: 1981, Nature, 292, pp. 721-4.

(45) Lepping, R.P., Burlaga, L.F., and Klein, L.W.: 1981, Nature, 292, pp. 750-3.

(46) Behannon, K.W., Connerney, J.E.P., and Ness, N.F.: 1981, Nature, 292, pp. 753-5.

(47) Brown, L.W.: 1975, Astrophys. J. 198, pp. L 89-92.

(48) Jones, D.E., Tsurutani, B.T., Smith, E.J., Walker, R.J., and Sonett, C.P.: 1980, J. Geophys. Res. 85, pp. 5835-40.

(49) Dolginov, Sh. Sh., Yeroshenko, Ye. G., Zhuzgov, L.N., and Pushkov, N.V.: 1961, Geomagn. Aeron. 1, pp. 18-25.

(50) Colburn, D.S., Currie, R.G., Mihalov, J.D., and Sonett, C.P.: 1967, Science 158, p. 1040.

(51) Ness, N.F., Behannon, K.W., Scearce, C.S., and Cantarano, S. C.: 1967, J. Geophys. Res. 72, pp. 5769-78.

(52) Runcorn, S.K.: 1975, Phys. Earth. Planet. Inter. 10, pp. 327-35.

(53) Brown, L.W.: 1976, Astrophys. J. 207, pp. L 209-12.

(54) Russell, C.T.: 1980, Rev. Geophys. Spa. Phys. 18, pp. 77-106.

LUNAR PALAEOMAGNETISM

S.K.Runcorn

School of Physics, The University,
Newcastle upon Tyne, NE1 7RU, England.

Abstract

Although the Moon possesses no general magnetic field today,
it is inferred from lunar palaeomagnetism, studied both in
returned samples and by magnetometers and satellites around the
Moon, implies the existence of an early lunar field generated
by a core dynamo.   This field decays from 1 G 4 Ga ago to
0.02 G 3.2 Ga ago: raising profound questions about the early
heat sources in the solar system.   Palaeomagnetic directions
are interpreted in terms of polar wandering abought about by
the impacts producing the multi-ring basins.

A. Coradini and M. Fulchignoni (eds.), The Comparative Study of the Planets, 291–294.
Copyright © 1982 by D. Reidel Publishing Company.

The discovery of lunar palaeomagnetism was not predicted.   It was
a result not at all easy to fit into the picture of the Moon which
had gradually come to be accepted by the beginning of the Apollo
project: a body formed by accretion but which being small had
never melted so that a convecting iron core in which a magnetic
field could have been generated was though inconceivable.   The
mean density 3.34 of the Moon was also close enough to that of
the basic silicates to make the existence of a core now quite
unnecessary.   Further, Luna 2 had detected no field (the magnetometer sensitivity was about 100γ).   However, the Apollo 11 mission
returned crystalline rocks (lavas) and high grade breccia which
possessed remanent magnetization with components with stable
properties not dissimilar to that of terrestrial rocks.   It was,
therefore, concluded that the rocks had indeed possessed a magnetization on the Moon and in the case of the lavas when they were in
the flows from which they had been excavated by relatively recent
impacts.   Confirmation of the widespread magnetization of the
lunar crust came from the analysis of magnetic fields in the
shadow the Moon casts in the solar wind from magnetometer
measurements from the very successful Explorer 35 satellite.
These observations were attributed to deflection of solar wind by
lunar magnetic anomalies at the limb (10γ on 100γ scale).   These
magnetic anomalies more frequent over the highlands than over the
mare were at first interpreted as lunar magnetization, perhaps
the result of meteorite impact but the fact that all the Apollo
landings found magnetized rocks presented a paradox.   This was
resolved because a uniformly magnetized flat plate extending to
infinity has no magnetic field external to itself.   Consequently
it was calculated that a general magnetization of the lunar
crust was compatible with the localisation of magnetic anomalies:
lines of magnetic force escape where there is an edge, either the
result of cratering or contrasts of intensity between adjoining
strata.

The subsatellites launched in the Apollo 15 and 16 mission
detected lunar anomalies and three component magnetic field maps
have been produced of limited regions of the Moon - limited
because except when the Moon is in the magneto-tail of the Earth,
the noise from the solar wind obscures the anomalies.   The
determination of the dipole field of the Moon from the subsatellite
magnetometers yielded a value (0.05γ) much too small to be
reconciled with a uniformly magnetized crust of reasonable thickness.
This second paradox of the subject was resolved when it was shown
that a uniforml spherical shell magnetized by an internally
generated field much later disappears and produces no external
magnetic field at all.   Thus the null value of the present lunar
dipole field argues in favour of an early dipole field.

The existence of an iron core in which such a lunar field could
once have been generated is now seriously debated.   It was first

suggested in connection with the modern explanation of the non-
hydrostatic shape of the Moon. The existence of this bulge
along the Earth-Moon line was inferred from Cassini's laws of
Moon's rotation and explained by Laplace in terms of an early
distortion retained by the finite strength of the solid Moon.
Jeffreys elaborated this idea by supposing the present figure
was a fossil tidal bulge. Such explanations were only
questioned as a result of the need to explain continental drift
by invoking convection in the Earth's mantle consequent on the
mantle having the property of solid state creep. To explain the
anomalous difference between the moments of inertia of the Moon
a too slow convection pattern was postulated and according to the
marginal theory of stability this demanded the presence of a
small core 300-500 km in radius. The most recently determined
value of the moment of inertia factor, 0.3905, is compatible with
the existence of such a core and one seismological observation
obtained by a meteorite hit on the far side is compatible with
such a core.

The original orientation of the Apollo rocks is not known but
the palaeointensity studies have resulted in interesting and
some surprising results. The Thellier-Thellier method of de-
magnetizing the NRM by successive heating to higher temperatures
in zero field and experiments giving the demagnetized rock
partial thermoremanent magnetizations have been used on many
samples. On more rocks analagous palaeointensity method has
been used which does not require heating and the consequent
danger of chemical change. In this method the NRM is removed
by AC demagnetization using higher and higher alternating
magnetic fields and the demagnetized specimen is given
anhysteresic magnetization by placing it in a small laboratory
field in the presence of higher and higher alternating fields.
In some cases both methods have been used on the same sample
giving similar results. The general result is that palaeo-
intensities for rocks of about 4 Ga are remarkably high, about
10^{-4} T and diminish exponentially by a factor of about 50 at
3.2 Ga. Some support for the high palaeo fields is now
available from an entirely different method - by the study of
isothermal magnetization (IRM).

The study of the magnetic anomaly maps of the Apollo 15 and 16
subsatellites has yielded important information on the
direction of magnetization of areas of the lunar crust (a few,
(100 km) across). This has allowed the hypothesis that the
Moon possessed an early dipole magnetic field to be tested. As
for other planets the dominance of the Coriolis force on the
core hydrodynamics would have resulted in the mean dipole lying
along the axis of rotation. Consequently, the palaeomagnetic
poles calculated from the directions of crustal magnetization
are ancient poles of rotation. These are found to fall into

three groups.   In each of which poles are antipodally
arranged along an axis, the mean poles being at 180° apart.
These "reversals" may be reversals of the polarity of the lunar
dynamo or they may arise because anomalies can be produced by demagnetization of already magnetized crust, e.g. by meteorite
craters or by excess magnetization, e.g. by breccia layers from
the great impacts.   The three groups of poles seem likely to
represent three different ages: those sources on the far side
older than 4 Ga, sources in the centre and' eastern longitudes of
the near side 4 Ga and sources mainly in Oceanus Procellarum
3.9 Ga.   The palaeoequators of the two latter groups fall near
the basins of Nectarian and Imbrium age respectively.

It is concluded that displacements of the pole are brought about
by the great impacts removing circular areas of the anorthositic
crust and changing the axis of moments of inertia.   Such polar
wandering is entirely reasonable mechanically provided that
solid state creep in the interior allows the hydrostatic
equatorial bulge to continually re-orientate itself so that it
remains perpendicular to the axis of rotation.   The proximity of
the multi-ring basins on both the near and far side to the
palaeoequators of their age succests that the bodies which
impacted the Moon were not in helio-centric but in geo-centric -
small moons of the Earth/Moon system.   It seems physically
reasonable that the gravitational attraction of the Moon along
with dissipation could result in small nearby moons moving in
orbits exactly co-planar with that of the Moon.

The existence of a core early in the lunar history requires the
entire melting of the Moon dating from highland samples and from
the middle ages of the mare basalts at 4.4 Ga.   Accretion looks
increasing the implausible heat sources for the differential
outer shell of the Moon: for deep melting it has never been
plausible.   The very early melting of the Moon - and widespread
melting of other small bodies of the solar system, e.g. the
parent bodies of meteorites seems one of the most puzzling
problems.   Extinct radioactivity, possibly superheavy elements,
must be a prime factor.   However, the short lived radio active
isotopes such as Al^{26} are implausible as a melting agent in
bodies like the Moon which probably took 10^8 years to accummulate.
Moreover, a heat source is required which will generate the early
magnetic field of the Moon which only disappeared since 3.2 Ga.
Siderophile or calcophile properties are likely in the superheavy
elements with atomic numbers around 100 and 14:   quite a
reasonable abundance (a fifty of that of U) and half life (100 Ma)
seem capable of melting the Moon and running the dynamo.

THE PRESENT PICTURE OF IO'S ELECTRODYNAMIC COUPLING WITH THE
MAGNETOSPHERE OF JUPITER

M.Dobrowolny

Istituto Plasma Spazio, CNR, Frascati (Italy)

The recent Voyager 1 measurements have greatly added to
our knowledge of Io's electrodynamic interactions with the
magnetosphere of Jupiter.We discuss here the present physical
picture of these interactions and its relation to the decametric radiation from Jupiter.

1.INTRODUCTION

The main motivation for the numerous studies of the
electrodynamic interactions of Io with Jupiter's magnetosphere
has been,and is still now,the fact,discovered by Bigg (1),that
Io exerts a remarkable modulating effect on the decametric
radiation (DAM) which we receive at Earth from Jupiter.After
more than twenty years of radio observations this effect has
been at lenght discussed in the literature (see reviews (2)
and (3)).The strong effect of Io on the radio emissions
suggests in fact that Io plays a fundamental part and,eventual-
ly,is the energy source for at least part of these emissions.
The studies of electrodynamic interactions of Io with the
Jovian magnetosphere must therefore be viewed in the context
of understanding the origin and physical mechanisms of the
decametric emissions from Jupiter.To this we must add that
these studies are also of a more general astrophysical interest in that they refer to the phenomenology consequent to the
motion of any conducting body through a magnetized plasma.
In spite of the long efforts,it is fair to say that a
unified theory of Io,s control of DAM does not exist so
far,the different studies having rather concentrated on diffe-

295

A. Coradini and M. Fulchignoni (eds.), The Comparative Study of the Planets, 295–309.
Copyright © 1982 by D. Reidel Publishing Company.

rent aspects of the overall problem.

The recent in loco measurements of Voyager 1,whose trajectory was passing in the vicinity of Io's magnetic flux tube
(IFT),at a minimum distance of 20000 Km from the satellite,be-
sides revealing an astonishing geological activity of the
surface of Io,have also greatly added to our knowledge of Io's
electrodynamic coupling with the magnetosphere of Jupiter.The-
se observational results,which are stimulating a great deal of
theoretical developments,lead,on the one hand,to a revision of
some previous ideas, and,on the other hand,give new guidelines
for the development of a unified theory of Io's control over
Jupiter decametric radiation.

In this paper we will first briefly review some of the
main ideas on Io's electrodynamic interactions which have been
put forward prior to the Voyager exploration (Sect.2).Then we
will recall those observations of the Voyagers which are most
significant for the electrodynamic coupling of Io with the
Jovian magnetosphere (Sect.3).Finally,we will present in
Sect.4 the physical picture of the phenomenology associated
with the Io's interaction which emerges from the observa-
tions.This picture indicates also which should be the plausible lines of future theoretical developments.

2. THEORETICAL MODELS OF IO'S ELECTRODYNAMIC INTERACTIONS

Out of the many theoretical studies on electrodynamic
interactions of Io with the Jovian magnetosphere we will
recall in this Section only those ideas which,in our opinion,
remain most significant in the light of the recent Voyager 1
observations.For a much more comprehensive and critical review
of theoretical models of Jovian decametric radiation (up to
the Jupiter encounters of the Pioneers),we refer the reader to
reference (4).

As first introduced in the work of Piddington and Drake
(5) and Goldreich and Lynden-Bell (6),if Io is idealized as a
perfect conductor (moving across the Jovian magnetic
field),one would see,from a reference frame at rest with
respect to the Jovian plasma far away from Io,a polarization
electric field

$$\underline{E}_0 = - \underline{V}_0 \times \underline{B} \tag{1}$$

to develop across its diameter (V_0 being Io,s velocity with
respect to Jupiter's magnetosphere).Conversely,from the Io's
frame,one would not have any electric field inside Io.
These ideas were prior to the noticeable discovery by Pioneer
10 (7) that Io is surrounded by an atmosphere and a conducting
ionosphere above. However, the analogy with the physics of a
conducting body moving through a magnetic field can still be
used if we consider as the interacting body the solid Io plus

its atmosphere-ionosphere system.
Using V_0 = 56.8 km/s and B=1900   for the Jovian field at Io, we obtain

$$E_0 \sim 0.11 \text{ volt/m} \tag{2}$$

and, correspondingly, a voltage across Io's diameter (R_{Io} 1820 km)

$$\Delta \phi \sim 400 \text{ Kvolts} \tag{3}$$

This large electromotive force is clearly the basic energy source coupling Io with the Jovian magnetosphere.

In the reasonable hypothesis of infinite conductivity of the magnetospheric medium along magnetic lines and zero transverse conductivity across, the magnetic lines are equipotentials. Then the perturbation represented by the Io's applied electric field is transmitted along IFT towards conjugate regions of the Jovian ionosphere as an Alfvén wave. This mechanism of Alfvén wave generation, which had been investigated earlier (8), has been recently reconsidered both for applications in the Earth's magnetosphere (9) and in relation to the Io's measurements of Voyager 1 (10).
The charge separation electric field sets the plasma within IFT in motion with velocity V_0 (for a perfectly conducting Io and infinite parallel conductivity of the medium). On the other hand the Alfvén wave carrying this perturbation, after reaching the dense layers of the Jovian ionosphere is likely to be at least partially reflected, as the massive layers resist the imposed motions (11).

Upon assuming a propagation time $2\tau_A$ of the Alfvèn wave down to the Jovian ionosphere and back to Io, smaller than the time required for Io to cross a distance equal to its own diameter, one arrives at the concept (6) of an IFT frozen with Io: the plasma in the flux tube intercepted by Io moves rigidly with Io with respect to the Jovian medium while its feet slide along the two conjugate areas of the Jovian ionosphere. A parallel current system is associated with the Io's generated Alfvèn wave, and, therefore, the concept of frozen-in IFT implies the existence of a DC current system (in Io's frame) within IFT which closes in the Jovian ionospheres where the tranverse Pedersen conductivity becomes important (see Fig. 1). Thus, in this idea, Io is coupled to the ionospheric layers of Jupiter and, actually, the value of the current along IFT was calculated by Goldreich and Lynden-Bell (6) by applying Ohm's law to the E layer portion of the circuit and then using current continuity.

An important consequence of the idea of a frozen-in IFT is that, because of the fact that the electric field across the flux tube (Io's frame) vanishes, the current down to Jupi-

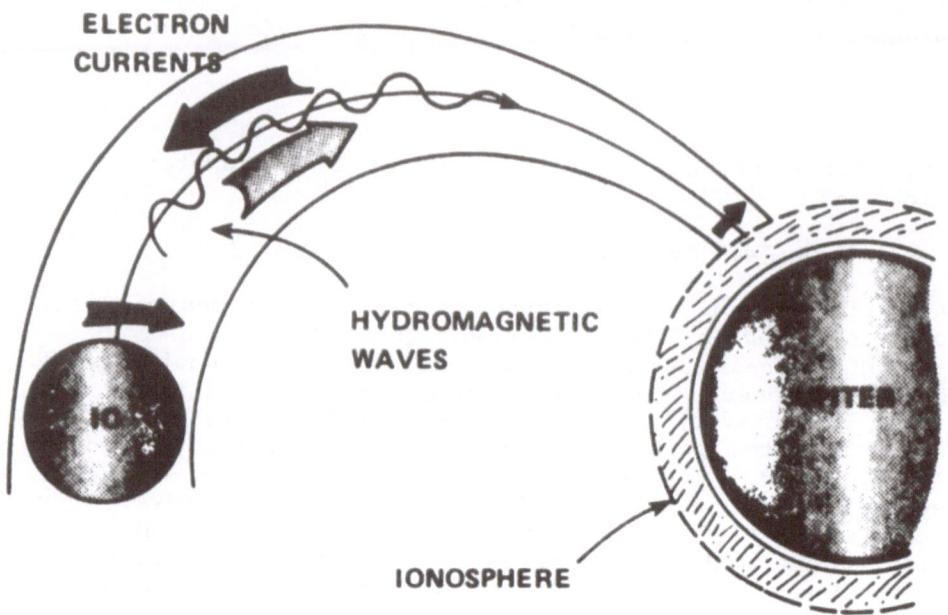

Fig. 1 : Schematic view of interaction of Io with the Jovian
         ionosphere.

ter's ionospheres must be carried in thin sheets along the
boundaries of IFT.This is a main point because it implies
(together with the rather low value of electron density in the
magnetosphere) that,in order to obtain the required current (I
10^8 amps), one needs rather high energy electrons (2.5 Kev in
the estimates of Goldreich and Lynden-Bell).Although these
authors did not show how the particles would attain these
large energies,the presence of these concentrated electron
beams is essential for their subsequent ideas as to the
generation of decametric emissions.

     The Io's controlled DAM appears to occur at frequencies
close to the electron cyclotron frequency (at a level close to
Jupiter's ionosphere) (2,3,12,13).Correspondingly Goldreich
and Lynden-Bell (6) proposed a coherent cyclotron instability
produced by the concentrated electron beams as the mechanism
generating such radiation and actually used the localization
of the beams in the thin current sheets to justify some
beaming properties of the observed radiation.

     For a discussion of this instability and other possible
instability mechanisms mostly based on having accelerated electron beams) which have appeared in the literature,we refer to
reference (4).

     The importance of having accelerated electrons to generate instabilities producing DAM has in fact led to several other works which specifically adressed the question of the mechanism of electron acceleration along magnetic field lines.Before the Pioneer 10 discovery of Io's ionosphe- re,Gurnett (14) pointed out to the possible existence of a charged particle sheath around the solid Io through some parts of which electrons could in fact be accelerated toward Jupi- ter.This model was subsequently adapted (15) to the new disco- very of an Io's ionosphere,the idea being that of placing the sheath as a transition between the top of the Ionian ionosphe- re and the magnetosphere of Jupiter.As pointed out by Smith and Goertz (16), an atmosphere-ionosphere system is something quite different from a solid surface and,consequently,the ori- ginal sheath concept (and structure) is likely not to hold anymore.These same authors proposed instead the existence of double layers (i.e. regions of electrostatic potential drops) along the magnetic lines treading Io,as possible sources of particle acceleration. Although double layers have been consi- dered in some recent theoretical literature (17, 18) and used to explain observed electron acceleration along auroral field lines of the Earth, no direct experimental evidence has been so far obtained that these transitions do exist under geophy- sical or astrophysical conditions.

3.   RECENT OBSERVATIONAL RESULTS FROM VOYAGER 1

Of the numerous discoveries on the Jupiter's satellite Io which have come out of the Voyager 1 flyby past Jupiter, we will recall here only those measurements which are directly relevant to the electrodynamic coupling of Io with the Jovian magnetosphere.

3.1 Magnetic field measurements

The magnetometer of Voyager 1 (19) measured a distinct magnetic field perturbation indicative of intense electrical currents at about the time of maximum approach to Io (minimum distance of 20000 Km south of the satellite at 1505 UT on March 5).

Figure 2 (from Ref. 19) indicates the result of a prelimi- nary study of such perturbation.The figure refers to a right handed coordinate system,centered at Io and moving with Io,with the z axis parallel to the background (Jovian) magne- tic field and the x axis located in the plane of the z axis and the direction of corotation of the magnetospheric flow at Io.The dots represent more precisely what is left of the magnetic measurements after subtraction of the Jovian field and therefore indicates first of all that the perturbation lies approximately in a plane transverse to the background field (as it should be for the field of an incompressible

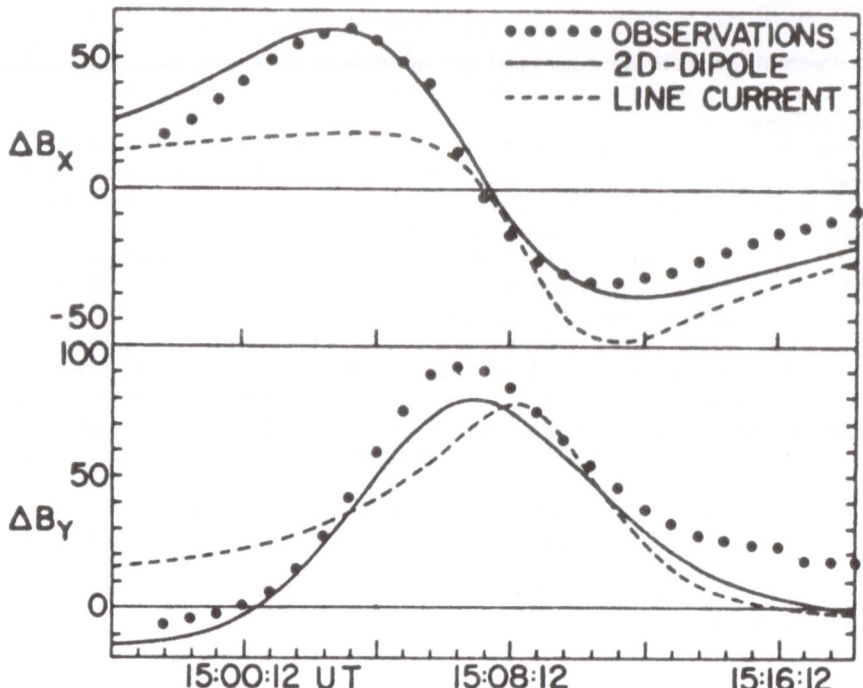

Fig. 2 : Magnetic field pertubation measured from Voyager 1 at
closest approach to Io (from Ness et al. (19)).

Alfvèn wave).The best fitting to tha data (solid line in the
figure) is obtained from a two dimensional magnetic dipole,which is the lowest order expansion of the far field of a
system of currents parallel and antiparallel to the background
field with zero average current.The dipole vector can be
expressed as

$$\underline{m} = \quad (\underline{x} - \underline{x}_D) \times \underline{j} \; dS \qquad\qquad (4)$$

where $\underline{j}$ is the current density, $\underline{x}$ the position vector and $\underline{x}_D$
the location of the dipole in the plane perpendicular to $\underline{j}$
with area element dS.As the observations give the value of m
(m=(1.2±0.4) 10^{10} Amp.Km), further assumptions or modelling
are necessary to guess a current distribution.The experimental
value of m is,for example,consistent, according to the analysis of Ness et al. (19) with a current I= 4.8x10^6 Amps
flowing along the outer surface of a cylinder of one Io
diameter and with a sinuisoidal variation with respect to the
azimuth in the plane perpendicular to the background
field.However,increasing the average distance between the antiparallel parts of the current distribution implies decreasing
the total current consistent with the same determination of m
(10).

Very recently Neubauer (10) has set up the proper theoretical framework for interpreting the magnetic field observations of Voyager 1.The interpretation is in fact that the perturbation is an Alfvèn wave which transmits the transverse polarization electric field (1) (or rather a part of it) away from the conductor.Transmission occurs along the Alfvèn wave characteristics given by

$$\underline{V} \pm \frac{\underline{B}}{\sqrt{4\pi\rho}} = const = \underline{V}_0 \pm \frac{\underline{B}_0}{\sqrt{4\pi\rho}} \tag{5}$$

where $V_0, B_0,$ represent the background velocity,magnetic field and density (and V,B are the velocity and field of the wave).These characteristics,and therefore also the current associated with the conductor, are therefore at an angle with respect to magnetic field lines given by

$$\Theta_A = arctg \frac{V_A}{V_0} \tag{6}$$

V_A being the Alfvèn speed.Fig.3 gives a schematic picture of the perturbation leaving the conductor (so called Alfvèn wings).

The parallel current density is found to be related to the transverse electric field (9,10) by

$$J_{||} = \sum_A \underline{V}_L \cdot \underline{E}_\perp \tag{7}$$

with

$$\sum_A = \frac{1}{\mu_0 V_A} \tag{8}$$

Thus the Alfvènic current tube is analogous to a transmission line with a conductance $\sum_A (\sum \sim 2.2$ taking a value for V_A appropriate for the Io's plasma torus).

Assuming the current j to be distributed over the surface of a cylinder at $r=R_{Io}$ (and thereby supposing a costant electric field inside the cylinder)and also taking a $\sin\varphi$ variation for such surface current,Neubauer (10) derives for the magnetic moment

$$m_x = 2\pi R_{Io}^2 (E_0 - E_i) \sum_A \tag{9}$$

with E_0 the Lorentz field (1) and E_i (Io's frame) the electric field inside the cylinder.The case of a complete short circuit of the Ionian current path $(E_i=0)$ gives rise to a maximum magnetic moment $m_x =5.6\times 10$ $Amp^1 Km$ which is smaller than that obtained in the analysis of Voyager data (19).Neubauer suggests that,even for a complete short circuit (infinite conductivity of Io's ionosphere) the current carrying cylinder must have a radius greater than R_{Io} and even more so for incomplete short circuiting.The current,in each direction,at the cylinder surface has the value

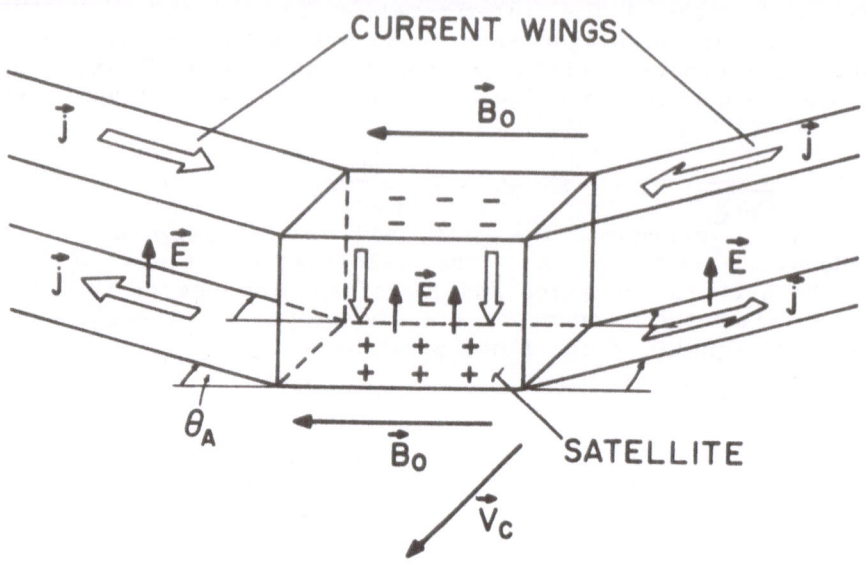

Fig. 3 : Schematic picture of Alfvén wings.

$$I = 4(E_O - E_i)R_{Io}\Sigma_A \qquad\qquad (10)$$

and,therefore,even for the ideal case $E_i = 0$,is limited to a maximum value

$$I_{max} = 4 E_O R_{Io}\Sigma_A \qquad\qquad (11)$$

due to the finite conductance Σ_A of the Alfvènic wings (and quite independently from the characteristics of the ionosphe- ric closure of the circuit).This theoretically computed maxi- mum value is

$$I_{max} \cong 1.74 \times 10^6 \text{ amps} \qquad\qquad (12)$$

which is of the same order as the value reported from the Voyager 1 data analysis.

3.2 The Io's plasma torus

Several Voyager 1 experiments (12,20,21,22) have given evidence of a torus of dense plasma centered around the magnetic equator of Jupiter at a distance of $6R_J$.The torus

has a diameter of 2R$_J$ so that Io's orbit is always within this
dense plasma region.Wave measurements (22)have shown that the
electron density decreases from up to 4000cm^{-3} in the center of
the torus to less than 20 cm^{-3} outside.The chemical composition
of the torus plasma appears to be quite different from that of
the Jovian magnetosphere,with considerable percentage of massive S^{++} ions.The upper limit torus composition envisaged by
Broadfoot et al.(21) corresponds to an average ion mass per
electron of 9 proton masses.Assuming this composition and a
density of 1500 cm^{-3} would give an average Alfvèn speed in the
torus of 356 Km/sec (and an Alfvènic Mach number M$_A$=0.18).Out-
side the torus,on the other hand,the Alfvèn speed,because of
the much lower densities,increases by an order of magnitu-
de.Clearly therefore the existence of this dense torus of plasma (quite apart from the problem of its origin) is of fundamental importance for the electrodynamic coupling of Io with the
surrounding medium.A recent paper by Goertz (23) considers in
detail some features of the interaction of Io with the plasma
torus with the main aim of a completely self-consistent determination of the transverse electric field in Io's vicinity.

3.3 Some features of the radio measurements

    Structure in the decametric emissions has been known from
ground based observations even before the space measurements
(2,3).The space measurements,in particular the high sensitivity
radio experiments on the Voyagers (12,13) have however considerably precised the structural features of the radio emis-
sions.All the decametric radiation appears to consist of discrete narrow band structures extending from a few MHz to ∿ 40
MHz.The Io's control of the emission intensity is on the other
hand confined to the higher frequencies (above∿ 20 MHz).Fig.4
gives some example of the decametric arc structure observed by
Voyager 1 in a frequency versus time diagram (12,13,14)
    Two main features should be clearly pointed out among the
complicated morphologies shown by the observations.The first is
the curvature of the arcs,i.e. the slow drift of frequency
versus time . (occurring over periods typically of a few
hours).The second feature is the multiplicity of the arc patterns and their regularity.During one Jupiter rotation one can
sometimes distinguish up to one hundred arcs or more. The
spacing between the arcs appears to be not at random.Groups of
close arcs are separated from one another by time intervals
ranging from some minutes up to ∿ 30 m.The width of each arc is
∿ 3m and essentially independent from frequency.
    The measurements of frequency and polarization of the
radiation (12,13) are consistent with emissions at frequencies
sligtly above the electron gyrofrequency and thereby determined
by the position along magnetic lines.The emission takes place
in a cone according to the model represented in Fig. 5 and is

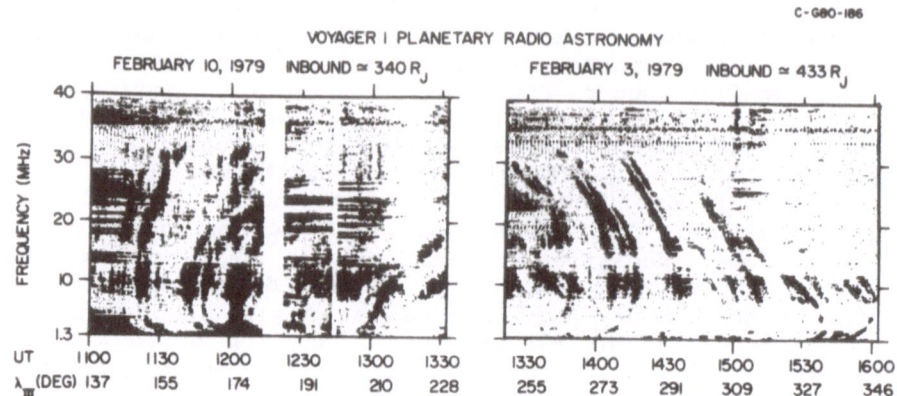

Fig. 4 : Frequency time spectrograms of decametric arcs observed by Voyager 1 during the approach to Jupiter by
the radio astronomy experiment (Warvick et al. (12)).

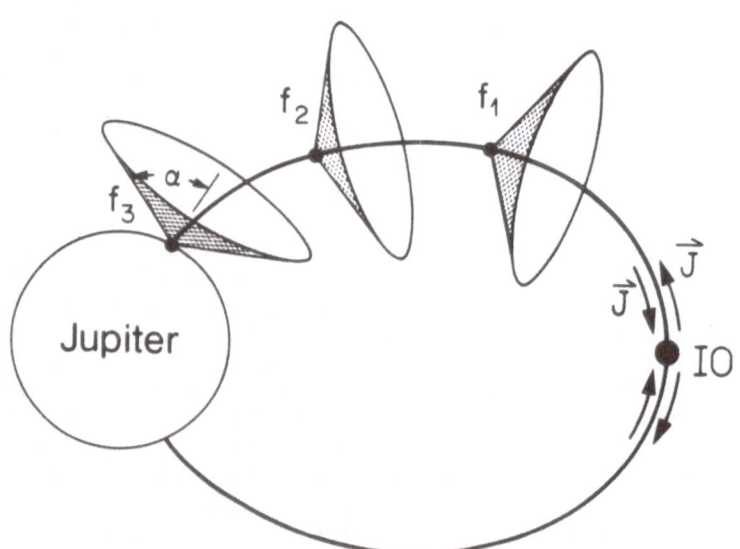

Fig. 5 : Explanation of the structure observed for decametric
arcs (taken from Gurnett and Goertz (25)).

received when the cone volume intersects the observer's position.The variation of frequency as a function of position along a given field line indicated in Fig. 4,together with the longitudinal motion of the source field line gives a natural explanation (12) of the frequency drift which is indicated by the observations.

4.PHYSICAL PICTURE OF IO'S ELECTRODYNAMIC COUPLING

Let us see what is a plausible physical picture of the electrodynamic coupling of Io with Jupiter's magnetosphere which follows from the recent Voyager measurements and point out,at the same time,in which way previous ideas must be modified.

The first point is that the magnetic measurements of Voyager 1 have proved the actual existence of an Alfvèn wave current system associated with Io.The current flows along the Alfvèn wings which are at an angle with respect to the magnetic field lines of IFT.The current circuit is closed,on the one hand,within the Iovian ionosphere and,on the other hand,in two conjugate regions of the Jovian ionosphere.

This resembles the original ideas of Piddington and Drake (5) and Goldreich and Lynden-Bell (6) with,however,some important differences.

A coupling of Io with the Jovian ionosphere (and the corresponding concept of an IFT frozen with Io's motion) would occur if the transit time of the Alfvènic disturbance from Io to the ionosphere of Jupiter and back would be smaller than the time required for Io to cross its diameter during its motion.The discovery of the dense plasma torus surrounding Io's orbit,with the corresponding low value of the Alfvèn speed, denies in fact this possibility.The time required for Io to cross its own diameter is

$$\tau_{Io} = 2R_{Io}/V_0 \sim 64 \text{ s}$$

On the other hand,with an average Alfvèn speed in the torus $V_A \sim$ 356 Km/s and a torus diameter of $2R_J$,the time of traversal across the torus is already $\sim$ 400 sec.There may be partial reflections of the Alfvèn perturbation within the torus boundaries (10) depending also from the position of Io within the torus. However,as indicated by a recent analysis of Goertz (23) of wave propagation through the torus,it is likely that the greatest part of the Alfvènic power is trasmitted across such boundaries and therefore reaches the Jovian ionosphere.There some dissipation is expected (through Ohmic heating of the ionospheric currents) but,at the same time,there will be an almost complete reflection of the wave disturbance.When the Alfvèn wave reaches back again Io's orbit it does not however find the satellite anymore.

The reflection coefficient at the ionospheric boundary can be calculated (24) to be

$$R = \frac{\Sigma_J - \Sigma_A}{\Sigma_J + \Sigma_A} \qquad (13)$$

where Σ_A ,given by (8),is the Alfvènic conductance and Σ_J is the Pedersen conductivity of the ionosphere.This gives,using reasonable values for tha various parameters,a number very close to unity.Recent estimates by Gurnett and Goertz (25) lead to the conclusion that a great number of reflections of the Alfvèn wave back and forth between conjugate ionospheres is possible (from 50 up to a thousand reflections according to the authors).Fig. 6,taken from Ref. (25),gives the corresponding representation of the three dimensional Alfvènic current system produced by Io's motion in Jupiter's magnetosphere.As it is seen,it is something considerably different from the original idea of frozen in IFT.

Concerning the current distribution within the Alfvènic flux tube, no information can be obtained from the magnetic data (and there are no d.c. electric field measurements from Voyager).However the idea (of both the old theories and the recent Neubauer's work)of a constant electric field across the tube and,consequently,a current confined on the surface boundary of the wings,is certainly too simplified.Owing to variations in the conductivity of Io's ionosphere,there will be corresponding variations of the transverse electric field and therefore it is more reasonable to expect a current structure across the wings (on a scale at least of the order of the scale height of the Ionian ionosphere).The type of current distribution across the wings, and hence the perpendicular wavenumber of the Alfvènic perturbation, is important for future studies on the evolution of such Alfvèn waves and their role in the DAM process.

Whatever the distribution of current in the wings,one can give an upper estimate to the power carried by the Alfvèn wave (in the hypothesis of infinite Ionian conductivity)as

$$P_{max} \sim \pi R_{Io}^2 \frac{E_o^2}{2\mu_o V_A} \qquad (14)$$

which leads,referring to the Jovian magnetosphere (outside Io's torus) to $P_{max} \sim 10^{12}$ watts.The recent calculations by Goertz (23) of the interaction of the Alfvèn wave with the plasma torus have precised this estimate considerably (see Fig.4 of Ref. 23).Even allowig for some reduction of the tranverse electric field,due to finite conductivity of Io's ionosphere and also a possible loss whitin the plasma torus, the power in the Alfvènic disturbance reaching Jupiter's ionosphere appears orders of magnitude greater than the average power in the Io's modulated decametric bursts : $P_{DAM} \sim 10^8$ watts.

This is a first fact that makes it reasonable to suppose that the DAM emission originates from the Alfvèn current system of Fig.4. There are however other points as well,recently developed by Gurnett and Goertz (25),which tend to confirm this idea.

The first one comes from the very narrow width of the decametric arcs. This corresponds to a longitudinal width of the emitting source of 0.42° which agrees quite well with the longitudinal width of the Io's flux tube (0.45°) and hence of the Alfvèn wings. This fact gives therefore strong evidence that the sources of Io's controlled DAM are within the Alfvén wings.

The second point made by Gurnett and Goertz (25) is a close correlation between the multiple system of Alfvénic wings (see Fig. 6) generated by Io and the multiplicity and rather regular spacing of the decametric arc structure recalled in Sect. 3. By simple geometrical considerations these authors infer a longitudinal separation between the consecutive Alfvén wings of Fig. 6 (i.e. a separation between successive reflections of Alfvén waves at the ionospheric layers) of 5.8° (corresponding to 123 reflections to cover one Jupiter's rotation). Using Io's rotational period (T_{Io} =1.77 days), one derives correspondingly a temporal separation of $\sim$ 40 m. As this periodicity varies according to the position of Io within the thick plasma torus, all temporal separations between 0 and $\sim$ 40 m appear to be possible. Although a strictly regular separation cannot be seen between successive decametric ars, the separation which are found are within this range.

Thus,to summarize the two points (and we refer to the paper of Gurnett and Goertz for a more exhaustive discussion),the width of an individual arc compares well with the presumable width of IFT (or the corresponding Alfvèn wings).On the other hand the multiplicity of the arcs during one Jupiter's rotation is likely to be correlated with the multiple Alfvèn wave system which is installed by Io around Jupiter the separation between successive Alfvèn waves falling in the same range as the temporal separation between arcs.

Therefore,on the one hand, the picture of the Alfvèn wave system shown in Fig.6 acquires considerable weight,and,on the other hand,these Alfvèn waves appear to be clearly related to the Io's controlled DAM. These associations are clearly phenomenological in character and do not constitute a theory of the Io's controlled decametric emissions. However they indicate quite strongly that such a theory must have as a basic ingredient the Alfvèn waves originating from Io.

As the current itself associated with the wings cannot produce the observed radio emissions,which are likely to

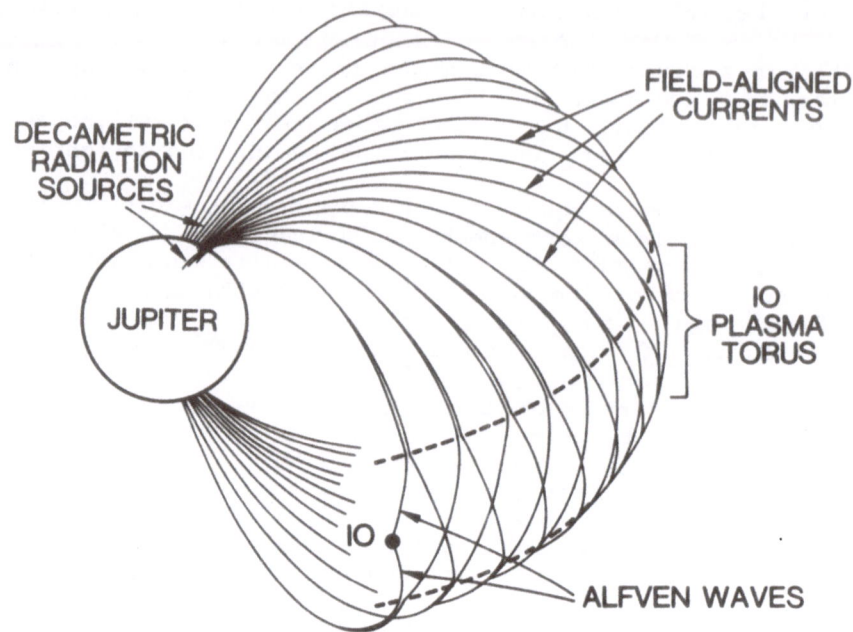

Fig. 6 : Three-dimensional Alfvénic current system produced by
         the motion of Io in the Jovian magnetosphere (taken
         from Gurnett and Goertz (25)).

require   instabilities   generated   by   energetic   electron
beams,one must now look for physical mechanisms through which
each Alfvèn wave,in the course of its propagation down to
Jupiter's ionosphere, leads to electron acceleration and then
to the generation of electron distributions unstable for generation of waves at about the electron cyclotron frequency.This
study, although not yet developed,is strongly indicated by the
recent space observations as the necessary step of investigation to obtain a unified picture of the Io's controlled
decametric radiation.

REFERENCES

1) Bigg E.L.,1964.Nature 203, pp.1008-1009
2) Warwick J.W.,1967.Space Sci.Rev. 6, pp.841-891
3)   Carr   T.D.   and   Desch   M̄.D.,1976.in"Jupiter",Ed.by
   T.Gehrels,The University of Arizona Press, pp.693-737
4) Smith R.A.,1976.in "Jupiter",Ed. by T.Gehrels,The University of Arizona Press, pp.1146-1189
5) Piddington J.H. and Drake J.F.,1968.Nature 217, pp. 935-937

6) Goldreich P. and Lynden-Bell D.,1969.Astroph.J. <u>156</u>,
    pp.59-78
7) Kliore A.,Clain D.L.,Fjeldbo G.,Seidel B.L. and Rasool
    S.I.,1974. Science <u>183</u>, pp.323-324
8) Drell S.D.,Foley H.M. and Ruderman M.A.,1965.J.Geophys.Res.
    <u>70</u>, pp. 3131-3145
9) Mallinckrodt A.J. and Carlson C.W.,1978.J.Geophys.Res. <u>83</u>,
    pp.1426-1432
10) Neubauer F.M.,1980.J.Geophys.Res. <u>85</u>, pp.1171-1178
11) Goertz C.K. and Deift P.A.,1973.Planet.Spa.Sci. <u>21</u>,
    pp.1399-1415
12) Warwick J.W.,Pearce J.B.,Riddle A.C.,Alexander J.K.Desch
    M.D.,Kaiser M.L.,Thieman J.R.,Carr T.D.,Gulkis S.,Boischot
    A.,Harvey C.C. and Pedersen M.B.,1979.Science <u>204</u>,
    pp.995-998
13) Warwick J.W.,Pearce J.B.,Riddle A.C.,Alexander J.K.,Desch
    M.D.,Kaiser M.L.,Thieman J.R.,Carr T.D.,Gulkis S.,Boischot
    A.,Leblank Y. Pedersen M.D.,Staelin D.H.,1979.Science <u>206</u>,
    pp.991-995
14) Gurnett D.A.,1972.Astrophys.J. <u>175</u>, pp.525-533
15) Shawhan S.D.,1976.J.Geophys.Res. <u>81</u>, pp.3373-3379
16) Smith R.A. and Goertz C.K.,1978. J.Geophys.Res. <u>83</u>,
    pp.2617-2627
17) Block,L.P.,1972.Cosmic Electrodynamics <u>3</u>, pp.349-376
18) Knorr G. and Goertz C.K.,1974.Astrophys. Spa.Sci. <u>31</u>,
    pp.209-223
19) Ness N.N.,Acuna M.H.,Lepping R.P.,Burlaga L.F. and Behan-
    non K.W., 1979.Science <u>204</u>, pp.982-987
20) Bridge· H.S.,Belcher J.W.,Lazarus A.J.,Sullivan J.D.,Mc
    Nutt R.L., Bogenal F.,Scudder J.D.,Sittler E.C.,Siscoe
    G.L.,Vasyliunas V.M., Goertz C.K. and Yeates
    C.M.,1979.Science <u>204</u>, pp.987-991
21)   Broadfoot   A.L.,Belton   M.J.S.,Takacs   P.Z.,Sandel
    B.R.,Shemansky D.E.,Holberg J.B.,Ajello J.M.,Atreya
    S.K.,Donahue T.M.,Moos H.M., Bertaux J.L.,Blamont
    J.E.,Strobel D.F.,McConnell J.C.,Dalgarno A.,Goody
    R.,McElroy M.B.,1979.Science <u>204</u>, pp.979-982
22) Gurnett D.A.,Shaw R.R.,Anderson R.R.,Kurth W.S.,Scarf
    F.L.,1979. Geophys.Res.Letters <u>6</u>, pp.511-514
23) Goertz C.K.,1980. J.Geophys.Res.<u>85</u>, pp.2949-2965
24) Scholer M.1970. Plan.Spa.Sci. <u>18</u>, pp.977-1004
25) Gurnett D.A. and Goertz C.K., 1980. Internal Report
    PPG-472, University of California Los Angeles

STRUCTURE OF SMALL METEOROIDS DEDUCED FROM TWO STATION LOW LIGHT LEVEL TV OBSERVATIONS OF METEOR TRAILS

M.A. Hapgood and P. Rothwell

Physics Department, University of Southampton

We report   on a study of meteor trails in the earth's upper atmosphere, which were imaged with low light level TV cameras simultaneously from two stations, and recorded on video tape at 50 frames/sec. This technique has several advantages over photographic techniques employed previously for recording meteor trails. Firstly the temporal resolution is much better.   Secondly, television cameras can detect meteors of apparent magnitude +7 or brighter, whereas photographic cameras can detect only those meteors brighter than magnitude +4. Thirdly, it is much easier to spot faint meteors on the moving video record than on photographic stills. The true position of the trails in space were determined by triangulation against the fixed star background, and hence we could deduce the meteoroid velocities, zenith magnitudes, trail lengths, beginning and end heights of trails, and the heights of maximum brightness of the trails.

We determined the characteristics of 98 meteor trails, 39 recorded in the South of England during the Perseid meteor showers of 1977 and 1978, and 59 recorded in Northern Norway, in October/ November 1977, as a by product of two station auroral observations. The information we have obtained on small meteoroids from studying these characteristics can be summarized as follows.

1.   Both sporadic and showers meteors were recorded in Norway in the autumn of 1977, in the zenith magnitude range +5 to -1. They were separated into two classes (compact and dustball) using the Ceplecha (1) technique of plotting meteor velocity against beginning height of the luminous trails. At any given velocity, the beginning heights of compact meteors are lower than the beginning heights of dustball meteors. All meteors identified

A. Coradini and M. Fulchignoni (eds.), The Comparative Study of the Planets, 311–315.

as shower meteors from their radiant points lay in the dustball
group (Fig. 1). The average length of the vertical component of
the compact meteoroid trails ($\sim 7 \pm 1$km) was shorter than the
average length of the dustball trails (12 ± 2km).

2. The shorter compact meteoroid trails were bright and/or slow,
while the longer trails were faint and/or fast. This implies
that the larger compact meteoroids had shorter trails than the
smaller ones. This is in agreement with the theory of Jones &
Kaiser (2) who suggested that larger compact meteoroids would
fragment during ablation because of large temperature gradients
within the body of the meteoroid and hence have shorter trails
than the smaller meteoroids which could heat up and ablate more
uniformly.

3. The dustball ablation theory of Hawkes & Jones (3) was tested
using data from 39 Perseid meteor trails in the zenith magnitude
range M_V = +4 to −2. This theory assumes that individual dustball grains are of similar size and are held together by a
relatively low melting point "glue".
Dustball meteoroid ablation therefore occurs in two stages.
First the glue melts releasing the individual dust grains, later
the dust grains reach a sufficiently high temperature to ablate
and produce light emission. Thus all dustball meteoroids of the
same velocity should become visible at the same height,
independently of their size. The glue of small dustballs should
have melted completely before ablation of the grains begins but
larger dustballs would continue melting and releasing additional
grains after the onset of ablation. Hence the length of the
dustball meteor trails should be independent of size for small
meteoroids but should increase with increasing size for
meteoroids larger than a critical mass. A meteoroid of critical
mass should have just fragmented completely at the onset of
ablation. The apparent magnitude of a meteor trail depends upon
the mass, velocity, and zenith distance of the meteoroid but the
last two variables are nearly constant for meteoroids from one
particular shower. In this case, meteor magnitudes and masses
should be nearly proportional.

The Perseid meteor trail beginning heights (h_B) were found
to be indipendent of meteor magnitude; they ranged from 105 to
115 km with a mean value of h_B = 110 $\pm$ 1 km. The trail end
heights (h_E) and heights of maximum brightness (h_M) were
independent of meteor magnitude up to a critical magnitude
M_V = 0. Their mean values for magnitudes fainter than M_V = 0
were h_E = 99 $\pm$ 1km and h_M = 103 $\pm$ 1km respectively. For meteors
brighter than M_V = 0, h_E and h_M decreased with increasing meteor
brightness. Thus the lengths of the dustball meteor trails were
independent of magnitude for meteors fainter than zenith
magnitude M_V = 0, but increased with increasing brightness for

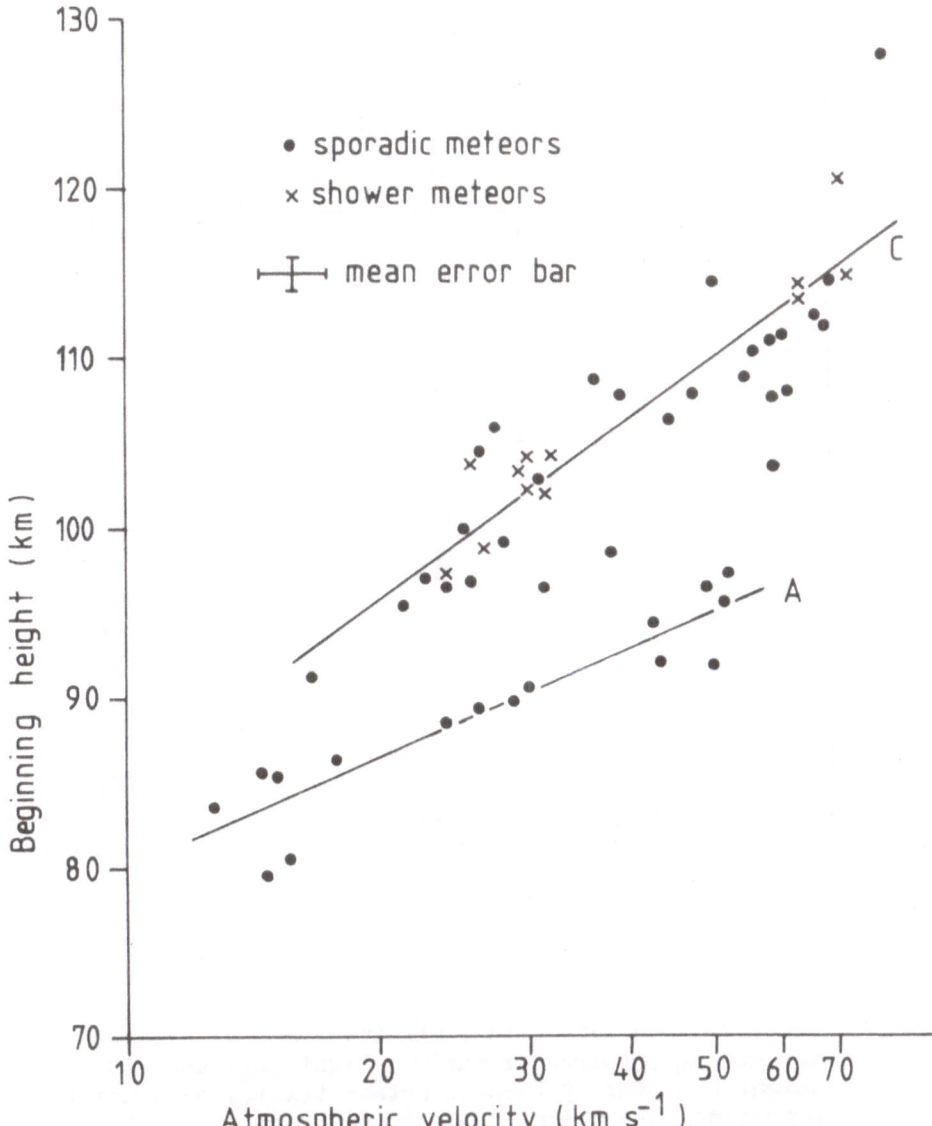

Fig. 1. The variation with meteor velocity of the beginning
height of meteoroid luminosity trails. Meteoroids
corresponding to points grouped around the lower line A
are identified as "compact", meteoroids corresponding
to points grouped around the upper line C are identified
as dustballs.

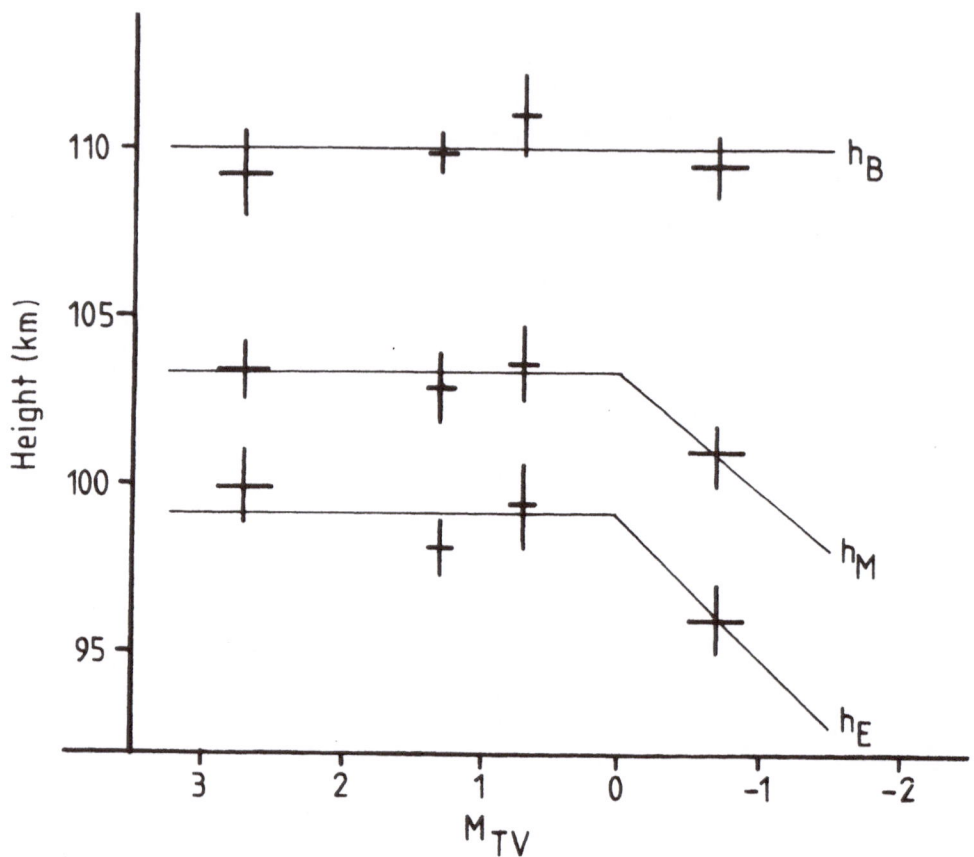

Fig. 2. The variation with meteor magnitude M_{TV} of beginning
height (h_B), height of maximum light (h_M) and end
height (h_E) for 39 Perseid meteor trails. Each point
represents the average height and average magnitude
for all trails within the zenith magnitude ranges
+4 to +2, +2 to +1, +1 to 0 and 0 to -2.

meteors brighter than $M_V = 0$, in good qualitative agreement
with dustball theory (Fig. 2).

4. The energy required to disintegrate Perseid dustball material
has been estimated from our value of critical magnitude $M_V = 0$
as $3 - 8 \times 10^5$ joule kg $^{-1}$, assuming that at Perseid velocities
of $\sim$ 63 km/sec., the corresponding critical mass of the meteroid
is 2×10^{-4} kg. This energy is several times lower than the
energy postulated by Hawkes and Jones ($1 - 6 \times 10^6$ joule kg $^{-1}$)
suggesting that Perseid dustball material may be weaker than
had been supposed previously.

References

(1) Ceplecha Z. (1968) Smithsonian Astr.Obs. Special Report 279.

(2) Jones J. and Kaiser T.R. (1966) Mon.Not. R. Astr. Soc. 133,
p. 411.

(3) Hawkes R.L. and Jones J. (1975) Mon.Not. R. Astr. Soc. 173,
p. 339.

NEAR INFRARED SPECTROSCOPY OF COMET BRADFIELD 1980t.

C. Barbieri, C. Bonoli, F. Bortoletto

Istituto di Astronomia, Università di Padova

Abstract. Three spectra of Comet Bradfield 1980t taken with the RETI-
CON system at Asiago Observatory are discussed; we present an estimate of the absolute fluxes and a discussion of the principal features
in the spectra.

1. INTRODUCTION

Comet Bradfield 1980t was observed with the RETICON spectrophotometer of the 182-cm telescope the evenings of Jan. 14 and 15, 1981.
On both dates the spectra were taken with the slit oriented E-W, 32
arcsec long and 3 arcsec wide, centered on the bright nuclear condensation. The area of comet entering the spectrophotometer was therefore
approximately $100 \ (arcsec)^2 = 2.35 \times 10^{-9}$ sr.
The spectra of Jan. 14 were obtained with the 600 gr/mm grating rota-
ted in order to cover the spectral region $\lambda\lambda\, 6000\text{-}9000$ Å, with a resolution of 5.85 Å/diode. The following evening the 1200 gr/mm grating
had to be mounted so that the region 5900-7300 was observed with a
resolution of 2.93 Å/diode.
The standard procedure for RETICON observations, which are entirely
computer-controlled, consists of two exposures, one on the object followed by an equally long exposure on the adjacent sky; between the two
exposures the object is removed off the slit by tilting a transparent plate in front of the spectrograph. However this movement was not sufficient
to exit from the extended coma of the comet, so that a different procedure had to be quickly implemented, namely to move the telescope in
Dec by some 0.5 degrees north. We shall discuss in the following the
three spectra obtained in this second way, disregarding those obtained
with the standard procedure.
Also in the reduction of the data we had to introduce a modification,

A. Coradini and M. Fulchignoni (eds.), The Comparative Study of the Planets, 317–322.
Copyright © 1982 by D. Reidel Publishing Company.

because the low elevation of the comet (Zenith distance $\approx 70°$ deg) prevented the observation of nearby standard stars.
As a consequence the chromatic sensitivity of the system, including in it the atmosphere, could not be determined in the usual direct way. To simplify the treatment of the problem it has been decided to remove from the row data a polynomial of low degree best-fitting the instrumental continuum; the effect is the artificial horizontal "continuum" present in Figs. 1 and 2. The discussion of the true fluxes has been confined to few wavelenghts, and is presented in Sect. 3.
The rest of image processing followed the usual flow which includes wavelenght calibration and linear filtering of the noise.

2. THE SPECTRA

Fig. 1 shows the two spectra obtained the 14th at UT $17^h 10^m$ and $17^h 15^m$ respectively, with 1 min exposure each. To aid in the identification of telluric and solar features, the lower part of the figure displays the spectrum of the A5-III star α Oph, obtained with the same equipment; the spectral regions strongly affected by the telluric features will not be discussed further. The spectrum obtained the following night at UT $16^h 38^m$ is shown in Fig. 2.
The spectra show good agreement; the strongest common features are reported in Table 1 together with their identifications and approximate fluxes. A closer examination however reveals few discrepancies which appear too large to be entirely due to the system noise or to the digital filters applied to the raw data. The largest differences are at $\lambda\lambda 6140$, 6242, 6440, 6807, 8325, 8894. All these features are fairly weak.
Part of them might be attributed to an imperfect subtraction of the OH bands from a still bright and rapidly changing sky. A second explanation could be sought in slightly different settings of the nucleus along the slit, causing different proportions of coma vs nucleus emissions in the spectra; this reason might be invoked for the $H_2 O^+$ features.
A comparison of our spectra with those of Comet Bradfield 1979l and obtained with an equipment very similar to ours published by Danks and Dennefeld (1981, in the following DD) doesn't help to resolve the discrepancies. The two comets had very similar spectra indeed judging from the strongest features, but no detailed comparison is possible with the small scale figure of that paper.

3. THE FLUX IN THE CONTINUUM AND IN THE BANDS

We discuss in this section the difficult determination of the absolute fluxes; only tentative estimates will be reached because of the lack of a certain calibration and of the difficulty to locate the continuum.
That the continuum has been detected is demonstrated by the strength of the telluric absorption bands, namely the B and A band of O_2 at 6860 and 7630, and the bands of H_2O at 7200, 8230 and 9020 respectively. More uncertain is the presence of the solar absorption lines: in practice

TABLE 1

Wavelength $\overset{\circ}{A}$	Identification, Remarks, Fluxes f($\times 10^{-5}$erg/sec cm^2 sr)
6109	strong,wide,complex,peaks at 6101 and 6118,NH_2,f=6
6190	complex H_2O+?, f=3
6300	strong, narrow, [OI] , f=11
6331	NH_2, f$\approx$2
6364	[OI] f=4
6415	weak
6501	weak, NH_2
6541	weak
6596	complex, H_2O+?
6623	strong, wide, complex, NH_2, f=7
6677	complex, NH_2
6729	weak, wide peak on wing of following feature
6758	strong, complex, f$\approx$6
6851	strong, narrow, at the edge of telluric B band
6958	wide, complex, edge of B band, f$\approx$4
7130	weak
7351	narrow, f$\approx$3
7380	narrow, f$\approx$2
7505	weak
7580	weak, uncertain, edge of A-band
7684	weak, uncertain, edge of A-band
7760	broad, complex feature, peaks at 7749 and 7769,C_2,f$\approx$4
7813	weak, C_2
7849	weak, C_2
7910	peak of a strong, wide, complex feature,C_2+CN?, f$\approx$10
7951	secondary peak, C_2+CN?, f$\approx$5
8007	weak, C_2 + CN?
8039	weak, C_2
8072	peak on a wide feature, CN
8110	strongest peak of same feature, CN, f$\approx$4
8143	weak,CN
8417	weak
8458	weak
8491	weak
8526	weak, possibly spurious, due to solar CaII 8542 absorption
8580	weak, possibly spurious, due to solar CaII 8542 absorption
8606	weak
8722	weak, possibly spurious, due to solar MgI 8736 abs.
8817	weak

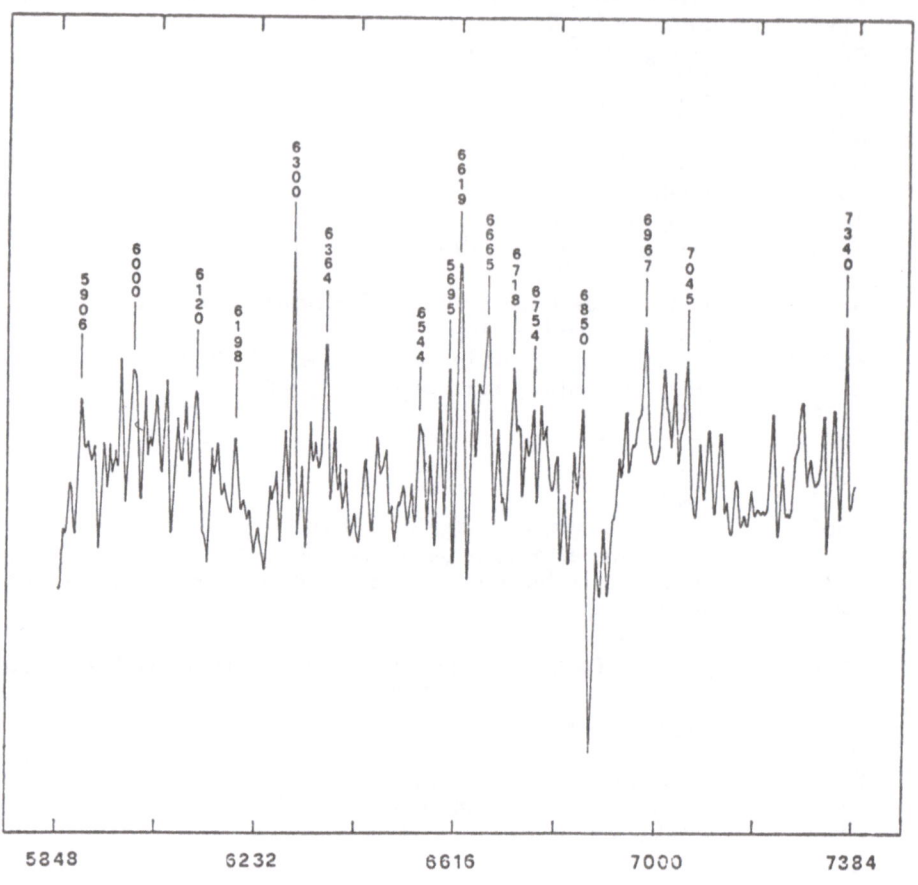

Fig. 1 – The two spectra taken on Jan 14th with the 600 gr/mm grating. The bottom spectrum is that of the comparison star   Oph, with the identification of telluric and solar absorptions. The horizontal trend of the continuum is an artifact of the data processing.

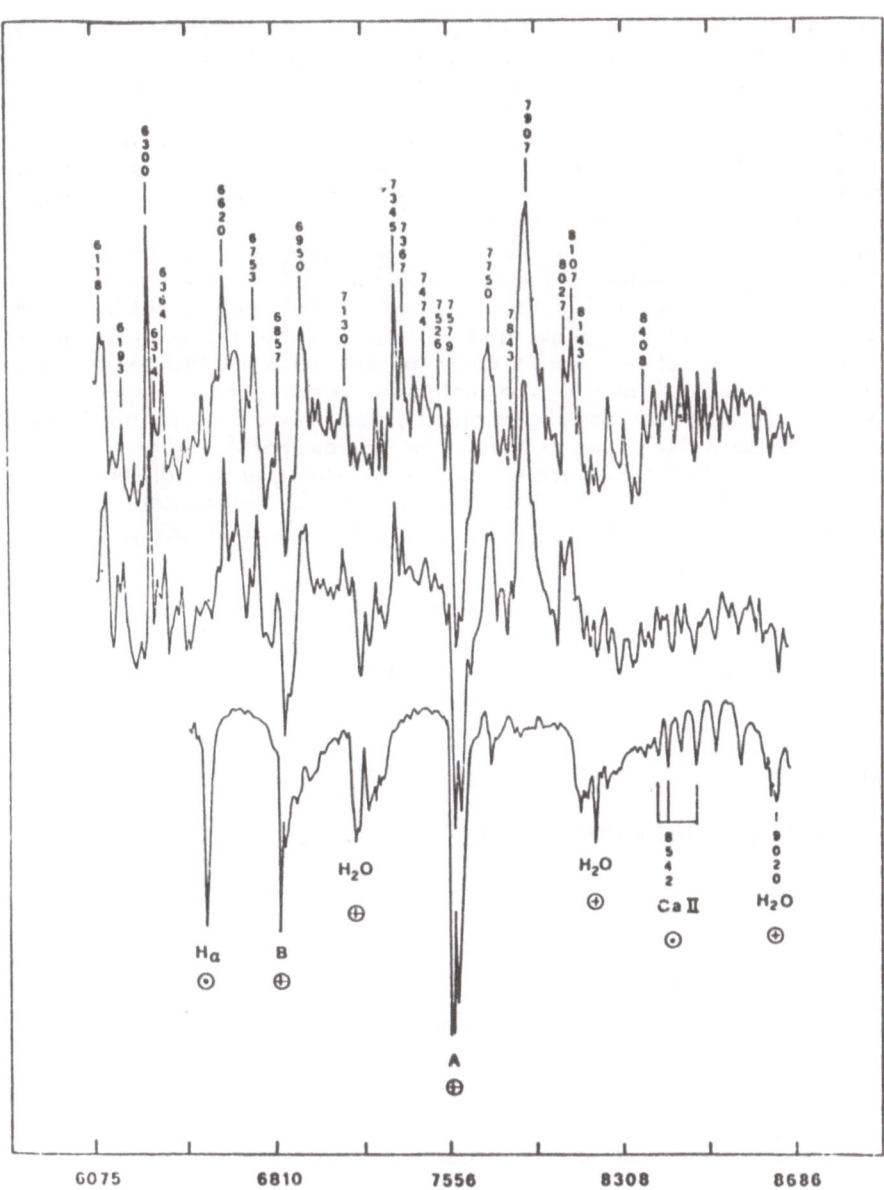

Fig. 2 – The spectrum taken on Jan 15th with the 1200 gr/mm grating. The horizontal trend is an artifact of the data processing. The telluric B-band is indicated.

we see the 8542 line of the CaII triplet, the blend of SiI and CaII at 8650, the MgI at 8740 and possibly H .

The instrumental units have been converted to flux units using average figures determined by the several programs of spectrophotometry carried out with that equipment. The result is a flux in the continuum of about 4×10^{-14} erg/sec cm^2 Å $(arcsec)^2$, or 1.7×10^{-5} erg/sec cm^2 Å sr. This value approximately obtains for the entire range of observed wave-lenghts; its formal error doesn't exceed $\pm 15\%$ so that our data are con-sistent with an essentially flat continuum. This result is again in good agreement with DD, who found a decrease of a factor 1.3 between 6100 and 10000 Å for Bradfield 1979l.

Although we stress that the absolute flux might be in error by as much as two times, the value 1.7×10^{-5} erg/sec cm^2 Å sr is the basic unit we have used to determine the line and band fluxes in Table 1. These fluxes are those at the peaks and not those under the area of the band, because the resolution doesn't allow a separation of the overlapping mo-lecules. The total fluxes in the molecular bands can be appreciably hi-gher, say three times for the strong C2 features at 7910 and 8100; the ratio C2 8100/7910 comes out however approximately correct, ≈ 0.40, not different within the errors from DD's. The greatest difference with DD is the strenght of the atomic [OI] lines in comparison with the mo-lecular features; our value is approximately half than DD's.

Aknowledgment. The observations were performed under suggestion of Prof. C. Cosmovici. A fuller discussion of the Asiago data on Comet Bradfield 1980t will be presented elsewhere.

REFERENCE

Danks A.C., Dennefeld M., 1981, Astron. J. **86**, 314

IMPACT CRATERING MECHANICS

Jean Pohl

Institut für Allgemeine und Angewandte Geophysik
Theresienstr. 41, D-8000 München 2, FRG

ABSTRACT

Impact cratering by high-velocity projectiles is an important
process in the evolution of planets. Observations on planets and
satellites with solid surfaces show that small impact craters are
bowl-shaped and deep. Larger impact structures are shallow, they
have overall flat interiors with several characteristic
morphological elements, such as central peaks or rings. In
addition to observations the cratering process is investigated by
impact and explosion experiments and by theoretical continuum
mechanics calculations. So far the formation of transient crater
cavities produced by ejection and by subsurface flow of target
material is relatively well understood. The subsequent
modifications of deep transient cavities to shallower final
bowl-shaped craters and especially to flat large complex
structures by wall failure and centripetal displacements of the
crater area and its surroundings are still controversial.

1. INTRODUCTION

Impact collision is one of the fundamental processes in the solar
system. Detailed knowledge of the impact process is essential for
example for understanding accretion of larger solid bodies,
evolution of planetary surfaces, interpretation of the origin of
lunar rocks and the use of crater frequencies for age dating.
Craters are formed on solid bodies in the solar system by
collision with smaller objects such as asteroids, meteoroids or
comet nuclei. Impact velocities vary from a few hundred m/s to
several tens of km/s. The cratering process depends on numerous
parameters which lead to a very large variety of cater morphology,

323

A. Coradini and M. Fulchignoni (eds.), The Comparative Study of the Planets, 323–331.
Copyright © 1982 by D. Reidel Publishing Company.

subsurface structure and ejecta formations. For the projectile,
composition, density, velocity, kinetic energy and impact angle
and for the target body, composition, density, geologic structure,
strength, gravitational acceleration and the nature of its
atmosphere are of major importance. Efforts to quantify the role
of the different parameters in the cratering process can be
divided into three groups: (i) Observations of caters on all
solid bodies in the solar system, (ii) Impact and explosion
experiments. (iii) Theoretical calculations.

2. MORPHOLOGIC AND STRUCTURAL INVESTIGATIONS OF IMPACT CRATERS

Morphologic observations on many planets and satellites and
structural investigations in terrestrial impact structures led to
the fundamental distinction between the smaller simple craters
and the larger complex craters.

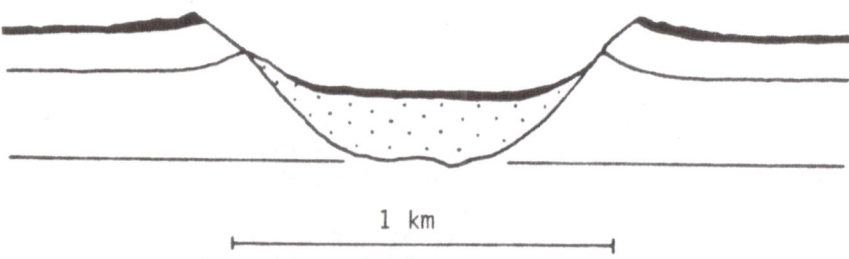

1 km

Fig. 1. Schematic cross section of a simple crater showing bowl
        shape, uplifted strata in the rim area, breccia lens
        (stippled), fall-back and fall-out ejecta (black)

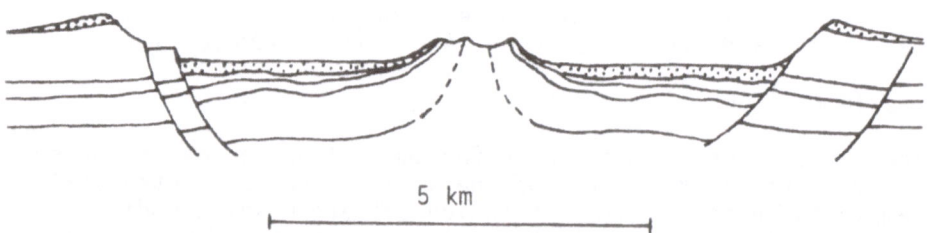

5 km

Fig. 2. Schematic cross section of a relatively small complex
        crater in a stratified sedimentary target showing flat
        floor, central uplift, terrace (left), subsidence in the
        rim area. Breccia in the crater and ejecta are stippled.

<u>Simple craters</u> are bowl-shaped, of high circularity and have
generally a rather smooth interior (Fig. 1). The depth to
diameter ratio d/D is roughly equal to 0.2 for all simple craters.

<u>Complex craters</u> (Fig. 2) have overall flat floors on which
several characteristic morphological elements can be observed,
such as central peaks or inner rings. Rims can be scalloped and
terraces indicate extended failure in the rim area. Terrestrial
complex structures show important subsurface displacements. The
largest complex structures are multiring basins with diameters up
to more than thousand km. Depth to diameter ratio varies from 0.2
to less than 0.01 in very large structures.

The transition from simple to complex craters on several planets
is shown in Fig. 3 in a depth versus diameter diagram (1, 2).
Transition diameter and depth of complex craters decrease with
increasing planetary gravity.

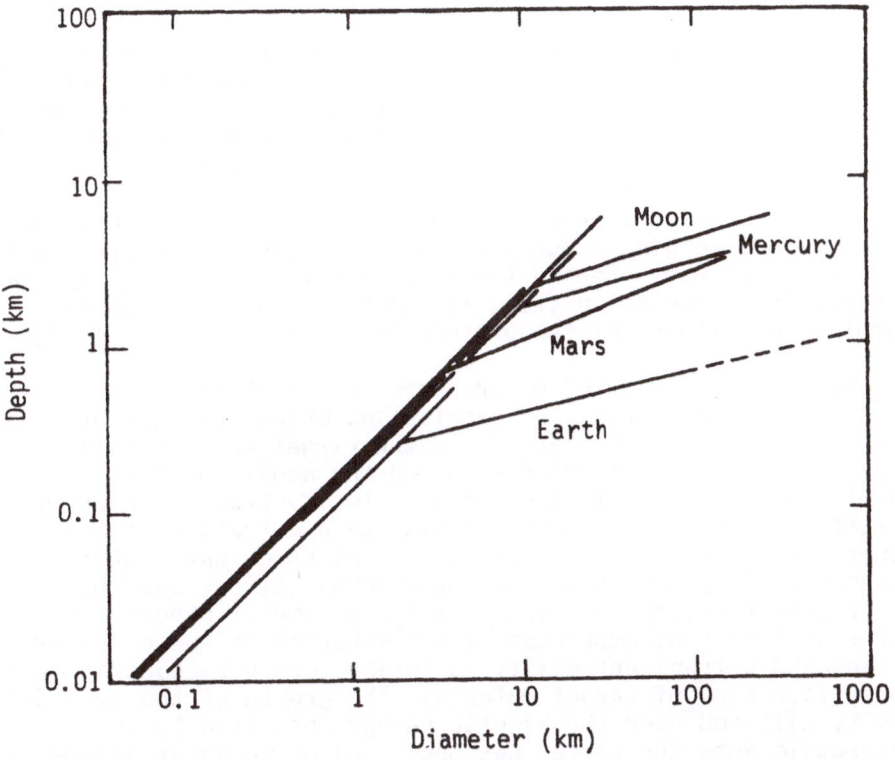

Fig. 3. Depth versus diameter diagram for different planets
        (1, 2). Least squares fits to the data. Steep slope is
        for simple craters.

3. FORMATION OF IMPACT CRATERS

3.1 Formation of transient cavities

In the following a short description of the impact process in the case of a hypervelocity projectile is given. The description is based on experimental impact and explosion cratering and on numerical modelling. Experimental data are limited to craters of moderate size (3, 4, 5) and numerical modelling is required to simulate the formation of craters larger than a few hundred m. Calculations are generally done with finite difference techniques. They necessitate knowledge of the equations of state of the material involved under the high pressure and temperature generated by the shock wave in the initial state and a realistic description of the rheological behaviour of shock-processed geological materials in late stages. Computations for the formation of impact craters, in some cases of km-size, have been carried out to fairly late times and give a detailed insight into the physics of cratering (3, 6, 7, 8, 9).

Compression stage. When a hypervelocity projectile hits a target surface, a shock wave is driven forward into the target and rearward into the projectile. Peak pressure and temperature in the shock-compressed material may reach several Mbar and several thousand degrees, depending on the projectile velocity and the projectile and target material. Both the projectile and compressed target material move downward into the target. In the target the diverging shock wave assumes rapidly an approximately hemispherical form. Behind the shock front the material motion is radial. The compression stage is essentially terminated at complete engulfment of the projectile by the target (Fig. 4).

Excavation stage. Relief of pressure begins where the shock wave reaches the free surface by rarefaction waves propagating into the projectile and the target. In the target the rarefaction produces a rearward acceleration and the addition of the rarefaction velocity to the radial velocity results in an upward directed velocity component, leading to a curvature of the flow lines. As a result of the passage of the shock wave, which becomes rapidly detached from the growing cavity, and the subsequent rarefaction waves, a large volume of shock-processed, but essentially decompressed target material has been set into motion and a transient cavity is beeing formed by ejection and subsurface flow of target material. The growth of the transient cavity will end when the kinetic energy deposited by the projectile into the target has been used up in deformational work and in gravitational work in the planetary gravity field. Crater scaling considerations for terrestrial planets show that in the case of target material with high strength, the crater depth scales as $E^{1/3}$ for impact energies E less than about 10^{12} J.

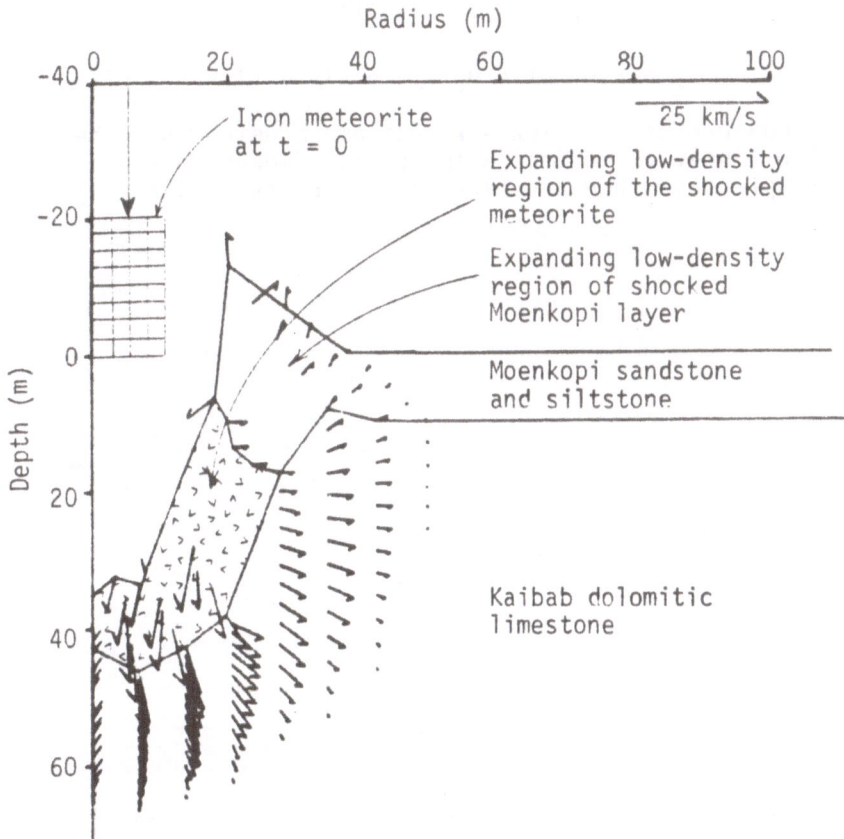

Fig. 4. Example of numerical finite-difference calculation of th
        formation of Meteor Crater, Arizona (after 7). Time is
        3.137 ms after the impact, shortly after the end of the
        compression phase. Arrows give the velocity vector field.

In the case of cohesionless material and for impact energies
greater than about 10^{22} J, gravity is the dominant parameter
limiting the crater growth and the crater depth scales as $E^{1/4}$
(gravity scaling). Between these two extreme regimes scaling
exponents range between 1/3 and 1/4 (5, 10, 11).

3.2 The Maxwell z-model

A first order analytical approximation of the cratering flow
field, which is much used at present, is the Maxwell z-model
(12, 13, 14). In the z-model an incompressible flow is assumed,
in which the radial and tangential velocities are given by (Fig. 5)

$$u_r = dR/dt = \alpha(t,\theta) \, R^{-z(t,\theta)}$$

$$u_\theta = R \, d\theta/dt = u_r \, (z-2) \sin \theta/(1 - \cos \theta)$$

α is a measure of the flow strength and z determines the shape of the flow lines. In general α and z will depend on t and θ. In the most simple case they are taken as constants $(2 \leqslant z \leqslant 4)$.

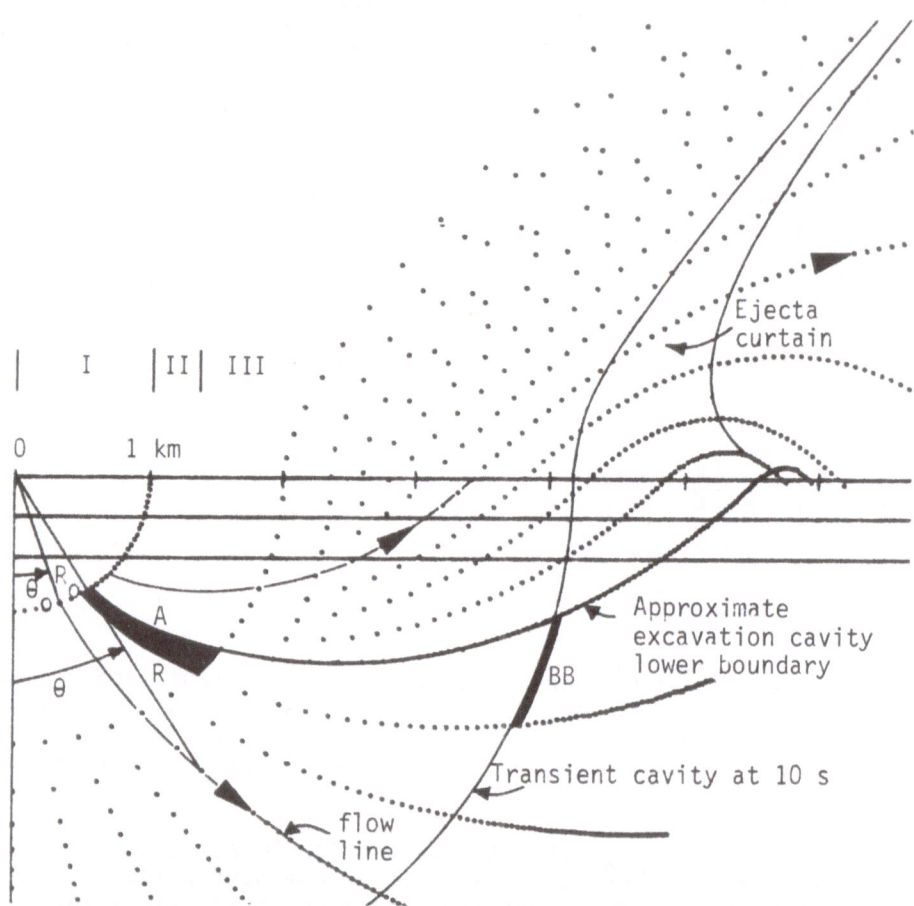

Fig. 5. Z-model flow field and ballistic ejection trajectories calculated with R_0 = 1 km, α = 7.5 km^4/s, z = 3. Dot spacing is 0.5 s.
Target material contained in volume A at 0 s will be contained in volume B at 10 s.
I, II and III indicate schematically zones of vaporization, melting and brecciation, corresponding to the decreasing peak pressure in the shock wave.

The flow lines are given by

$$\left(\frac{R}{R_0}\right)^{z-2} = \frac{1 - \cos\theta}{1 - \cos\theta_0}$$

Fig. 5 shows an example of the z-model subsurface flow field and ballistic ejection trajectories in $g = 9.81$ m/s^2 and in vacuum. Calculation of subsurface motions was stopped when the stream velocity was less than 50 m/s. The transient cavity at $t = 10$ s and the material contained in the ejecta curtain at this moment are also shown.

The z-model explains many important features observed in impact craters, such as for example uplifted rims and ordered ejection leading to an inverted stratigraphy in impact formations. Fig. 5 also shows the excavation cavity (13, 15, 16), from which the ejected material originates. Target material originally below the excavation boundary will not leave the crater cavity, but will be displaced downward and outward along the flow lines shown in Fig. 5. In order to compensate for the subsurface displacements a transient uplifted rim will be formed. Material incorporated in the subsurface flow will undergo strong shearing deformation, resulting in an important shortening of the initial distance between high-shock and low shock zones (Volumes A and B in Fig. 5). The simple z-model has severe   limitations, which are currently investigated by comparison with numerical modelling to late stages of crater formation (e.g. 6, 8).

3.3 Modification of transient cavities

In the case of simple bowl-shaped final craters with a depth to diameter ratio of 0.2 the transient cavity will be modified by centripetal slumping of brecciated material from the cavity wall and by subsidence of the rim area (Fig. 1). For the formation of larger complex structures much more important modifications of a primary transient cavity are required in order to obtain the final overall shallow morphology and the characteristic inner structures such as central peaks, terraces, ring grabens etc.

The formation of central peaks and central uplift structures requires an inversion of the motions involved in the excavation process, i.e. a centripetal displacement with a vertical component in the center. The uplift is necessarily associated with subsidence displacements in the outer regions. There is evidence from observations on several impact craters, from large explosions and from numerical modelling, that such a velocity reversal can begin near to the end of the excavation phase. The rebound

mechanism does not exclude deep base failure in the planetary
gravity fields, leading to flattened structures and to a large
increase of the outer diameter. Both mechanisms can operate
together and transform a deep transient crater into a larger
shallow complex structure.

Several other factors may play an important role in the formation
of complex craters. Target layering with varying shock wave impe-
dance can under certain conditions lead to extended shallow
excavation of near surface material and produce a nested crater
type with an overall small final depth to diameter ratio. However
subsequent modifications as described above are also required in
this case to explain the characteristic inner features. Another
possible explanation for the formation of some shallow structures
is that the the transient crater itself was much shallower.
Shallow transient cavities can be produced by low-density projectiles, such as comet nuclei. It is however difficult to imagine
that all complex structures were produced by low-density projectiles, since this would imply two completely different populations
of projectiles for simple and complex craters. In very large
structures the transient crater will be shallower because of the
dominating role of the planetary gravity field in the excavation
process.

The problem of the formation of complex structures is not solved
and further detailed ·investigations especially on terrestrial
structures have to be made as well as realistic numerical
modelling.

REFERENCES

1. Pike, R.J., 1980a: Formation of complex craters: Evidence
        from Mars and other planets. Icarus 43, 1-19

2. Pike, R.J., 1980b: Control of crater morphology by gravity
        and target type: Mars, Earth, Moon. Proc. Lunar Planet.
        Sci. Conf. 11th, 2159-2189

3. Roddy, D.J., Pepin, R.O., Merrill, R.B. (eds.), 1977: Impact
        and Explosion Cratering. Pergamon, New York, 1301 p.

4. Stöffler, D., Gault, D.E., Wedekind, J., Polkowski, G., 1975:
        Experimental hypervelocity impact into quartz sand:
        Distribution and shock metamorphism of ejecta.
        J. Geophys. Res. 80, 4063-4077

5. Gault, D.E., Wedekind, J.A., 1977: Experimental hypervelocity
        impact into quartz sand II, Effects of gravitational
        acceleration. In: Impact and Explosion Cratering (D.J.
        Roddy, R.O. Pepin, R.B. Merrill, eds.). Pergamon,
        New York, 1231-1244

6. Thomsen, J.M. et al., 1979: Calculational investigation of
        impact cratering dynamics: Early time material motions.
        Proc. Lunar Planet. Sci. Conf. 10th, 2741-2756

7. Roddy, D.J. et al., 1980: Computer code simulations of the
        formation of Meteor Crater, Arizona: Calculations MC-1
        and MC-2. Proc. Lunar Planet. Sci. Conf. 11th, 2275-
        2308

8. Austin, M.G. et al., 1980: Calculational investigation of
        impact cratering dynamics: Material motions during the
        crater growth period. Proc. Lunar Planet. Sci. Conf.
        11th, 2325-2345

9. Orphal, D.L., Borden, W.F., Larson, S.A., Schultz, P.H.,
        1890: Impact melt generation and transport. Proc. Lunar
        Planet. Sci. Conf. 11th, 2309-2323

10. Gault, D.E. et al., 1975: Some comparisons of impact craters
        on Mercury and the Moon. J. Geophys. Res. $\underline{80}$, 2444-2460

11. O'Keefe, D.J., Ahrens, T.J.,1981: Impact cratering: The
        effect of crustal strength and planetary gravity. Rev.
        Geophys. Space Phys. $\underline{19}$, 1-12

12. Maxwell, D.E., 1973: Cratering flow and crater prediction
        methods. Tech. Memo TCAM 73-17, Physics International,
        Calif., 50 p.

13. Ivanov, B.A., 1976: The effect of gravity on crater formation:
        Thickness of ejecta and concentric basins. Proc. Lunar
        Sci. Conf. 7th, 2947-2965

14. Maxwell, D.E., 1977: Simple z model of cratering, ejection,
        and the overturned flap. In: Impact and Explosion
        Cratering (D.J. Roddy, R.O. Pepin, R.B. Merrill, eds.).
        Pergamon, New York, 1003-1008

15. Pohl, J., Gall, H., 1977: Bau und Entstehung des Ries-Kraters.
        Geologica Bavarica $\underline{76}$, 159-175

16. Croft, S.K., 1980: Cratering flow fields: Implications for the
        excavation and transient expansion stages of crater
        formation. Proc. Lunar Planet. Sci. Conf. 11th, 2347-2378

EXPERIMENTAL STUDY OF EFFECTS ASSOCIATED WITH MACROSCOPIC HYPERVELOCITY IMPACTS

G.Martelli[+],R.Bianchi[°],P.Cerroni[+],M.Coradini[°],
R.Flavill[x],P.Hurren[+],P.N.Smith[+],F.Waldner[^]

[+] University of Sussex,Brighton,UK
[°] Istituto di Astrofisica Spaziale,Rome,Italy
[x] University of Kent.Canterbury,UK
[^] Università di Bari,Bari,Italy

We present and discuss results from an ongoing series of hypervelocity impact experiments performed using explosive shaped charges. Primary and secondary cratering have been studied both at atmospheric pressure and "in vacuo" (p <1 mm Hg) using basalt-like and clay targets. The primary impacts have been recorded using fast framing photography, and studied with the help of computer enhanced false colour analysis; the secondary impacts, on glass and quartz samples, using a scanning electron microscope. Magnetic phenomena associated with the impact produced plasma have also been recorded.

1. INTRODUCTION

American and Soviet space missions to the planets of the Solar System have shown that hypervelocity impact phenomena have played a fundamental role in the formation and evolution of the bodies of the Solar System.

Although many theoretical and, to a lesser extent, experimental studies on hypervelocity impacts have hitherto been carried out, much is still to be learned about the physics of the hypervelocity impacts and related phenomena. This is basically due to a lack of systematic studies of "true" hypervelocity experiments

A. Coradini and M. Fulchignoni (eds.), The Comparative Study of the Planets, 333–357.
Copyright © 1982 by D. Reidel Publishing Company.

in the laboratory, since the techniques currently used seldom reach impact velocities higher than, say, 8 km/sec On the other hand, a deeper understanding of phenomena such as the formation of primary and secondary craters, the mass and velocity distribution of the ejecta, the thermal and mechanical response of various types of impacted surfaces (rock, ice, permafrost, etc.), the distribution of angular momentum between the parent body and the ejecta, etc., can be obtained only by performing experiments where the impact velocities are truly representative of meteoritic velocities.

In the pursuit of the acceleration of macroscopic projectiles (m≳0.01 gr) to very high velocities, the only available technique to date is the explosive shaped charge, suitably modified as described below. Apart from the obvious advantage of accelerating relatively large masses to meteoritic speeds, this technique suffers from some drawbacks, in that the choice of the projectile material is limited, and each test involves the detonation of a substantial amount of explosive, which in turn makes it necessary to conduct the experiments in special (and not always easily available) installations.

In the present paper we relate how, in spite of the above difficulties, shaped charges have been used successfully in a series of macroscopic hypervelocity impact experiments designed to study various aspects of cratering phenomena. We describe first in some detail the experimental techniques and diagnostics employed, then report and discuss the results obtained. More detailed discussions and interpretation will be found in other papers in the present volume (1,2) together with a description of the latest technical advances on the production of hypervelocity projectiles (3).

2. DESCRIPTION OF THE EXPERIMENTS

2.1 The shaped charge

The acceleration of macroscopic (∿1gr) projectiles to velocities of the order of 10 km/sec was achieved using the metal lined explosive shaped charge technique, suitably modified as described in (4). This technique consists essentially of a cylinder of explosive having a high detonation velocity v_d (typically 7-8 km/sec). A hollow metal cone is inserted at one end of the cylinder, their axes being rigorously aligned (Fig. 1).

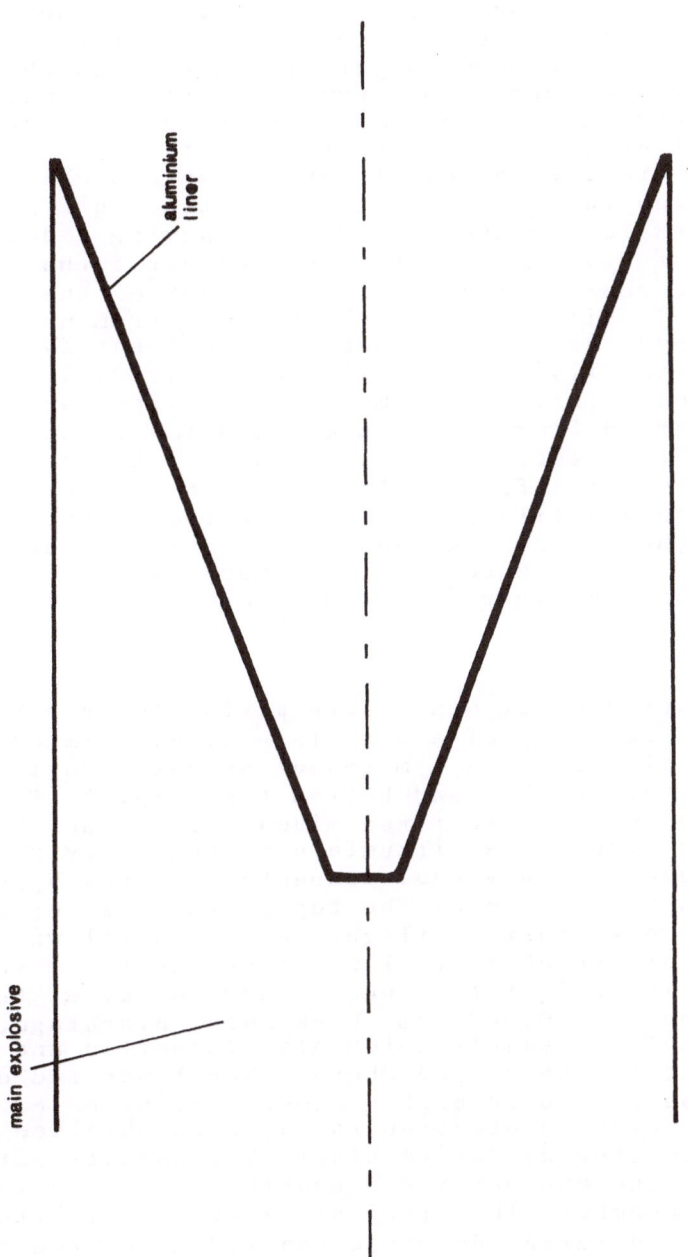

Figure 1. Sketch of the shaped charge.

When detonated at one end, the detonation wave collapses the cone, which is forced inside out and then
ejected with a velocity larger than v_d. As long as the
opening angle α of the cone exceeds some critical value,
the collapsed cone forms a jet, the tip of which moves
faster than the rest. A subsequent suitably timed detonation sweeps away the slower part of the jet and the tip
alone is allowed to reach the target. In the experiments
reported here the conical liner was made of HE 30 Aluminium alloy (density 2.73 gr/cm^3), the angle α being
41°. The explosive used was Comb B, having a detonation
velocity of 7.9 km/sec. It should be noted that digiti-
sation and reproduction in false colour of the pictures
of the projectiles taken with the fast framing camera
before impact has made it possible to identify the position of the flying projectile and its trail, and to
put an upper limit to its mass, which previously had
to be inferred from rather uncertain theoretical considerations (5). Thus, Fig. 2, which is a black and white
reproduction of a false colour picture of the flying tip,
shows the projectile bounded by the two isobars which
identify the compression and decompression front of the
blast wave (6) produced by its supersonic motion through
the residual gas in the vacuum chamber (see below).

2.2 Experimental set-up

Most cratering experiments were performed in a vacuum
chamber, consisting of a stainless steel cylinder $\sim$1 m
in inside diameter and 1 m inside height, provided with
domed ends at the top and bottom (see Fig.3). Four ports
of 40 cm diameter were fixed midway along the chamber,
equispaced around the circumference and carrying plexiglass windows and a vacuum connector flange. Hypervelocity projectiles entered the top of the chamber after
passing down a plastic "flight tube" coaxial with the
main body of the chamber, 1.2 m long and 10 cm inside
diameter. The flight tube was vacuum sealed at the up-
permost end by a 0.0025 mm thick mylar diaphragm, clam-
ped in a nylon assembly which also supported and located accurately the shaped charge. The lower end of the
flight tube passed through a close fitting hole in a
4 cm thick steel protection table which shielded the
chamber from the explosive blast. The targets were hou-
sed inside the chamber and supported by the inside base of the chamber. In a typical experimental situation
the charge to target distance was 2.2 m and the residual pressure, when the chamber was evacuated, was
$\sim$ 1 torr (hereafter referred to as "vacuum").

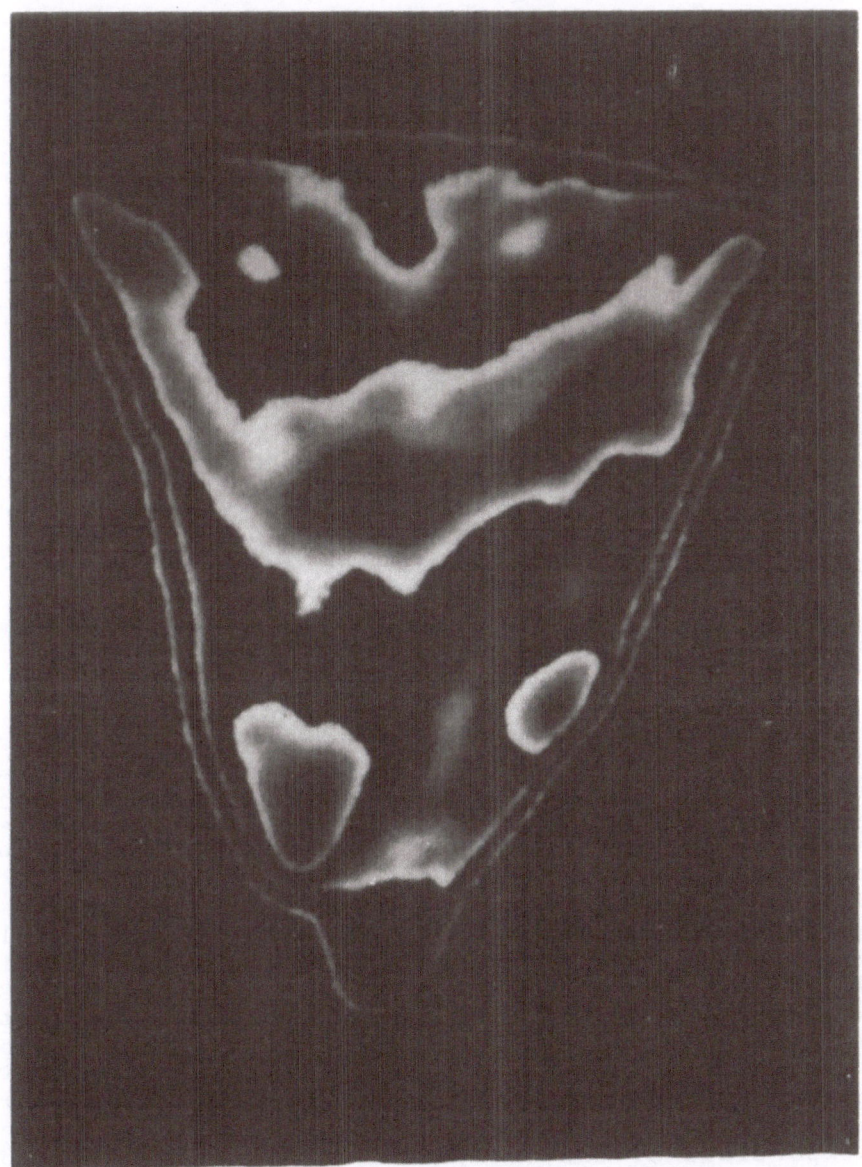

Figure 2.   Black and white reproduction of a false
colour picture of the flying tip

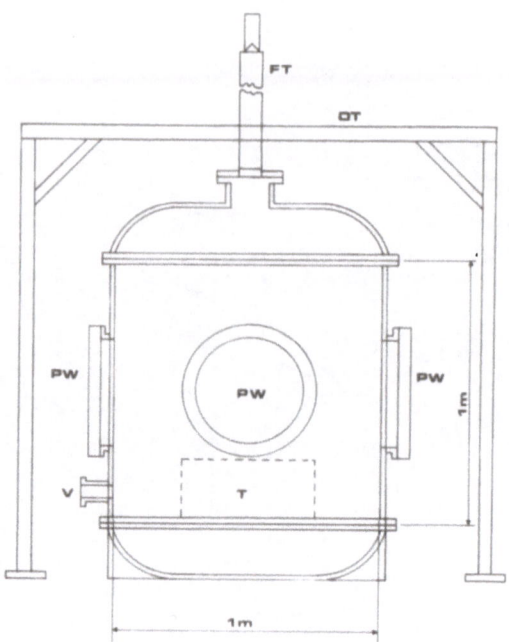

Figure 3. Sketch of the vacuum chamber: FT - fligth
tubes; DT - deflection table; PW - perspex windows;
T - target; V - to the vacuum pump.

        In order to measure the distribution in size, ele-
vation, angle and velocity of secondary particles ejected from the impact craters, collection arenas were
placed around the impact area in two experiments. The
arenas consisted of a conical outline of height 27 cm
and base diameter 37 cm, constructed from Aluminium and
plastic bars with top and bottom supported by stainless
steel rings. Plasticene was used for potentially stressed joints, thus achieving a structure which was highly
transparent to blast waves and at the same time contained intentional mechanical failure points to enhance
sample survival. Glass and optically polished quartz
samples were mounted on 13 mm scanning electron microscope (SEM) stubs and arrayed along the bars to inter-
cept portions of the ejecta flux at various elevation
angles.

        The impact events were photographed using an Ima-
con 790 image converter camera (7). The camera viewed
the test area through a 45° mirror located inside a
right angled aperture set into the wall of the blast
proof bunker and closed with a 14 cm diameter plexiglass

window. The chamber was positioned so that the camera
looked down on the impact area at an angle of 15° to the
horizontal with an optical separation of 4 m. Thus,
with a 76 mm lens, the chamber window just filled the
camera field of view. The impact events are highly luminous, making very short exposure times possible. Typically, 10 to 20 exposures per impact were made, each
lasting 1.8 μsec at 9 μsec intervals.

The trigger pulses needed to start the camera framing sequence were generated by an "active" electronic
timing system, which made it possible to define the projectile's distance from the target when the first exposure was made. In a typical experiment the camera was
preset to operate when the projectile was about 20 cm
from the target.

Electromagnetic signals generated by the flight of
the projectile and its impact were measured using 50-turn
pick-up coils having a diameter of 11 cm, mounted in or-
thogonal planes near the impact point, and recorded, via
buffer amplifiers positioned inside the vacuum chamber,
on magnetic tape. The overall sensivity of the magnetic
pick-up system was ∿100 mV per 3.4 milligauss at 100 KHz.

2.3 Target material

Most cratering experiments were performed using basalt-
-like targets, which were cut from large boulders and
encased in duraluminium boxes, with walls 15 mm thick
and having dimensions 50 x 50 x 30 cm^3.

The boulders were obtained from the ancient lava
field of Monte Falcone, a small cone 203 m above s.l.,
in the volcanic region of Colli Albani near Rome. The
thin section analysis shows an effusive rock with larger
crystals (phenocrysts) of leucite and augite set in a
finer-grained ground-mass (porphyritic texture); in the
hypocrystalline grain matrix, close to the most crystal-
lized material, very limited glassy areas, probably generated by late remelting, are present. Most of the leuco-
cratic components are formed by leucitites with crystals
in tetragonal system, thickly geminate and pseudopara-
morph in respect to the cubic system; in the ground-mass
sanidine crystals, sometimes Carlsbad geminate, are present. Among the coloured components the clinipyroxenes
(augitic type) prevail, but also some isolated Mg-olivine
(Fo_{92}) crystal is present. Besides the grain matrix contains a lot of opaque crystals, also included into the
leucite phenocrysts with the tipical crown structure

NIGGLI VALUES

si	al	fm	c	alk	k	mg	ti	p	qz	Si°	Magmatic Type
99.0	19.1	35.7	28.0	17.2	0.70	0.26	1.7	1.1	-70	0.59	Normal sommaitic

MINERALOGICAL ANALYSIS

SiO_2	TiO_2	Al_2O_3	Fe_2O_3	FeO	MnO	MgO	CaO	Na_2O	K_2O	H_2O^+	H_2O^-	P_2O_5	Sum
44.69	1.03	14.57	5.56	3.69	0.15	5.83	11.68	2.37	8.62	0.60	0.11	1.16	100.06

Table I. Chemical composition and the Niggli values of the target material.

(Schlackenkreuzchen Struktur). Interstitial magnetite
is also present, togheter with rare crystals of nephe-
line and mililite. Interstitial calcite was placed by
late filling. Table I shows the chemical composition
and the Niggli values according to Washington (8).

This material cannot be considered very close in
comparison to lunar rocks, in spite of some chemical and
mineralogical analogies to some lunar basalts. However,
it does reproduce some of the magnetic features of lunar
materials, since it is magnetically soft but not particularly viscous.

A few other tests were carried out using layers of
clay of different colours resting upon a concrete base
also encased in dural boxes.

The leucitite targets were chosen in order to reproduce, as closely as possible, the impact of meteorites
on a planetary solid surface. The experiments with clay
were designed to reproduce the reversal of the stratigraphy in the layers forming the ejecta blankets.

3. RESULTS

3.1 Primary cratering

We have performed a statistically significant number of
macroscopic cratering experiments, at velocities of about
10 km/sec, both "in vacuo" (8 tests) and at atmospheric
pressure (7 tests). The development of the crater, and
the formation of the ejecta plume and of the impact produced plasma, were recorded photographically by the Imacon camera at 9 μsec intervals over a period of 100 μsec
(for the experiments "in vacuo") or about 200 μsec for
the experiments in air. Figures 4, 5 and 7 show typical
sequences of fast framing camera pictures, the order of
the individual frames being as follows:

$$
\begin{array}{ccc}
\ldots & 5 & 3 & 1 \\
\ldots & 6 & 4 & 2
\end{array}
$$

The use of the fast framing camera has made it possible to study the evolution of the cratering processes
in some detail. Figure 4 shows a typical sequence taken
at atmospheric pressure for a projectile impacting at
right angles onto leucitite target. The first three frames show the "meteoroid" and meteor trail before impact.
The fourth frame reveals the first contact, the total

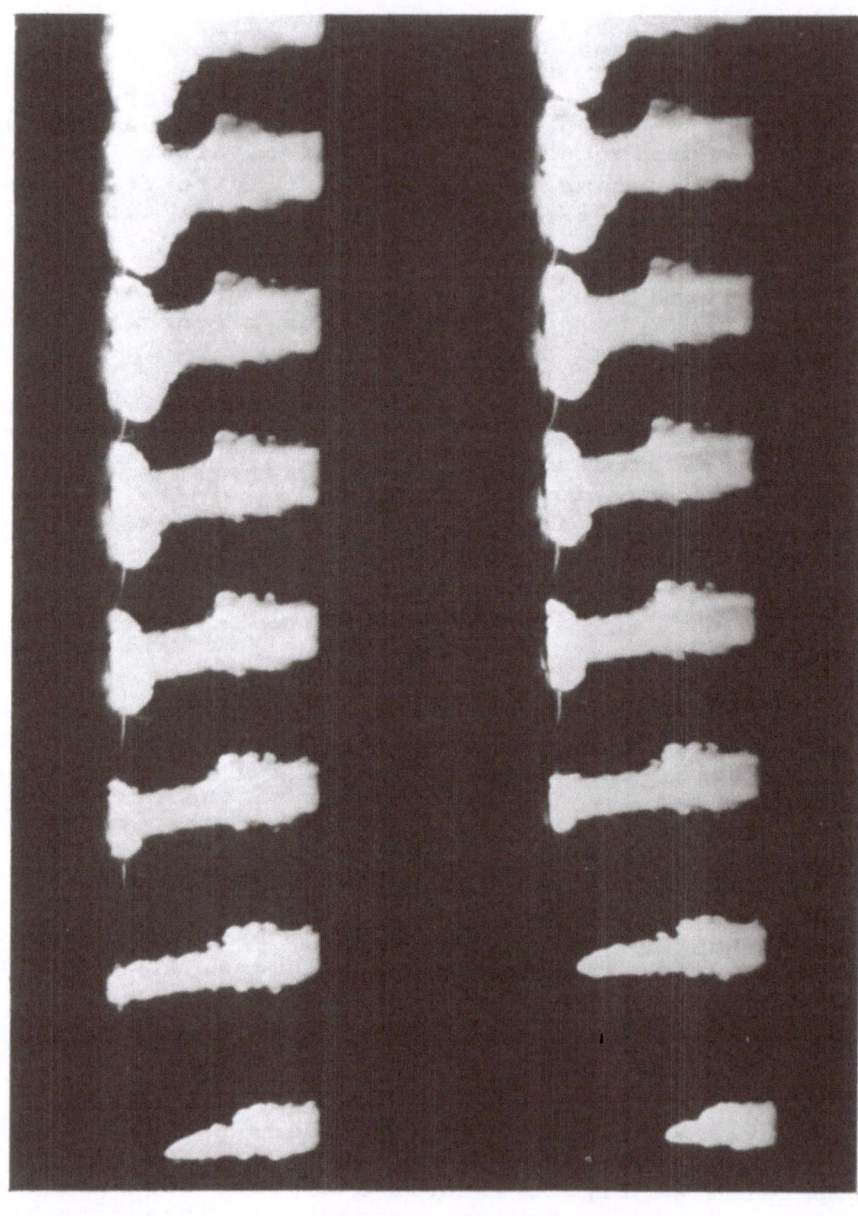

Figure 4. Sequence of cratering at atmospheric pressure.

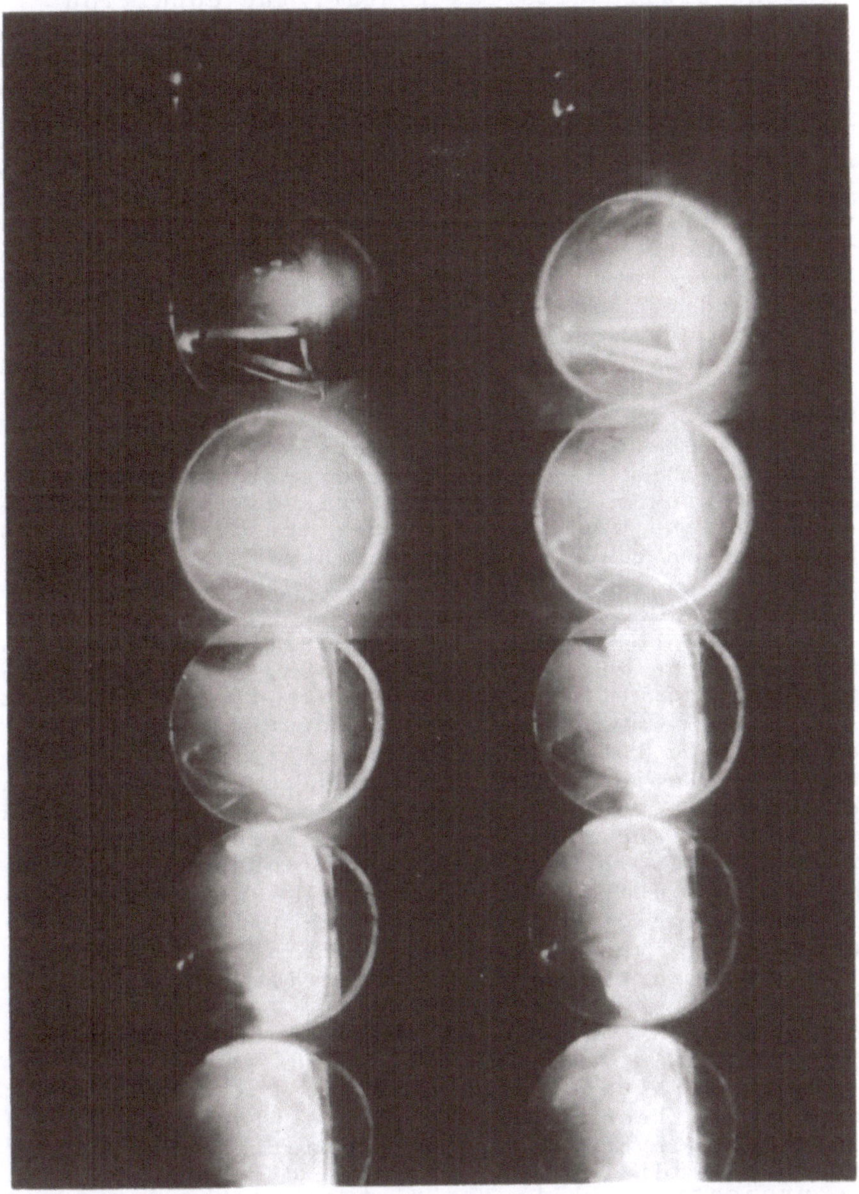

Figure 5.   Sequence of cratering "in vacuo".

length of the bright trail seen in the frame being approximately 50 cm. In the first three frames the shock
wave in the trail satisfies roughly the condition
sin α = 1/M, where α is the angle of the shock wave and
M is the Mach number. Its energy is probably dissipated
rapidly in the dense medium, through excitation, dissociation and ionization processes. After impact the formation of the plume and the ejection of material is visible. From frame 10 onwards, individual large ejecta
can be followed from frame to frame, and this gives an
indication of the time taken for crater formation.
From an estimate of their velocity (a few hundred m/sec)
it is possible to go back to the time they were impact
expelled, indicating that the large masses detached themselves from the targets at a few tens of μsec after impact. In a previous set of experiments performed under
similar conditions it was observed that no microscopic
secondary craters were formed on the scanning electron
microscope stubs placed approximately 20 cm from the
impact area. This indicates a severe ablation of the
very fast moving micron sized particles over relatively
short paths, and would account for the limited expansion
of the plume in air, as shown in the last frames of
Fig. 4.

The sequence in Fig. 5 shows cratering "in vacuo".
In this case the first two frames show a well developed
blast wave, indicating that the shock is not attenuated
over the first few tens of centimetres of the trail. A
comparison with the previous set of pictures shows a
different evolution of the post-impact gases and dust.
(Note the arena housing the SEM stubs). In this picture
the rapidly expanding cloud of fine secondary ejecta is
clearly visible. These features are enhanced in Fig.6,
which is a computer processed enlargement of frame 8 of
Fig.5. This type of processing of the pictures has consistently shown that the cloud of finer ejecta moves
much faster than the primary projectile, with velocity
of the order of 20 km/sec. These findings have been confirmed by an SEM analysis of the secondary craters.

Figure 7 illustrates one of the tests in which the
surface of the leucitite targets was placed at 45° to
the direction of the impacting projectile. The larger
ejecta moved radially away from the direction of arrival
of the projectile, while the plasma and the finer ejecta plume developed at right angles to the target surface. Post-impact examination showed that the crater
was roughly circular. The post-impact motion of the large ejecta and the plume development illustrate clearly

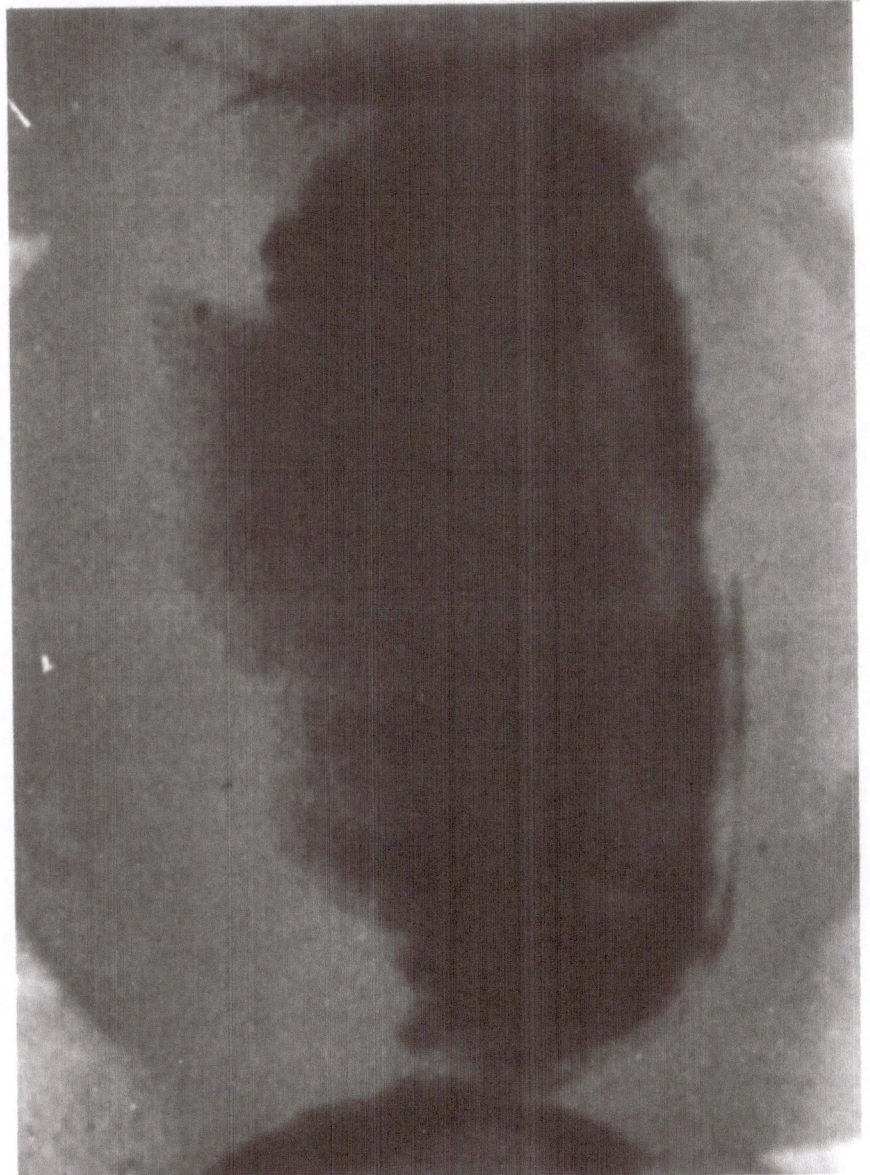

Figure 6. Black and white reproduction of a computer processed enlargement of frame 8 of Fig. 5.

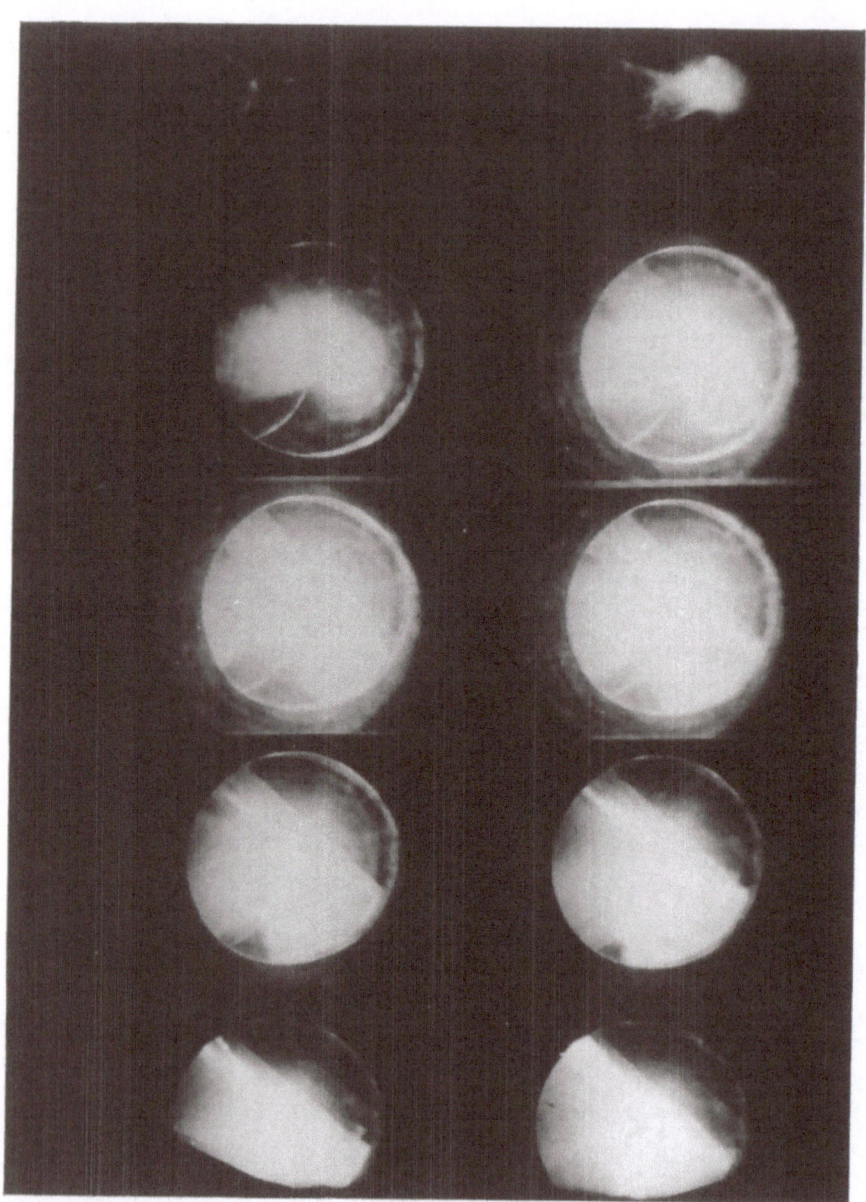

Figure 7. Sequence of cratering at an angle of 45°.

the different roles played by the momentum transfer and the explosive energy release respectively. All the craters in leucitite targets, both in air and "in vacuo", had diameters ranging from a few centimetres to 20 centimetres or so.

A morphological analysis of the craters was carried out. Plaster moulds were produced in order to obtain lighter and more handy crater samples and to make measurements of the morphological parameters. The vertical profiles of each crater at different depth were determined by cutting the plaster moulds into a number of parallel sections 1 cm wide (Fig. 8). The depth was measured on the vertical profile obtained from the central section. The crater diameter has been taken to the average of the maximum and minimum diameters. The craters, although approximately circular in planform, are in fact somewhat irregular in shape, due to erratic spalling of the larger pieces of ejecta fragments. The volume of the target ejected material was determined by filling the waterproofed moulds with water at room temperature. The depth matrixes, shown in Fig. 9, were computed by digitizing the vertical profiles of the craters and were used as data base to obtain contour lines plots representing the general trend of the crater walls. In Fig. 10 the contour lines of some craters are shown. Two distinct topographic parts were produced by the impact: the inner part of the crater with steeper walls, and the outer part with shallow sloped walls. The first one represents the part of the target where most of the impact energy was concentrated, producing finer-grained ejecta and sublimation of target material. The second can be interpreted as the part excavated by spallation of large fragments. This interpretation is supported by the similarity in morphology and roughness of the surface of the larger pieces the ejecta and the conchoidal surface of the crater walls. In some cases the contour lines of the craters show that the inner and the outer portions are characterized by the same slope. This is due to the lack of spallation during the impact. Fractures, nearly perpendicular to the surface of the basalt block and roughly radial with respect to the crater axis, appear to delimit the boundaries of the larger pieces of the ejecta. The morphology of the craters does not show reversal of the melted material, as the one shown in the microparticle impact craters, up to a centimeter in diameter, found in lunar rocks. Although there is evidence for shock melting in the laboratory experiments, the absence of reversed morphologies and the presence of spalla-

Figure 8.   Vertical profiles of a typical crater.

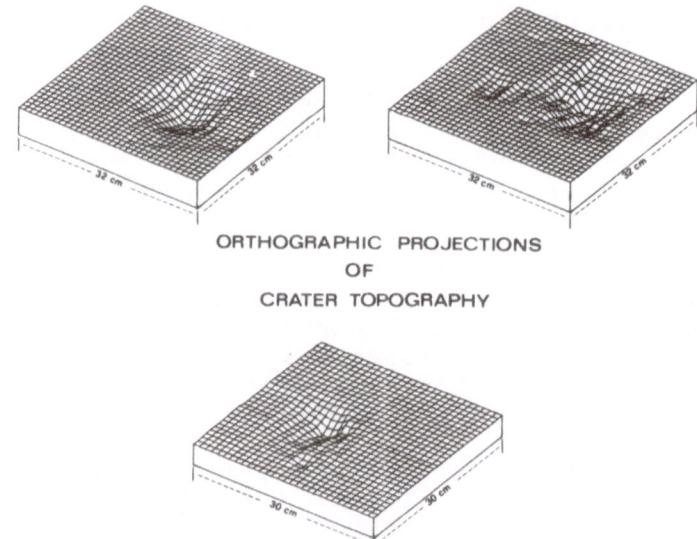

ORTHOGRAPHIC PROJECTIONS
OF
CRATER TOPOGRAPHY

Figure 9.   Depth matrixes , computed by digitising the
           vertical profiles of the craters.

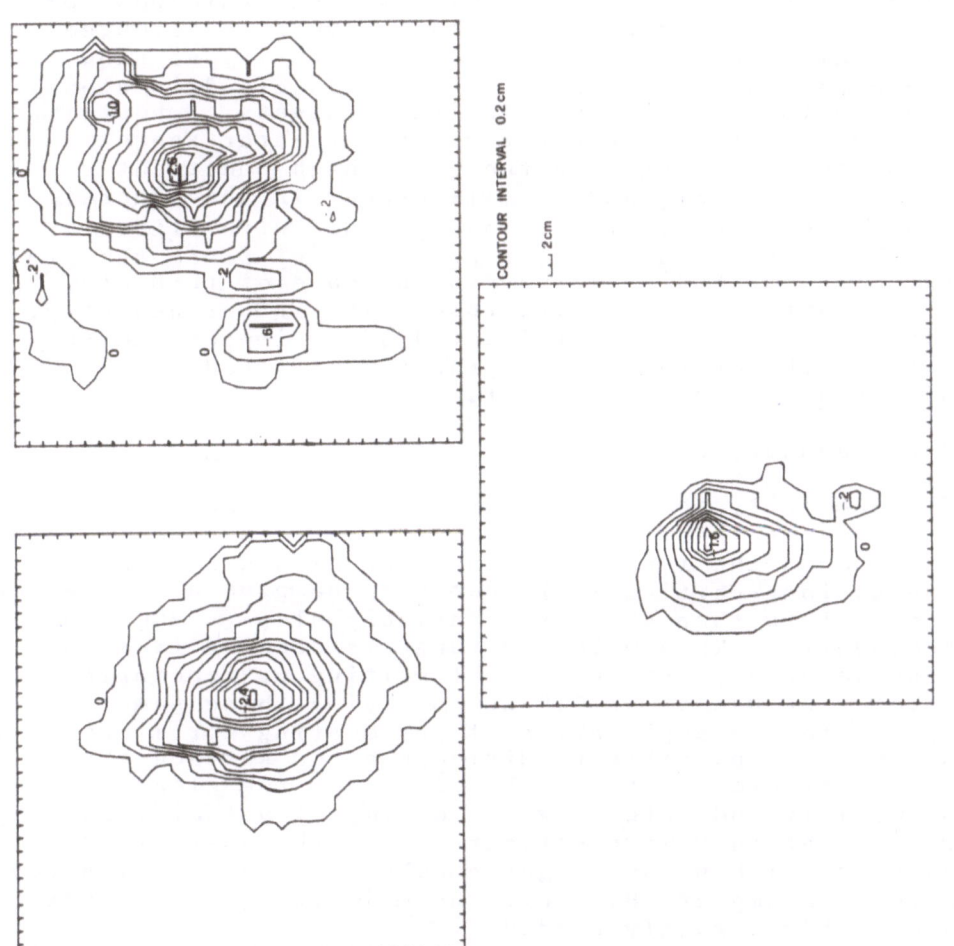

Figure 10. Contour lines of some craters.

tion of rock plates around the craters may be due to
the very high strength of the crystalline target material.

One exploratory experiment was done using a soft
multicoloured layered clay target, on a concrete foundation. Three layers of clay with a total thickness of
10 cm were separated by two thin layers of coloured
material. The photographs of the impact event show that
the expansion of the secondary ejecta cloud is slower
than in the case of the leucitite targets. A post-impact
inspection of the target revealed a crater approximately 39 cm in diameter in the clay which underwent stratigraphic reversal, and a small crater in the concrete
bed. More experiments are now planned with different
depths of clay and smaller projectiles, since the present results indicate that in the case of thin clay
over a hard bed of finite thickness, in our geometry,
most of the energy is reflected after impact and the
site of the crater rim is probably affected by the finite geometry of the target.

3.2 Secondary cratering

The SEM analysis of secondary cratering caused by the
finer ejecta hitting the small targets (SEM stubs) placed around the impact area has revealed a surface encrustation (particurlarly dense on samples at elevation
angles less than 40°) of relatively slow (< 200 m/sec)
secondaries. This velocity range is deduced from the
absence of any obvious impact melting or fracturing features. The use of an SEM in observing these samples
(prior to the application of any coating for microscopy)
has made it possible to dislodge a few surface strains
purely by electrostatic charging from the electron
beam. This indicates a very low impact velocity and suggests that they were collected from the dust cloud circulating within the target chamber for several minutes
after the impact. However, the majority (>95 %) were
apparently strongly bonded.

SEM studies of the glass and quartz samples have
shown that very different terrains have resulted on these materials due to the many secondary impacts. In the
experiment with glass samples the flux of secondary
particles was apparently much higher than in the case
of the the quartz samples, and the entire top surface
has become very tightly bound to the slow ejecta blanket (Fig. 11).

Figure 11. Secondary microcraters on glass SEM samples.

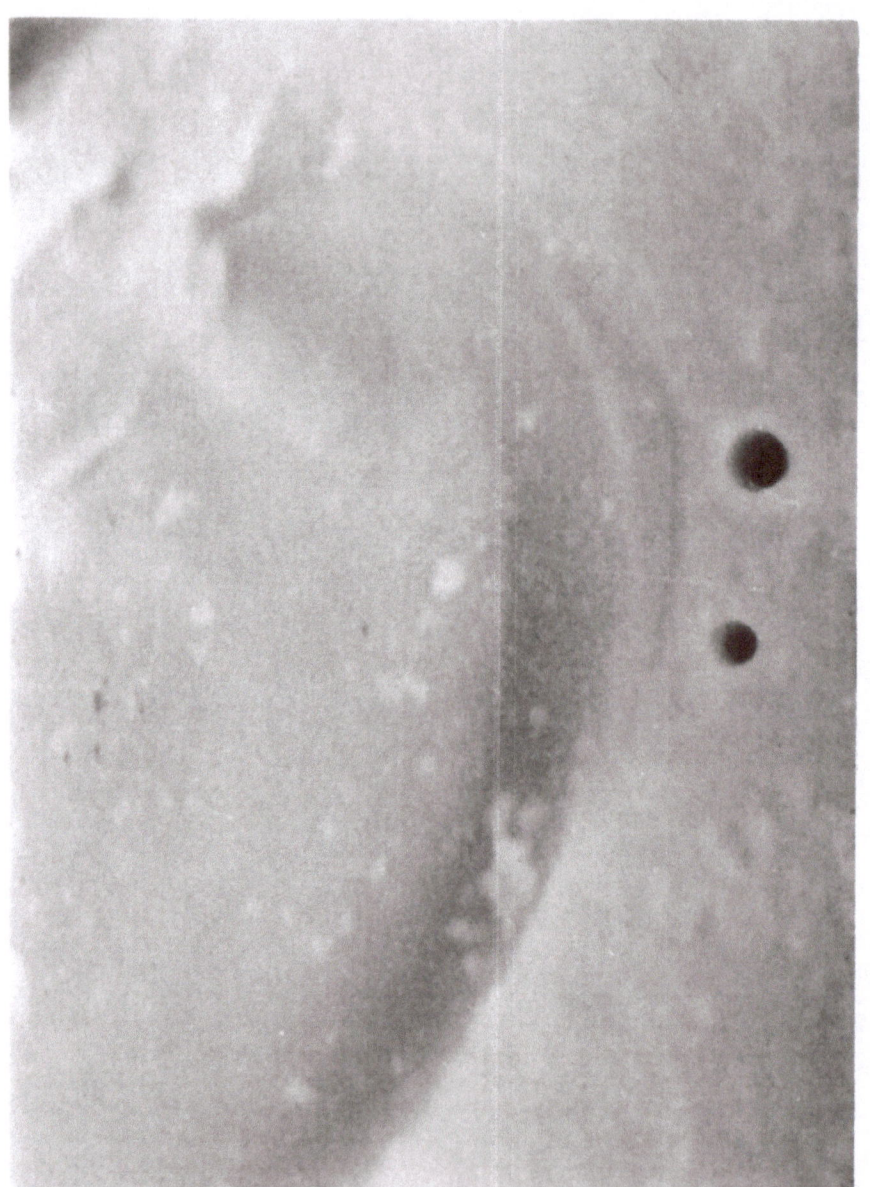

Figure 12. Microcraters on quartz with features suggesting very high impact velocities.

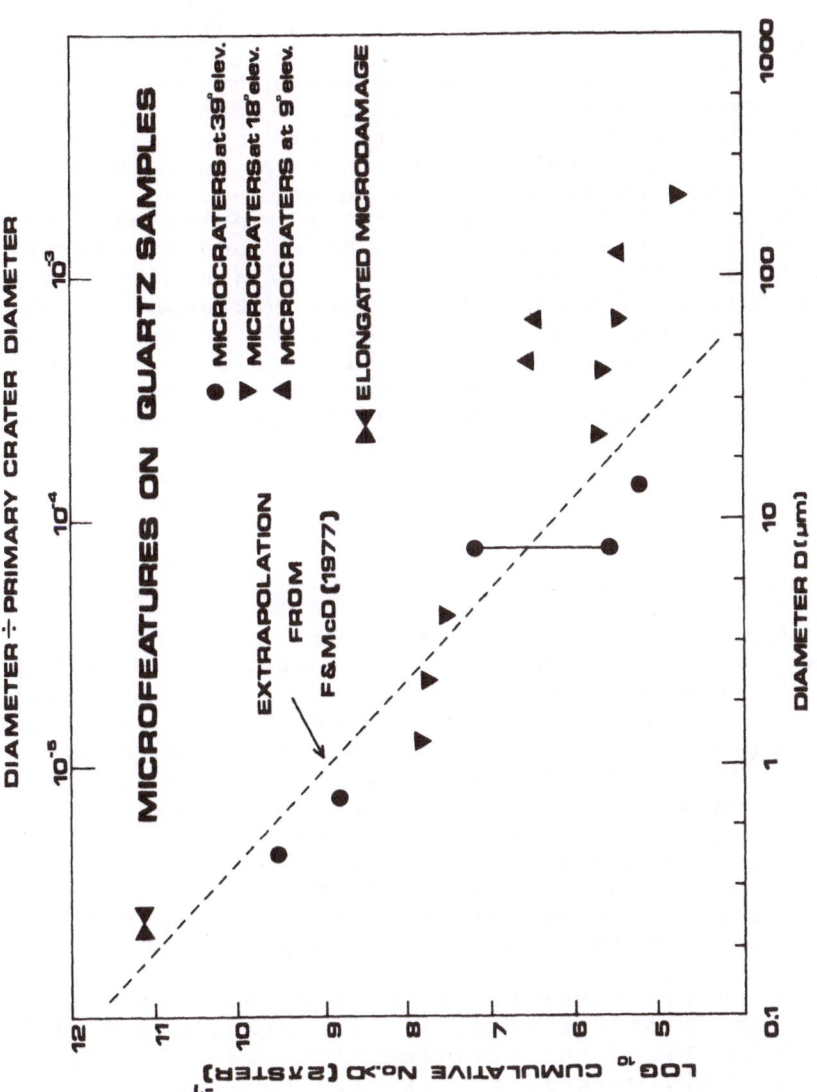

Figure 13. Cumulative number versus diameter for secondary microcraters on quartz SEM samples.

On the quartz samples, removal of the low velocity encrustation reveals an areal concentration per square centimetre of approximately 10 to 50 craters of diameter of approximately 100 micrometres. We have also
discovered a population of similar craters of smaller
diameter. Microcraters of diameter ∿1 micrometre with
features suggesting higher impact velocities have also
been observed on the quartz (Fig. 12). These microcraters could be due to the impact of the very fast ejecta
particles identified from the computer enhanced photographs of the moving ejecta cloud. Because of the satu-
ration of the glass samples, with subsequent loss of
information on the small sized craters·, only the results from the quartz samples have been used in our a-
nalysis of the secondary microcrater population. We
have found that the distribution of secondary microcraters is in general agreement with previous microscale
work (9) derived from impact of iron microspheres with
velocities in the range 3-6 km/sec onto lunar crystalline rocks. In this work primary craters of 14 μ spall
diameter were formed and the secondary microcrater
population was measured for secondary to primary diameter ratios down to 0.01. The present results extend
this work over several orders of magnitude, as shown
in Fig.13 where the points from the present work are
compared with the extrapolated results from (9).

3.3 Magnetic measurements

In the previous series of experiments attempts have
been made to study magnetic effects associated with the
impact produced plasma in a magnetic field free environment (<50γ). This was achieved using a 2 m^3 enclosure,
shielded with μ-metal sheets and a triple system of
Helmholtz coils. The basalt target was also demagne-
tised, but only to a depth of a few centimetres. We
believe that for this reason the results were not conclusive, since the magnetic perturbation observed in the
neighbourhood of the craters could be equally interpreted in terms of edge effects associated with the remaining bulk magnetisation of the basalt block.

In the current series of tests we have concentrated our efforts on the study of the magnetic and electromagnetic effects in the presence of a pre-existing magnetic field. This work is preparatory to another series
of experiments which will be performed in the near future in a field free environment using totally demagneti-
sed basalt targets. Pick-up coils surrounding the impact area have been used to detect the impact associated

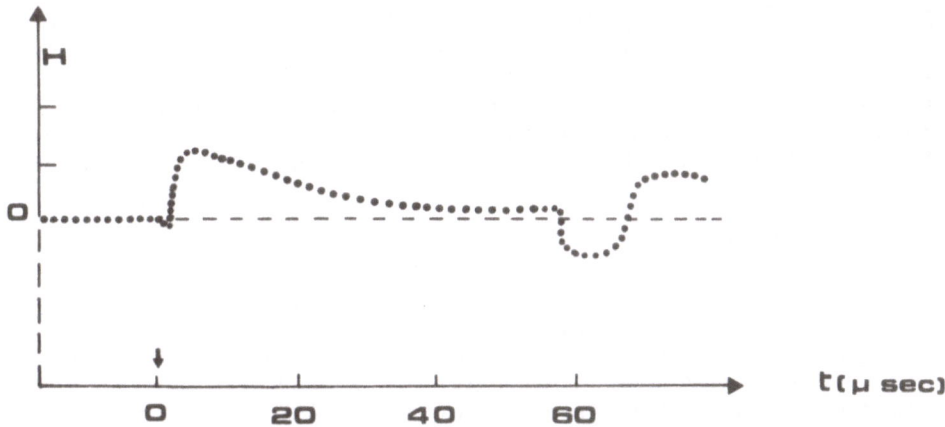

Figure 14. Electromagnetic signal picked up by a horizontal coil. The vertical scale is 1 division per .01 Gauss.

effects described below. However, since these coils have a finite cross section(∿1 cm ^{2}) and thus interfere with the expansion of the ejecta, we have confined this type of measurements to only three main experiments "in vacuo". A typical signal picked up by a horizontal coil is shown in Fig. 14. It can be seen that before impact, when the projectile is at a distance of a few tens of centimetres from the target, the magnetic field starts increasing, reaching a value of approximately 0.5 mG before undergoing a rapid reversal at impact, reaching a minimum of the order of a few mG, followed by a slow recovery towards its initial value. This behaviour is common to all three experiments. We limit our analysis to the first 100 µsec after the impact because the signals which follow this recovery phase are erratic and are probably associated with the motion of the coils invested by large flying ejecta. We interprete the pre-impact enhancement of the magnetic field in terms of paramagnetic currents circulating in the plasma trail which accompanies the projectile, where thermal and density gradients having a component perpendicular to the external magnetic field can drive azimuthal currents and thus enhance the field (10). The subsequent decrease of the signal, which is probably limited by the bandwidth of the tape recorder, corresponds to the expulsion of the field by the rapidly expanding high-β impact produced plasma. This situation

lasts a few tens of μsec, i.e. for a time which is comparable with the propagation time of the shock through
the impacted block.   This mechanism could account for
the relatively strong magnetic perturbation observed in
the neighbourhood of the craters, which appears as an
enhancement of the pre-existing magnetisation of the
rock.   These results confirm the findings of (4).

4. CONCLUSIONS

In the series of experiments reported we have confirmed
that a metal lined explosive shaped charge is a very effective technique for accelerating macroscopic projectiles (with masses of the order of 1 gr) at velocities of
approximately 10 km/sec.   The use of fast framing photography in association with computer processing and reproduction in false colour of the images has made it
possible to identify the position of the projectiles during the flight and to follow the phenomena associated
with the impact on a given target, i.e. the formation
of the plume, of the plasma cloud and of the ejecta.
The results obtained hitherto have already provided u-
seful information on the morphology of the craters for
both normal and oblique impacts, and on the modes of
ejection of the material during the first 100 μsec of
the crater formation.   SEM analysis of the glass and
the quartz targets placed around the impact area have
revealed some aspect of the secondary cratering process
which takes place when particles are ejected at very
high speed during the impact.   Comparison of the labo-
ratory-formed secondary craters with microcraters found
in lunar samples confirms that secondary microparticles
can move faster than the primaries.   Magnetisation of
the impacted surfaces has also been investigated, as
well as the generation of magnetic fields in the plasma
trailing the projectile during the flight.

ACKNOWLEDGEMENTS

The experiments were carried out at the Detonics Centre
of the Italian firm "Difesa e Spazio" (formerly SNIA
VISCOSA) at Colleferro near Rome, Italy.   We are particularly indebted to the Director of the Centre, Dr.
F. Scorretti, and his team for their co-operation, and
to Dr. L. Menegazzoli for the preparation of the charges
and many useful discussion.   Dr. R. Orfei, of the Istituto Plasma Spazio, Frascati, Italy, has helped with

the magnetic measurements. Mr. B. Blackman, of the
University of Sussex, has given invaluable help in the
preparation and execution of the experiments. We thank
Prof. A.Mottana for the target mineralogical analysis.

REFERENCES

1. Coradini,M.,Flamini, E., Martelli, G., Hurren,P.,
    Smith,P.N.: Effect of microparticle hypervelocity
    impacts on polished surfaces for the choice of the
    Halley M.C.M. (this volume), p. 367.

2. Cerroni, P. and Martelli,G.: A possible mechanism
    for the impact magnetisation of cratered surfaces
    (this volume), p. 363.

3. Martelli,G.: The trumpet charge, a new technique
    for the production of hypervelocity macroscopic
    projectiles (this volume), p. 359.

4. Martelli, G. and Newton,G.: 1977, Nature 269,pp.
    478-480

5. Bond, J.W., Keyse, R.J., and Newton,G.: 1980, Pla-
    net. Space Sci. 28, pp. 599-608

6. Wallace, J.E. and Burke, A.F.: 1965, Proc. 4th
    Symposium Rarefied Gas Dynamics, pp.487-507

7. Technical Report: 1980,"Imacon 790: the image con-
    verter system" Hadland Photonics Ltd., Bovingdon,
    Herts H13 OEL, UK (and references therein)

8. Washington, H.S.: 1970, American J. Sci. IV, pp.
    50-60

9. Flavill, R.P., and McDonnell, J.A.M.: 1977, Meteo-
    ritic 12, pp. 222-225

10. Cerroni, P. and Martelli, G., submitted for publi-
    cation to Planet. Space Sci.

THE "TRUMPET CHARGE", A TECHNIQUE FOR PRODUCING MACROSCOPIC
HYPERVELOCITY PROJECTILES

G. Martelli

School of Mathematical and Physical Sciences,
University of Sussex, Brighton BN1 9QH, U.K.

A novel technique is proposed for the acceleration of macroscopic
projectiles to velocities approaching 20 km/sec using explosive
propulsion.

Most hypervelocity phenomena of astrophysical and astronomical
interest, such as cratering, structural modification under shock
loading, etc., could be simulated more realistically than has
been possible hitherto if higher velocities could be achieved on
a routine basis. The limit of 10-12 km/sec for macroscopic
projectiles ($m \sim 0.1$ gr) has so far proved an insurmountable
barrier, whether light gas guns or conventional explosive
techniques are used.

In the fastest technique to date, the explosive shaped
charge,[1] the maximum attainable velocity v_t for the tip of the
jet is ultimately limited by the detonation velocity v_d of the
explosive (8-9 km/sec, say), the relation between the two
velocities being $v_t = qv_d$, with $q < 2$ and generally not exceeding
1.3 - 1.4.

We propose here another method, which would increase the
"apparent velocity" of the explosive (i.e. the velocity at
which the liner "sees" the advancing detonation front), and
hence the maximum attainable velocity of the projectile. We
suggest initiating the detonation of a cylindrical shaped charge
simultaneously along a circumference, and shaping the liner in
such a way that the detonation wave moves at a constant angle α
to it. Figure 1 illustrates how this can be achieved. The
figure shows a cross-section containing the axis of the charge,

359

A. Coradini and M. Fulchignoni (eds.), The Comparative Study of the Planets, 359–362.
Copyright © 1982 by D. Reidel Publishing Company.

the profile of the liner (which resembles a trumpet) being shown
by the thicker line. The propagating detonation front, initiated
at OO, is shown by the circles drawn as broken lines, each circle
corresponding to a different time after detonation. Let C be a
point on the liner, and let α be the angle between the tangent to
the liner at C and the tangent to the detonation wave at C. A
simple calculation shows that, if α does not depend on the choice
of C, the profile of the liner, in polar coordinates ρ and θ,
is given by

$$\rho = \rho_o \exp \theta \tan \alpha$$

where ρ_o is the length of the segment OA, and the origin of θ is
the line OP. When the detonation is started along the circumference OO, the velocity of the detonation front along the liner
is $v_d/\sin \alpha$. For our purposes, we are interested in the interval
$\alpha < \theta < \pi/2$, so that $\alpha = \theta$ at OA. The region where $\theta < \alpha$ is of
no interest, since the walls of the liner diverge in the opposite
direction.

The angle α cannot be made too small, because the transverse
momentum transfer from the detonating explosive to the liner
decreases with decreasing α. Thus, if we define the mass ratio
(i.e. the ratio of the mass of the explosive behind an anulus of
the liner to the mass of the anulus of the liner itself) as
$\mu = \rho_{ex}/d_1\rho_1$, where ρ_{ex}, ρ_1, d_{ex} and d_1 are the density and
thickness of the explosive and liner respectively, it is obvious
that to maintain the mass ratio of a conventional shaped charge,
in the trumpet charge μ must be increased by $(\sin \alpha)^{-1}$, to maintain a similar momentum transfer in a direction perpendicular to
the liner. Such an equivalent mass ratio will be indicated by
$\mu' = \mu/\sin \alpha$. Therefore, the length OA must be chosen in such a
way that the mass ratio μ' is not less than some preset minimum
at the point A (in our drawing the mass ratio increases with θ,
but this is irrelevant to the present discussion).

Now, in a conventional shaped charge having a conical liner
and filled with, say, Comb B, which has a detonation velocity
$v_d = 7.9$ km/sec, if the liner is made of Copper, velocities of
8-9 km/sec are obtained, and 10-12 km/sec if the liner is made of
Aluminium. Furthermore, a velocity gradient is established along
the jet, so that the highest velocity is achieved only by the tip
of the jet.

In a trumpet charge, realistic figures are $\alpha = 30°$, OA = 3
cm, $d_{ex} = 3$ cm, $\rho_{ex} = 1.79$ g/cm^3 (for Comb B), $d_1 = 1.2$ mm and
$\rho_1 = 8.9$ g/cm^3 (for Copper). This gives a mass ratio $\mu' = 5$ at
A. Then, by virtue of the fact that all the elements of the
liner are equally accelerated, a thin "rod" should be produced
moving at 16-18 km/sec, and exhibiting no (or a very small)

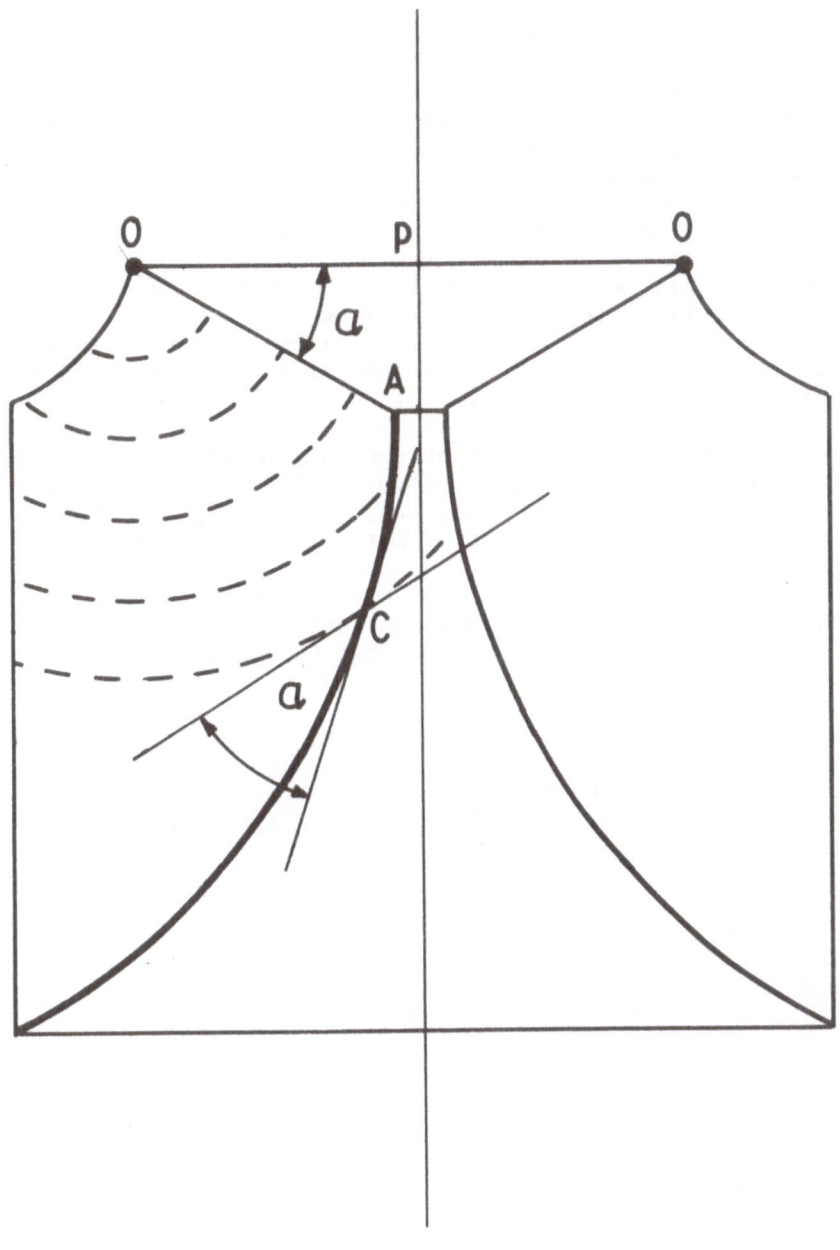

Fig. 1

A cross-section containing the axis of a charge, the profile of
the liner being shown by a thicker line. The detonation front,
initiated at OO, is shown by the circle drawn as broken lines.

velocity gradient.   If the liner is made of Al having the same
thickness, one would expect speeds in excess of 20 km/sec.

The computational results by (2) suggest that the jet should
become unstable if the speed of the detonation front exceeds by
some units the speed of sound in the liner, for a given angle of
incidence of the detonation front onto the liner.   Thus we would
expect the accelerated liner to break up into a shower of
relatively small particles (order of 0.1 gram or so).   Preliminary tests with Al liners have shown this to be indeed the case,
and velocities of the order of 20 km/sec have been recorded
using time of flight techniques.   Further work is now being
carried out to establish how best to isolate individual fast
particles and to deflect away from the target the debris from the
detonation.

In conclusion we suggest that the fast particles produced
with this technique can be used as very suitable projectiles for
hypervelocity impact experiments, when relatively large masses
and speeds truly representative of meteoritic speeds are
required.

The author wishes to acknowledge useful discussions with
Dr. P. Cerroni and Mr. P. Hurren of this University.

REFERENCES

1.   Martelli, G. and Newton, G.:   1977, Nature 269, p. 478.
2.   Chou, P.C. and Carleone, J.:   1977, J. Appl. Phys. 48, p.
     4187.

A POSSIBLE MECHANISM FOR IMPACT MAGNETISATION OF CRATERED SURFACES

P. Cerroni and G. Martelli

School of Mathematical and Physical Sciences,
University of Sussex, Brighton BN1 9QH, U.K.

A mechanism is proposed for the generation or amplification of
magnetic fields during the expansion of impact-produced plasma
clouds, the temperature gradient providing the thermal force
required to excite azimuthal currents at the surface of the
plasma clouds.

We present here some theoretical considerations which suggest a
possible mechanism for amplifying a pre-existing magnetic field
in the neighbourhood of a crater formed by a plasma producing
impact. Should the impacted target material be ferromagnetic the
amplified field would be imprinted as a permanent perturbation.
It is well known that if a dense hot plasma is produced and is
left to expand in a volume permeated by a magnetic field B and if
the plasma pressure nkT exceeds the magnetic pressure $2B/\mu_0$, two
effects relevant to the present discussion will occur. Firstly,
the magnetic field will be excluded from the volume occupied by
the plasma and will subsequently diffuse back into the plasma at
a rate which depends on the plasma resistivity. Secondly, an EMF
will be induced along the "equator" of the expanding plasma
spheroid. This EMF can, under certain circumstances, maintain a
transient current which generates an additional toroidal magnetic
field B_t.

We propose to prove that, in many impact produced plasmas,
the transient magnetic field can be substantial, the energy for
its production being supplied by the thermal energy on the hot
core of the plasma. Finally we will indicate that the various
time scales involved (for the growth of the magnetic field, for
the propagation of a shock through the target and for the spin-

A. Coradini and M. Fulchignoni (eds.), The Comparative Study of the Planets, 363–366.
Copyright © 1982 by D. Reidel Publishing Company.

lattice relaxation) are all compatible.

From Maxwell equations and the equation of continuity we can write for the rate of change of the magnetic field:

$$\partial \underline{B}/\partial t = \underline{\nabla} \times (\underline{v} \times \underline{B}) + (1/\mu_0 \sigma) \nabla^2 \underline{B} \tag{1}$$

where μ_0 is the magnetic permeability of free space, σ is the conductivity of the plasma and the first and second term on the right hand side of (1) are the convective and the diffusive terms respectively. We assume that $B = B(0,0,z)$ and that the component B_x and B_y during the deformation of the field by the expanding plasma cloud are negligible. For the convective term to predominate over the diffusive term we require the dimensionless number $R = \mu_0 \sigma vL \gg 1$, where L is the radius of the expanding plasma cloud at a given time. The existence of a temperature gradient at the boundary of the cloud ensures that a thermal force will drive azimuthal electric currents. The situation is similar to that which exists for a conductor with a radial temperature gradient $\underline{\nabla}T$ embedded in a magnetic field at right angles to $\underline{\nabla}T$ (Nernst-Ettinghausen effect). It can be shown that in this case the action of the thermal force can be described through a vector $\underline{G} = Q\underline{\nabla}T_e$ which has the dimension of a velocity. As long as the dimensionless number (Hibberd number) $R_H = \mu_0 \sigma GL \gg 1$, Eq. (1) reduces to

$$\partial \underline{B}/\partial t = \underline{\nabla} \times (\underline{G} \times \underline{B}) \tag{2}$$

It can be shown (see, e.g. (1)) that for a plasma Q is always positive; a pre-existing magnetic field will therefore be amplified by this mechanism.

In order to put some realistic figures in these formulae we use data from hypervelocity impact experiments performed under controlled conditions (2,3). The plasma was formed by an Aluminium projectile ($m \sim 1$ gr, $v \simeq 10$ km/sec) impacting on basalt-like blocks. The plasma cloud was roughly spherical in shape, with a radius $L \sim 10$ cm. This assumption is supported by the analysis of the pictures of the impact event taken with an image intensifying fast framing Imacon camera (3). The plasma was assumed to have a temperature of a few eV, say 3 eV, per particle. The particle density of the plasma, assuming complete ionisation, is $n \sim 6 \times 10^{23}$ m^{-3}.

Figure 1 illustrates the relevant geometrical parameters of the plasma. We note here a flat central plateau AA where we assume the electron temperature T_e and the plasma density n_e to be constant (or to vary little with respect to the variations in the toroidal shell AB). In the shell AB, which is not drawn to scale, we take T_e and n_e to vary a few orders of magnitude, so that the largest value of $\underline{\nabla}T_e$ is to be found here.

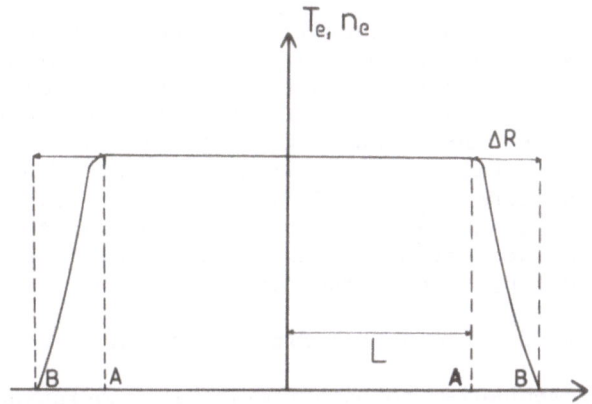

Figure 1. A possible distribution for the electronic temperature and density, T_e and n_e, in the plasma cloud.

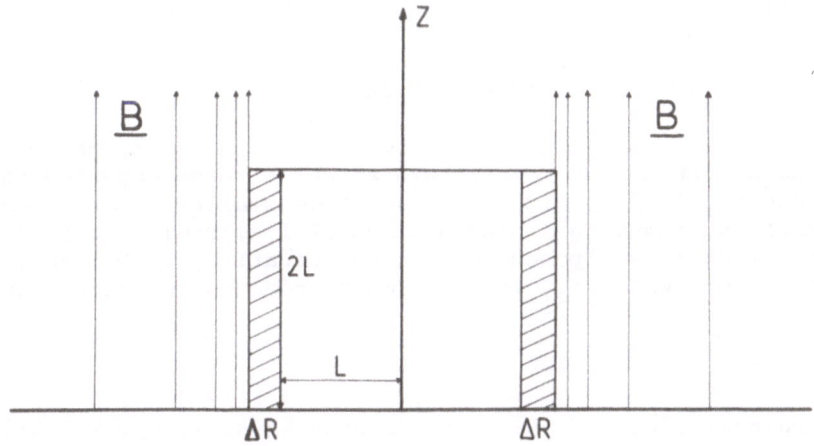

Figure 2. The cylindrical equivalent of the plasma spheroid. The shaded area represents a shell of inner radius L and thickness ΔR. The vertical lines represent the lines of force of the magnetic field compressed by the plasma.

These assumptions are justified both by observations of laser-produced plasma (4) and by the analysis of the pictures of the cratering events.

For the sake of simplicity we assimilate the plasma spheroid to a cylinder of radius L and height R = 2L (see Figure 2); the thickness ΔR of the external cylindrical is taken to be a few millimetres or so.

The temperature gradient ∇T_e over the region of thickness ΔR will then be $\nabla T_e = 4.8 \times 10^{-16}$ Jm^{-1}. The coefficient Q, in the expression which defines $\underline{G}$, and the conductivity of a fully ionised plasma, evaluated according to (1), are respectively $Q = 3.5 \times 10^{20}$ (SI units) and $\sigma \simeq 5.6 \times 10^3$ (SI units). The expression for $\underline{G}$ then yields $\underline{G} = 1.68 \times 10^5$ ms^{-1} and the value of the Hibberd number R_H is 118. We are thus satisfied that the convective term in Eq. (1), which is responsible for the toroidal magnetic field associated with the presence of a strong thermal gradient, predominates over the diffusive term.

The value of the magnetic field thus generated is $B_t \sim 10^{-5}$ T.

For this mechanism to be effective in imprinting this additional magnetic field on a ferromagnetic target such as basalt, the time τ_m, typical of the growth of B_t, must be equal to, or larger than, the time τ_s taken by the shock to propagate to the rim of the crater and this in turn must be larger than the spin-lattice relaxation time τ_{sl}, i.e. the hierarchy $\tau_m > \tau_s \gg \tau_{sl}$ must be obeyed.

This is indeed the case both in the laboratory experiments (3) and for a planetary surface. It can be easily shown that, for the laboratory experiments, $\tau_m \sim 10^{-5}$ sec, $\tau_s \simeq 10^{-5}$, while it is well known that at the temperature at which these experiments are performed τ_{sl} is many orders of magnitude smaller. For a planetary surface, assuming a plasma cloud of 1 km radius to be formed and the mechanism proposed here to be operative, $\tau_m \sim 1$ sec, $\tau_s \simeq 10^{-2}$ sec, while τ_{sl} is still many orders of magnitude smaller.

REFERENCES

1. Braginskii, S.I.: 1965, Rev. Plasma Phys. 1, pp. 205-312.

2. Martelli, G. and Newton, G.: 1977, Nature 269, pp. 478-480.

3. Martelli, G., Cerroni, P., Hurren, P., Smith, P.N., Coradini, M., Bianchi, R., Flavill, R. and Waldner, F.: 1982, Experimental study of effects associated with macroscopic hypervelocity impacts; this volume, pp. 333-357.

4. Fedosejevs, R., Tomov, I.V., Burnett, N.H., Enright, G.D. and Richardson, M.C.: 1977, Phys. Rev. Letts. 39, pp. 932-935.

THE EFFECTS OF MICROPARTICLE HYPERVELOCITY IMPACTS ON POLISHED
SURFACES: TESTS FOR THE CHOICE OF THE HALLEY MULTICOLOUR CAMERA

M. Coradini,[*] E. Flamini,[*] P. Hurren,[+] G. Martelli[+] and
P.N. Smith[+]

[*]I.A.S., Reparto di Planetologia, Viale dell'Universita'
11, 00185 Roma, Italy

[+]University of Sussex, Brighton BN1 9QH, UK

In order to test the response to hypervelocity impacts of materials
to be used in the construction of the mirror to be flown on the
ESA GIOTTO probe as part of the multicolour camera which will
image the Halley comet, we have bombarded reflecting surfaces with
quartz particles of 10^{-6} - 10^{-7} gr travelling at velocities well
in excess of 8 km/sec.

Metallic and glassy mirrors were tested. Crater counts,
morphological analysis of the impacted surfaces and theoretical
considerations allowed us to evaluate the response of different
materials to microparticle hypervelocity impacts.

We have found that nickel plated aluminium mirrors are the
most resistant to hypervelocity impacts, but their optical
properties may be destroyed more rapidly than those of the
polished Al mirrors, because of the irregular morphology of the
crater rims.

1.   INTRODUCTION

Several hypervelocity bombardment experiments have been
performed in order to test some of the materials being considered for the reflecting surface (hereafter referred to as "the
mirror") of the Halley multicolour camera optical system. This
will provide high resolution photographs of the Halley comet
during the GIOTTO fly-by mission.

A. Coradini and M. Fulchignoni (eds.), The Comparative Study of the Planets, 367–387.

The experiments were carried out at the proving grounds of the Italian firm DIFESA E SPAZIO (formerly SNIA VISCOSA) at Colleferro, near Rome, Italy, within the framework of a collaboration between the Reparto di Planetologia of the Istituto di Astrofisica Spaziale and the Space and Plasma Physics Group of the University of Sussex. The materials to be tested were provided by the OFFICINE GALILEO (Italy) which is responsible for the construction of the mirror. Professor Barbieri's group from the University of Padua (Italy) is in charge of the design and assembly of the mirror and mounting flange.

The goal of the experiments was to check:
a)   the resistance of different materials to microparticle hypervelocity impacts;
b)   possible deformations and/or destruction of the mirrors;
c)   the degradation of the optical properties of the mirrors.

2.   EXPERIMENTAL TECHNIQUE

The principle underlying the present technique is as follows: during the hypervelocity impact of a primary projectile on a hard target, the vaporisation of the projectile produces a blast wave of neutral and ionised gases travelling faster than the primary projectile. If the impact area is covered by a layer of granulous material of known density and composition, one would expect some of these grains to be drag accelerated by the blast wave. Fast framing photography makes it possible to measure the velocity of the rapidly expanding spherical shell.

The principle has been experimentally verified in the series of tests described below.

The tests were performed in vacuo, the primary projectile consisting of an aluminium pellet of mass approximately 0.5 gr moving at 10.0 ± 0.5 km/sec (1), the loose granulous material consisting of pre-selected quartz grains. In the tests we used two sizes of quartz grains corresponding to masses of 10^{-2} and 10^{-4} gr.

Because of the extremely high acceleration experienced by the grains we expected them to break up into smaller sizes during flight.

Energy balance considerations indicate that a precise knowledge is required of the thickness and density of the loose target layer. For this reason we have paid particular attention to the selection and preparation of the "sand" grains. The grains were composed of quartz crystals with a negligible amount of calcite. The non-quartz components were eliminated by dissolving them in high concentration hydrochloric acid. After

accurate measurements of the particles in groups of 1000, 100, 10
and 1 elements it was found that the most representative average
values of the particle weights were 1.1×10^{-4} gr and 1.3×10^{-2}
gr respectively. Figure 1 shows quartz grains of 10^{-4} gr against
a 1 mm grid and Fig. 2 quartz grains of 10^{-2} gr after treatment
(the reference line is 2 mm long).

3.   EXPERIMENTAL SET UP

In order to avoid ablation of the secondaries, the tests were
performed in a vacuum tank (see Fig. 3) at a residual pressure of
10^{-1} - 10^{-2} torr. A steel cylinder (30 cm in diameter and 40 cm
in height, Fig. 4) was placed inside the vacuum tank.

The quartz particles were placed on a steel disc (15 cm in
diameter, 3 cm thick) hereafter referred to as the "tray". This
was welded to the supporting cylinder to prevent it from shifting
its position prior to the firings. Figure 5 shows the tray loaded
with 10^{-2} gr quartz particles.

The loads consisted of either a monolayer of 10^{-2} gr grains
or a two diameter thick layer of 10^{-4} gr grains. All the tests
were performed in the vacuum tank shown in Fig. 3; the primary
projectiles were accelerated by the modified shaped charges
placed on top of a vertical flight tube (not shown in the figure),

Figure 1. Quartz particles of approximately 10^{-4} gr. The scale
is provided by the background 1 mm grid.

Figure 2.   Quartz particles of approximately 10^{-2} gr.   The white
reference line is 2 mm long.

Figure 3.   The explosion tank ready for the test.   The primary
projectile flight tube is not shown.

Figure 4. **The supporting steel cylinder** before the assembly of the explosion tank.

Figure 5. **The sand tray loaded with** 10^{-2} **gr grains.** The thickness of the sand layer is approximately one particle diameter.

vacuum sealed at one end by a mylar diaphragm. The top of the
horizontal trays was located approximately halfway up the windows.

The impact events were monitored via a mirror by an IMACON
790 fast framing camera placed inside the bunker, the length of
the optical path being 470 cm. Figure 6 shows the post-impact
evolution of the grain cloud.

Preliminary tests were performed by placing the mirrors beyond
a steel plate provided with horizontal slits viewing the impact
point at different elevation angles. However, the first two tests
indicated that the most energetic grain ejection occurred at very
low angles and consequently very few particles could make their
way through the horizontal slits. For this reason it was
decided that in subsequent tests the mirrors should be exposed
unprotected to the whole of the expanding grain cloud from 0° to
30°.

Aluminium witness plates mounted alongside the impact area
during the first two experiments showed no traces of impacts of
large and slow particles, which were to be avoided in the philo-
sophy of the experiments. This was expected, since the thick
steel disc was chosen specifically to avoid the formation of spall
plates travelling at low speed.

When slow particles impacted the mirrors the features formed were

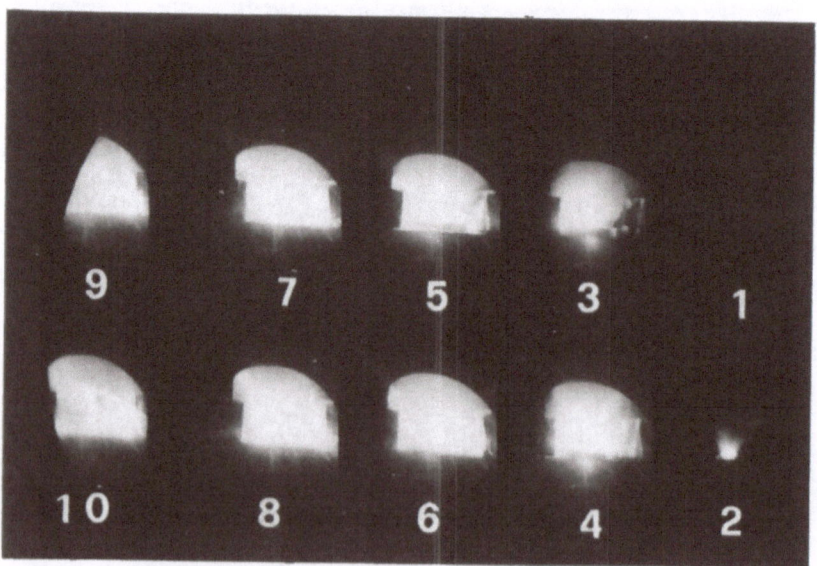

Figure 6. Evolution of the post-impact grain cloud. The framing
rate was one frame every 9 µsec.

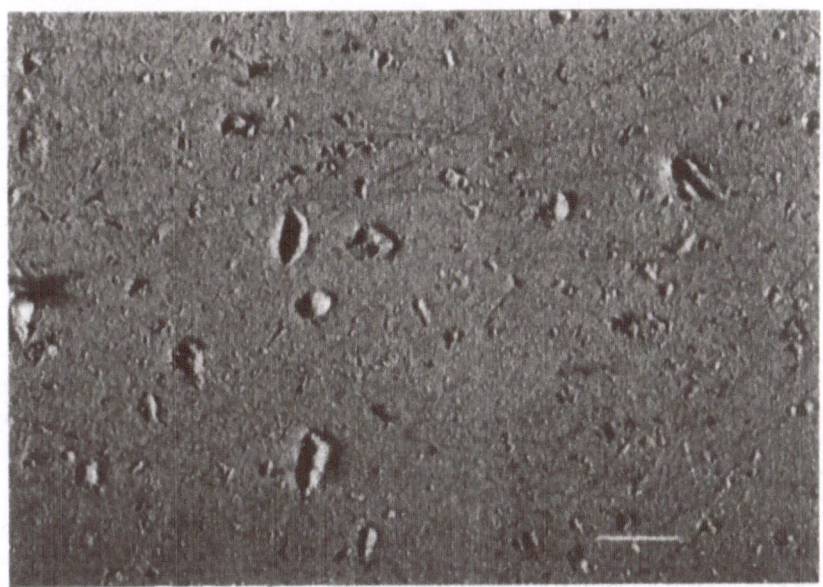

Figure 7. Irregular depressions formed by the impact of slow particles are clearly visible. The white reference line is 1 mm long.

easily identifiable because their morphology is quite irregular, as shown in Fig. 7.

During the shots the IMACON camera was triggered in such a way that the first frame showed the flying projectile and the second the impact phase. In the third frame the expanding grain cloud has already reached the mirror; simple geometrical considerations indicate that the front velocity of the cloud must exceed 8.3 km/sec, and is probably well in excess of 10 km/sec.

This is to be expected, since it is well known that the material ejected at low elevation angles can travel much faster than the primary projectile (2), although the total mass ejected is of course much smaller than that of the primary projectile.

4. MIRROR CLEANING

The detonation of the shaped charges shatters the flight tube and fumes produced outside the vacuum tank are sucked into it. This results in a coating of the target by carbonaceous detonation products (in future experiments the explosion tank will be provided with a fast shutter to seal the vacuum immediate-

ly after detonation). In order to remove the thin carbon coating
from the mirrors they were placed in an ultrasonic cell containing a 6% sodium hydroxide solution. This solution mildly
corroded the aluminium surfaces causing a partial disappearance
of the polished layers (1 - 2 μm). Thus, we could not determine
the response of the aluminised mirrors to the thermal and
mechanical stresses caused by the microparticle bombardment.

On the other hand, the 40 μm nickel layer on aluminium mirrors
was not affected by the cleaning process.

5.   CRATER COUNTS

The counting of the craters formed after the impacts was
performed on a 6 cm^2 rectangular area selected on each mirror as
shown in Fig. 8. The area with the maximum exposure to the
particle impacts was chosen, as shown in Fig. 9.

The selected surfaces were photographed with slant angle of
illumination of approximately 45°, so as to enhance the crater
concavity. The illumination conditions were kept constant during
all photographic operations. Crater counts were operated on
enlarged prints of the pictures. Craters were identified on the
basis of the presence of a hemispherical shadow inside the
inspected feature. This criterion also provides some information

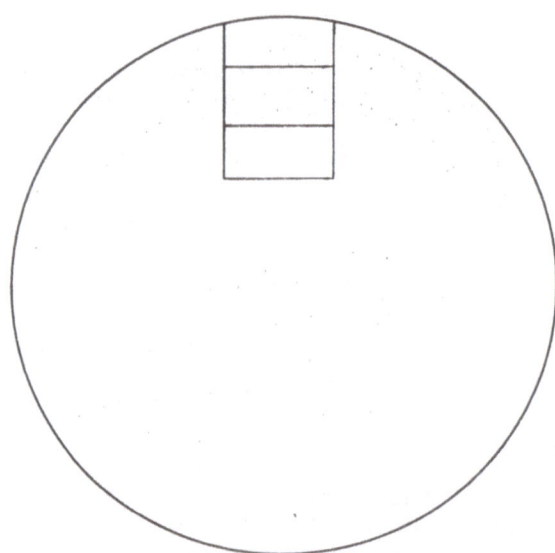

Figure 8.   Rectangular area selected for the crater counts is
shown. The area is approximately 6 cm^2.

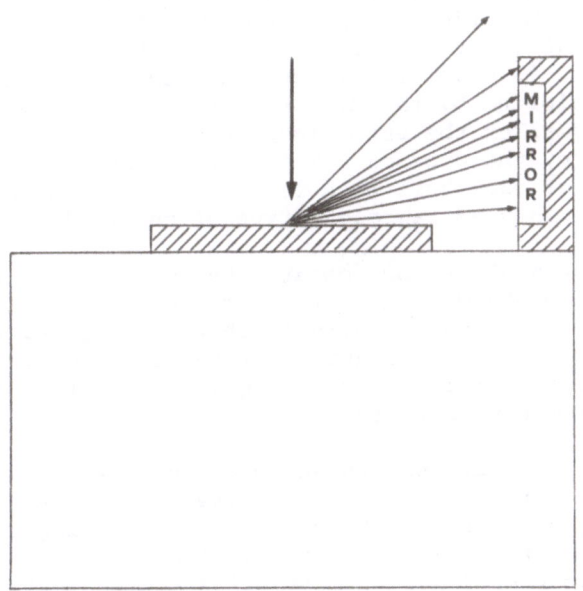

Figure 9. Paths of the particles ejected by the impact of the primary projectile.

about its depth. Because the masses accelerated were relatively large (approximately 10^{-6} gr) and the resolution was limited, craters with diameters smaller than 10 μm were ignored.

6. RESULTS

Crater densities and morphologies were similar in both series of tests performed with grains of different masses (10^{-4} and 10^{-2} gr). This was to be expected in view of the extremely high acceleration (10^8 G) imparted to the grains, which caused them to fragment into subparticles characterised by similar mass distributions which were independent of the initial grain sizes.

According to (3) the dependence of the ratio of the crater diameter D to the particle diameter d is $D/d \propto v^{2/3}$. Using this empirical law, we find that in our case the values of the impacting masses ranged between 10^{-9} and 10^{-5} gr, the largest number of craters corresponding to impacting masses of the order of 10^{-6} gr.

During the counts allowance had to be made for crater obliteration due to the chemical erosion caused by the cleaning

procedure. Assuming that the craters are characterised by a
diameter to depth ratio constant and equal to tan α, where α is
the opening angle, the smaller craters can be erased completely
if their radius is equal to or less than the chemically eroded
thickness multiplied by tan α. Obviously, the larger the crater
diameter the smaller the error introduced into the crater diameter
determination by the erosion. This fact allows us to accept the
determination of diameters much larger than the eroded thickness
multiplied by tan α as representative of the real crater diameter.

Let us now analyse and compare the cumulative size frequency
distributions obtained from four mirrors with different compositions and/or structures. Figure 10 shows the cratering curve
(diameter classes versus logarithm of the ratio of the number of
craters with diameter larger than a given value to a given area)
of polished aluminium mirrors.

The curve is quite smooth and characterised by a continuous
increase of the crater number with decreasing diameters. This
trend suddenly changes and the curve flattens owing to the very
low number of craters with diameter smaller than 140 μm. This is
probably due to the limited number of particles with masses lower
than approximately 10^{-7} gr impacting the mirror.

The solid line visible in Fig. 10 is the linear best fit of
the cumulative distribution; its slope is very low, indicating
the presence of relatively very large craters.

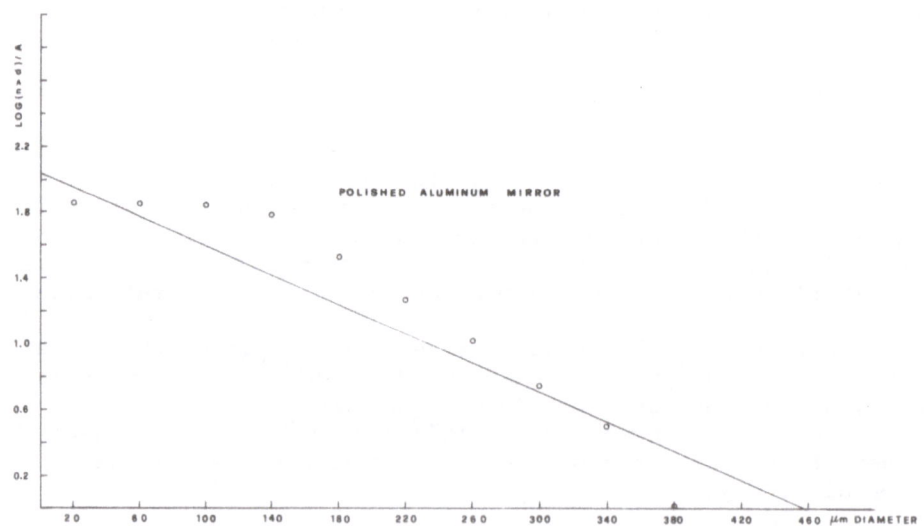

Figure 10.   Cumulative size-frequency distribution obtained from
polished aluminium mirror.

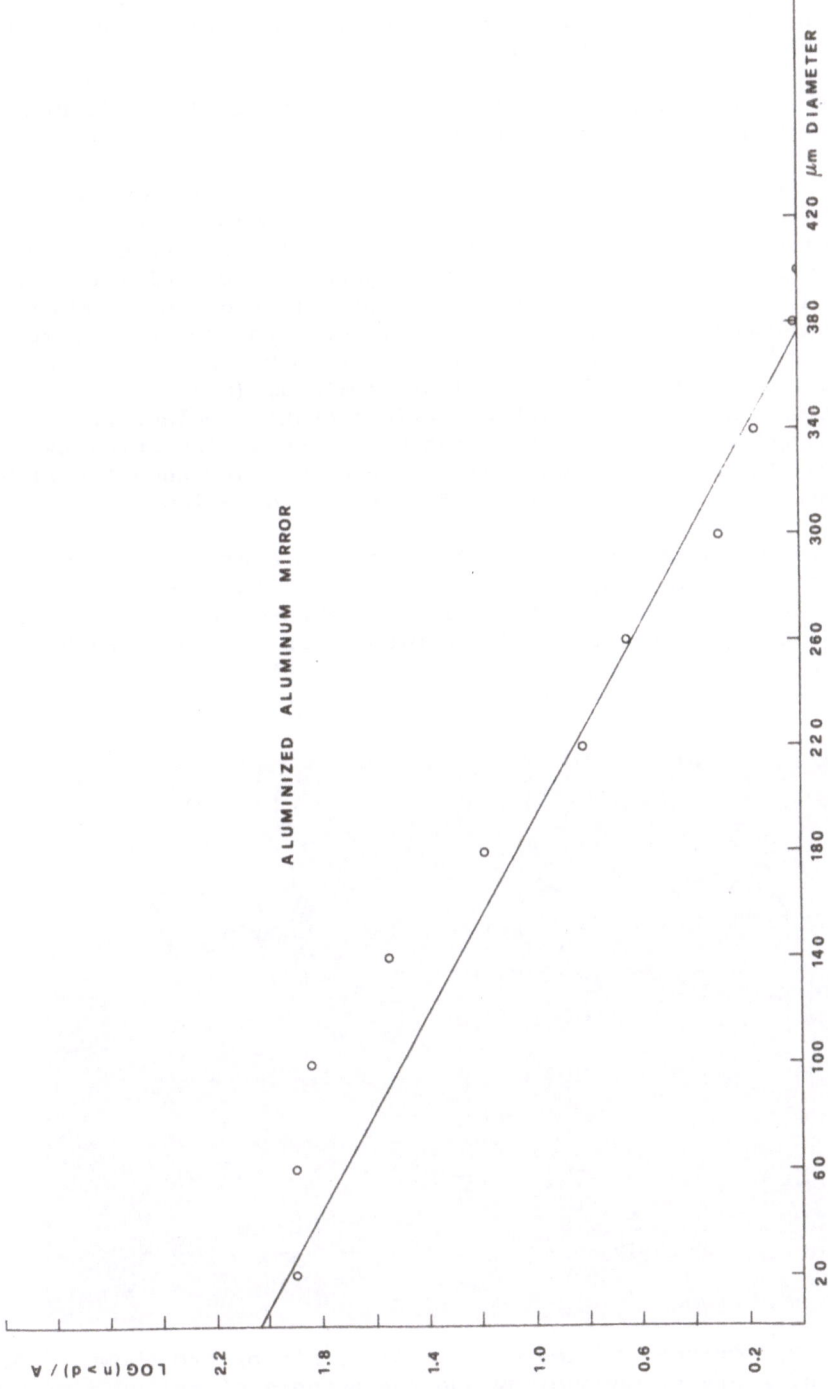

Figure 11. Cumulative size-frequency distribution obtained from aluminised aluminium mirror.

The cumulative size-frequency distribution (see Fig. 11) shows
that the aluminised aluminium mirror responded differently to
the microparticle bombardment.

A sharp decrease in the number of craters with a diameter
larger than 180 μm makes the best fit line slope steeper.

Craters with diameter of 100 μm are the most numerous. From
this we could erroneously infer that the population of impacting
particles had a different mass distribution. On the other hand,
the total number of craters and the general shape of the curve
are similar to those obtained with the polished aluminium mirror.
This suggests that the aluminised mirror withstood the hypervelocity bombardment better. This is probably due to the presence
of a very thin layer (1 - 2 μm) of aluminium (evaporated onto the
aluminium substratum) which can affect crater emplacement.
Indeed the surface of contact can behave as a discontinuity
surface which reflects part of the shock energy, thus lowering
the amount of energy available for crater formation.

The morphology of the craters on both aluminium mirrors is
very regular. In Fig. 12 craters formed on the aluminised mirror
are clearly visible. Craters are bowl shaped, their rims are
sharp and no identifiable ejecta blanket is present. Such a
regular morphology is to be attributed to the target material
ductility which favours projectile penetration and plastic

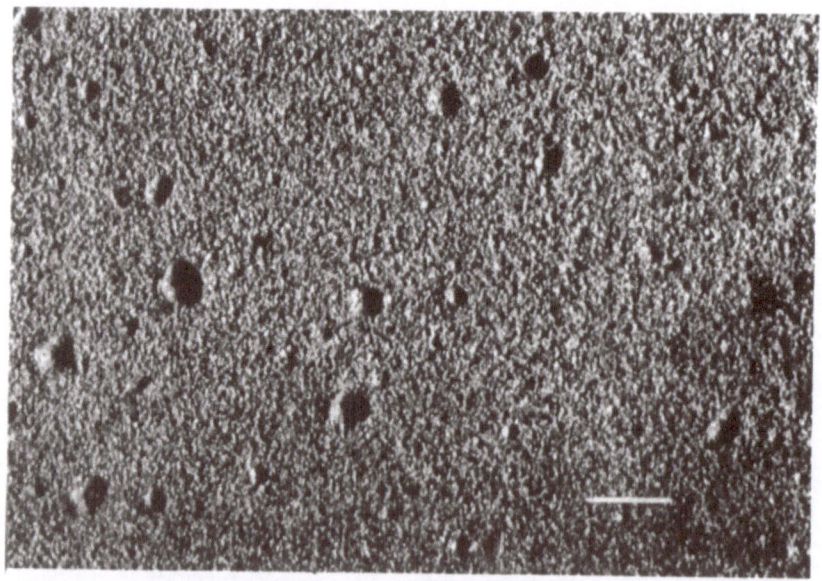

Figure 12. Craters formed on aluminised mirror are shown. Note
the regular crater morphology and the absence of raised rims.
The line is 2 mm long.

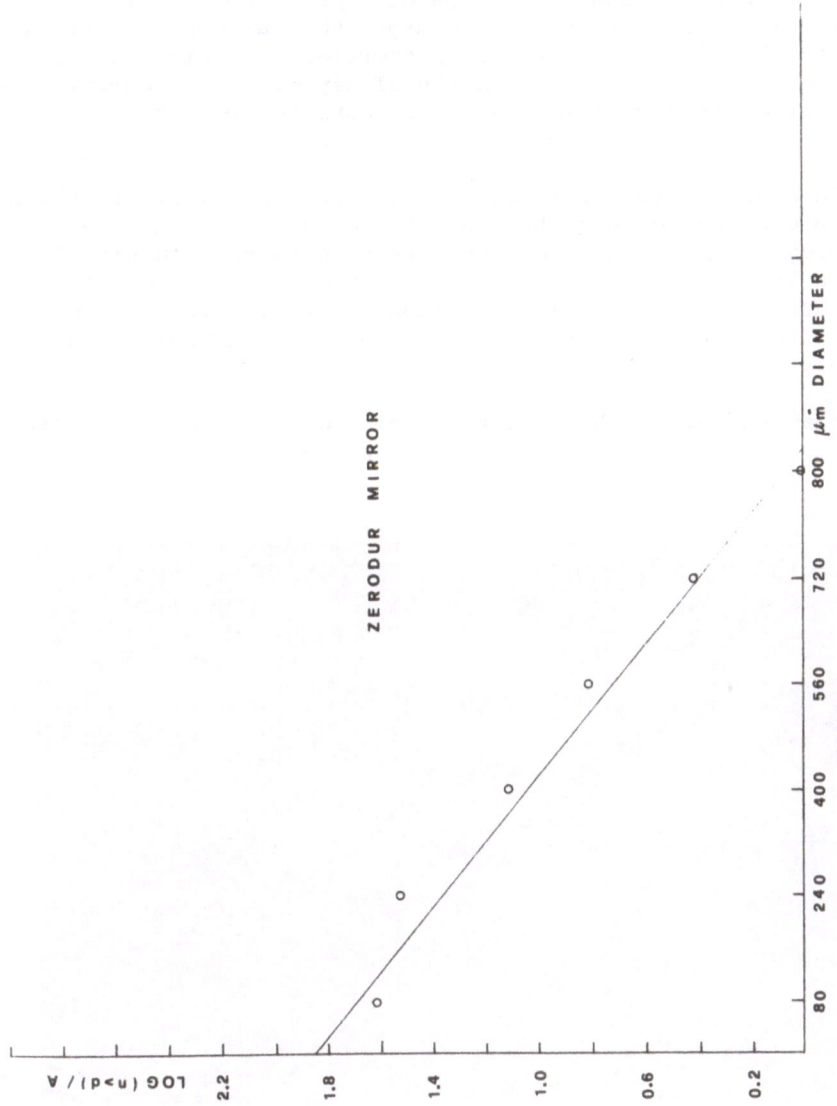

Figure 13.   Cumulative size-frequency distribution obtained from zerodur mirror.

deformation during impact.

The zerodur mirrors presented a totally different response
to the microparticle bombardment. The fragility and the
crystalline nature of the material were responsible for the
formation of craters with diameters much larger than those formed
on metallic mirrors. This phenomenon can be explained in terms
of the fracturing processes taking place along preferential
planes, with consequent formation of typical concoidal morphologies. This process severely damages the reflecting surface
which totally loses its optical properties. In Fig. 13 the
cumulative curve obtained from the glassy mirror is shown. Note
the great difference from the curves obtained with the aluminium
mirrors.

In Fig. 14 craters formed on the zerodur mirror are shown.
It should be noted that the scale factor does not influence the
morphology of craters formed on glassy surfaces. In Fig. 15
craters formed on a quartz target (1) are shown. The concoidal
structures are quite similar to those shown in Fig. 14 and
support the hypothesis that even microscopic particles can
severely damage a glassy mirror.

When analysing the effects of microparticle hypervelocity
impacts on the nickel plated mirror, we should bear in mind the
double layer structure, formed by a 40 μm nickel layer super-

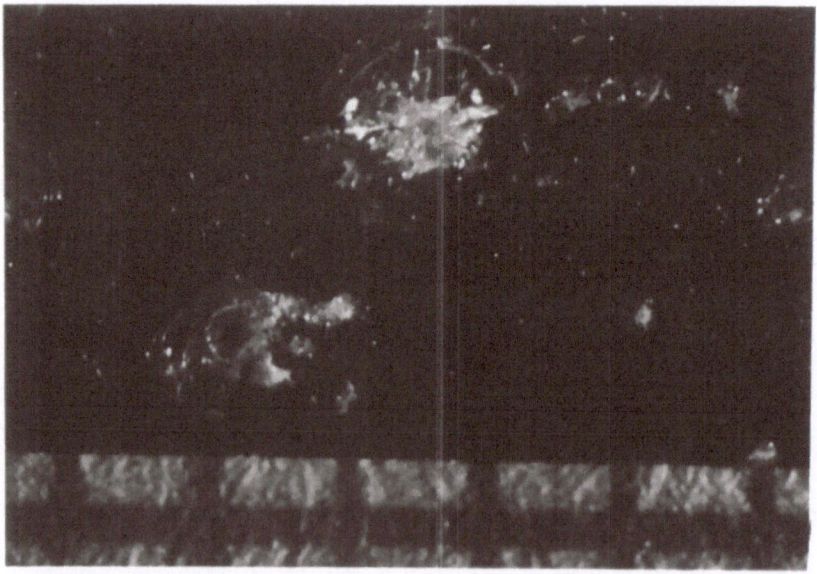

Figure 14.   Concoidal craters on zerodur.   The scale is given by
the 1 mm grid.

Figure 15. Concoidal craters on quartz (1).

imposed on an aluminium substratum approximately 1 cm thick. The slope of the linear best fit of the cumulative cratering curve, shown in Fig. 16, is steeper than those obtained from the other metallic mirrors.

This is due to a lack of large craters (D > 260 μm) and to a high density of craters with diameter of ∿ 100 μm. The formation of smaller craters could of course be attributed to the high density and hardness of the nickel. However, a striking difference is observed between the morphology of the craters formed in the nickel plated mirrors and of those formed in non-protected aluminium.

Craters in nickel are characterised by a depth to diameter ratio equal to 1/8, while for craters in non-protected aluminium this ratio is 1/2. The formation of such shallow craters in the nickel plated mirror can be explained as the result of the reflection of shock waves and the dispersion of energy caused by the interface nickel-aluminium. Craters on nickel also present irregular raised rims, and sometimes a central peak has been identified. These features make the craters on nickel very similar to the craters on the surfaces of heavenly bodies, regardless of the scale factor.

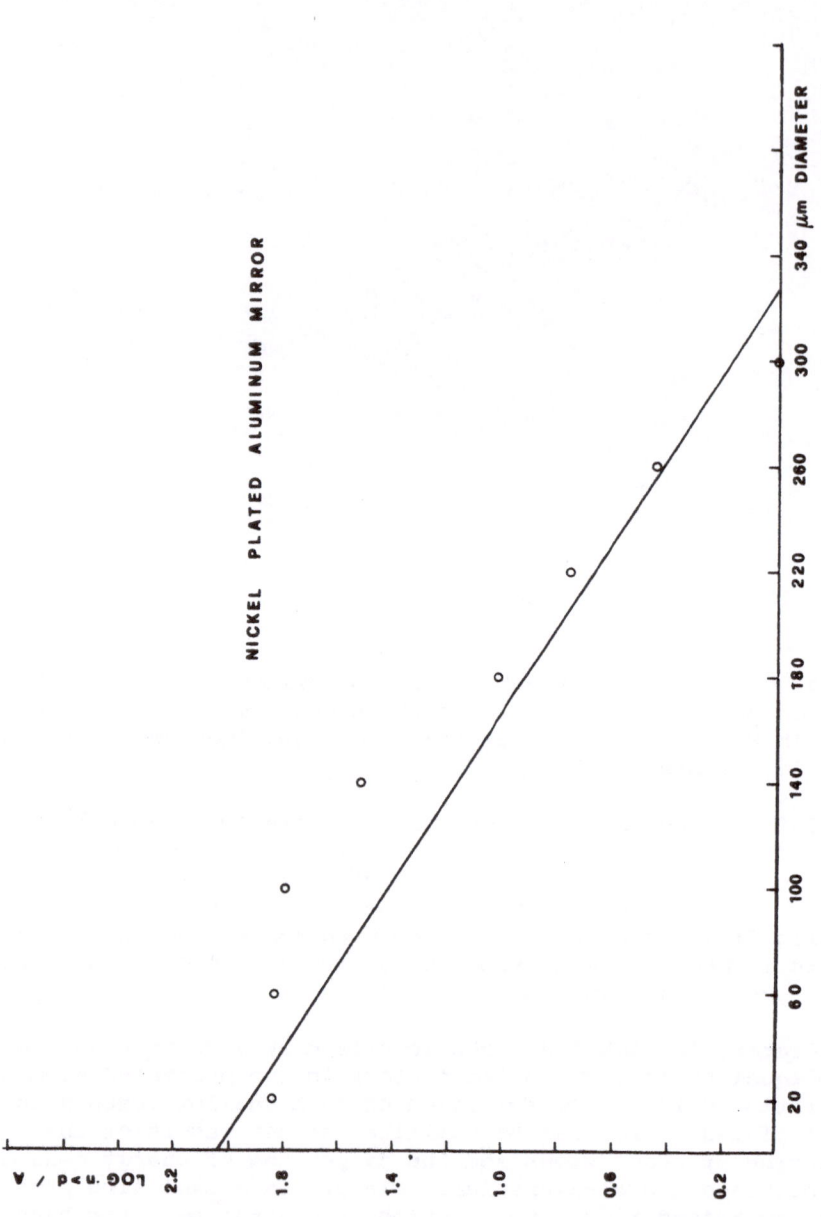

Figure 16. Cumulative size-frequency distribution obtained from a nickel plated aluminium mirror.

The above considerations show clearly that the zerodur mirror is totally unsatisfactory for our purposes. This conclusion is supported both by the large dimension of the craters formed by hypervelocity impacts and by the poor resistance of this material to the mechanical stresses which it undergoes during bombardment. In fact, we found that the breakage of the mirror, as shown in Fig. 17, was due not to mechanical vibrations caused by the impact of the primary projectile on the steel cylinder, but to the impacts of the grains.

This fact is demonstrated by the lack of carbon fume coating along the fracture planes, indicating that the fractures were formed in the mirror body, the various pieces being held together by short range forces while the fumes deposited. Subsequent mechanical vibrations caused a final separation of the parts along the fracture planes.

The nickel plated aluminium showed the best response to the hypervelocity impacts: craters exhibit smaller diameters and lower depths. The reflecting surface can be damaged severely if the nickel layer wrinkles or partly detaches. In Fig. 18 a zone of the nickel layer detached during bombardment is shown.

It is not certain whether the nickel fragmented because of particle impacts or as a result of other processes. The presence of a number of large and regular craters on the aluminium sub-

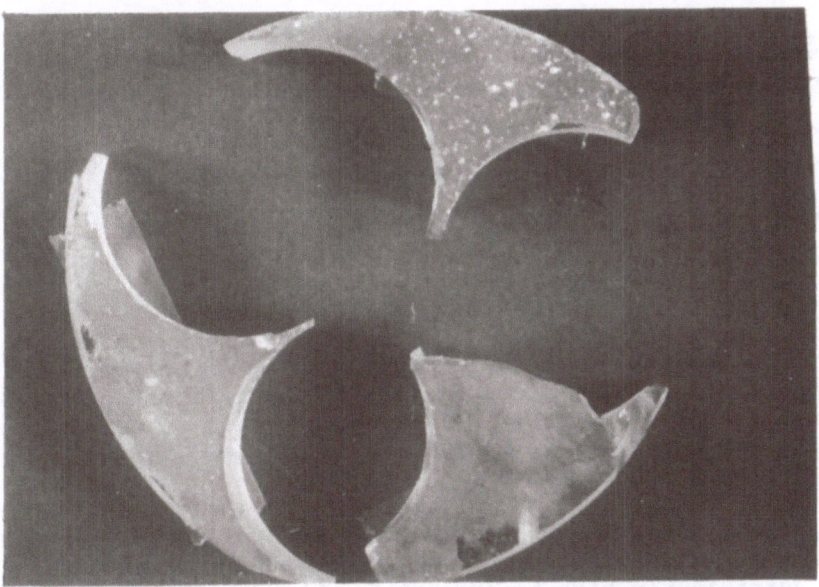

Figure 17. Remnants of the zerodur mirror after hypervelocity particle impacts.

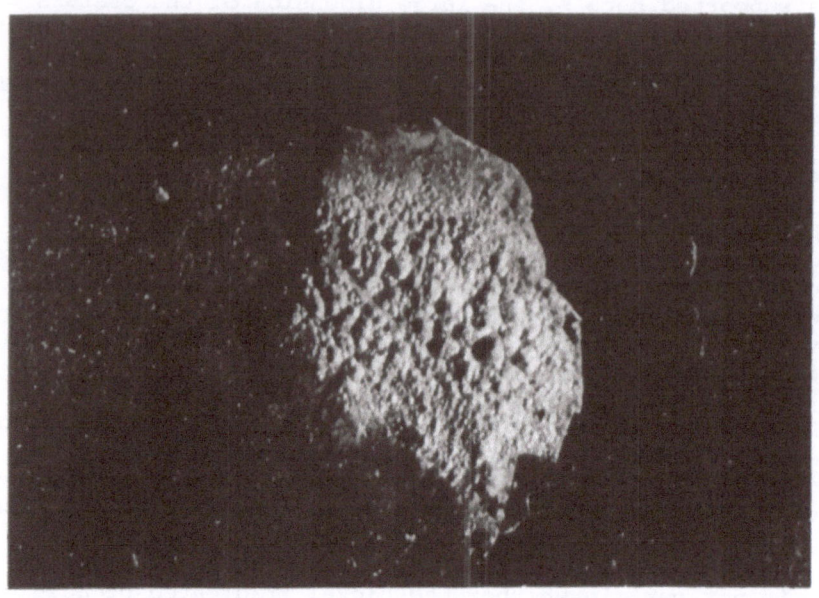

Figure 18.   Zone of the nickel plated mirror where the top layer
detached.   Note the craters formed on the aluminium substratum.

stratum suggests that the nickel spallation had taken place
during an early phase of the bombardment.

     This is shown in Fig. 19 where we have plotted the ratio of
the transmitted to reflected stress given by:

$$\frac{C_t}{C_r} = \frac{2G_2}{G_2 - G_1} \qquad (1)$$

where $G_1 = \rho_1 C_1$ and $G_2 = \rho_2 C_2$, C_1, C_2, ρ_1, ρ_2 being the speed
of sound in, and the density of, two materials labelled 1 and 2
respectively, forming a double layer; t and r refer to transmission and reflection.   Eq. (1) gives the ratio of the transmitted to reflected stress as a function of the ρ and C of the
materials in the double layer.

It can be seen that if the top layer is made of nickel and the
substratum of aluminium, the magnitude of the reflected stress
can be sufficiently high to account for the detachment of the
top nickel layer.

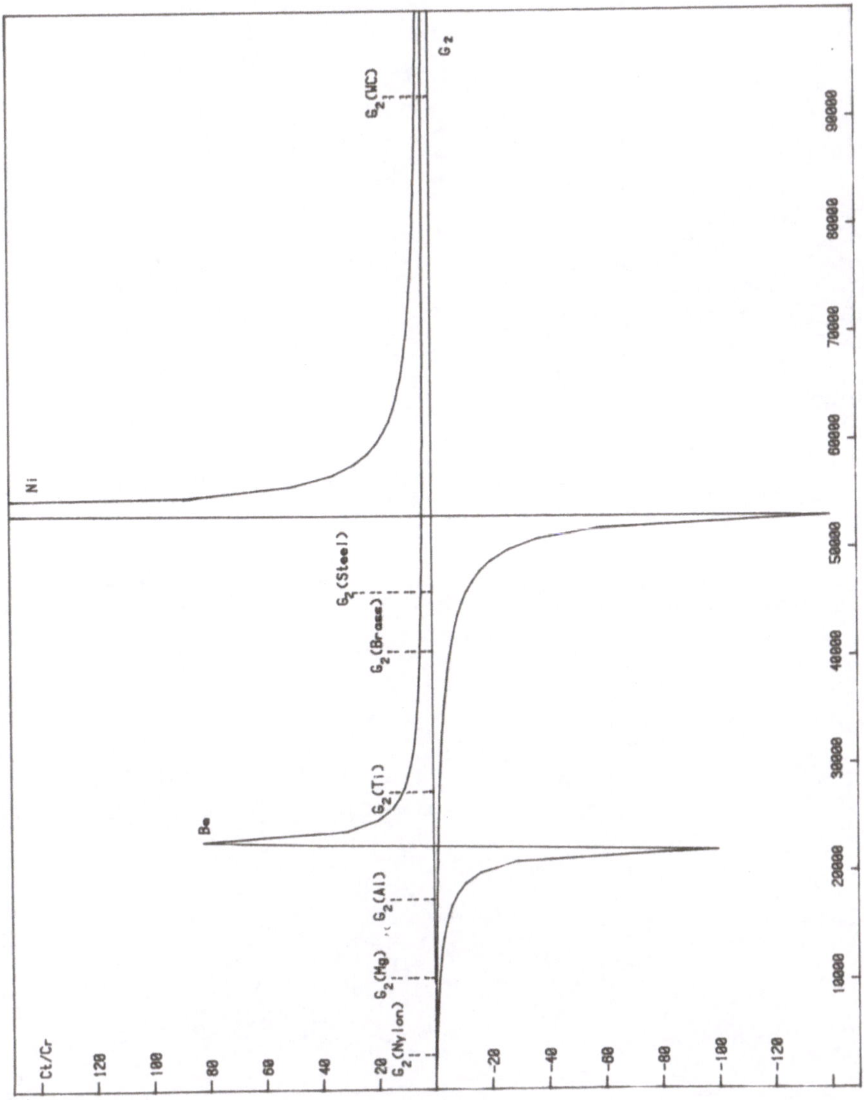

Figure 19.   Plot of the ratio of the reflected to transmitted stress vs acoustic resistance G.

Figure 20. Craters on the nickel plated mirror. Note the irregular morphology.

ACKNOWLEDGEMENTS

The authors are indebted to B. Blackman, for providing technical support during the experiments, to S. Pozio and R. Salvatori for contributing to the crater counting and morphological analysis. R. Bianchi operated particle selection and classification, and F. Capaccioni made helpful contributions to the theoretical study of the double layer.

REFERENCES

1. Martelli G., Bianchi R., Cerroni P., Coradini M., Flavill R., Hurren P., Smith P.N. and Waldner F.: Experimental study of effects associated with macroscopic hypervelocity impacts (this volume), pp. 333-357.

2. Gault D.E., Shoemaker E.M. and Moore H.J.: Spray ejected from the lunar surface by meteoroid impact. NASA Technical Note D-1767, 1963.

3. Neukum G., Schneider E., Mehl A., Stozer D., Wagner G.A. Fechtig H. and Bloch M.R.: Lunar craters and exposure ages derived from crater statistics and solar flare tracks. Proceedings of the Third Lunar Science Conference, pp. 2793-2810, Vol. 3, 1972.

PART III

GEOLOGY OF PLANETARY BODIES

GEOLOGIC OVERVIEW OF THE TERRESTRIAL PLANETS

Raymond E. Arvidson

McDonnell Center for the Space Sciences, Department of
Earth and Planetary Sciences, Washington University,
St. Louis, Missouri 63130

1. ABSTRACT

The terrestrial moons and planets of our solar system are
composed largely of rocky and metallic materials, and preserve
geologic records. The moons of the giant planets, especially the
larger moons, can be thought of as terrestrial bodies, albeit
exotic ones. The terrestrial bodies are close enough in bulk
physical and chemical parameters to make meaningful comparisons
between them. On the other hand, they are different enough so
that each one has evolved a unique geological record. Planetary
geology is devoted to deciphering those records and using the
information in a comparative sense to better understand the
physical and chemical processes that are related to the origin
and evolution of planetary interiors and surfaces.

The geological records for the terrestrial bodies are
currently the subject of a great deal of research. The geology
of the Earth has, of course, been discussed in detail in a number
of references, and will not be treated here. In this paper, the
Moon, our most primitive object is discussed first, followed by
Mars, then Venus, and the paper ends with a discussion of the
evolution of the Galilean Satellites.

2. GEOLOGY OF THE MOON

A vast data base exists for studying the lunar surface.
That base includes: Images acquired in visible light from Earth,
from lunar orbit, and from the lunar surface; Earth-based
spectrophotometric observations in the visible through the

A. Coradini and M. Fulchignoni (eds.), The Comparative Study of the Planets, 391–407.

thermal infrared; Orbital observations in the thermal IR;
Earth-based and Orbital observations with radar; and Orbital
observations of surface radioactivity and major element
concentration (see: Head et al., 1978). In addition, samples
have been returned from eight distinct geologic provinces. Lunar
research is presently in an advanced state, where most
researchers utilize a diverse range of remote sensing data to
understand the origin of given features (Soderblom et al., 1977).

The classic work of Shoemaker and Hackman (1962) using
Earth-based photographs, established the basis for a lunar
stratigraphic column. Their work centered about the crater
Copernicus and the Imbrium basin. They were able to show that
the ejecta from Copernicus lies stratigraphically above the dark,
smooth maria and that these ejecta also partially cover a number
of other post-mare craters. They also mapped the extent of
ejecta deposits that formed as a result of the impact event that
produced the multiringed Imbrium basin and showed that these
deposits overly most of the surrounding terra materials. The
lunar stratigraphy has been considerably revised since Shoemaker
and Hackman's (1962) initial work. Wilhelms and McCauley (1971)
have synthesized a number of lunar geologic maps and other data
into a map of the lunar frontside, produced at a scale of
1:5,000,000. Their most important data source was the contiguous
photographic coverage produced by the Lunar IV Orbiter, which
provided a spatial resolution increase that was an order of
magnitude better than can be obtained from Earth. The
stratigraphic column can be divided into four systems
(Pre-Imbrian, Imbrian, Eratosthenian, Copernican), where a system
represents a given period of lunar history during which given
rock units were emplaced.

The Pre-Imbrian System includes materials emplaced before
the impact event that produced the Imbrian basin and ejecta
deposits. Based on shock metamorphic ages of Apollo 14 breccia
samples that are thought to have been emplaced during the Imbrian
impact event, the Imbrian event occurred about 3.9 billion years
ago (Tera et al., 1974). The Pre-Imbrian system thus includes
rocks older than 3.9 billion years. Rocks older than 3.9 billion
years are abundant on the lunar surface and comprise most of the
lunar terra. The terra, which occupy 80% of the lunar surface
and almost the whole of the lunar backside, consist of heavily
cratered surfaces composed of an anorthosites, norites, and
troctolites that have been intensely brecciated and mixed by
numerous impact events (Taylor, 1975). Shock metamorphic ages
for such rocks range from about 3.8 to 4.4 billion years (Tera et
al., 1974). It is thought that the terra crust formed during
last stages of lunar accretion by progressive crystallization of
a magma ocean that was produced by the intense heating of the
outer parts of the moon due to accretion. The abundance of
craters records either: (a) the sweeping-up of the last of the
debris during accretion or (b) a late heavy bombardment of the

lunar surface by projectiles perturbed into the inner solar system from orbits beyond Jupiter (Wetherill, 1975). In either case, the shock ages demonstrate that such intense bombardment begin to decrease about 4 billion years ago. Combining the crater abundances at the landing sites with sample ages shows that the torrential bombardment exponentially decayed with time and that the cratering flux has been constant for about the past 3.0 billion years

The Imbrian System begins with rocks that were emplaced during the Imbrian impact event. Ejecta deposits extend for several hundred kilometers from the margins of the Imbrium basin. Oberbeck et al., (1975) showed that the kinetic energy contained in the ejecta deposits from such events could allow more mass to be excavated by the secondary craters produced by impact of the ejecta than the mass of ejecta originally excavated by the impact of the asteroid or comet that formed the basin. The material excavated from the secondaries could form part of a ballistic debris surge moving radially outward from the primary crater. For basin-sized events, they postulate that regions close to the basin rim would be scoured and striated by such a process, while regions further away would be mantled as the debris was deposited. In fact, the ejecta deposits associated with the Imbrian event form a linearly sculpted terrain (the Imbrium sculpture) close to the basin, while numerous smooth areas (terra light plains) exist further away. Another consequence of the debris surge is that a great deal of local material would be stirred-up and included in the resultant ejecta deposits. Considering the vast numbers of terra craters and basins, such processes must have thoroughly disrupted and mixed the terra crust.

The Imbrian System also includes most of the materials that comprise the lunar maria. These deposits are composed of basaltic lavas that preferentially filled the larger basins of the lunar nearside (Mutch, 1972; Taylor, 1975). Photogeologic analyses demonstrate that the maria exhibit the following landforms that are diagnostic of emplacement by low viscosity lava flows: lobate scarps, sinuous rilles, domes, cones, dark haloed craters, collapse craters, kipukas, lava terraces, mare ridges, and volcanic complexes (Head, 1975). Based on crystallization ages of samples returned from the maria, extensive volcanism began about 3.8 billion years ago and extended over about 1.0 billion years in duration (Taylor, 1975). The first deposits are thought to consist of the dark, mantled regions outcropping near the edges of a number of maria regions. These dark deposits were sampled at the Apollo 17 landing site on the eastern edge of Mare Serenitatis (Lucchitta and Schmitt, 1974). The low albedo has been shown to be due to the presence of 2 to 18% of orange and black droplets of glass in the impact-generated regolith (Heiken et al., 1974). Fire fountaining has been suggested as the origin of the droplets.

Head (1975) argues that this early phase of volcanic activity
also produced the basalts collected at the Apollo 11 and 17
sites. A number of researchers have shown that the extent of
these early volcanics can be mapped based on their bluish color,
which has been shown to be indicative of a relative Ti-enrichment
(see: Pieters, 1978). Based on crystallization ages of Apollo
11 and 17 basalts, together with color and geologic mapping, this
early volcanic phase lasted from 3.8 to 3.5 billion years and
covered most of the eastern maria. Age estimates have been
obtained for mare regions beyond the landing sites by comparing
the extent of crater rim and wall degradation relative to the
extent of degradation of craters at the landing sites (Soderblom
and Lebofsky, 1972; Boyce et al., 1974).

Head (1975) recognized a second period of volcanism, ranging
from 3.5 to 3.0 billion years ago (middle to late Imbrian), and
covering much of the remaining maria regions. These deposits are
characteristically redder and Ti-poor. A third period of
volcanism can also be recognized (3.0 to about 2.5 billion years
ago; early Eratosthenian in age) that covered parts of Mare
Imbrium and the western maria and was again rich in titanium.
The Eratosthenian system also includes materials associated with
craters superimposed on the maria that are still fresh-appearing,
but lacking extensive ray systems. The youngest system on the
Moon is the Copernican, which includes craters with extensive ray
systems, such as Copernicus and Tycho.

3.   GEOLOGY OF MARS

Photogeologic interpretations of the Martian surface began
in 1965 with imaging data acquired during the Mariner 4 flyby.
Mariner 4 acquired 22 images, covering about 2% of the surface,
with a best resolution of about 5 km/pixel. The image data
showed an ancient, highly cratered surface that appeared to have
been modified to a greater extent than the lunar terra (Leighton
et al., 1965). In 1969, Mariner 6 and 7, also flyby spacecraft,
acquired near-encounter data over about 20% of the Martian
surface. Most of these data also pointed to an ancient, heavily
cratered crust, although some of the images pointed to a
collapsed, chaotic terrain that hinted at a variegated and
complex surface (Leighton et al., 1969). In 1971-72 Mariner 9,
an orbiter, acquired about 7000 images of the atmosphere and
surface, with nearly 100% coverage at a resolution of 1 to 3
km/pixel and about 1% coverage at 0.1 to 0.3 km/pixel (Masursky
et al., 1973). The increase in coverage and in resolution
profoundly altered our conception of the Martian surface.
Mariners 4, 6, and 7, unfortunately, were targeted to acquire
imaging data largely in the southern hemisphere, a region
dominated by an ancient, heavily impacted crust. The northern
hemisphere, as shown by Mariner 9, consists of a wide variety of
terrain types, mostly sparsely cratered and thus younger than the

ancient crustal units exposed in the south. The Viking Orbiter
missions (2 spacecraft) appreciably increased our understanding
of the origin of various terrain units, in addition to allowing
the characterization of a complex of surface phenomena that were
not evident in the Mariner data (Snyder, 1979; Arvidson et al.,
1980). Part of the reason is that the Viking Orbiter cameras,
returning about 50,000 images over their 4 year lifetime, had a
somewhat better resolving power than the Mariner cameras. In
addition, the periapsis was lowered to 100 km for the Viking
Orbiters and scan platform motion was used as an image smear
compensation system (Snyder, 1979). Thus, resolutions of as
small as 8 meters/pixel were obtained in limited areas.

A fairly coherent picture of the global geologic evolution
of Mars can now be constructed, largely by combining new data
from the Viking mission, with the global distribution of geologic
units inferred from the Mariner 9 data (Carr, 1980; Arvidson et
al., 1980). Martian geologic evolution has involved extensive
differentiation of the interior, extrusion of major quantities of
volcanic materials, both extensional and compressional tectonics,
the formation of geomorphic features by periglacial processes,
and wind action. Approximately half of the Martian surface is
occupied by ancient cratered terrain, a surface that has a crater
abundance per unit area roughly similar to the highly cratered
surfaces of the Earth's moon and Mercury (Oberbeck et al., 1977).
The lunar terra, based on the radiometric ages of samples
returned during the Apollo missions, seems to have formed about
4.65 billion years ago and to have been heavily bombarded until
up to about 4.0 billion years ago (Taylor, 1975). The broader
scale topography on the Martian cratered terrain is probably of
similar age. This ancient cratered terrain is separated from the
extensive, sparsely cratered plains to the north by a great
circle inclined to the equator by about 35 degrees (Mutch et al.,
1976). The boundary between the cratered terrain and the plains,
where not directly covered by volcanic materials, consists of a
complex of faulted escarpments. One of the major problems of
Martian geologic evolution is the reason for or the processes by
which this major hemispheric asymmetry came about (Wise et al.,
1979). The northern hemisphere is covered with plains, largely
of volcanic origin, and exhibiting a variety of crater abundances
and thus a variety of ages (Soderblom et al., 1974; Neukum and
Wise, 1976). Age ranges probably occupy several billions to
several hundreds of millions of years or younger. A reasonable
hypothesis for formation of the asymmetry is that, before or
during core formation, convective overturn within the interior
led to extension in the north, thereby fracturing and thinning
the northern cratered terrain crust. The crustal thinning and
extension eased the rise of magmas, leading to a massive covering
with volcanic materials. Most of the southern cratered terrain
proper was partially mantled by a relatively thin cover of
volcanic materials during the time period corresponding to the
maximum volcanic resurfacing rate in the northern plains.

The Tharsis plateau of Mars, occupying about 2000 km in breadth, 8 to 10 km in height, and straddling the equator, is covered with a vast expanse of relatively young volcanic materials. The plateau also dominates the lower order terms of the Martian gravity field (Phillips and Saunders, 1975). The topography and the gravitational signature make the plateau the second major asymmetry on Mars. Photogeologic mapping shows that volcanic activity in this region extended over a large fraction of geologic time (Plescia and Saunders, 1979). Furthermore, exposures of ancient crust within the plateau, together with mapping of lava thickness based on largest craters that are nearly filled to their brims, show that the plateau height may be largely caused by relief within the lithosphere and asthenosphere, rather than by thick sequences of lava flows (Plescia and Saunders, 1979). The plateau may be underlain by a thin crust and by a mantle that is slightly less dense than typical (Sleep and Phillips, 1979).

A great surprise from photogeologic studies was the discovery of a number of valleys or channels cut into the Martian surface, especially for the older plains and the cratered terrain regions. Most investigators, based on comparison to terrestrial systems, together with consideration of the physics involved in channel formation, feel that the major channels were cut by torrential floods of water or by water-mud slurries (see: Masursky et al., 1977; Baker, 1979). The most likely source of the water would have been catastrophic break-out from underground reservoirs. Morphological comparisons between the distinctive and much smaller integrated Martian valley networks and terrestrial fluvial systems demonstrate that the Martian systems need not have formed by rainfall and runoff (Pieri and Sagan, 1979). In constrast to terrestrial systems, the Martian networks do not exhibit junction angle (angle between branch and trunk) patterns that are indicative of downhill flow of water. Pieri and Sagan (1979) suggest that the Martian networks are due to groundwater sapping. An alternative hypothsis is that the integrated channels systems may have formed early in Martian history, when an ammonia and methane-rich atmosphere supported a greenhouse heated atmosphere that was warm enough to allow rainfall and runoff (Pollack, 1979). Other geologic features that suggest the presence of large reservoirs of water, more than predicted from scaling the abundances of noble gases in the atmosphere (Arvidson et al., 1980), include: (a) the fretted and chaotic terrains, which are reminscent of collapse following geothermal melting of ice bodies (Sharp, 1973); (b) the peculiar multi-lobed ejecta deposits surrounding many craters that may indicate the combined effects of impact-melted crustal ices and interaction of the ballistic ejecta with the atmosphere (Carr and Schaber, 1977); (c) patterned or polygonal ground covering much of the northern high latitudes, which may be similar to freeze-thaw polygons or to dessication features (Carr and Schaber, 1977), and (d) numerous lines of evidence, both

theoretical and experimental, that suggest that much of the
Martian soil consists of weathered materials consisting of
chemically bound hydroxides and water, carbonates, together with
adsorbed and absorbed water and carbon dioxide (Huguenin,
1976; Baird et al., 1977; Soderblom and Wenner, 1978; Singer et
al., 1979).

The presence of a thin atmosphere on Mars provides for
another set of geologic processes that involve transfer of
condensates to and from the surface, together with mechanical and
chemical coupling of the surface and the atmosphere. The Martian
atmosphere consists primarily of about 5 to 10 mb of CO_2,
together with smaller amounts of H_2O, argon, and nitrogen (Owen
et al., 1977). Water is thermodynamically stable only as a solid
or gas, although recent radar observations of the surface suggest
the presence of minute amounts of liquid water at certain times
of the day and year (Zisk and Mouginis-Mark, personal communication).

During the winter season in a given hemisphere a mixture of
carbon dioxide and water ice condenses onto the surface near the
poles, forming a thin (meters) seasonal polar cap. The temporary
deposits completely evaporate in the northern polar cap area
during the ensuing spring, exposing a 1 to 3 km thick sequence of
layered dust and ice deposits (Cutts et al., 1976). The Viking
Orbiter thermal mapper recorded brightness temperatures as high
as 210 degrees K over these deposits during the northern summer
(Kieffer et al., 1976). In addition, the Viking Orbiter Water
Vapor Mapper showed an atmosphere saturated with water vapor over
the northern summer cap (Farmer et al., 1976). Both results are
consistent with water-ice as the dominant volatile in these
deposits, since carbon dioxide or clathrate would have evaporated
at 148 degrees K and 151 degrees K, respectively.

Seasonal CO_2 and H_2O condensates were seen on the surface
from the Viking 2 Lander, which appears to have landed just on
the inside of the southernmost extent (48 degrees N. Lat.) of
the northern polar cap (Jones et al., 1979). The condensates
formed during the two winter seasons of Viking observations. The
deposits remained on the surface for about 200 Mars days, finally
evaporating as the spring season progressed. After the
condensates had evaporated, the surface was found to be
considerably brighter and redder, probably because of the
accumulation of a thin (several microns) layer of red dust
(Guinness et al., 1979). It seems likely that formation of the
condensate and the dust accumulation are linked - the condensates
probably formed in the atmosphere by accumulation of water and
carbon dioxide ice around tiny dust particles. The increasing
particle size as more ice condensed led to higher settling
velocities and thus a greater chance of being deposited onto the
surface. Such a process would be even more enhanced over the
colder polar regions, providing a plausible mechanism for
accumulation of the layered deposits of dust and ice.

A thick sequence of dust and ice deposits also exists in the south pole, although the composition of the ice is uncertain because global dust storms obscured the deposits during the southern summer seasons during which the Viking Orbiters conducted observations (James et al., 1979). The normal albedo of the south polar deposits has been estimated to be about 0.7, which may make it difficult for all the CO2 and H2O deposited during the winter to evaporate during the following spring and summer seasons (James et al., 1979). Thus, the south polar cap deposits may be composed of a mixture of carbon dioxide, water, clathrates, and dust.

Global dust storms generally occur during the summer season in the southern hemisphere. The reason seems to be that Mars is at its perihelion position in its orbit during the southern summer and, as a consequence, receives about 50% more solar energy then it does when it is in its aphelion position (Mutch et al., 1976). However, periodic changes in the orbital obliquity and eccentricity, together with precession of the pericenter and spin axis, should lead to, among other effects, cyclic alternations of which hemisphere is in its summer season during perihelion (Ward, 1979). The net result of such orbital variations may be that the asymmetry in polar deposits between the water-ice dominated northern deposits and the carbon dioxide-ice dominated southern deposits may cyclically exchange over 10**5 and 10**6 year timescales.

4. GEOLOGY OF VENUS

Venus is a prime target for radar observations since its surface is obscured from optical and infrared observations by a thick atmosphere and dense clouds. Earth-based imaging of the surface of Venus makes use of the delay-Doppler interferometry mapping techniques similar to techniques used to construct lunar radar images (Pettengill et al., 1974).

Earth-based observations are only acquired in the near-equatorial portion of the planet, centered near 320 degrees East Longitude. The reason is that this area of Venus is always toward Earth at inferior conjunction. Earth-based, near-equatorial radar images of Venus have been produced by both the Arecibo and Goldstone Facilities. These data are somewhat complementary since Arecibo maps extend to temperate latitudes but miss the equatorial regions, which Goldstone maps in great detail. Typical spatial resolutions for the Earth-based data are several km/resolution element. Pre-Pioneer Venus geologic interpretations of Venus radar echos are given by Malin and Saunders (1977).

Earth-based radar image of Venus from Arecibo data is shown in Campbell and Burns (1980). Perhaps the most dramatic features

are the dark Lakshmi Plateau and the bright Maxwell area. The
center of Maxwell contains a dark circular depression which may
be a caldera, which in turn would make this feature a rather
large shield volcano. Beta is another prominent broad rise,
about 1000 km across, that may be a shield volcano. Several
areas of valley and ridge topography that are similar to the
surface expression of terrestrial basin and range faulting can be
seen. Alpha is the most obvious example. A probable fracture
zone, some 1000 km long, and 90 km wide, can be seen just to the
north of the equator. Finally, a number of circular features
that may be craters can be detected. In fact, the abundance of
circular features larger than 80 km in diameter, if they are
impact craters, suggest that the Venusian surface is at least a
billion years old.

The Pioneer Venus mission carried a small radar altimeter
(Pettengill et al., 1979, 1980). This instrument measured the
distance to the surface along the ground track of the
spacecraft's orbit, along with measuring the scattering
characteristics of the surface out to a distance of about 64 km
on either side of the beam. The Pioneer-Venus spacecraft is in
elliptical orbit that is inclined at an angle of 74 degrees to
the equator. The altimeter experiment provided the data to
construct the first global (93% of surface) topographic map of
Venus. This map has a vertical resolution of 200 meters and a
horizontal "footprint" of about 100 kilometers, about one degree
on the surface of Venus (Pettengill et al., 1980). The
topographic map of Venus shows a planet that has a topography
which is quite different from the Earth's. Perhaps the most
fundamental difference consists of the difference in the
equatorial bulge for Venus as opposed to the Earth. Due to
rotation the equatorial radius of the Earth is 42 km larger than
the polar radius. Venus, in contrast, is a perfect sphere,
albeit somewhat lumpy, to within the accuracy of the Pioneer
altimeter. A comparison of the hypsometric curve for the Earth
and Venus illustrates the differences in the distribution of
topography. Earth data are shown as deviations from a reference
geoid while Venus data are referred to a sphere with 6051.4
kilometer radius. Eighty percent of the Venusian topography
falls within a topographic range of two kilometers, while about
5% of the surface is occupied by plateaus. These relationships
can be clearly seen on a shaded relief map of the distribution of
topography, based on PV altimetry measurements. Venus lacks the
bimodality between continental and oceanic crust found on Earth,
suggesting fundamental differences between the interiors of the
two planets. Gravity maps, obtained from tracking spaccraft
Doppler accelerations, combined with results for the topography,
suggest a correlation between gravity and topography. Such a
correlation does not exist on Earth, again suggesting differences
in the interiors of these two bodies.

5. GEOLOGY OF THE GALILEAN SATELLITES

Four of the satellites of Jupiter, known as the Galilean
Satellites, are roughly the size of the Earth's moon. Proceeding
outward from Jupiter, they are Io, Europa, Ganymede, and
Callisto. Except for Io, the bulk density of these bodies, known
from mass inferences based on orbits together with volume
estimates, indicate that substantial quantities of water exist
within their interiors. In fact, both Callisto should have about
40% by weight of water and Ganymede should contain about 50%
water by weight (Lupo and Lewis, 1979). The remaining materials
are thought to consist of alumino-silicates. Europa should have
about 5% by weight of water. Pre-Voyager thermal evolution
models suggested that Europa, Ganymede, and Callisto should have
differentiated in such a manner to produce rocky cores surrounded
by water or water ice mantles and icey crusts. Earth-based
spectral reflectance observations supported this interpretation
in that all three bodies showed strong absorption features in the
IR due to water (Pilcher et al., 1977). Observations of Io,
which was known from bulk density considerations to be a rocky
body, showed an anhydrous surface with a reflectance spectrum
characterized by an extremely sharp drop-off in the UV, leading
to a yellow-reddish coloration to the surface (Fanale et al.,
1979). In addition, Earth-based imaging of Io and Jupiter in the
sodium ion emission part of the spectrum showed a cloud of singly
and doubly ionized sodium atoms that formed a torous in Io's
orbital wake (Matson et al., 1977). It was thought that the ions
were sputtered from the surface by the influx of charged
particles from Jupiter's magnetosphere. The sharp UV drop-off in
the reflectance spectrum, together with the sodium cloud data,
led some researchers to suggest that radiation damaged salts,
left over from Io's degassing, dominated the surface materials
(see: Fanale et al., 1979).

The Voyager 1 and 2 encounters with Jupiter and its
satellites in 1979 provide the data necessary to begin to
understand the geologic histories of the Galilean satellites.
The histories of these four bodies is of special interest since
their internal composition and structure are quite distinct from
the other terrestrial bodies. The Voyager imaging system
acquired data in 4 broad bands from the UV to the red parts of
the spectrum (Smith et al., 1979). In addition, a radiometer and
an interferometer were utilized to acquire visible albedo and
thermal IR spectra (Hanel et al., 1979).

Callisto and Ganymede: Callisto is the outermost satellite
and has the lowest density of the four bodies. The surface is
also the darkest of the satellites, although the absolute albedo
is still a factor of 2 higher than the Earth's moon. Voyager
observations show this body to be the most heavily cratered
surface in the Galilean system. The abundance of craters is
similar to that found on the lunar terra, indicating that the

icey crust has remained fairly intact for about 4 billion years.
One of Callisto's hemispheres seems to be dominated by a large
multi-ringed structure, although that feature is distinctivly
different in detail from impact basins on the Earth's moon.   In
addition, a number of other large (°200 km) circular, very
shallow ring structures can be seen.   The very shallow appearance
of these features is consistent with viscous relaxation of crater
structures, with larger features relaxing to a greater extent
than smaller features.

Ganymede is Jupiter's largest Galilean satellite and Voyager
observations show it to have rather unique surface (Smith et al.,
1979).   Darker regions on Ganymede are occupied by a highly
cratered crust, much like the icey crust on Callisto.   One
section of the crust even exhibits evidence for the presence of a
large ring structure.   Brighter striped regions were found to be
cutting across the darker crustal regions.   The stripes are
characterized by a complex of closely spaced, shallow grooves
that form networks running roughly parallel to the boundaries of
the stripes.   The grooved terrain divides the heavily cratered
crust into polygons that range in size from several hundred to
approximately one thousand kilometers across.   The data strongly
suggest that the older crust has been disrupted and covered,
gradually being replaced by the grooved terrain.   The grooved
terrain may have formed by injection of water magma along dikes.
The method of consumption of the older crust, either by
subduction or assimilation, or covering remains controversial.
The low temperatures of both bodies may explain why a dominantly
icey crust has remained intact for so long, although the marked
absence of topographic relief (nowhere greater than 1 km) does
imply relaxation by slow flow of ice.

Europa and Io: Europa displays no obvious features related
to either ring basins or to impact craters.   Relief is very
subdued.   The dominant surface features are a complex set of
fractures (Smith et al., 1979).   Increasing resolution reveals
fractures at an increasingly finer scale.   The lack of craters
implies a youthful surface, e.g., a surface that reflects an
active interior capable of destroying the ancient crusts
preserved on Callisto and partially preserved on Ganymede.

Io is perhaps the most interesting of the satellites in that
Voyager discovered a number of active volcanic vents, spewing gas
and ash up to 270 km above the surface (Morabito et al.,
1979; Strom et al., 1980).   The reflectance spectrum, with the
sharp UV drop-off and a deep absorption at 4.0 microns, is now
thought to be dominated by sulfur dioxide, together with
allotropes of sulfur (Fanale et al., 1979: Hapke, 1980).   The
surface exhibits more than 100 calderas, some as much as 200 km
in width.   A number of the calderas have narrow (10's km wide),
long (100's km) flows extending from them.   The absence of relief
along the flows implies lavas of rather low viscosity.   Much, if

not all, of the surface of Io is covered by volcanic materials.
About 10% of the surface was beneath the seven volcanic plumes
seen by the Voyager spacecraft. In addition, IR observations of
an area with a particularly dark caldera are best explained by
having the dark area be about 150 degrees kelvin hotter than the
surrounding regions (Hanel et al., 1979). This interpretation
suggests that warm volcanic materials were sensed. The
resurfacing rate calculated from the observed volcanic activity
corresponds to about 10**-1 to 10**-4 cm/year, making the surface
activity rate comparable to the Earth's (Johnson et al., 1979).
The reason that both Io and Europa are so active may be because
of a forced tidal resonance with Jupiter, during which a
considerable amount of tidal energy is transferred from Jupiter
to the interiors of these two satellites (Peale et al., 1979).
In fact, it is calculated that tidal heating may dominate the
thermal histories of these two bodies, making them unique among
solar system objects observed thus far.

A number of escarpments can be seen extending over the
surface of Io. Voyager showed a number of discrete white clouds,
thought to be composed of sulfur dioxide crystals, extending from
the escarpments (McCauley et al., 1979). These clouds are
probably due to crystallization of SO2 gas emanating from the
interior and escaping along zones of weakness, such as fractures.
This observation also suggests that the energetics of this
process may provide a unique mechanism for mass movements on Io.

REFERENCES

Arvidson, R.E., K.A. Goettel, C.M. Hohenberg, 1980, A post-Viking view of Martian geologic evolution, Reviews Geophysics Space Physics, 18, 565-603.

Baird, A.K., and others, 1977, Viking X-ray fluorescence experiments: Sampling strategies and laboratory simulations, J. Geophys. Res., 82, 4595-4624.

Baker, V.R., 1979, Erosional processes in channelized water flows on Mars, J. Geophys. Res., 84, 7985-7993.

Boyce, J.B., A.L. Dial, L.A. Soderblum, 1974, Ages of lunar nearside light plains and maria, Proc. Lunar Sci. Conf. 5th, 11-23.

Campbell, D. and B. Burns, 1980, Earth-based imagery of Venus, J. Geophys. Res., 85, 1110-1119.

Carr, M.H. and G.G. Schaber, 1977, Martian permafrost features, J. Geophys. Res., 82, 4039-4054.

Cutts, J.A., and others, 1976, North polar region of Mars, Imaging results from Viking 2, Science, 194, 1329-1337.

Fanale, F.P., R.H. Brown, D.P. Cruikshank, R.N. Clarke, 1979, Significance of absorption features in Io's IR reflectance spectrum, Nature, 280, 763-766.

Farmer, C.B., and others, 1977, Mars: Water vapor observations from the Viking orbiters, J. Geophys. Res., 82, 4225-4268.

Guinness, E.A., R.E. Arvidson, D.C. Gehret, L.K. Bolef, 1979, Color changes at the Viking landing sites over the course of a Mars year, J. Geophys. Res., 84, 8355-8364.

Hanel, R. and others, 1979, Infrared observations of the Jovian system from Voyager 1, Science, 204, 972-976.

Hapke, B.A., 1979, Io's surface and environs: A magmatic-volatile model, Geophys. Res. Lett., 799-802.

Head, J.W., 1975, Lunar volcanism in space and time, Reviews of Geophysics and Space Physics, 14, 265-300.

Head, J.W., C. Pieters, T. McCord, J. Adams, S. Zisk, 1978, Definition and detailed characterization of lunar surface units using remote observations, Icarus, 33, 145-172.

Heiken, G.H., D.S. McKay, R.W. Brown, 1974, Lunar deposits of
   possible pyroclastic origin, Geochem. et. Cosmochem. Acta, 38,
   1703-1718.

Johnson, T.V., A.F. Cook, C. Sagan, L.A. Soderblom, 1979, Volcanic
   resurfacing rates and implications for volatiles on Io, Nature,
   280, 746-750.

Jones, K.L., R.E. Arvidson, E.A. Guinness, S.L. Bragg, S.D. Wall,
   C.D. Carlston, D.G. Pidek, 1979, One Mars year: Viking Lander
   imaging observations of sediment transport and H2O condensates,
   Science, 204, 799-806.

James, P.B., G. Briggs, J. Barnes, A. Spruck, 1979, Seasonal
   recession of Mars' polar cap as seen by Viking, J. Geophys. Res.,
   84, 2889-2922.

Kieffer, H.H., and others, 1976, Martian north pole summer
   temperatures: Dirty water ice, Science, 194, 1341-1344.

Leighton, R.B. and others, 1965, Mariner IV photography of Mars:
   Initial results, Science, 149, 627-630.

Leighton, R.B. and others, 1969, Mariner 6 and 7 television
   pictures: Preliminary analysis, Science, 166, 49-67.

Lucchitta, B.K. and H.H. Schmitt, 1974, Orange material in the
   Sulpicius Gallus Formation at the southwestern edge of Mare
   Serenitatis, Proc. Lunar Sci. Conf. 5th, 223-234.

Lupo, M.J. and J.S. Lewis, 1979, Mass-radius relationships in icy
   satellites, Icarus, 40, 157-170.

Matson, D.L., T.V. Johnson, F. Fanale, 1977, Sodium D-line emission
   from Io: Sputtering and resonant scattering hypotheses,
   Astrophys. J., 192, 43-46.

McCauley, J.F., B.A. Smith, L.A. Soderblom, 1979, Erosional scarps
   on Io, Nature, 280, 736-737.

Malin, M.C. and R.S. Saunders, 1977, Surface of Venus: Evidence of
   diverse landforms from radar observations, Science, 196, 987-990.

Masursky, H., 1973, An overview of results from Mariner 9, J.
   Geophys. Res., 78, 4009-4030.

Masursky, H. and others, 1977, Classification and time of formation
   of Martian channels based on Viking data, J. Geophys. Res., 82,
   4016-4038.

Morabito, L.A., S.P. Synnott, P.N. Collins, 1979, Discovery of active extraterrestrial volcanism, Science, 204, 321.

Mutch, T.A., 1972, Geology of the Moon: A stratigraphic View, Princeton Univ. Press, 400 p.

Mutch, T.A., R.E. Arvidson, J.W. Head, K.L. Jones, R.S. Saunders, 1976, The Geology of Mars, Princeton Univ. Press, 400 p.

Neukum, G. and D.U. Wise, 1976, Mars: A standard crater curve and possible new timescale, Science, 194, 1381-1387.

Oberbeck, V.R., F. Horz, R.H. Morrison, W.L. Quaide, D.E. Gault, 1975, On the origin of the lunar smooth plains, The Moon, 12, 19-54.

Oberbeck, V.R., W.L. Quaide, R.E. Arvidson, H.R. Aggarwal, 1977, Comparative studies of lunar, Martian, and Mercurian craters and plains, J. Geophys. Res., 82, 1681-1689.

Owen, T.B. and others, 1977, The composition of the atmosphere at the surface of Mars, J. Geophys. Res., 82, 4635-4639.

Peale, S., P Cassen, R. Reynolds, 1979, Melting of Io by tidal dissipation, Science, 20, 892-894.

Pettengill, G.H., S.H. Zisk, T.W. Thompson, 1974, The mapping of lunar radar scattering characteristics, The Moon, 10, 1-16.

Pettengill, G.H., and others, 1979, Venus: Preliminary topographic and surface imaging results from the Pioneer Orbiter, Science, 205, 90-93.

Pettengill, G.H., D.B. Campbell, H. Masursky, 1980, The surface of Venus, Scientific American, 242, 54-65.

Phillips, R. and R.S. Saunders, 1975, The isostatic state of Martian topography, J. Geophys. Res., 80, 2893-2898.

Pieri, D. and C. Sagan, 1979, Origin of Martian valleys, NASA TM-80339, 349-352.

Pieters, C., 1978, Mare basalt types on the frontside of the moon: A summary of spectral reflectance data, Proc. Lunar Planet. Sci. Conf. 8th, 2825-2849.

Pilcher, C.B., S.T. Ridgeway, T.B. McCord, 1972, Galilean satellites: Identification of water frost, Science, 178, 1087-1090.

Plescia, J.B. and R.S. Saunders, 1979, Geologic evolution of the
    Tharsis volcanoes (abstract), Proc. Lunar Planet. Sci. Conf.
    10th, 989-991.

Pollack, J.B., 1979, Climatic change on the terrestrial planets,
    Icarus, 37, 479-553.

Sharp, R.P., 1973, Mars: Troughed terrain, J. Geophys. Res., 78,
    4063-4072.

Shoemaker, E.M. and R.J. Hackman, 1962, Stratigraphic basis for a
    lunar timescale, in The Moon, IAU Symposium, No. 14 (Z. Kopal and
    Z.K. Mikhailov, eds.), Academic Press, 290-300.

Singer, R.B., T.B. McCord, R.N. Clark, J.B. Adams, R.L. Huguenin,
    1979, Mars: Surface composition from reflectance spectroscopy: A
    summary, J. Geophys. Res., 84, 8415-8426.

Sleep, N.H. and R. Phillips, 1979, An isostatic model for the
    Tharsis Province, Mars, Geophys. Res. Lett., 6, 803-806.

Smith, B.A., and others, 1979, The Jupiter system through the eyes
    of Voyager 1, Science, 204, 951-972.

Snyder, C.W., 1979, The extended mission of Viking, J. Geophys.
    Res., 84, 7917-7933.

Soderblom, L.A. and L.A. Lebofsky, 1972, Technique for rapid
    determination of relative ages of lunar areas from orbital
    photography, Proc. Lunar Sci. Conf. 3rd, 1191-1199.

Soderblom, L.A., C.D. Condit, R.A. West, B.M. Herman, T.J.
    Kreidler, 1974, Martian planetwide crater distributions:
    Implications for geologic history and surface processes, Icarus,
    22, 239-263.

Soderblom, L.A., J.R. Arnold, J.M. Boyce, R.P. Lin, 1977, Regional
    variations in the lunar maria: age, remanent magnetism, and
    chemistry, Proc. Lunar Sci. Conf. 8th, 1191-1199.

Soderblom, L.A. and D.B. Wenner, 1978, Possible fossil H2O
    liquid-ice interfaces in the Martian crust, Icarus, 34, 622-637.

Strom, R.B., 1979, Mercury: a post-mariner 10 assessment, Space
    Sci. Rev., 24, 3-70.

Strom, R.B., R.J. Terrile, H. Masursky, 1979, Volcanic eruption
    plumes on Io, Nature, 280, 733-736.

Tang, C.H., Boak, T.I.S., Grossi, M.D., 1977, Bistatic radar measurements of electrical properties of the Martian surface, J. Geophys. Res., 82, 4305-4315.

Taylor, S.R., 1975, Lunar Science- A post Apollo View, Pergamon Press, 372 p.

Tera, F.D., A. Papanastassiou, G.J. Wasserburg, 1974, Isotopic evidence for a terminal lunar cataclysm, Earth Planet. Sci. Lett., 22, 1-21.

Ward, W.F., 1979, Present obliquity oscillations of Mars: Fourth-order accuracy in orbital E and I, J. Geophys. Res., 84, 237-241.

Wetherill, G.W., 1975, Late heavy bombardment of the moon and terrestrial planets, Proc. Lunar Sci. 6th, 1539-1561.

Wilhelms, D:E. and J.F. McCauley, 1971, Geologic map of the nearside of the moon, U.S. Geological Survey Map I-703.

Wise, D.U. and others, 1979, Tectonic evolution of Mars, J. Geophys. Res., 84, 7934-7939.

FAULTING AND FRACTURING OF PLANETARY SURFACES

Donald U. Wise

Department of Geology and Geography, University of
Massachusetts, Amherst, Massachusetts 01003 USA

1. ABSTRACT

Investigation of faulting and fracturing of planetary
surfaces involves the determination of past stress systems and
their effect on the materials of those surfaces and to the upper
layers of the planet. The stress tensor and its relationship
to fault types using theories of E.M. Anderson are described.
Fault systems, stress systems, and possible ways of dating and
generating these stresses on regional and planetary scales are
discussed. These principles are applied to the Alba Volcano of
Mars. Examples of topographic lineaments may be related to the
sigma 1 – sigma 2 plane using domain swarm analysis. Finally
some implications for planetary tectonic models are presented.

2. INTRODUCTION

A fractured planetary surface indicates past concentrations
of stress in excess of the stength of materials comprising that
surface. The fractures may represent simple separation or
extension along a plane or they may have suffered slippage
parallel to the fracture planes, in which case they are termed
faults. The fractures commonly are severely modified by erosive
processes and etched into the topography as subtle lines or
lineaments.

The types, orientations, and patterns of the fractures indicate past stress orientation and character. If the mechanical
properties of the material are known, the stress magnitudes
can be estimated. Conversely, if the major driving stress

A. Coradini and M. Fulchignoni (eds.), The Comparative Study of the Planets, 409–418.
Copyright © 1982 by D. Reidel Publishing Company.

is gravity of the planet, then the fractures give some clues to
the strength of the near-surface materials of the body. The
relative ages of the fractures also can be determined in many
instances as being younger than any material or structure
that they cut and older than any rock unit or structure which
is superimposed upon them or cuts them.

   Thus, the fracture patterns can yield detailed data on
major tectonic episodes and events in the history of the planet,
details which when coupled with geophysical and geochemical
observations, volcanic data, and distribution of major rock
units, define much of the inner workings of a planetary machine.
They are a significant component in devising a tectonic model
of a planet's evolution.

   The fractured "surface" of a planet may have several meanings.
On earth, a transition from generally brittle to more or less
ductile behavior at typical terrestrial strain rates occurs at
depths of about 10 km. At greater depths, much of the deformation may be by ductile flow or ductile faults, that is on
fault planes without sharp discontinuity and not necessarily of
rapid enough motion to create significant seismic waves.
Nevertheless, with higher rates of strain and less silicic
materials, faults with measurable seismic energy release do
occur on earth at depths up to 700 km. The word "crust"
implies the material above the earth's mantle, namely the
granitic continents with thicknesses of about 35-40 km and the
oceanic floor with thickness of about 5 km. The word
"lithosphere" implies the relatively rigid plate of plate
tectonics moving on an "asthenosphere" of weaker, more ductile
mantle material at depths on the order of 150 km. On other
planetary bodies the "regolith" or meteorite impacted and
broken material may constitute the upper, 1-10 km; this
regolith may have a coating of relatively unbroken lava flows;
on colder bodies a few km of permafrost may impregnate and
lithify the upper part of the regolith. In brief, the concept
of fracturing of planetary surfaces may include consideration
of many types of materials and mechanical layering, many of
which are capable of changing as a function of time.

3.   PRINCIPLES

   The foundation of much of fracture analyses is the deduction
of stress from the observed strains, namely the fractures.
Three dimensional stress is a tensor quantity as illustrated
in Fig. 1. It is composed of 9 elements, three acting normal
to the faces(sigma zz, sigma yy, sigma xx)and six acting as
shears along the faces. Each of the shears, indicated by tau,
has a two component subscript: tau zy means a shear acting on
a z face in the y direction, etc. The stress tensor is commonly

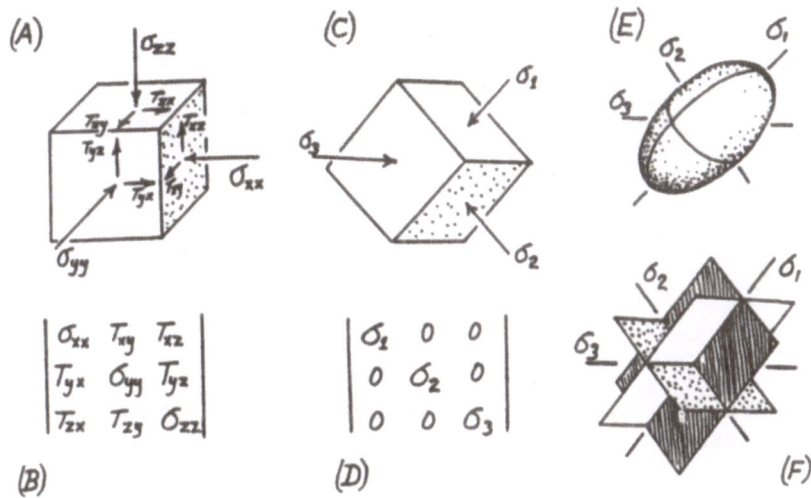

FIGURE 1.   SIX DIFFERENT WAYS OF EXPRESSING THE STRESS TENSOR.
PRINCIPAL STRESS AXIS AND PRINCIPAL PLANES WITH NO SHEAR
ALONG THEM ARE ILLUSTRATED IN (E) AND IN (F) RESPECTIVELY.

written as in (1B) but may always be recast into its eigenvalues
as in (1D). Thus, for every stress tensor there is always
some coordinate system (1C) in which three mutually perpendicular
stresses act normal to the faces of a unit cube in that system
and in which there are no shears on those faces.   Direction
cosine orientations of these three directions are given as their
eigenvectors.   The three directions are called principal
stress axes: the largest compressive principal stress is
termed sigma 1; the intermediate is sigma 2; and the least
compressive stress is sigma 3.   Their magnitudes and orientations
are commonly represented as the stress ellipsoid of 1E.
Figure 1F illustrates the fact that for any stress system there
are three and only three principal planes, each normal to one
of the principal stresses and having no shear components along
it.

Two of the major relationships of fracture directions to
principal stresses appear in Fig. 2 A and B.   If sigma 3 is
negative or tensional (or effectively tensional as a result of
partial support of overlying rock by fluid pressure) an
extensional fracture is produced in the sigma 1 – sigma 2 plane
(Fig. 2A).   Alternatively with compressional failure, paired
or conjugate fault planes are produced (Fig. 2B) intersecting
in sigma 2 and lying about 30 degrees on either side of sigma 1.
The motion on the fault planes is such that the actue angled
wedge is forced inward causing the rock mass to expand in the

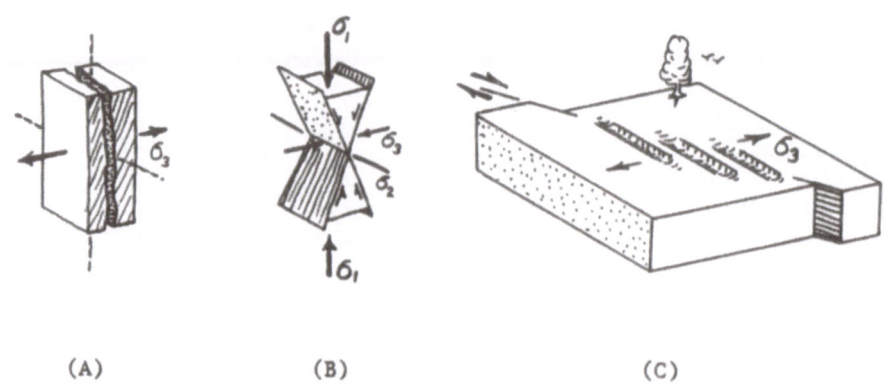

(A)   (B)   (C)

FIGURE 2.   RELATIONSHIP OF SOME FRACTURE TYPES TO PRINCIPAL
     STRESS AXES

sigma 3 direction.

     These basic relationships were exploited by E.M. Anderson
(1) in a theory widely applicable to planetary faulting.
Anderson noted that the surface of the earth is a plane of no
shear, that is, a principal plane, requiring that one of the
three principal stresses be vertical near the earth's surface.
These three possibilities result in the three major classes of
faults (Fig. 3):   compressional thrust faulting with sigma 3
vertical, extensional normal faults producing horsts (up) and
grabens (down) with sigma 1 vertical, and strike-slip or
lateral displacement with sigma 2 vertical.   If the strike-slip
displacement is distributed across a zone of shearing (Fig. 2C)
a zone of en echelon horst and graben can be produced with long
axes normal to sigma 3.   In that the sigma 3 orientation is related
to the direction of shearing, a zone of en echelon horst and
graben may be used to determine the motion sense of shearing
along the zone itself.   An alternative interpretation is also
possible, as discussed in the next paragraph.

     Ideally, these principles apply to isotropic rock masses.
Most planetary surfaces have, at a minimum, some layering or
vertical differences in strength and mechanical properties.   In
addition there is commonly an array of pre-existing fracture
orientations on most planetary surfaces, producing a strength
anisotropy favoring some fracture directions over others.
Discussion of all the ramifications of these anisotropies is
beyond the space limits of this paper although one needs
emphasis.   If a regional fracture anisotropy has been produced
on a surface by a stress system similar to 2A, a slight shift
in sigma 3 stress orientation should produce fracture directions

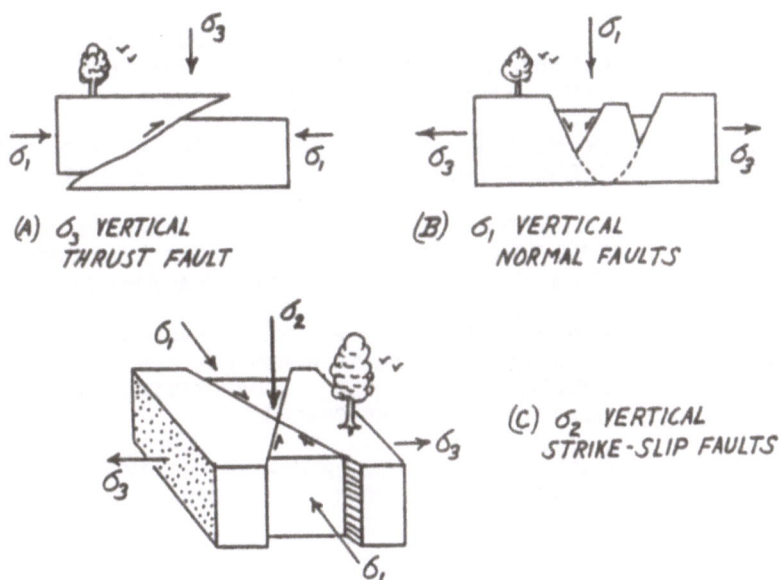

FIGURE 3. THE THREE MAJOR CLASSES OF FAULTS RELATED TO THE
    THREE ORIENTATIONS OF THE STRESS ELLIPSOID ACCORDING TO
    THE MODEL OF E.M. ANDERSON.

normal to the new sigma 3. The trace of these new fractures
may be marked by lines of horst and graben, with individuals
oriented parallel to the older fracture trends. The result
is an _en echelon_ system similar to 2C which may be interpreted
incorrectly as evidence for strike-slip faulting, lateral
shifting of the crust, and even a tendency toward incipient
plate tectonics.

4. AGE OF FRACTURING

Determination of the relative and absolute age of fracturing is an important part of the analysis. The faulting and
fracturing must be younger than any rock unit or feature which
they cut and older than any unit or feature superimposed upon
or cutting them. The absolute ages of the stratigraphic units
are obtained, for the most part, by crater density studies
linked to radiometric dates on lunar surfaces (see Neukum,
this volume). Photo interpreted relative ages of the fractures
are obtained by some of the relationships illustrated in Fig. 4.

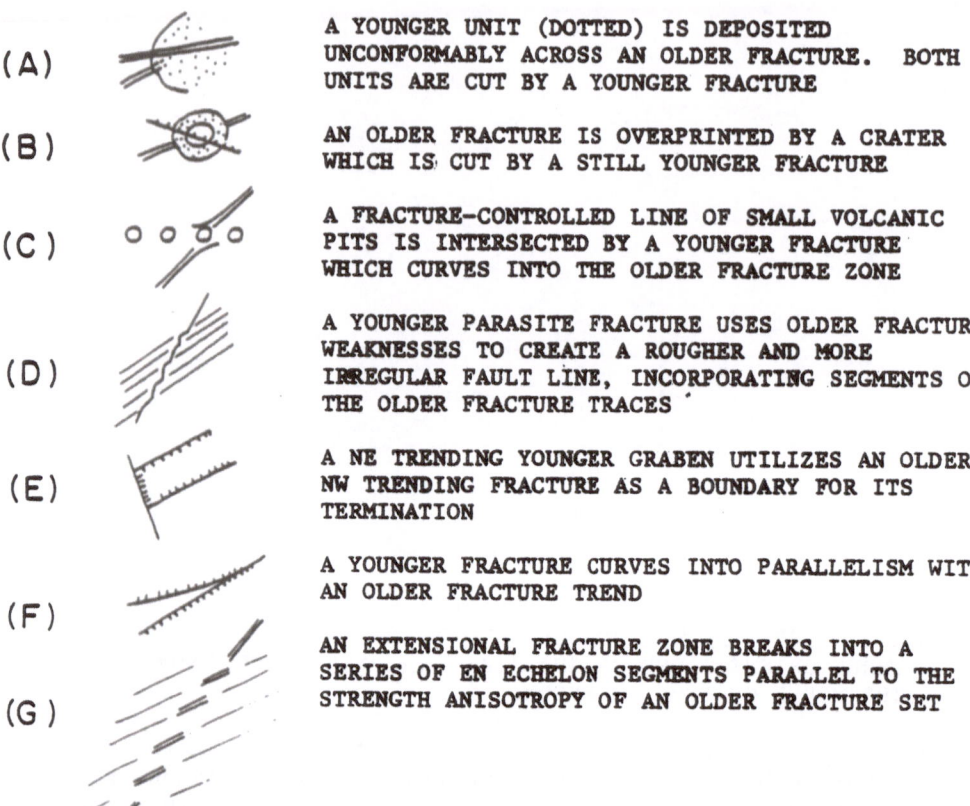

(A)   A YOUNGER UNIT (DOTTED) IS DEPOSITED
      UNCONFORMABLY ACROSS AN OLDER FRACTURE. BOTH
      UNITS ARE CUT BY A YOUNGER FRACTURE

(B)   AN OLDER FRACTURE IS OVERPRINTED BY A CRATER
      WHICH IS CUT BY A STILL YOUNGER FRACTURE

(C)   A FRACTURE-CONTROLLED LINE OF SMALL VOLCANIC
      PITS IS INTERSECTED BY A YOUNGER FRACTURE
      WHICH CURVES INTO THE OLDER FRACTURE ZONE

(D)   A YOUNGER PARASITE FRACTURE USES OLDER FRACTURE
      WEAKNESSES TO CREATE A ROUGHER AND MORE
      IRREGULAR FAULT LINE, INCORPORATING SEGMENTS OF
      THE OLDER FRACTURE TRACES

(E)   A NE TRENDING YOUNGER GRABEN UTILIZES AN OLDER
      NW TRENDING FRACTURE AS A BOUNDARY FOR ITS
      TERMINATION

(F)   A YOUNGER FRACTURE CURVES INTO PARALLELISM WITH
      AN OLDER FRACTURE TREND

(G)   AN EXTENSIONAL FRACTURE ZONE BREAKS INTO A
      SERIES OF EN ECHELON SEGMENTS PARALLEL TO THE
      STRENGTH ANISOTROPY OF AN OLDER FRACTURE SET

FIGURE 4.   RELATIONSHIPS INDICATING RELATIVE AGE OF FRACTURES

5.   STRESS GENERATION

A variety of structural conditions on planetary surfaces
can produce stress differentials of a few hundred bars or levels
adequate to produce faulting. On earth, the driving forces of
plate tectonics and the associated stress fields of the plates
are generally regarded as including the gravitative sinking of
densifying oceanic plates beneath island arcs and the sliding
of plates off upraised, less dense, hotter ridges of oceanic
mantle materials. This process, in that the density contrasts
are thermally powered, is a form of convective drive even though
the mantle is a relatively passive element for the system.
It differs from many refinements of ideas of the driving
process which include large, thick cells of deep mantle actively
overturning or of convective mantle plumes dragging passive

surface materials along on their tops. The relative contributions of these processes to Earth's moving tectonic plates is
uncertain. Their operation at other times or on other planets
with differing energy levels, lithospheric thickness, and mantle
characteristics is even less certain.

Many other processes are capable of generating stress
fields of regional or planetary scale. Impacting bodies
create massive but short-lived stress fields. Thermal changes
producing expansion or contraction at differing depths can be
generated by long-term cooling, concentrations of radioactive
material, or gravitative energy associated with core formation.
Volumetric changes associated with permafrost regions of Mars
are probably causing many local fractures. Changes in surface
or subsurface loading can be produced by volcanic piles, polar
deposits, erosion at surficial or subcrustal depths, or lateral
magmatic movements. Gravitational sliding of surface materials
off topographic highs creates many local stresses. Any loads
not supported by isostatic equilibrium must be supported by
stresses maintained by the strength of the planet; for example,
the lunar mascons create strong local stress fields in their
immediate vicinity. Even with isostatic equilibrium, differences
in density at the same depths in adjacent columns will cause
stress differences. In addition, tidal torques are stressing
the surfaces of the bodies; changing rotation rates or "despinning" of planets alter their oblateness expanding the polar
crustal regions, compressing the equatorial regions, and deform-
ing the middle latitudes by pure shear. There appears to be
no shortage of potential stress generators for the observed
planetary strains.

An example of stress fields generated by gravitative
loading combined with a regional sigma 3 is illustrated in
Fig. 5 for Alba Volcano of Mars. The map pattern shows a system
of horsts and graben wrapping around the volcano. Calculations
integrating the various stress tensors and simplifying them
by eigenvalues and eigenvectors yield cross sections similar
to Fig. 5 and map patterns duplicating the observed fault systems.
The differential stress magnitude seems too low (10 to 20 bars)
to produce regional faulting suggesting a strength anisotropy
of a pluton beneath the volcano deflecting the faulting around
it.

6. LINEAMENTS

Among the most widespread and least understood of all
paleostress indicators are large-scale fracture traces of
lineaments etched by erosion into the topography of almost
all areas of the earth. These are readily visible on LANDSAT
and SEASAT imagery of the earth. Similar linear features appear

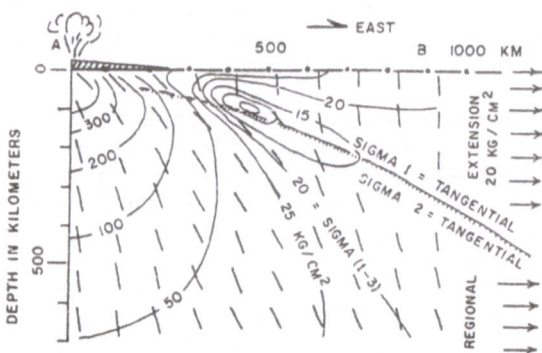

FIGURE 5.   STRESS TRAJECTORIES CALCULATED FOR ALBA VOLCANO ON
    MARS.   THE STRESS TENSORS ARE CALCULATED FOR MARTIAN
    GRAVITY AND A POISSONS RATION OF .4 USING NUMERICAL
    INTEGRATION OF THE BOUSSINESQ EQUATIONS UNDER THE CONE
    AND COMBINATION WITH A REGIONAL EXTENSILE STRESS.   FROM
    WISE (1976).

on almost all solid bodies of the solar system imaged
to date and have been the basis of proposed "lunar grids",
"martian grids" and "mercurian grids". It is generally agreed
that these represent some type of fracture traces related in
some way to stress systems, but the exact relationsips are
hotly debated. This field, in spite of its importance for
planetary work, is a shambles. The reproducibility of where
the lines are drawn by different observers is open to question;
there is no generally agreed upon statistical methodology
to treat the lines; the nature of the origin of the linear
features where they can be observed on earth is not clear.
These linear features, as the second most common topographic
element on planetary surfaces, have considerable potential for
interpretation of the stress history of those surfaces. Unfortunately, in order to realize that goal, significant advances
in understanding their origin need to be made.

     One study attempting to relate lineaments swarms to stress
fields is that of Wise et al., (2,3) based on fracture
domain analysis of Italy. The work suggests that most of these
topographically etched lineaments represent zones of extension
and local jointing rather than true faults. They seem to
represent the outcrop traces of sigma 1 – sigma 2 planes
produced by a variety of overlapping structural events. Some
of their typical tectonic settings are illustrated in Fig. 6.

POSSIBLE TECTONIC SETTINGS FOR LINEAMENT SWARM DEVELOPMENT

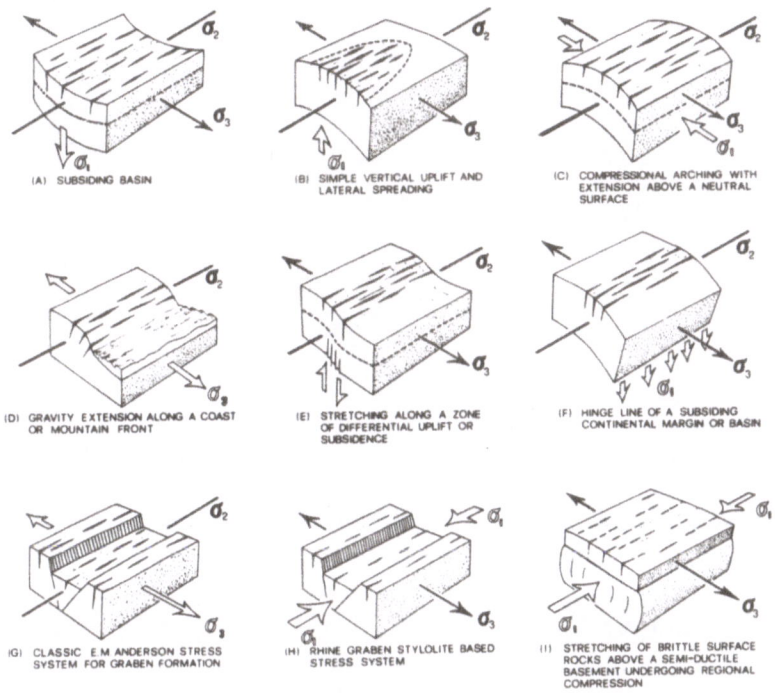

FIGURE 6. WISE, FUNICIELLO, PAROTTO, AND SALVINI (1980)

Similar lineament analysis applied to other planetary surfaces may provide evidence of subtle warpings and stress field changes not readily recognized by most other techniques.

7. REFERENCES

(1) Anderson, E.M., 1951, *The Dynamics of Faulting*, London: Oliver and Boyd, 206 pp.

(2) Wise, D., Funiciello, R., Parotto, M., and Salvini, F., 1980, Origins of regional scale lineament swarms as suggested by fracture domain analysis of Italy, NASA Technical Memorandum 81776, Reports of Planetary Geology Program 1979-1980, pp. 83-85.

(3) Wise, D., 1976, Faulting and stress trajectories near Alba
    Volcano, Northern Tharsis Ridge of Mars, *Geologica Romana*,
    XV:403-433.

AEOLIAN MODIFICATION OF PLANETARY SURFACES

Ronald Greeley

Department of Geology and Center for Meteorite Studies, Arizona State University, Tempe, Arizona 85287, U.S.A.

ABSTRACT

Any planet or satellite having a dynamic atmosphere and a solid surface has the potential for aeolian processes. Survey of the Solar System shows that wind plays an important, and in some regions, the key role in surface modifications. Most deserts and many coastal areas on Earth are subject to aeolian processes. Seasonal dust storms sweep the surface of Mars, where aeolian activity appears to dominate. Measurements of wind speeds on Venus, observations of its surface, and estimates of particle threshold wind velocities in the venusian environment suggest that aeolian processes operate there as well. Recent discoveries of the predominantly nitrogen atmosphere of Titan raise the possibility of wind activity on this, the largest satellite of Saturn and the only moon known to have an appreciable atmosphere. From the extremely dense, hot atmosphere of Venus to the low atmospheric density of Mars and the extremely cold environment of Titan, there is the opportunity to study a single geological process under a wide range of environments to derive fundamental knowledge of how aeolian processes operate.

1.0 INTRODUCTION

Many physical and chemical processes modify planetary surfaces. Aeolian is defined (Gary et al., 1972) as 'Pertaining to the wind; esp. said of rocks, soils, and deposits (such as loess, dune sand, and some volcanic tuffs) whose constituents were transported (blown) and laid down by atmospheric currents, or of landforms produced or eroded by the wind, or of sedimentary structures (such as ripple marks) made by the wind, or of geologic processes (such as erosion and deposition) accomplished by the wind.' Thus, any planet or satellite having a dynamic atmosphere and a solid surface is subject to aeolian or wind processes. A survey of the Solar System shows that Earth, Mars, Venus, and possibly Titan meet these criteria (Table 1). These planets afford the opportunity to study a basic geological process—aeolian activity—in a comparative sense with each planet being a vast

A. Coradini and M. Fulchignoni (eds.), The Comparative Study of the Planets, 419–434.
Copyright © 1982 by D. Reidel Publishing Company.

Table 1. Relevant properties of planetary objects potentially subject to aeolian processes

	Venus	Earth	Mars	Titan
Mass (Earth = 1)	0.815	1	0.108	0.02
Density (water = 1)	5.2	5.5	3.9	1.4
Surface gravity (Earth = 1)	0.88	1	0.38	.11
Surface gravity (cm s^{-2})	890	978	371	110
Atmosphere (main components)	CO_2	N,O	CO_2	N
Atmospheric pressure at surface (millibars)	9×10^4	10^3	7.5	$\sim 2 \times 10^3$
Mean temperature at surface (^{O}C)	480°	22°	-23°	-200°

natural laboratory having strikingly different environments. Because terrestrial processes and features have been studied for many years, Earth is the primary data base. However, because surface processes are much more complicated on Earth—primarily because of the presence of liquid water and vegetation—many aspects of aeolian processes that are difficult to assess on Earth are more easily studied on other planets.

Wind blowing across a planetary surface has the potential for directly eroding material and redistributing it to other areas. Winds transport sediments via three modes: *suspension* (mostly silt and clay particles, i.e., smaller than about 60 μm), *saltation* (mostly sand size particles, 60 to 2000 μm in diameter), and *surface creep* (particles larger than about 2000 μm in diameter). Wind threshold curves (Fig. 1) define the minimum wind speeds required to initiate movement of different particles for given planetary environments (1). The ability of wind to attain threshold is a function primarily of atmospheric density, viscosity, composition, and temperature. Thus, the very low density atmosphere on Mars (surface pressure is about 1/200 that of Earth) requires wind speeds that are about an order of magnitude stronger than on Earth. It can be thought of partly in terms of the number of gas molecules impinging on and passing over the particles to be moved; for the same amount of work to be done in a low density atmosphere (fewer molecules) the wind must be moving faster to achieve the same effective flux of molecules. Although this is an oversimplification, it demonstrates to a first order the relationship between atmospheric density and wind velocity for particle threshold.

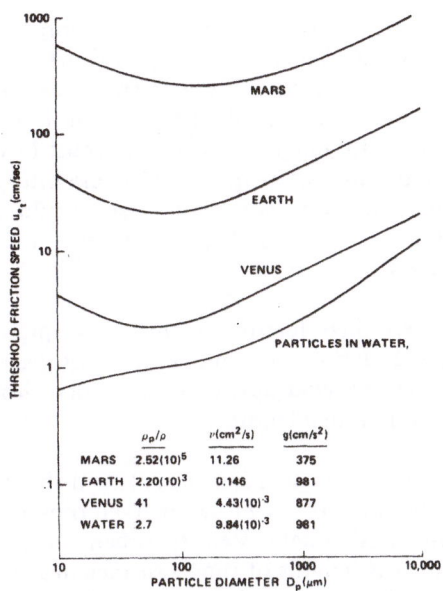

Figure 1: Comparison of threshold friction speed versus particle diameter for Mars, Earth, and Venus and in liquid; ρ_p/ρ is ratio of particle density to fluid density, ν is kinematic viscosity, g is gravitational acceleration, U_F is terminal speed, u_{*_t} is friction speed at threshold (after Ref. 1).

1.1 Relevance to Planetary Geology

Aeolian processes are capable of redistributing enormous quantities of sediment over planetary surfaces, resulting in the formation of landforms large enough to be seen from orbit and deposition of windblown sediments that can be hundreds of meters thick. Any process capable of effecting these changes is relevant to understanding the geological environment of planets so involved. Furthermore, because aeolian processes involve the interaction of the atmosphere and lithosphere, an understanding of aeolian activity sheds light on meteorological problems. Aeolian activity can be considered in terms of large scale modifications, small scale modifications, and as an observable active process.

Large scale modifications involve features that can be observed from distances of orbiting spacecraft. One of the most useful types of features for interpretation of surface processes is the dune, a depositional landform. Both the planimetric shape and cross sectional profile of dunes can reflect the prevailing winds in a given area. Thus, if certain dune shapes and/or slopes can be determined from orbital data, local wind patterns can be determined. Repetitive viewing of the same dunes as a function of season may reveal seasonal wind patterns.

On Earth great quantities of silt and clay are transported in dust storms and eventually deposited as loess. Thick loess deposits are found throughout the geological column. Even where relatively young and well-exposed on the surface, loess deposits are nearly impossible to identify by remote sensing methods. Yet, identification of such deposits could be very important in understanding planetary surfaces. For example, substantial areas of Mars are interpreted to be mantled with aeolian sediments. However, other processes could also lead to similar appearing terrains. Thus, some definitive means for identifying fine-grained aeolian sediments is needed.

Large-scale aeolian erosional features include: 1) pits and hollows (called *blowouts*) that form by deflation, or removal of loose particles, 2) wind sculptured hills called *yardangs*, 3) *windgaps*, or wind-eroded 'notches' in ridges, and 4) general unclassified wind-eroded landforms.

Small-scale aeolian features include *ventifacts* (wind-shaped rocks) and aeolian sedimentary structures–features that can be observed only directly on the ground or inferred from remote sensing data. Ventifacts can provide information about local wind directions and the lengths of time a surface has been exposed. Identification of ventifacts has relevance to other aspects of planetology. For example, rocks at the Viking landing sites that show pitted surfaces have been interpreted as vesicular igneous rocks and are part of the basis for identifying the surrounding plains as volcanic; alternatively, the pitted rocks could be the result of aeolian erosion and are not igneous.

Observations of active aeolian features provide direct information on the atmosphere. For example, crater streaks on Mars are albedo surface patterns that show surface wind direction; they occur in great number over much of Mars. Repetitive imaging of these and other variable feature patterns has shown that many of them disappear, reappear, or change their size, shape, or position with time. Mapping the orientations of variable features has been used to derive a near-surface atmospheric circulation model.

Impact crater frequency distributions are widely used in planetary geology to obtain relative· dates for different surfaces. On planets having active aeolian processes, the erasure of craters by erosion or burial by aeolian sediments can drastically alter the crater record and invalidate crater-derived ages. Thus, knowledge of rates of aeolian erosion and deposition for a wide range of planetary environments is required in order to assess the possible effects on the impact crater record.

1.2 Approach for Investigating Aeolian Processes

Aeolian processes incorporate elements of geology, meteorology, physics and chemistry. A unified study, therefore, requires a multidisciplinary approach. The approach commonly used is to isolate parts of the aeolian process for detailed study. The pioneering work of Bagnold who analyzed the physics of windblown sand is an example of this approach. Once the fundamental principles are understood, then it is possible to extrapolate to a wide range of conditions, i.e., other planetary environments. Before this can be done, however, it is necessary to fit the results from studies of isolated parts of the problem back into the system: for example, Bagnold's (2) work on threshold winds for particle movement were carried out principally in wind tunnel studies; before making generalizations, however, he field tested the results using natural sands.

In the study of planetary aeolian processes there is seldom the opportunity to field test extraterrestrial predictions. Thus, we must rely on a somewhat different method, as follows:

1. Identification of the general problem and isolation of specific factors for study (e.g., wind threshold speeds for particles of different sizes on Mars).
2. Investigate the problem under laboratory conditions where various parameters can be controlled for the 'Earth case' (e.g., wind tunnel test of particle threshold).
3. Field test the laboratory results under natural conditions to verify that the simulations were done correctly (e.g., threshold tests in the field).
4. Correct, modify, and/or calibrate the laboratory simulations to take into account the field results.
5. Carry out laboratory experiments for the extraterrestrial case duplicating or simulating as nearly as possible the planetary environment involved (e.g., threshold tests under martian conditions).
6. Extrapolate the results to the planetary case using the laboratory results and theory (for parameters that cannot be duplicated).
7. Field test the extrapolation via spacecraft observations and apply the results toward the solution of problems involving aeolian processes.

A benefit of this approach is not only to provide a logical means for understanding extraterrestrial problems, but also a contribution toward solving aeolian problems on Earth as well.

2.0 MARS

Windstorms were suspected to occur on Mars even before Mariner 9 returned conclusive evidence of aeolian activity in 1971. Earth-based observations made over the last 100 years showed albedo patterns that were attributed to a variety of processes. Some of the earliest interpretations of these patterns as dust storms were those of Dean McLaughlin, as reviewed by Veverka and Sagan (3).

As knowledge of the composition and density of the martian atmosphere became better defined, predictions based on theory were made as to the wind velocities required to set particles in motion. Because of the low atmospheric density on Mars, the estimated minimum windspeeds are about an order of magnitude higher than on Earth (4). Wind tunnel tests conducted under low atmospheric pressure in a martian simulation substantiate these estimates (5,6,7).

When Mariner 9 and the Soviet spacecraft Mars 2 and 3 arrived at Mars in 1971 during a major global dust storm, the speculations and predictions of martian aeolian processes were amply verified. After the dust cleared, the Mariner 9 cameras revealed abundant features attributed to aeolian activity, including dunes, yardangs, and various pits and grooves considered to be deflation features (8). The Viking mission (1976-1981) added substantially to the catalog of martian aeolian features and provided details not previously observed (Figs. 2, 3). And for

Figure 2: High resolution image of cratered terrain south of Elysium Planitia at 14.4° S and 190° W showing numerous transverse dunes on plains between mountains; the position and orientation of the dunes appear to be controlled by the surrounding topography; large crater is about 2 km in diameter (Viking Orbiter frame 763A10).

Figure 3: Dune field on the floor of an ancient impact crater in the region south of Sinus Sabaeus, centered at 46° S, 339° W. Craters and other topographic depressions are natural traps for windblown sediments and the crater shown here is typical of many that have been photographed from orbit. Crater is about 150 km across (Viking Orbiter frame 94A42).

the first time images were seen of the surface of Mars at two landing sites. The surface consists of rock fragments up to several meters in diameter and very fine-grained material described as aeolian drift deposits (9). Some of the rocka are pitted and sculpted, suggestive of ventifacts.

2.1 Dunes

The vast erg of the circumpolar region of Mars is one of the most impressive discoveries of the Viking mission. The field covers more than 7×10^5 km^2, larger than Rub Al Khali in Arabia (the largest active erg on Earth). All of the dunes are either transverse or barchan (Fig. 4). Mapping the dune morphologies (10,11) and other indicators of wind directions have enabled regional maps of the wind circulation pattern to be derived. Two major wind directions are suggested, off-pole winds that become easterly due to coriolis forces during summer, and on-pole winds that become westerly during winter. These wind patterns compare favorably with those based on models of the atmosphere. The low albedo (i.e., dark) appearance of the dunes suggests a composition other than quartz, an observation fitting with the apparent lack of silicic materials on Mars. Because basaltic lavas are very common over much of Mars, including the smooth plains south of the dune field, it is suggested that the north polar dunes are composed of windblown basaltic particles.

Figure 4: Viking Orbiter view of the north polar region of Mars showing isolated dunes and dune complexes; image covers an area about 30 km wide (Viking Orbiter frame 544B07).

2.2 Yardangs

Yardangs are primary wind erosional features observed on Mars. Most yardangs occur in equatorial regions, notably in the Amazonis region, Aeolis region, Ares Valles, and Iapygia. Some of the largest features are interpreted to be early-stage yardangs; they are 50 km long, 1 km wide, and 200 m high, and appear to have developed from the erosion of mesas. From studies of terrestrial yardangs Ward (12) concludes that the martian features are geologically young (on a martian time-scale) and probably are composed of friable rocks such as ignimbrites (many of the yardang localities are near known volcanic craters), or indurated regolith (regolith in this sense being fragmental debris generated by impact cratering). On Earth yardangs develop by erosion of grains that are loosened by weathering processes involving liquid water; Ward suggests that on Mars (in the absence of liquid water) exfoliation, salt weathering, or freeze-thaw processes may operate, but that the net weathering rate would be slower than on Earth.

2.3 Variable Features

The most abundant aeolian features on Mars are *variable features*, so named from the Mariner 9 mission (13) for albedo patterns that changed their size, shape, and position with time (Fig. 5). Crater streaks are the most common of the variable features. They commonly occur as either light or dark forms, although 'mixed' forms are found in which both light and dark streaks occur in association with the same crater. The origin of crater streaks is a matter of debate, with several models having been proposed. Nearly all investigators agree that streaks represent a surface manifestation of windblown processes, such as relatively thin (~cm) deposits of particles that shift in response to winds.

Figure 5: Mariner 9 images of a 17 km crater in the Daedalia region near Solis Lacus imaged early in the orbiting mission, showing only a small dark zone (left) and the same feature 38 days later (right) showing the growth of the dark zone due to aeolian activity; these and similar surface patterns are termed variable features (from Ref. 13).

Several different models (14) of light and dark streak formation can be postulated as functions of wind characteristics (wind velocities, turbulence, etc.), particle sizes, and particle compositions. In addition, the origin of streak material (from a source within the crater or part of the general surface around the crater) may control the development of streaks. Nearly all models must take into account the flow patterns generated by winds blowing over and around craters. Wind tunnel simulations and limited field studies show that a horseshoe vortex (Fig. 6) wraps around the crater rim and creates an erosive zone in the wake of the crater and a depositional zone in the immediate lee of the crater rim (15). The size and shape of zones of erosion and deposition are functions of crater geometry, wind speeds, time, and other parameters (16).

Some elongated bright streaks associated with craters in the south polar region appear to be accumulations of CO_2 frost. Based on their form and seasonal behavior, they have been interpreted to be accumulations of wind-transported frost in the lee of craters, similar to the formation of bright crater streaks involving dust and sand particles (17). Because frost particles would have a fairly low density, threshold wind speeds would be lower than for sand or dust, and frost streaks therefore would be more active.

Crater streaks of all types have been used to map surface wind patterns and applied to atmospheric circulation models (18). Comparisons of crater streaks observed from orbit with wind measurements obtained at the Viking Lander sites (14) shows that dark streak orientations coincide with directions of maximum winds (Fig. 7).

2.4 Dust Storms

More than one hundred years of observations of Mars have produced an impressive catalog documenting active dust storms (19). Spacecraft observations of Mars have greatly enhanced our understanding of these storms. Mariner 9 arrived in 1971 at a time when the planet's surface was obscured by dust (20), and observed the waning stages of a major global storm. The Viking spacecraft,

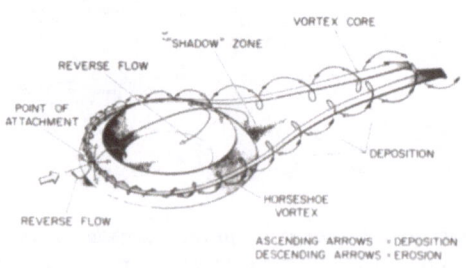

Figure 6: Flow field pattern for wind blowing across a raised-rim crater showing zones of preferential erosion and deposition (from Ref. 15).

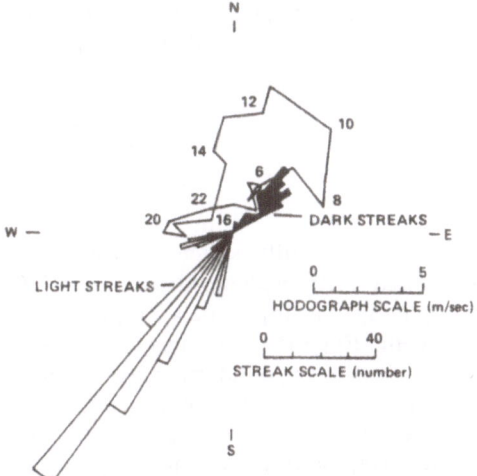

Figure 7: Rose diagram showing light and dark streak orientations in Chryse Planitia (region of Viking Lander 1) and hodograph of winds measured via the lander; dark streaks are oriented with the wind directions (from Ref. 14).

however, provided not only orbital viewing of all stages of dust storms (19), but also provided ground-based meteorological data (21). These observations show that major storms typically begin during the southern hemisphere spring or early summer (close to perihelion). In a given year the major storms vary in intensity and number from none to two. The major dust storms have been observed to originate from areas in both the southern (Hellespontus, Noachis, and Solis Planum) and northern hemispheres (Isidis Planitia).

The major storms appear to go through three phases. In phase I, numerous local dust storms occur in the southern hemisphere, associated with the sublimating south polar cap and strong surface thermal contrasts (22). These local storms contribute dust to the global atmosphere increasing diurnal and thermal tides. This increases the likelihood of a global storm because of the energy absorbed by the atmosphere from the suspended dust. Phase II is the expansion of a local storm into a global event. This occurs rapidly typically involving 3 to 7 days. Phase III marks the decay of the storm and lasts from 50 to 100 days. The first areas to clear are the poles and topographically high regions, such as the summits of the shield volcanoes.

Mariner 9 and Viking results show that average particle size in the atmosphere is less than 2 μm, or about the same as the particles carried over the Atlantic by major Saharan dust storms. The dust on Mars was found to be well mixed in the atmosphere to heights of 30 to 40 km and had the effect of raising the atmospheric temperature by as much as 50° K. Calculation of dust deposition from the storms suggests significant mantling of the surface of Mars over geological time.

In summary, aeolian processes appear to be the dominant process currently active on the surface of Mars, and has played an important role in the geological past.

3.0 VENUS

The atmosphere of Venus is composed primarily of CO_2 with minor amounts of hydrochloric, hydrofluoric, and sulfuric acids. With a surface pressure of more than 90 bar, it has the highest atmospheric density of all the terrestrial planets (Table 1). Venus is completely enveloped in a perpetual shroud of clouds that hide the surface from viewing. Repetitive pictures of the cloud tops obtained over a period of 8 days during the flyby of Mariner 10 in 1974 showed circulation patterns and allowed wind speeds to be determined for the upper atmosphere (23). Although speeds of about 100 m s^{-1} were obtained for the upper clouds in the equatorial zone, when extrapolations were made to the surface the winds were estimated to be very sluggish.

The Soviet landers, Venera 9 and 10, measured wind speeds near the surface for two sites on Venus of 0.5 to 1 m s^{-1} at the height of the windsensors (1 to 2 m above surface). More recent measurements of windspeeds obtained by the Pioneer-Venus atmospheric probes have been extrapolated to the surface and yield values of 1 to 2 m s^{-1} (24). These values are well within the range predicted for particle threshold (Fig. 1), based on a combination of theory (25,26) and extrapolations of wind tunnel experiments (1). Venera 9 and 10 images of the surface of Venus show rock fragments several cm and larger set in a mass of fine (<1 cm) material interpreted to be sand size or smaller (27). This bimodal size distribution is indicative of fluid transport and because liquid water cannot exist at the extremely high temperature on Venus, it is assumed that the fluid involved is the atmosphere, or wind. Thus, it is likely that aeolian processes are active at present on Venus, and probably have been active in the geological past.

4.0 TITAN

The Voyager spacecraft flew past Saturn and made observations of some of the satellites. One of the most important observations was the radio occultation of the atmosphere of Titan, Saturn's largest moon. The atmosphere was found to be predominantly nitrogen with a density that suggests surface pressures of 2 bar. Estimates of surface temperatures suggest solid nitrogen on the surface. Thus, there is the potential for aeolian processes on this, the only satellite known to have a substantial atmosphere. Whether the atmosphere is dynamic, and whether granular particles exist on the surface are unanswered questions. Dunes composed of ice particles being blown about by the dense, extremely cold nitrogen atmosphere borders on the realm of science fiction but remains a possibility.

5.0 SUMMARY

Aeolian processes play an important role in the modification of the surfaces of Earth and Mars. Indirect evidence suggests that Venus and perhaps Titan, one of the satellites of Saturn, also may experience aeolian activity. Study of aeolian activity in a planetary context thus affords the opportunity to examine a fundamental process under a wide range of environmental conditions—each planet can be viewed as a vast natural laboratory.

The key questions for future work can be posed as follows:

1. *How are dust storms initiated?* This question applies to both Earth and Mars. Dust storms involve very fine grained materials (typically a few μm in size); and examination of the threshold curve (Fig. 1) shows that extremely strong winds are required to move material this fine and it is unlikely that grains are placed into suspension by simply having a strong wind blow across the surface. The explanation frequently given is that sand size grains (relatively easily moved) placed into saltation impact dust particles, setting them into motion where they can easily be wafted aloft. This does not explain frequent dust storms on Earth that begin in areas lacking sand. Thus, there may be other factors involved, perhaps related to electrostatic effects, phase changes of absorbed volatiles, or other triggering mechanisms. This is a general problem that has received little attention, despite its importance on both Earth and Mars.

2. *What are rates of aeolian erosion in various planetary environments?* Despite much interest in aeolian erosion (both deflation and abrasion), relatively few studies have been undertaken even on Earth that provide quantitative data, much less that are relevant to other planets. There are two parts to the problem, one part deals with eroison on a small scale (e.g., individual rocks), the other part concerns erosion of landforms (e.g., crater rims).

3. *What is the evolution of aeolian landforms?* Many studies have been carried out that define stages of development for various aeolian features on Earth. In some cases, there is disagreement over the results and the interpretations that are involved. Various models of evolution should be tested on Earth and studies done to determine how to apply them to extraterrestrial environments.

4. *What are the characteristics of aeolian sediments on planets?* On Earth, most aeolian sediments are quartz sand and silts, various clay minerals, plus minor amounts of particles of other composition (gypsum sands, calcite sands, etc.), and most of the knowledge of the physics of particle motion and aeolian features such as dunes deals with these materials. On Titan, if windblown particles exist, they may be composed of nitrogen ice or other frozen volatiles; composition of aeolian particles on Venus is completely open to speculation; on Mars quartz is probably absent or is only a minor constituent because major source rocks for quartz, such as granite, are apparently absent. Basaltic sands have been proposed for

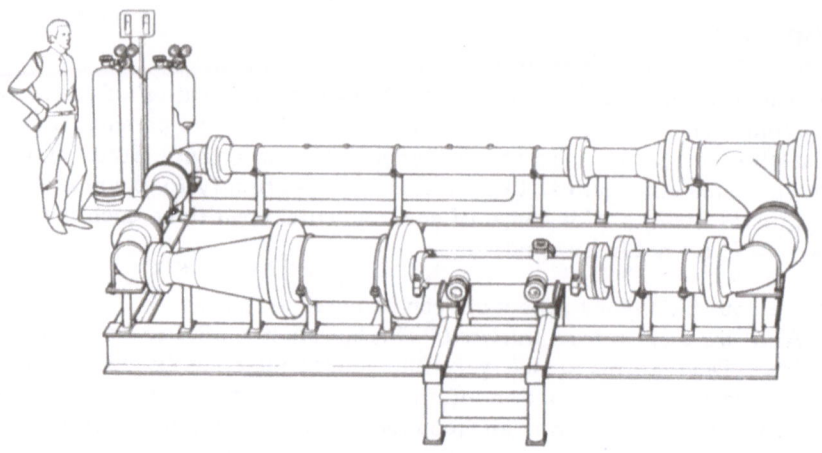

Figure 8: Diagram of the Venusian Wind Tunnel, a facility at NASA-Ames Research Center used to carry out research on aeolian processes under simulated venusian conditions.

Mars, given the widespread occurrence of basaltic lava flows, the low albedo of martian dunes, and the prevalence of mafic to ultra mafic compositions derived from spectral studies. How do windblown basaltic sands behave? What is their evolution? How efficient are they as agents of abrasion in the aeolian environment? Do they accumulate in deposits having morphologies different from quartz particles? These and other questions are intimately linked with understanding the general sedimentary cycle on Mars; similar sets of questions can be applied to Venus and possibly Titan as more information becomes available.

5. *What are aeolian conditions on Venus?* The physics of windflow and particle movement in the high temperature and high pressure and density atmosphere of Venus involves flow regimes not previously investigated. Even such fundamental questions as threshold wind speeds can only be estimated crudely. Other aeolian problems such as mass flux of windblown particles, rates of erosion, landforms, etc. require knowledge of threshold and additional study appropriate to the venusian environment (Fig. 8).

Because aeolian activity involves the interaction of the atmosphere with the surface, the analysis of various wind-related features can provide clues to the nature of the atmosphere. Understanding rates of aeolian erosion and analysis of ancient aeolian features have the potential for providing information on the climatic conditions of the past, and for helping to unravel the geological history of the Solar System.

ACKNOWLEDGMENTS

I thank M. Fulchignoni and A. Coradini for the opportunity to participate in the NATO-ASI Comparative Study of the Planets. This work was supported by the Planetary Geology Program, National Aeronautical and Space Administration, Washington, D.C.

REFERENCES

1. Iversen, J.D., Greeley, R., and Pollack, J.B.: 1976, Windblown dust on Earth, Mars and Venus. J. Atmos. Sci. 33, pp. 2425-2429.

2. Bagnold, R.A.: 1941, The Physics of Blown Sand and Desert Dunes. Methuen and Co., London, 265 p.

3. Veverka, J. and Sagan, C.: 1974, McLaughlin and Mars. Amer. Scientist 62, pp. 44-53.

4. Sagan, C. and Pollack, J.B.: 1969, Windblown dust on Mars. Nature 223, pp. 791-794.

5. Greeley, R., White, B., Leach, R., Iversen, J. and Pollack, J.: 1976, Mars: wind friction speeds for particle movement. Geophys. Res. Let. 3, pp. 417-420.

6. Greeley, R., White, B.R., Pollack, J.B., Iversen, J.D. and Leach, R.N.: 1977, Dust storms on Mars: considerations and simulations. NASA Tech. Memo, TM 78423, 29 p.

7. Greeley, R., Leach, R., White, B., Iversen, J. and Pollack J.: 1980, Threshold windspeeds for sand on Mars: wind tunnel simulations. Geophys. Res. Let. 7, pp. 121-124.

8. McCauley, J.F.: 1973, Mariner 9 evidence for wind erosion in the equatorial and mid-latitude regions of Mars. J. Geophys. Res. 78, pp. 4123-4137.

9. Arvidson, R.E., Binder, A.B. and Jones, K.L.: 1978, The surface of Mars. Sci. Amer. 238, No. 3, pp. 76-89.

10. Tsoar, H., Greeley, R. and Peterfreund, A.R.: 1979, Mars: the north polar sand sea and related wind patterns. J. Geophys. Res. 84, pp. 8167-8180.

11. Breed, C.S., Grolier, M.J. and McCauley, J.F.: 1979, Morphology and distribution on common 'sand' dunes on Mars: compariosn with the Earth. J. Geophys. Res. 84, pp. 8183-8204.

12. Ward, A.W.: 1979, Yardangs on Mars: evidence of recent wind erosion. J. Geophys. Res 84, pp. 8147-8166.

13. Sagan, C., Veverka, J., Fox, P., Dubisch, R., Lederberg, J., Levinthal, E., Quan, L., Tucker, R., Pollack, J.B. and Smith, B.A.: 1972, Variable features on Mars: preliminary Mariner 9 television results. Icarus 17, pp. 346-372.

14. Greeley, R., Papson, R. and Veverka, J.: 1978, Crater streaks in the Chryse Planitia Region of Mars: early Viking results. Icarus 34, pp. 556-567.

15. Greeley, R., Iversen, J.D., Pollack, J.B., Udovich, N. and White, B.: 1974, Wind tunnel simulations of light and dark streaks on Mars. Science 183, pp. 847-849.

16. Iversen, J.D., Greeley, R., White B.R. and Pollack, J.B.: 1976, The effect of vertical distortion in the modeling of sedimentation phenomena: martian crater wake streaks. J. Geophys. Res. 81, pp. 4846-4856.

17. Thomas, P., Veverka, J. and Campos-Marquetti, R.: 1979, Frost streaks in the south polar cap of Mars. J. Geophys. Res. 84, pp. 4846-4856.

18. Thomas, P. and Veverka, J.: 1979, Seasonal and secular variation of wind streaks on Mars: an analysis of Mariner 9 and Viking data. J. Geophys. Res. 84, pp. 8131-8146.

19. Briggs, G.A., Baum, W.A. and Barnes, J.: 1979, Viking orbiter imaging observations of dust in the martian atmosphere. J. Geophys. Res. 84, pp. 2795-2820.

20. Leovy, C.G., Briggs. G.A., Young, A.T., Smith, B.A., Pollack, J.B., Shipley, E.N. and Wildey, R.L.: 1972, The martian atmosphere: Mariner 9 television experiment progress report. Icarus 17, pp. 373-393.

21. Ryan, J.A. and Henry, R.M.: 1979, Mars atmospheric phenomena during major dust storms, as measured at surface. J. Geophys. Res. 84, pp. 2821-2829.

22. Peterfreund, A.R. and Kieffer, H.H.: 1979, Thermal infrared properties of the martian atmosphere 3, local dust clouds. J. Geophys. Res. 84, pp. 2853-2864.

23. Murray, B.C., Belton, M.J.S., Danielson, G.E., Davies, M.E., Gault, D., Hapke, B., O'Leary, B., Strom, R.G., Soumi, V. and Trask, N.: 1974, Venus: atmospheric motion and structure from Mariner 10 pictures. Science 183, pp. 21-29.

24. Counselman, C.C., Gourevitch, S.A., King, R.W., Loriot, G.B. and Prinn, R.G.: 1979, Venus winds and zonal and retrograde below the clouds. Science 205, pp. 85-87.

25. Hess, S.L.: 1975, Dust on Venus. J. Atmos. Sci. 32, pp. 1076-1078.

26. Sagan, C.: 1975, Windblown dust on Venus. J. Atmos. Sci. 32, pp. 1079-1083.

27. Florensky, C.P., Ronca, L.B., Basilevsky, A.T., Burba, G.A., Nikolaeva, O.V., Pronin, A.A., Trakhtman, A.M., Volkov, V.P. and Zazetsky: 1977, The surface of Venus as revealed by Soviet Venera 9 and 10. Geol. Soc. of Amer. Bull. 88, p. 1537-1545.

SOME THERMODYNAMIC RELATIONSHIPS GOVERNING THE BEHAVIOR OF PERMAFROST AND FROZEN GROUND

Duwayne M. Anderson

State University of New York at Buffalo

It is well established that water-ice occurs in the surface materials of Mars and that temperature regimes are such that ice-rich permafrost may be present in many localities. Water-ice is also thought to be a major constituent of the surface materials on Europa, Ganymede, Callisto, and several of the moons of Saturn.

When ice exists in intimate contact with silicate minerals and other solids a thin interfacial transition layer is present. Recent discussions of this topic in connection with planetology have centered on the nature, properties, and behavior of this layer (Anderson 1979, 1980). In brief, this transition layer possesses the properties of a two-dimensional fluid and is, therefore, referred to as "unfrozen water". The viscosity of this unfrozen, liquid-like water is considerably higher than that of pure water as inferred from nuclear magnetic resonance data and there is evidence of heterogeneity in this interfacial transition zone by phase changes in the range -40°C to -50°C. These major phase changes appear to be dependent upon the nature of the adjacent silicate surface or other solid phase. Below about -80°C the liquid-like behavior of the interface ceases and behavior more characteristic of solids is observed. The thickness of the unfrozen water interface varies with temperature, pressure, and concentration of dissolved substances. The relationships have been extensively studied and discussed and in general are exponential in nature.

The existence of a mobile interfacial fluid facilitates the transport of ionic species through frozen ground and also the response (by regelation) of the ice and mineral particles to changes in temperature and stress fields toward an equilibrium

A. Coradini and M. Fulchignoni (eds.), The Comparative Study of the Planets, 435–440.

or steady state configuration. The equilibrium among phases present in frozen ground can be described by appropriate thermodynamic relationships derived for heterogenious systems. The variables needed to define the thermodynamic state of this heterogenious mixture of ice, unfrozen water, and mineral matter are temperature, pressure, and composition. Other variables such as electrical gradients, gravitational fields, etc. are usually neglected but can be included if appropriate or needed. The defined thermodynamic functions required are: The partial molar free energy of the substance under discussion, in this case water ($\bar{F}$); the partial molar enthalpy ($\bar{H}$); the partial molar entropy ($\bar{S}$); and the activity (a). Also needed in the discussion are: The partial molar volume ($\bar{V}$); the partial molar heat capacity ($\bar{c}$); and the freezing point depression (θ). Although thermo-dynamic theory proceeds from a basis in first principles from which "ideal behavior" can be predicted the theory ultimately becomes empirical because of complex interatomic and inter-molecular behavior that as yet is too complex to have been sufficiently understood and described in terms ammenable to inclusion. Consequently, recourse is invariably made to empirical data which are then directly related to the partial molar quantities listed. When the empirical data are accurate and known with sufficient completeness, equilibrium relationships can be predicted with confidence and accuracy. Enough work has now been done to allow the ready determination or prediction of the following phenomena in permafrost and/or frozen ground: The freezing point depression (θ); the latent heat of freezing or thawing; the variation in thickness of the unfrozen water interface separating ice from mineral surfaces as a function of temperature, pressure, and concentration of dissolved substances; the direction of movement of water through frozen ground in response to temperature or electrical field gradients; the maximum, equilibrium pressure that may be developed in or sustained by frozen ground at varying temperatures below freezing.

Consider a mixture of ice and an assembly of silicate minerals for which adequate data on their surface properties, etc. is available (e.g. an assembly of clay minerals, quartz, etc.). Assume that such variables of state as the interfacial area, the composition of the water and other constituents present are constant and invarient. Under these conditions the equilibrium existing between the unfrozen water and the ice present can be described in terms of their respective partial molar free energies and the dependence of these quantities upon temperature and pressure. At equilibrium the partial molar free energies of the unfrozen water and ice present are equal. At slightly different temperatures and pressures, T + dT and P + dP:

$$\bar{F}_u + d\bar{F}_u = \bar{F}_i + d\bar{F}_i \ . \tag{1}$$

It follows that

$$d\bar{F}_u = d\bar{F}_i .\qquad(2)$$

Taking the total derivative of Eq. (2) and identifying the partial derivatives as the partial molar volume $\bar{V}$ and the partial molar entropy $\bar{S}$, respectively, we obtain

$$d\bar{F}_u = \bar{V}_u dp - \bar{S}_u dT\qquad(3)$$

and

$$d\bar{F}_i = \bar{V}_i dp - \bar{S}_i dT.\qquad(4)$$

Equating and rearranging Eqs. (3) and (4) results in

$$\frac{dp}{dT} = \frac{(\bar{S}_i - \bar{S}_u)}{(\bar{V}_i - \bar{V}_u)}\qquad(5)$$

By definition

$$\frac{\Delta \bar{H}_f}{T} = (\bar{S}_i - \bar{S}_u).\qquad(6)$$

in which $\Delta \bar{H}_f$ is the difference in partial molar enthalpy between ice and water. Substitution of this identity in Eq. (5) yields

$$\frac{dp}{dT} = \frac{\Delta \bar{H}_f}{T \Delta \bar{V}} .\qquad(7)$$

This is a form of the Clausius-Clapeyron equation describing the phase relationships for ice and the unfrozen water interface in a frozen mixture. The relationships are shown in schematic form in Figure 1 for water and ice in a montmorillinite clay mixture. Equation 7 describes the lines a-a", b-b", etc. in Figure 1.

Two principle factors determine the position of lines b-b", c-c", and d-d", that depict the liquid-solid equilibrium for three water contents. First is the lowered freezing temperature, T, of the drier material (thinner interfacial water), or in other words, a greater freezing point depression. Secondly, there is a reduction in the latent heat of freezing, $\Delta \bar{H}_f$ at lower temperatures. Available estimates of $\Delta \bar{H}_f$ compared with measured freezing point depressions indicate that, of the two opposing effects, the decrease in $\Delta \bar{H}_f$ at lower temperatures is the stronger function and predominates. For this reason Figure 1 is drawn to show progressive decreases in slope of the solidus-liquidus line with decreasing water content.

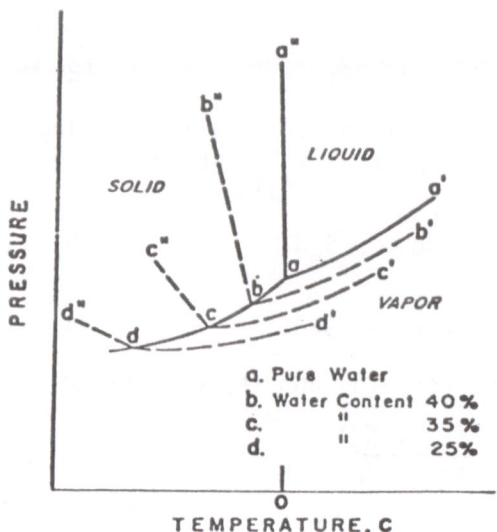

Figure 1.   Schematic phase diagram for water imbibed by
             montmorillonite showing the effect of clay
             surfaces on the vapor pressure, the freezing
             point, and the slope of the solid-liquid
             equilibria (not to scale).

The partial molar volume change associated with freezing the
interfacial water could be an important factor if, like T, it
can become very small. The specific volume of ice is about 10%
greater than that of water at 0°C, but the specific volume of the
unfrozen interfacial water next to montmorillonite surfaces is at
least 3% greater than that of pure bulk water so that $\Delta \bar{V}$ in this
case must be of the order of 7%. It appears that as the temperature is lowered, $\Delta \bar{V}$ for a given thickness of interfacial water
is further reduced. The rate of decrease and the final, limiting
value, however, have yet to be determined. It is believed,
however, that in comparison with the influence of $\Delta \bar{H}_f$ and T,
the effect is minor and that the slopes given in Figure 1 are
qualitatively correct.

The consequences of decreasing slopes of the solidus-
liquidus lines with decreasing water contents are several. The
most obvious is that ice pressed against montmorillonite surfaces
is more susceptible to pressure melting than normal, and that
the application of pressure at constant temperature leads to an
increase in the thickness of the unfrozen water interface. This
is important in the theory of ice segregation and frost heaving,
for at a given temperature below freezing, the build-up of
heaving pressures or the external application of pressure acts

to increase the thickness of the interfacial water above that which would otherwise prevail.

The movement of particles suspended in ice along thermal gradients has been demonstrated at temperatures near $0^{\circ}C$. The application of pressure is thus expected to enhance the rate of particle migration at a given temperature, and the application of sufficiently high pressure could induce particle migration at temperatures lower than ordinarily experienced.

In terrestrial environments subject to annual freezing and thawing, frost heaving results in the progressive elevation of the ground surface from 10 to 30 *cm* during winter. The process is terminated by spring thaw. If the length of the winter season were doubled the extent of frost heaving would be correspondingly increased. In regions of the earth where permafrost occurs, annual freeze thaw is confined to the upper few meters of the ground and frequently freezing occurs simultaneously from the top downward and the bottom upward. The resulting confinement of unfrozen soil in between results in considerable deformation and churning of the ground. Even when the soil is solidly frozen, redistribution of water and the growth of ice lenses occurs. Water is continually transported through the unfrozen liquid interfaces toward regions of lower and lower temperatures. Particle rearrangement and the continual deformation of the frozen ground result. The variety of geomorphic land forms characteristic of permafrost and frozen ground are formed by an interplay of processes involving the transport and redistribution of material in the unfrozen interfacial water and the processes associated with thermal expansion and contraction of the frozen earth and ground water. Similar phenomena are to be expected on Mars and the other planetary bodies where ice is found intermixed with silicate minerals. Unfamiliar land forms and surface features are expected on these planetary bodies where the extremes of temperature and pressure greatly exceed thosecommonly occuring on the earth.

REFERENCES

Anderson, D. M. Water in the Martian Regolith, Comparative Planetology, 219-224, Academic Press, Inc., 1978.

Anderson, D. M. The Role of Interfacial Water and Water in Thin Films in the Origin of Life. Proceedings NASA Conference on Life in the Universe, Ames Res. Center, Moffett Field, CA., June 1979. (in press)

Anderson, D. M. Tice, A. R., Low Temperature Phase Changes in Montmorillonite and Nontronite at High Water Contents and High Salt Contents. Cold Regions Science and Technology, 3: 139-144, 1980.

Tice, A.R., Anderson, D.M., Sterrett, K.F. Unfrozen Water
     Contents of Submarine Permafrost Determined by Nuclear
     Magnetic Resonance. Proceedings, The 2nd International
     Symposium on Ground Freezing, Norwegian Institute of
     Technology, U. of Trondheim, Trondheim, Norway, June 24-26,
     1980.

LAVA FLOWS ON ETNA, A MORPHOMETRIC STUDY

Rosaly Lopes and John E. Guest

University of London Observatory,
Mill Hill Park, London NW7 2QS

Abstract: Photogeological interpretations of volcanic products
on the terrestrial planets demand an understanding of the controls
of such factors as effusion rate, duration of eruption, viscosity
and yield strength of lava and topography on the shape of the
final flow field.

Using new data on historical flows from Mount Etna, all of
which are of similar composition, we examined the way in which
flow length and width are controlled by effusion rate, volume
and duration of eruption, and topography.

1. INTRODUCTION

Volcanic processes have played a major role in the development
of the surfaces of the Earth, Moon and the terrestrial planets.
On Earth, there is a diversity of volcanic landforms, each being
explicable in terms of a specific style or combinations of styles
of volcanic activity. Although most types of terrestrial eruption
have been described, it is only in recent years that physical
models of the different processes have been developed and many
more quantitative data on eruptive activity are required.

Volcanism has not been observed in action on the Moon and
other terrestrial planets, and apart from returned rock samples
from the Moon, we must interpret the style of eruption and
composition of materials erupted from the resultant landforms.
A valid approach is to try to quantify the morphology of landforms
on Earth and establish their relations to known parameters of
style of eruption and composition of materials. These studies

441

A. Coradini and M. Fulchignoni (eds.), The Comparative Study of the Planets, 441–458.
Copyright © 1982 by D. Reidel Publishing Company.

can then be used to interpret landforms on other planets with
the proviso that different environmental conditions are taken into
account.

The gross morphology of a volcano is controlled by a number
of inter-related factors such as ratio of lava to pyroclastic
materials, the lengths and thicknesses of lavas, distribution of
vents, and caldera formation. Each of these is controlled by
such factors as lava composition, gas content, lava rheology,
effusion rates, tectonism, volcano plumbing and rate of magma
output. It is important to determine the role of each of these
factors and their relative importance to the resultant landform.

As part of an endeavour to understand this problem, we
have studied the morphologies of historical lava flows of Mount
Etna. All these historical flows are similar in composition and
yet a variety of flow morphologies have been produced, from long
and thin flows to short and thick flows, showing a range of
widths. The majority of flows are aa but a few major flows have
pahoehoe surfaces. Because the composition of all lavas is
similar, we are able to eliminate this characteristic allowing
other specific factors to be examined in more detail. Good
records of eruptions are available (i),(ii),(iii) and flows were
mapped and dated where possible in the recently produced
Geological Map of Etna (iv), from which we obtained a significant
amount of the data used here. Data were obtained also from (v),
(vi), (vii), (viii), (ix), (x) and some thicknesses of flows were
estimated in the field.

Records of volcanic eruptions on Etna show that they have
have been of two types during historic times: persistent activity
at the summit and periodic flank eruptions (xi). We have studied
mostly the lavas produced by flank eruptions, since the small
summit flows tend to go unrecorded.

The morphology of individual lava flow fields is controlled
by a number of inter-related factors including:

- Viscosity and yield strength
- Effusion rate
- Underlying slope and topography
- Volume
- Duration of eruption.

We have examined, where possible, the relations between these
factors and the flow dimensions (length, width, thickness) with
the aim of determining whether there is a dominant factor
controlling any of the flow's dimensions, and how the interplay
of several factors controls flow development.

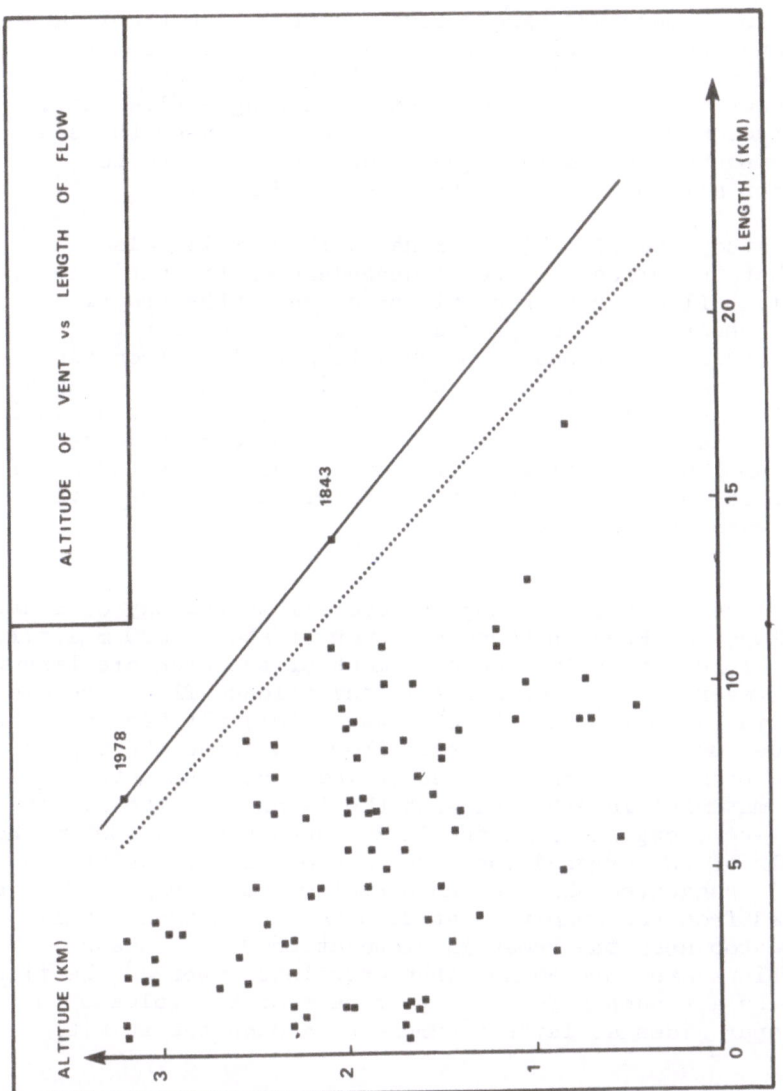

Figure 1: Altitude of start of flow plotted against flow's length, for 86 Etnean flows. The dots represent Walker's (xiii) maximum lengths with altitude of vent. The line represents our limit using new data.

2. LENGTH OF FLOWS

2.1. Effusion rate and altitude

Walker (xii) using data from Etnean flows argued that effusion
rate is the most important single factor controlling the length
of a flow, and that viscosity has at most an indirect control.
Hence the higher the effusion rate, the longer the resulting flow
with a tendency towards the development of a single flow unit.
With low effusion rates, lava tends to accumulate near the source
forming a compound flow field. These results appear to be
consistent for all the volcanoes examined by Walker.

From a study specifically of Etna, Walker (xiii) also
concluded that the effusion rate is dependent on the altitude of
vent possibly reflecting hydrostatic pressure of the erupting
magma. This observation may provide an explanation for the
tendency of Etnean eruptions to produce longer flows from the
lower altitude vents. Following Walker (xiii) we used new data
to examine the maximum length of flow with altitude. As shown in
Figure 1 these data confirm Walker's conclusion and indicate the
normal maximum length of flow erupted at different altitudes. It
should be noted, however, that not all eruptions produce lava
flows that reach this maximum.

The relation between maximum length of flow and altitude may
be of importance in understanding the overall morphology of Etna.
The lower slopes of Etna up to an altitude of about 1800 m a.s.l.
have slopes of less than 10° and over most of the area are less
than 5°. However, above 1800 m a.s.l. the volcano flanks become
steeper, having slopes of about 20° between this altitude and the
summit at just over 3,300 m. Steeper slopes towards the top of a
volcano like Etna are often ascribed to increasing amounts of
pyroclastic material interbedded with the lavas near the summit
crater. However, exposures in the lava sequence above 1800 m show
that the bulk of the exposed cone consists of lavas. Guest and
Murray (xvi) suggested that as the normal maximum length of lava
flows erupted from the summit is about 6 km, the longest summit
flows would stop near the break in slope at 1800 m. Because
summit eruptions are more common than eruptions lower on the flanks
there would be a tendency for the upper part of the volcano to
develop steeper sides as lavas accumulated around the summit
region.

One implication of the observation that length of flow is
altitude dependent is that the effusion rate is also altitude
dependent. Unfortunately, there are few data on the actual rate
at which lava was erupted and to examine this we must consider
an average value for effusion rate obtained by dividing the flow's
volume by the duration of the eruption. This introduces several

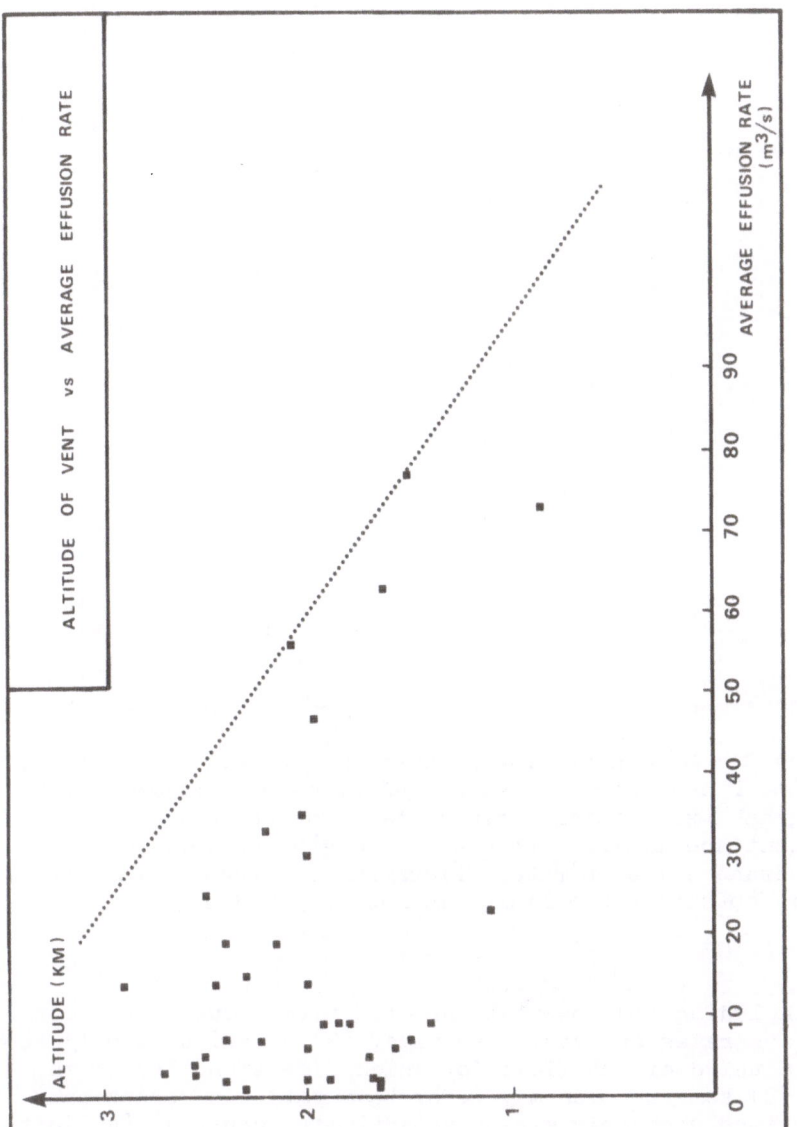

Figure 2: Altitude of vent plotted against average effusion rate for 37 Etnean flows, showing that flows erupted at lower altitudes tend to have higher effusion rates. The dotted line represents the highest average effusion rate for a given altitude based on the available data.

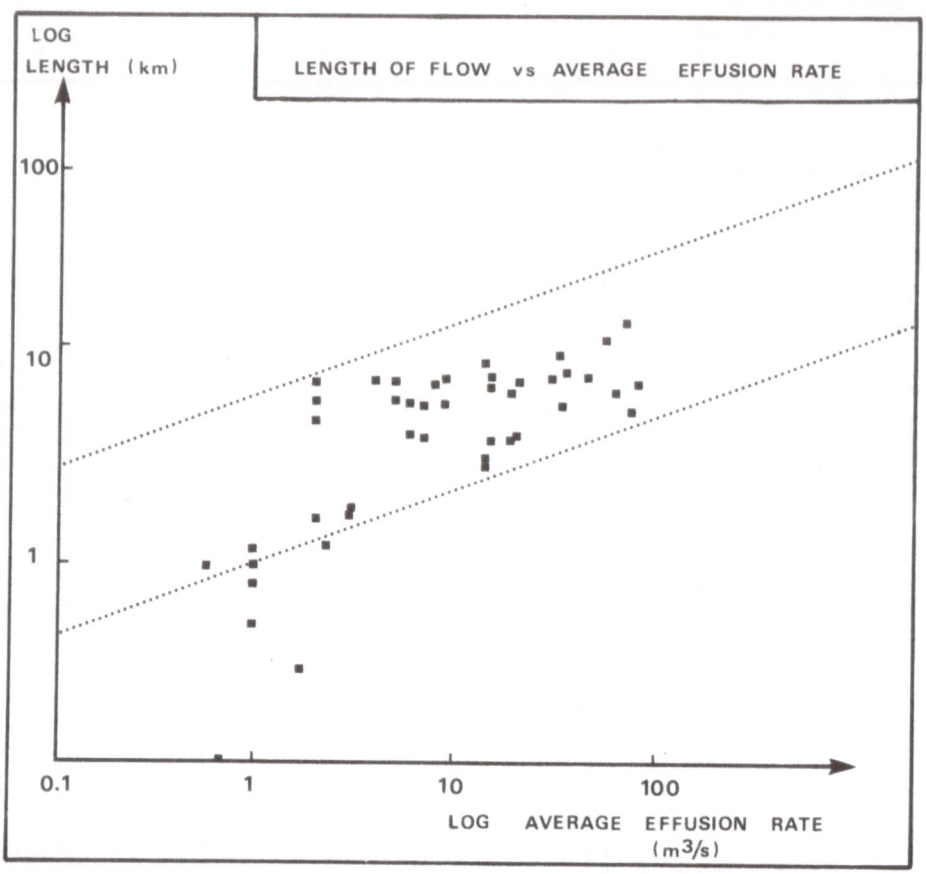

Figure 3: Length of flow plotted against average effusion rate for 44 Etnean flows. The dotted lines were reproduced from Walker's graph (xii) which included data for several volcanoes, and represent the limits within which length can vary given a certain average effusion rate. Flows shorter than about 1 km tend to lie outside these limits as noted by Wadge (xiv).

problems including (1) observations of active flows have shown that effusion rates can vary with time, (2) duration is only known for about a third of the flows for which dimensions have been measured, (3) some records may refer to duration of eruption including times when only explosive activity occurred, (4) initial phases of eruption are likely to have gone undectected and (5) average thicknesses of flows used in estimating volumes cannot be determined accurately and an error of up to 50% may be expected in some cases.

Using the available data we can show that there is no linear correlation between altitude and average effusion rate but that there is a maximum effusion rate at any particular altitude and that the rate is higher with decreasing altitude(Fig 2).

The implications of the data presented in Figures 1 and 2 depend on the conclusion of Walker (xii) that there is a relation between maximum length of flow and effusion rate. Using more recent data, we have again investigated this relation (Fig.3). We conclude that there is a correlation between flow length and average effusion rate, although there is more scatter than was apparent in Walker's plot (xiii) which was based on only eleven historic flank flows. However, nearly all the Etnean flows lie between the limits given by incorporating Walker's data from several other volcanoes (xii). There are some anomalous Etnean flows that lie outside the limits determined by Walker. These are all flows of less than about 1 km in length with relatively low effusion rates at about $1 m^3 s^{-1}$. This anomaly was previously noted by Wadge (iv) who concluded that the relation between maximum length and effusion rate was only valid for flows longer than 1 km. He also concluded that the results obtained from Etna could not necessarily be extrapolated to tube-fed pahoehoe flow fields of the type observed in Hawaii where much greater lengths can be achieved by comparison with Etnean aa flows of equivalent effusion rate.

2.2. Volume

From a study of historic Hawaiian flows, Malin (xv) found that there was a better correlation between length and volume of flow than between length and effusion rate. He concluded that factors such as cross-sectional area, effusion rate and volume, all play important roles in governing the emplacement of Hawaiian flows and that no single factor appears important. However Malin pointed out that most of his results were based on tube-fed pahoehoe flows including those erupted with low effusion rates, and thus a correlation with effusion rate might not be expected making a relation with volume more apparent.

For comparison with Malin's results we plot the lengths of Etnean flows against their volumes (Fig.4); although a correlation is apparent it is not as strong as that determined for Hawaii. Several Etnean flows lie noticeably outside the main trend, including that of 1763 which is an unusually thick flow by Etnean standards (average thickness 50 m) and is relatively short for its total volume.

The small volume flows below about $1 \times 10^6 m^3$ show a wide

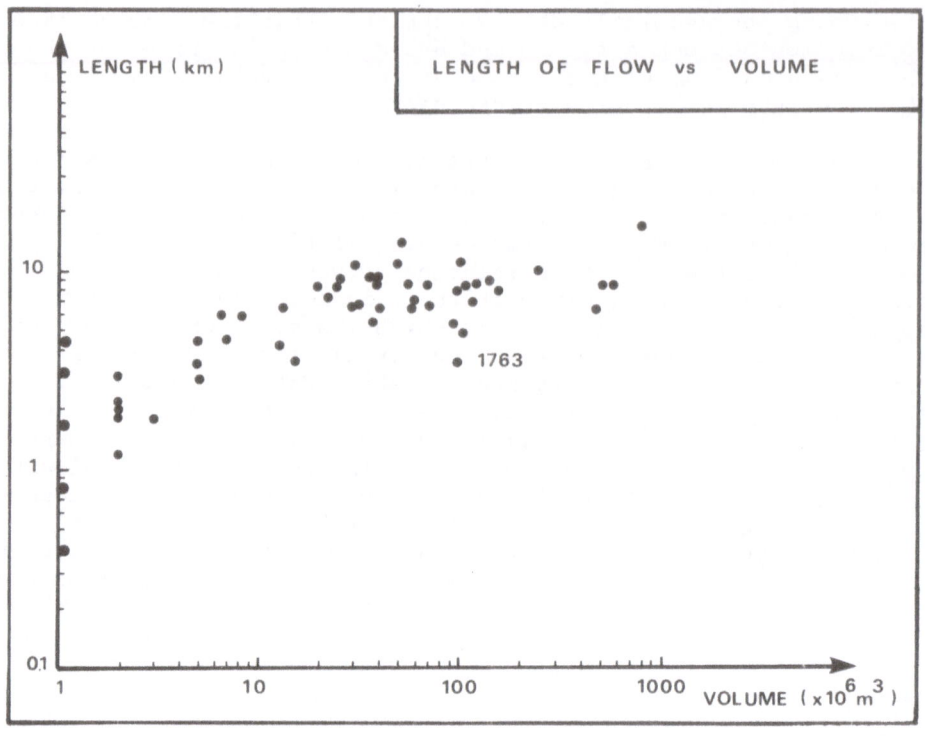

Figure 4: Length of flow plotted against volume of flow for 54 Etnean flows.

scatter with no clear relation between volume and length. Because there is clearly some relation between volume and length — if only because a flow cannot increase its length without increasing its volume — it is necessary to eliminate the volumetric consideration. To do this, we have plotted lengths of flows against their effusion rates, but have grouped the flows according to their volume. In Figure 5, which only shows flows of volume less than $5 \times 10^6 \text{m}^3$, there is good correlation and for flows longer than 1 km, length increases with effusion rate. Comparison with Figure 4 indicates that for most of these flows there was no volumetric correlation with length. In Figure 6 we show the same relations for groups of lavas of greater volume. Only one group (between 5 and $50 \times 10^6 \text{m}^3$) has enough flows to show a correlation between length and effusion rate, but only for flows less than about 11 km in length. Above this, there appears to be no increase in length with increasing effusion rate. Thus for a given effusion rate the maximum length can only be achieved if

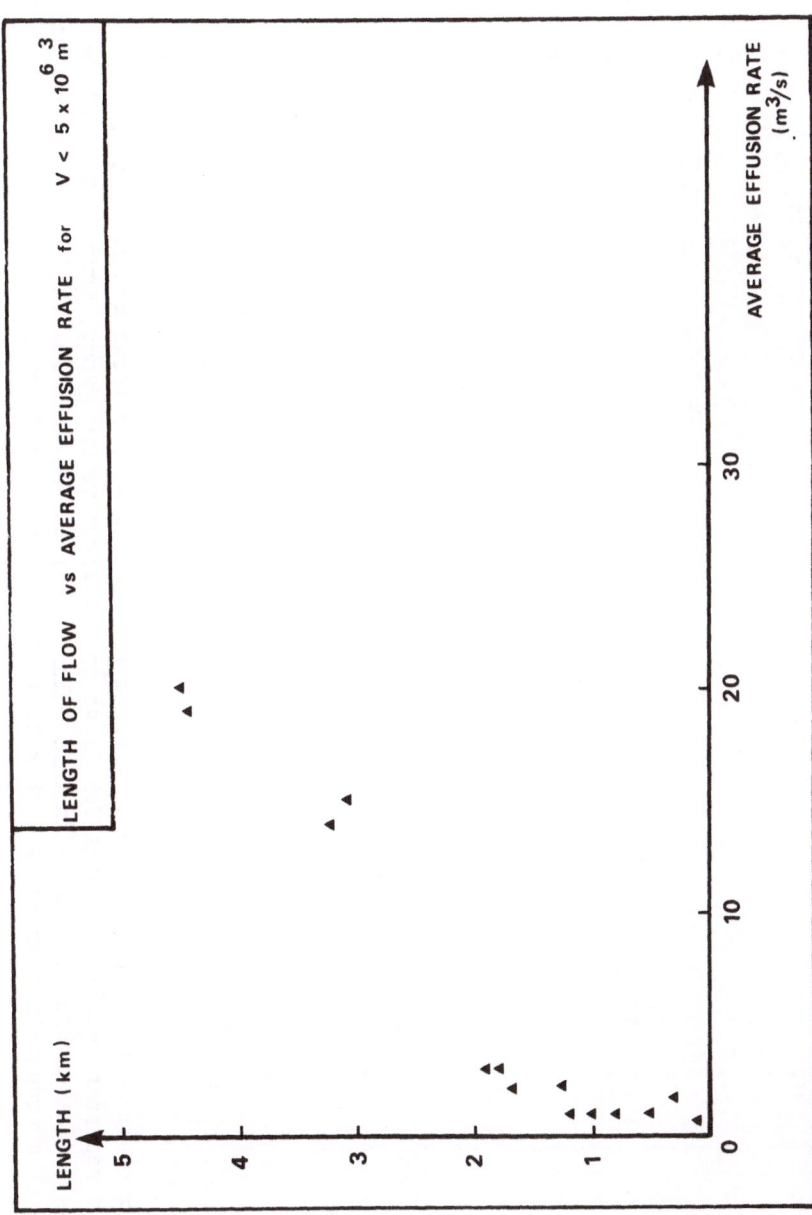

Figure 5: Length of flow versus average effusion rates for 14 Etnean flows of volumes less than $5 \times 10^6 m^3$.

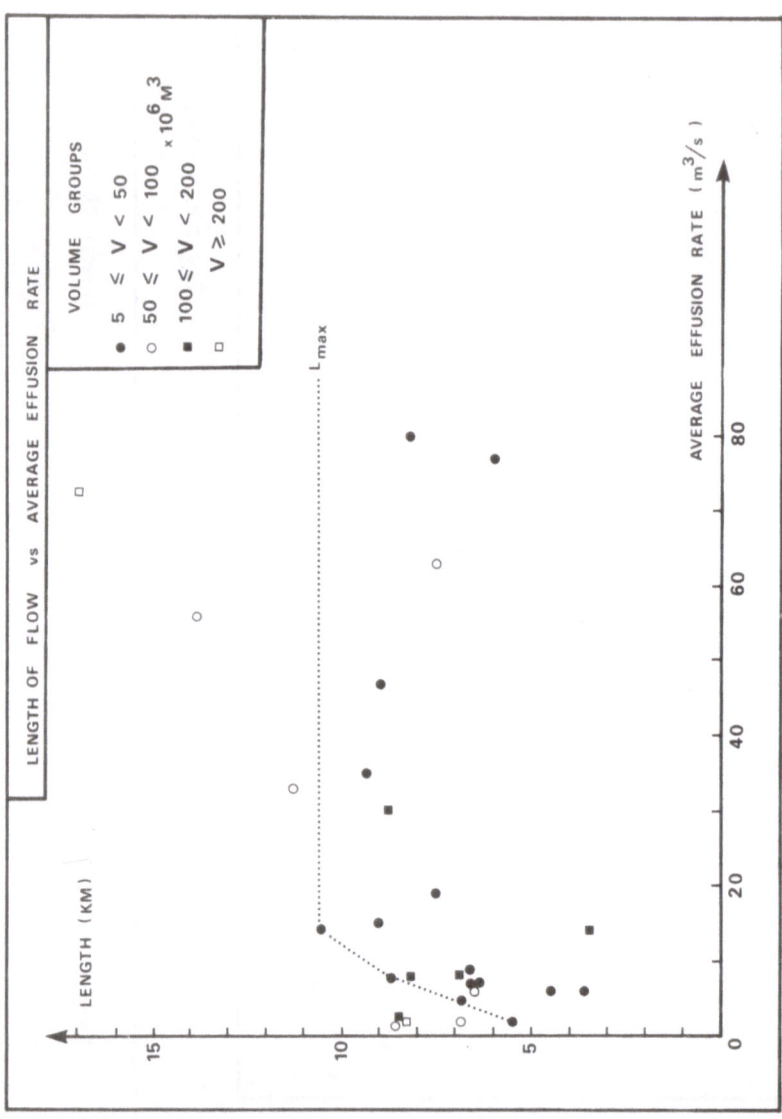

Figure 6: Length of flow versus average effusion rate with flows grouped according to volume. The dots represent maximum lengths for flows in the lower volume group, erupted at different rates. Our data indicates that for flows in this volume group the maximum length is about 11 km, and flows erupted at average effusion rates higher than about 20 m^3s^{-1} will not exceed this length.

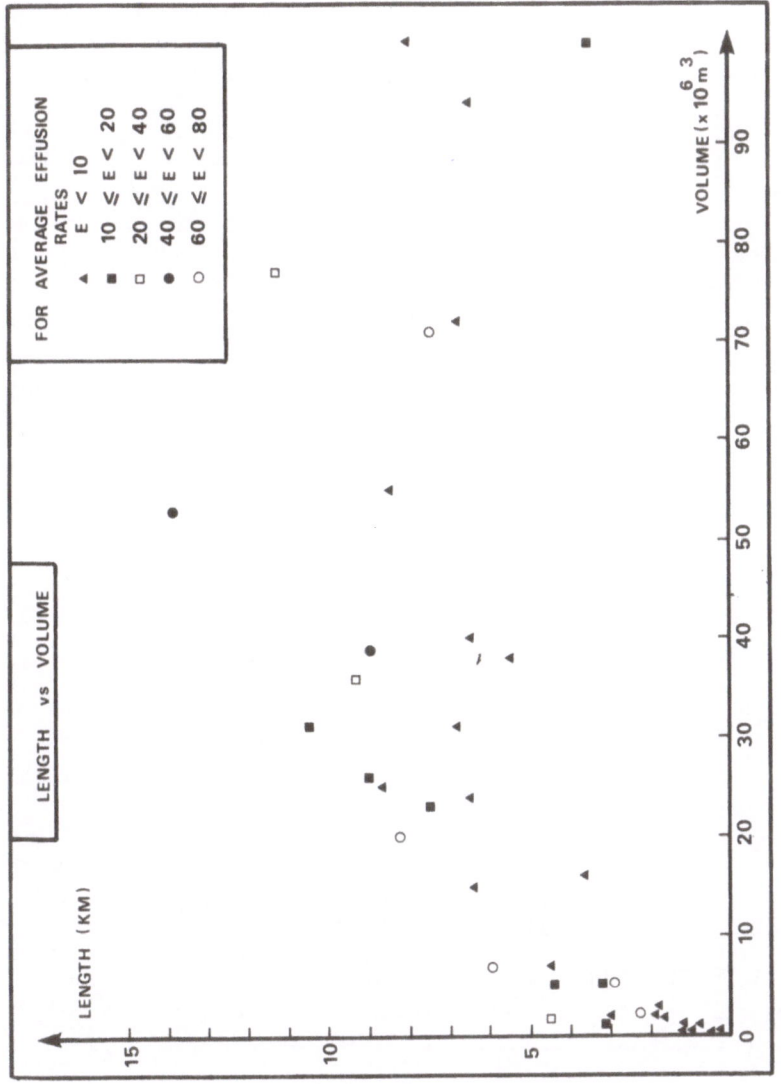

Figure 7: Length of flow plotted against volume with flows grouped by their average effusion rate. Flows having volumes greater than $100 \times 10^6 m^3$ are not included. Effusion rates are in $m^3 s^{-1}$.

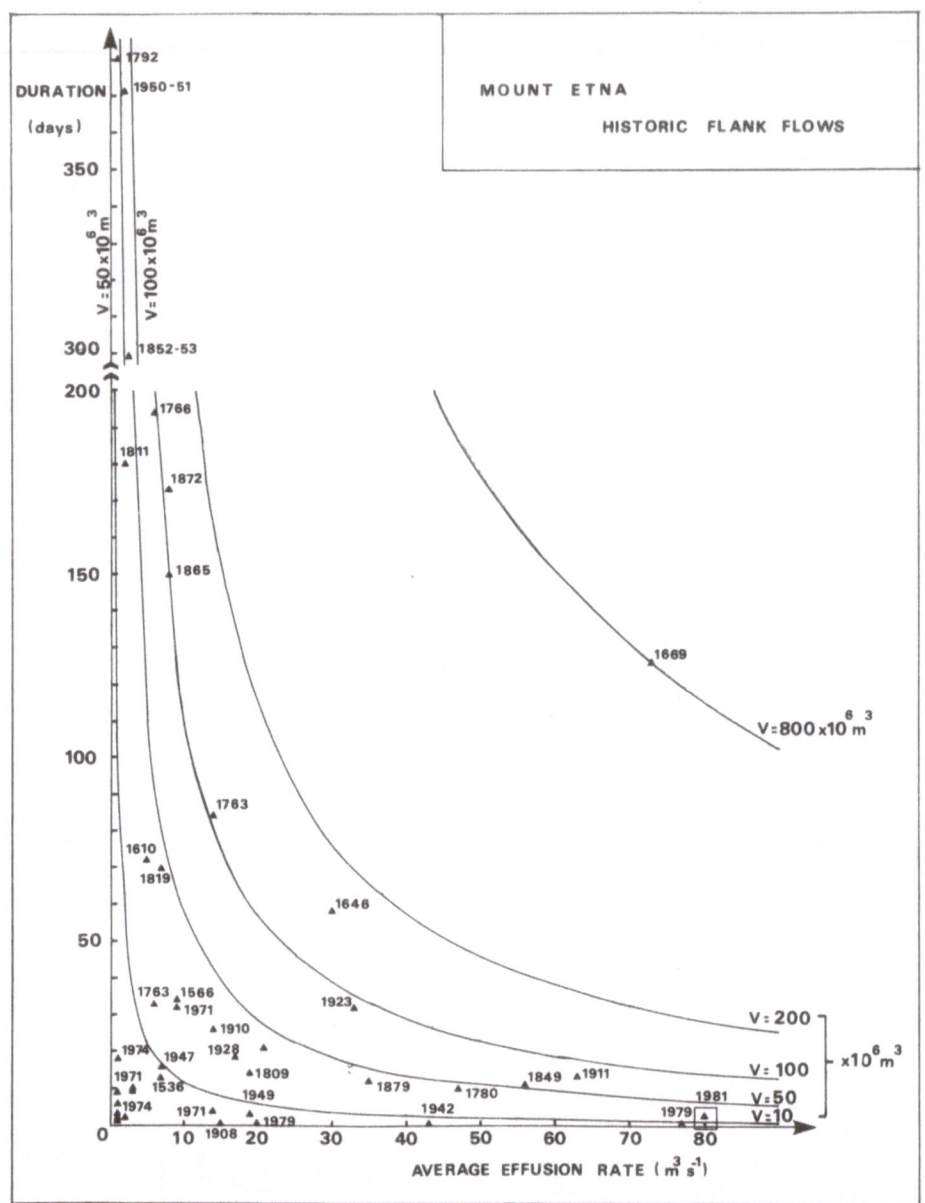

Figure 8:   Duration of eruption versus average effusion rate
for historical flank eruptions, lines represent constant volumes.
Etnean flows do not usually have volumes greater than 200 x 10^6m^3,
exceptions being the 1669 flow and the 1614-24 flow (not shown)
which has a volume of about 1500 x 10^6m^3 and lasted about 10 years.

there is sufficient volume (i.e. if the eruption lasts long
enough). Although the number of data points for flows in the
higher volume groups are limited, we can quantify to some extent
the limits on the maximum length of flows imposed by volume for a
given effusion rate. Therefore, for flows of volume up to
$5 \times 10^6 m^3$ the maximum length appears to be about 5 km (Fig.5); flows
of volume between 5 and $50 \times 10^6 m^3$ have a maximum length of about
11 km; and flows of volume between 50 and $100 \times 10^6 m^3$ have a
maximum length of at least 14 km (Fig.6). There is not enough
data for higher volume flow groups to indicate a maximum length.

A similar relation can be observed by considering volume and
length for flows erupted at similar effusion rates grouped
together (Fig.7). For average effusion rate below the median
value ($8 m^3 s^{-1}$), flows do not advance further than about 8 or 9 km
even if their volume is high. Thus once a flow has reached its
maximum length, continued eruption increasing the total volume of
the flow field and must either thicken or widen the flow. We do
not have the data to determine whether for high effusion rates
(above $20 m^3 s^{-1}$) there is also "excess" volume.·

3.3. Duration of Eruption

Effusion rate, volume and duration are clearly related to
one another. Figure 8 indicates that the majority of eruptions
on Etna do not involve more than a maximum of $200 \times 10^6 m^3$ of
lava; exceptional eruptions include those of 1614-24 (not shown)
and 1669. Those eruptions with high average effusion rates
(above $20 m^3 s^{-1}$) tend to be of short duration, but those of long
duration have low effusion rates (below $10 m^3 s^{-1}$).

Whether a flow is erupted at high rate over a short
period of time or a low rate over a long period appears to have
some control over the planimetric form of the final flow field
and eruptions of longer duration tend to produce flows that are
wider with respect to length than those of shorter duration
(Fig.9). However, this does not seem to be the case for small
volume flows (less than $3 \times 10^6 m^3$) which have both duration and
average effusion rates below the median values (i.e. 22 days and
$8 m^3 s^{-1}$).

3. WIDTH OF FLOWS

The factors that control the spreading of a flow have not
previously been considered quantitatively for Etna. As with lava
length, it is difficult to disentangle the various factors that
control the way the flow will spread. We have shown once a flow
has achieved its maximum length there may be excess volume which
either causes the flow to thicken or spread. We have also shown

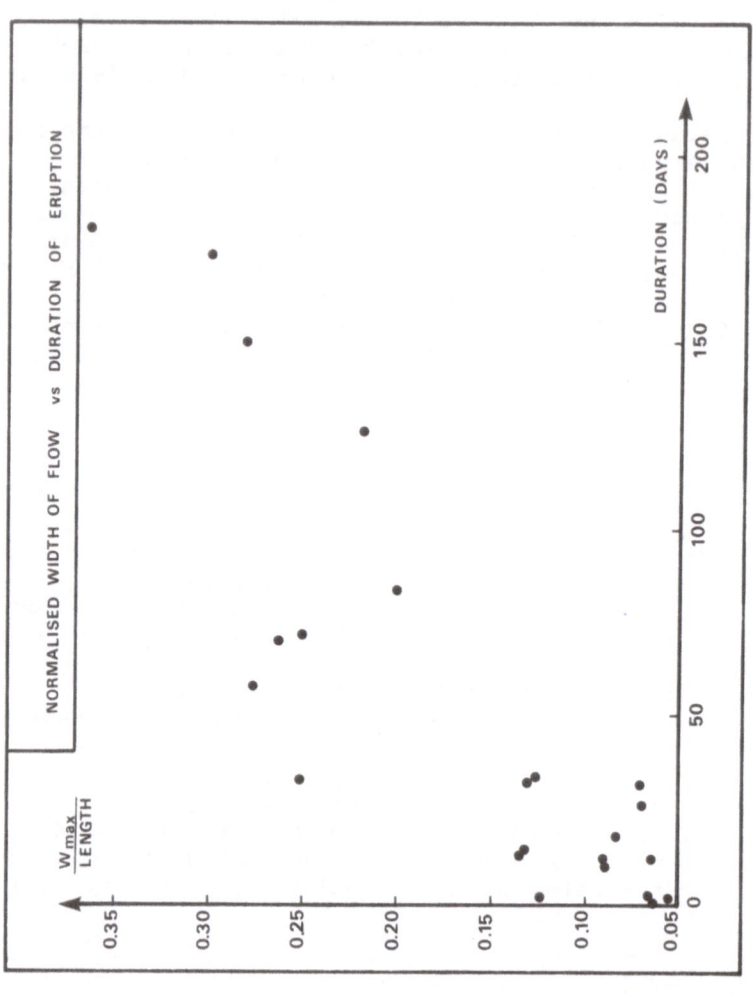

Figure 9: Maximum width/length plotted against duration of eruption for flows of volume greater than $3 \times 10^6 m^3$, showing the effect of duration on the final planimetric form of the flow field.

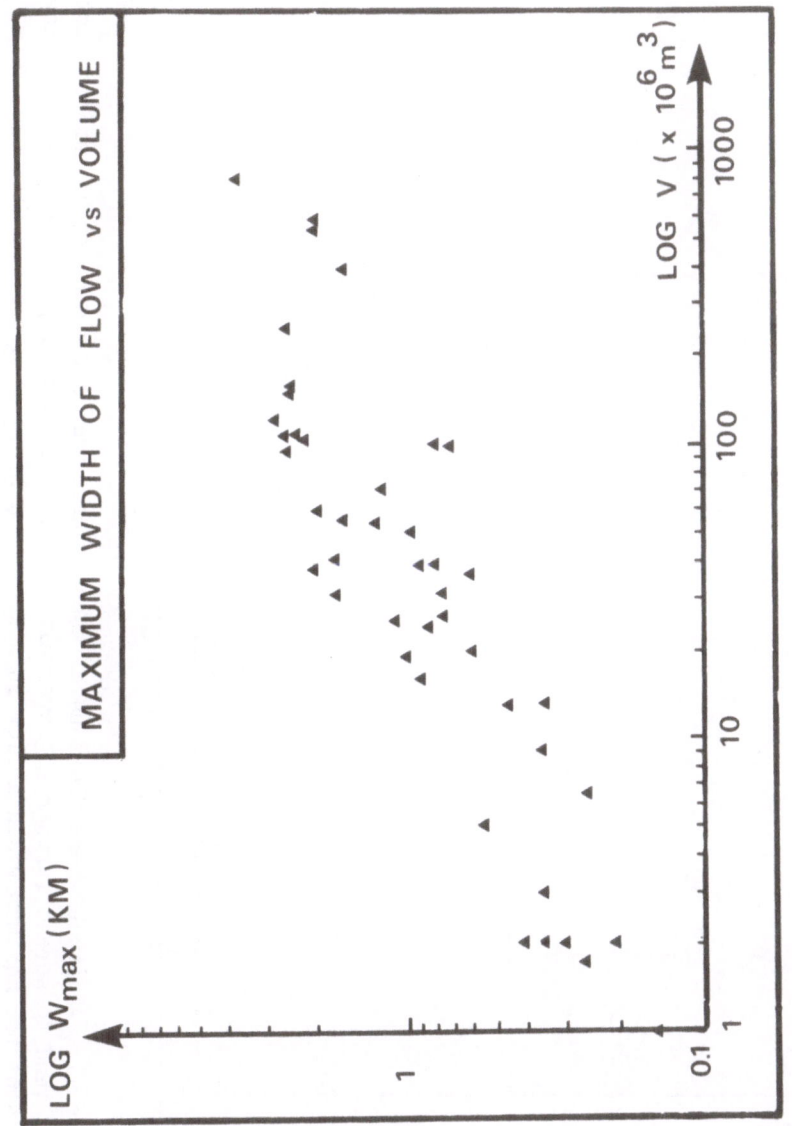

Figure 10: Maximum width of flow versus volume for 44 Etnean flows.

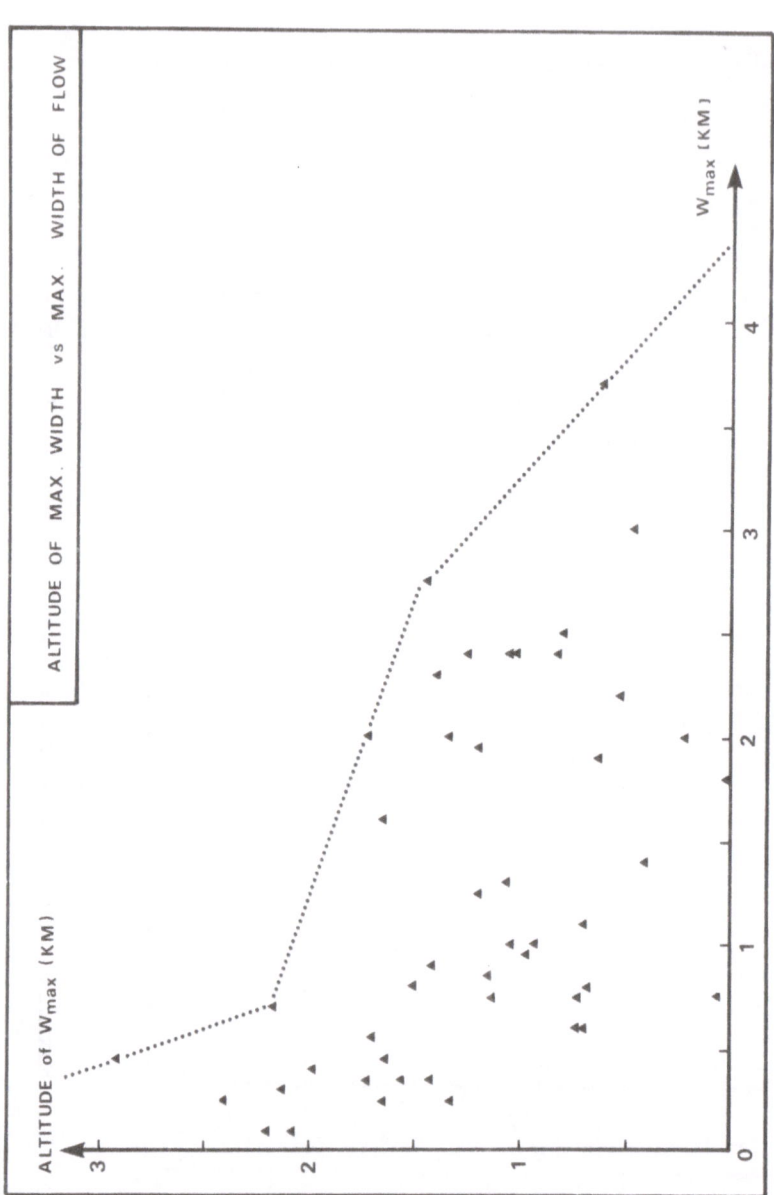

Figure 11: Altitude where flows have reached their maximum width versus maximum width, showing that flows tend to spread further at lower altitudes, where the slopes of Etna are shallower. The dots represent the maximum width a flow is likely to achieve at a certain altitude, based on data for 47 historical flows.

that if we normalise the length, the flows become wider with
increasing duration of eruption. The same relation may be seen
in Figure 10 which shows the maximum width versus volume and
again indicates that these two factors are related. The problem
of thickening of the flow with excess volume is harder to examine
on Etna as our knowledge of lava thickenesses is limited. However,
excessively thick flows are rare on Etna and tend to be associated
with unusual eruption conditions, such as spasmodic effusions
(e.g. 1763 eruption) or the controlling influence of lava tubes
and lava ponding (e.g. 1614-24 eruption). The evidence in
Figures 9 and 10 would tend to suggest that spreading occurs with
increasing duration of eruption assuming that there is sufficent
volume to accomplish this.

The influence of topography on maximum width of a flow is
clearly not negligible as shown in Figure 11 where the altitude
at which the flow reaches its maximum width is plotted against
this maximum width. Flows tend to be wider at lower altitudes
where slopes are less than 10°. There is a noticeable reduction
in the maximum width of flows above about 2,000 m where steeper
slopes occur. The maximum width for a given altitude indicated
on Figure 11 can only be considered as a generalisation as it
disregards special circumstances and it does not take into account
local variations in topography, such as valleys, or depressions
where ponding can occur.

4. CONCLUSIONS

The maximum length of a flow is determined by an interplay
of factors which is probably dominated by the effusion rate
assuming sufficient volume of lava is erupted. We are able to
determine to some extent the minimum volumes required for a
given effusion rate for the maximum flow length to be achieved.
For a given effusion rate, once the lava has reached its maximum
length, excess volume tends to contribute to flow widening rather
than thickening and greater widths of flow field are more likely
with increasing duration of eruption. Volume of flow exerts
some control on the maximum width reached, and the position on
the flow where maximum widths occur tend to be controlled by
topography.

The above conclusions are all based on examination of
historical flows using available documentary evidence for eruption
durations. To test the conclusions here, it is necessary to
compare flow length, width and thickness with actual effusion
rates during eruption and as the frequency of flank eruptions on
Etna is fairly high (about one every ten years) this volcano is
an ideal one for continued study of the controlling influences
on lava flow morphology.

REFERENCES

   (i) Von Walterhausen, W., Sartorius: 1880, 'Der Ätna',
vols.I and II. Leipzig: Engelman.

   (ii) Imbo, G.: 1965, 'Catalogue of the active volcanoes of
the world including solfatara fields', Rome: International
Association of Volcanology, Vol.18.

   (iii) Wadge, G.: 1977, J.Volcanol.Geotherm. Res.2, pp.361-84.

   (iv) Romano, R., et al: 1979, Carta Geologica del Etna,
Consiglio Nazionale delli Richerche (Italy).

   (v) Romano, R., and Guest, J.E.,: 1979, Boll.Soc.Geol. Italy,
98, pp.189-215.

   (vi) Romano, R., and Sturiale, C.: 1975, Boll.Soc.Geol.Italy,
94, pp.1121-1133.

   (vii) Guest, J.E., Huntingdon, A.T., Wadge, G., Brander, J.L.,
Booth, B., Carter, S., Duncan, A.: 1974, Nature, vol.250, pp.
385-387.

   (viii) Rittmann, R., Romano, R., Sturiale, C.: 1973, Atti
della Acc. Gioenia Sc. Nat. Ser.III, vol.3, pp.1-29, Catania.

   (ix) Bottari, A., Lo Giudice, E., Patani, G., Romano, R.,
and Sturiale, C.: 1975, Revista Mineraria Siciliana n.154-156,
pp.175-178.

   (x) Kilburn, C.: pers. comm.

   (xi) Duncan, A.M., Chester, D.K., and Guest, J.E., 1981,
The Geog. Journ. Vol.147, n.2, pp.164-78.

   (xii) Walker, G.P.L. : 1973, Phil.Trans.R. Soc.Lond. A.274.

   (xiii) Walker, G.P.L. : 1974, Geol.Soc.Lond.Misc. Paper 3,
pp.23-41.

   (xiv) Wadge, G.: 1978, Geology 6, pp.503-506.

   (xv) Malin, M.C.: 1980, Geology 8, pp.306-308.

   (xvi) Guest, J.E., and Murray, J.B. : 1979, J.Geol.Soc.Lond.
136, pp.347-354.

A PETROLOGICAL MODEL ON MAGMA EVOLUTION OF VULCANO ERUPTIVE
COMPLEX (AEOLIAN ISLANDS - ITALY)

G. Castellet y Ballarà, R. Crescenzi, A. Pompili and
R. Trigila

Istituto di Mineralogia e Petrografia, Università di Roma

ABSTRACT. A scheme of magmatic evolution, aimed at explaining lava
types occurring at Vulcano eruptive complex is presented. For this
purpose, data on chemistry, petrography and mineralogy of the lavas
have been treated on a quantitative basis by means of crystal-liquid
fractionation and mass-balance computer programs. $P-T-XH_2O$ of the
melts in equilibrium with the respective phenocrysts assemblages
have been estimated too. The results put in evidence the generation
of lc-tephritic, trachyandesitic, trachybasaltic melts from a single
magma of basaltic affinity. The problem on the origin of latitic
magmas and of the extreme differentiated ones, keeps ambiguous solutions still, possibly due to water and alkalies pre-eruptive loss.

1. INTRODUCTION

Vulcano is the southernmost of the Aeolian Islands. Together with
Stromboli (at the extreme north of the islands arch), it represents
an active eruption site. Both can be distinguished among the other
volcanic islands because of the marked alkaline character of their
recent products. This peculiarity has been explained as the senile
stage of an island arch volcanism (1). The petrological model herein
developed aims at the evaluation of the parameters which control
the magmatic evolution, also considering the recent results on Pb
and Sr isotopes which seem to leave out the hitherto accepted hypothesis on the origin of the different rock-types from a single
parental magma (2),(3). Details on geology, volcanology, mineralogy,
petrology and geochemistry of Vulcano island have been published
by Keller (4); we remind his paper for a complete list of references
too. Therefore only a brief account of the volcanic history of the
island is given below.

A. Coradini and M. Fulchignoni (eds.), The Comparative Study of the Planets, 459–476.

2. VOLCANIC HISTORY

The volcanic history of the island is shown by the geological setting of four main structural units: i) South Volcano (stratovolca-
no) and Piano Caldera; ii) Lentia Group; iii) Fossa Caldera and
Volcano; iiii) Vulcanello; (fig.1). South Volcano had origin during upper Pleistocene; it is composed by a cone (not quite dissimilar to the recent Stromboli) with a rather steep slopes built
up by alternating lava flows, pyroclasts beds and various types of
dikes. The collapse of the upper part of the stratocone (partially due to the magma increasing explosivity with time) produced
Piano Caldera. The following activity was concentrated around the
caldera rim with lava emissions from a number of secondary centers
(Gelso-Petrulla, Timpone del Corvo, M.Rosso, M.Molineddo, etc.)
but also of explosive type inside the caldera floor which was by
sectors subsiding gradually. Later on (upper Wurm) the lava domes
and viscous flows of Lentia Hills Group (probably representing the
relics of a former volcano collapsed to the north of Piano Caldera) were emplaced. An actual subsidence area which began to form
at the end of last glaciation (4) is represented by Fossa Caldera.
The respective products are mainly pyroclasts but also lava flows
around the southern border (M.Saraceno, P.Luccia) are present.
Fossa Volcano is the outstanding edifice occupating a large area
of the caldera floor. It is characterized by a mixed volcanic activity lasting to historical times. Pyroclasts are dominant and a-
mong them phreato-magmatic products have been recently recognized
(5). Finally Vulcanello grew up during 183 b.C. eruption as a separate island in the sea channel separating Vulcano and Lipari and
joined to the former during 16th century. It is composed by a large platform made up of a certain number of a very fluid lava flows,
dominated by a composite pyroclastic cone 100 mt. high from which
in a later stage a viscous flow of differentiated magma (P.del Ro-
veto) emerged.

3. CHEMISTRY, PETROGRAPHY AND MICROPROBE MINERALOGY OF LAVA TYPES

Forty chemical analyses (selected out of more than 100) of the different rock-types which constitute the above specified structural
units are presented in Table 1. A general inspection of the analytical data justifies the rock-types attribution to the shoshonite
association (6). A closer approach shows the presence of two "end
member" series (at different K_2O content)(1): the shoshonitic and
the lc-tephritic one. They seem to have origin from a common parental composition, separate in the range 51-55% SiO_2 and then
merge again towards the most acidic rock-types (fig. 2).

A comprehensive picture of the compositional ranges is shown
in the classificative Streckeisen's double triangle (fig.3) where
the results of 102 chemical analyses (as C.I.P.W. Norms modified

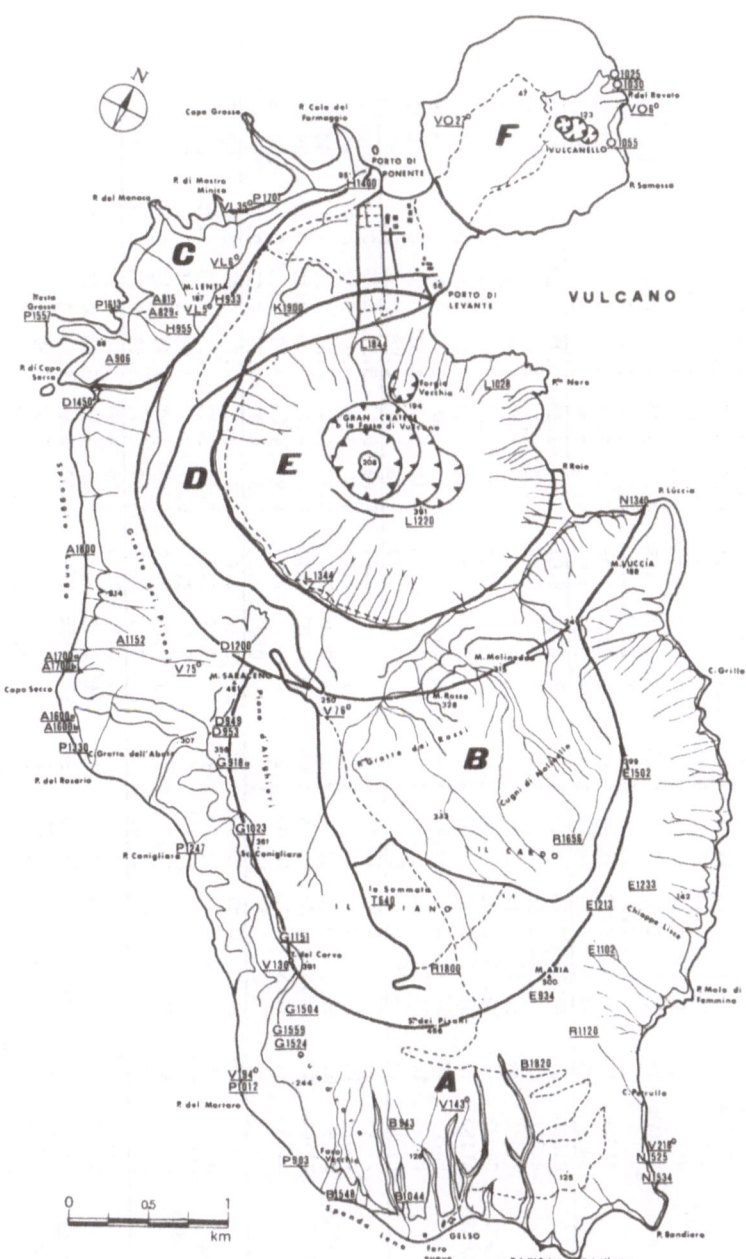

Fig. 1. Topographic domains of volcanic units of Vulcano eruptive
complex. A: South Volcano; B: Piano Caldera; C: Lentia
Group; D: Fossa Caldera; E: Fossa Volcano; F: Vulcanello.
Samples locations are also indicated.

TABLE 1 – VULCANO ERUPTIVE COMPLEX: SELECTED CHEMICAL ANALYSES

VOLCANIC UNIT	SAMPLE	TYPE	SiO_2	TiO_2	Al_2O_3	Fe_2O_3	FeO	MnO	MgO	CaO	Na_2O	K_2O	P_2O_5	H_2O+	TOT.
SOUTH VOLCANO	E1213	TB	50.43	0.64	15.78	3.90	4.59	0.16	7.69	9.64	2.86	2.38	0.24	2.06	100.37
	E1502	TB	51.82	0.73	15.54	5.53	3.78	0.21	6.06	10.09	2.50	2.71	0.25	0.81	100.03
	E1102	TB	51.87	0.72	16.08	4.57	4.39	0.16	5.42	9.32	3.34	2.82	0.25	1.00	99.94
	A1700a	TB	52.15	0.64	17.79	4.82	3.00	0.14	3.84	8.92	3.72	2.98	0.38	1.72	100.10
	A1600a	TB	52.15	0.64	17.49	4.92	3.25	0.13	4.21	8.39	3.66	3.62	0.40	1.10	99.96
	•V218	TA	52.3	0.75	15.5	5.1	3.75	0.12	6.4	9.7	2.8	2.6	0.35	1.2	100.57
	N1534	TA	53.19	0.67	15.00	5.51	3.27	0.17	6.50	9.63	2.62	2.78	0.30	0.38	100.02
	A1800a	TA	53.23	0.65	17.95	5.69	2.35	0.16	4.18	8.28	3.52	3.35	0.40	1.18	100.94
	•V143	TA	53.5	0.8	17.8	4.3	3.55	0.19	4.0	7.3	3.5	3.3	0.37	0.7	99.71
	B1548	TA	54.48	0.72	18.63	3.81	3.97	0.14	3.06	7.11	3.64	3.73	0.41	0.10	99.80
	B943	TA	54.85	0.71	18.88	3.53	4.26	0.17	3.09	6.72	3.97	3.44	0.40	0.47	100.49
PIANO CALDERA	R1800	AB	49.14	0.74	13.19	5.34	4.97	0.20	9.07	12.63	2.13	1.93	0.29	0.47	100.10
	R1656	TB	49.93	0.68	16.85	5.47	3.56	0.22	5.74	9.63	2.44	2.11	0.27	2.15	99.05
	V194	AB	50.2	0.7	12.8	3.7	5.1	0.13	8.8	13.0	1.9	1.75	0.3	1.4	99.78
	G1023	LTF	51.13	0.72	16.94	2.92	5.46	0.17	4.40	8.07	4.04	3.62	0.49	1.12	99.04
	A1152	TB	51.47	0.67	16.01	4.97	3.48	0.20	5.53	9.41	3.11	2.68	0.32	0.88	98.73
	•V130	TB	52.5	0.9	16.4	4.4	4.75	0.12	4.5	9.1	3.0	2.8	0.33	1.1	99.90
	B1044	TA	54.90	0.72	18.52	2.19	5.48	0.15	3.35	7.11	3.69	3.51	0.39	0.10	100.10
	P1012	TA	54.99	0.74	17.15	3.53	4.97	0.16	3.75	7.78	3.43	2.88	0.36	0.10	99.84
LENTIA GROUP	P1701	LT	56.22	0.54	13.47	4.53	2.74	0.13	6.31	8.63	3.32	3.57	0.29	0.74	100.49
	•VL5	TR	61.1	0.55	16.1	2.35	3.15	0.16	2.2	3.8	3.9	5.0	0.31	0.9	99.52
	•VL35	RY	66.9	0.32	13.9	2.3	1.6	0.05	2.4	3.9	3.4	4.6	0.36	1.0	99.73
	H1400	RY	69.35	0.23	14.44	1.86	1.38	0.07	0.90	2.04	4.00	4.94	0.10	0.53	99.84
	H933	RY	70.69	0.20	14.09	2.21	0.79	0.07	1.09	1.95	4.08	4.82	0.10	0.57	100.66
	VL6	ARY	72.0	0.15	13.2	1.10	1.45	0.07	1.2	1.3	4.0	4.7	0.07	0.70	99.94
FOSSA CALDERA	•V76	LTF	50.7	0.85	16.3	4.35	6.1	0.18	4.4	7.7	3.6	4.7	0.54	0.6	100.02
	D1200	TB	52.14	0.77	16.78	7.93	1.64	0.20	3.52	7.54	3.21	4.33	0.48	1.35	99.89
	•V75	TB/TR	55.0	0.75	17.4	2.7	4.8	0.16	3.3	6.0	3.6	4.9	0.45	0.8	99.86
	D1327	TB/TR	55.38	0.65	17.51	3.11	4.25	0.32	3.29	5.67	3.82	4.77	0.43	0.39	99.59
	D949	TR	67.07	0.26	14.83	1.24	2.26	0.12	0.98	2.39	4.71	5.63	0.14	0.38	100.00
FOSSA VOLCANO	N1340	LTF	55.17	0.60	18.28	5.34	1.36	0.16	2.38	4.86	4.97	5.71	0.54	0.57	99.94
	K1900	TR	58.88	0.48	16.38	5.05	0.94	0.13	2.21	4.27	4.11	5.66	0.33	0.90	99.33
	L1344	TR	59.13	0.49	16.48	4.39	1.57	0.12	2.52	4.50	4.45	5.53	0.28	0.63	100.09
	L1028	TR	64.46	0.35	15.25	1.59	2.82	0.11	1.71	3.27	4.50	4.38	0.21	0.44	100.09
	L1220	TR	66.69	0.29	15.13	1.42	2.31	0.10	1.15	2.47	4.57	5.43	0.18	0.37	100.11
	•VF23	TR/RY	67.2	0.26	14.3	2.2	1.7	0.06	1.2	2.4	3.9	5.2	–	1.8	100.22
	L1844	RY	73.14	0.14	14.05	0.88	1.40	0.08	1.2	1.22	4.21	4.99	0.05		100.81
VULCANELLO	VO27	LTF	52.6	0.68	15.7	2.95	5.45	0.12	5.05	7.9	3.6	4.75	0.45	0.75	99.70
	Q1025	LTF	52.69	0.65	15.83	3.71	4.21	0.15	4.54	7.35	3.87	5.06	0.44	0.30	98.80
	•VO6	TR	58.5	0.6	15.7	2.9	3.55	0.16	3.3	5.1	3.9	5.3	0.37	0.3	99.68

•chemical analyses taken from (4)

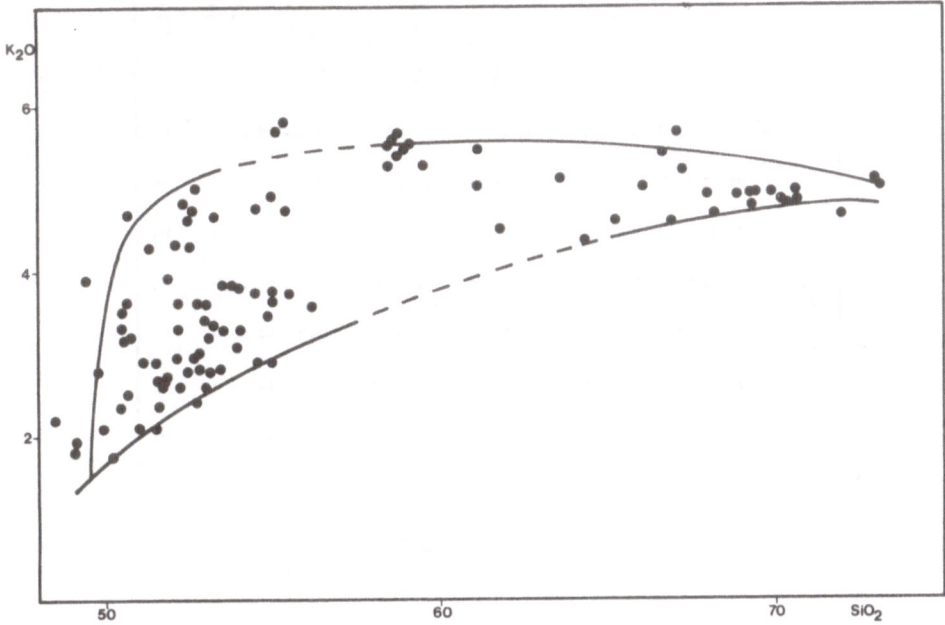

Fig. 2. K_2O-SiO_2 diagram. Points distribution puts in evidence
two "end member" series at different K_2O content: the
shoshonitic (LK) and the lc-tephritic (HK) one.

according to the salic minerals compositions) have been reported.
A prevalence of mafic and intermediate compositions at the initial
stages of volcanic activity i.e.: trachybasalts, trachyandesites
of the South Volcano stratocone and Piano Caldera, is shown. Intermediate and acidic rock-types (trachyandesites, trachytes, rhyolites) dominate the lava output of the more recent Lentia Group,
Fossa Caldera and Fossa Volcano. Lc-tephrites and trachytes occur
at Vulcanello typically. An exception to the adopted Steckeisen
grid concerns two compositions of Piano Caldera (R1800 and V194)
which have been classified shoshonitic alkaline basalts according
to the grid proposed by Cox et al.(7). Obviously the distribution
of the different compositions into the fields reported in fig. 3
cannot be used to evaluate the extent of magmatic relationships
because of distorsions resulting by the projection of the same compositions on the QAPF surface. Therefore this argument will be fa-
ced on a quantitate basis in a later section.

   In Table 2, 19 modal analyses representative of the major
rock-types have been reported. The South Volcano trachybasalts
(TB) and trachyandesites (TA) are strongly porphyritic rocks. The

TABLE 2 - VULCANO ERUPTIVE COMPLEX: SELECTED MODAL ANALYSES

VOLCANIC UNIT	TYPE	SAMPLE	PHENOCRYSTS %	ol	cpx	pl	san	op	%	GROUNDMASS ol	cpx	biot	pl	san	lc	op	gl	(VESH.) %
SOUTH VOLCANO	TB	A1800a	46.42	2.39	11.61	31.56	-	0.86	53.58	X	X	-	X	X	-	X	-	(10.83)
	TB	A1600a	46.35	1.90	14.57	28.46	-	1.42	53.65	X	X	-	X	X	-	X	-	(3.21)
	TB	A1700a	49.94	2.88	10.82	35.03	-	1.21	50.06	X	X	X	X	X	-	X	-	(2.35)
	TA	B1548	49.20	3.90	2.70	41.90	-	0.70	50.80	X	X	-	X	X	-	X	-	-
	TB	E1102	53.04	4.30	19.23	29.04	-	0.47	46.96	X	X	-	X	X	-	X	-	(1.37)
	TB	E1502	55.74	4.26	19.80	31.68	-	0.48	44.26	X	X	-	X	X	-	X	-	(17.80)
	TA	M1534	41.61	3.66	20.44	14.72	-	0.73	58.39	-	X	-	X	-	-	X	X	-
PIANO CALDERA	TA	P1012	7.47	0.12	3.31	4.04	-	-	92.53	-	X	X	X	X	-	X	X	-
	AB	R1800	8.60	1.65	6.69	0.26	-	-	91.40	-	X	-	X	-	-	X	X	(17.29)
	LTF	G1023	27.13	0.30	11.79	14.35	-	0.73	72.87	X	X	X	X	X	X	X	-	(1.80)
LENTIA GROUP	LT	*P1701	26.34	0.22	23.53	1.65	0.15	0.47	73.66	-	X	X	X	X	-	X	-	(17.21)
	RY	H1400	13.65	-	3.83	8.15	0.98	0.69	86.35	-	X	X	-	X	-	X	X	-
	RY	H933	7.32	-	2.73	2.93	1.07	0.59	92.68	-	X	X	-	X	-	X	X	-
FOSSA VOLCANO AND CALDERA	LTF	N1340	15.75	0.90	2.96	11.08	-	0.81	84.25	X	X	-	X	X	X	X	-	(13.21)
	TR	L1028	23.68	-	5.36	12.08	5.85	0.39	76.32	-	X	X	X	X	-	X	X	(3.30)
	TR	L1344	17.99	0.98	7.28	2.16	6.39	1.18	82.01	-	X	X	X	X	-	X	X	-
	TR	K1900	26.28	-	9.35	13.26	1.27	2.41	73.72	-	X	X	X	X	-	X	X	(30.59)
	TB/TR	D1327	13.00	0.13	8.65	3.36	-	0.87	87.00	-	X	-	X	X	-	X	-	(6.55)
VULCANELLO	LTF	Q1025	21.29	-	16.26	4.14	-	0.89	78.71	-	X	X	X	X	X	X	X	(0.89)

*including 0.33% biotite phenocrysts

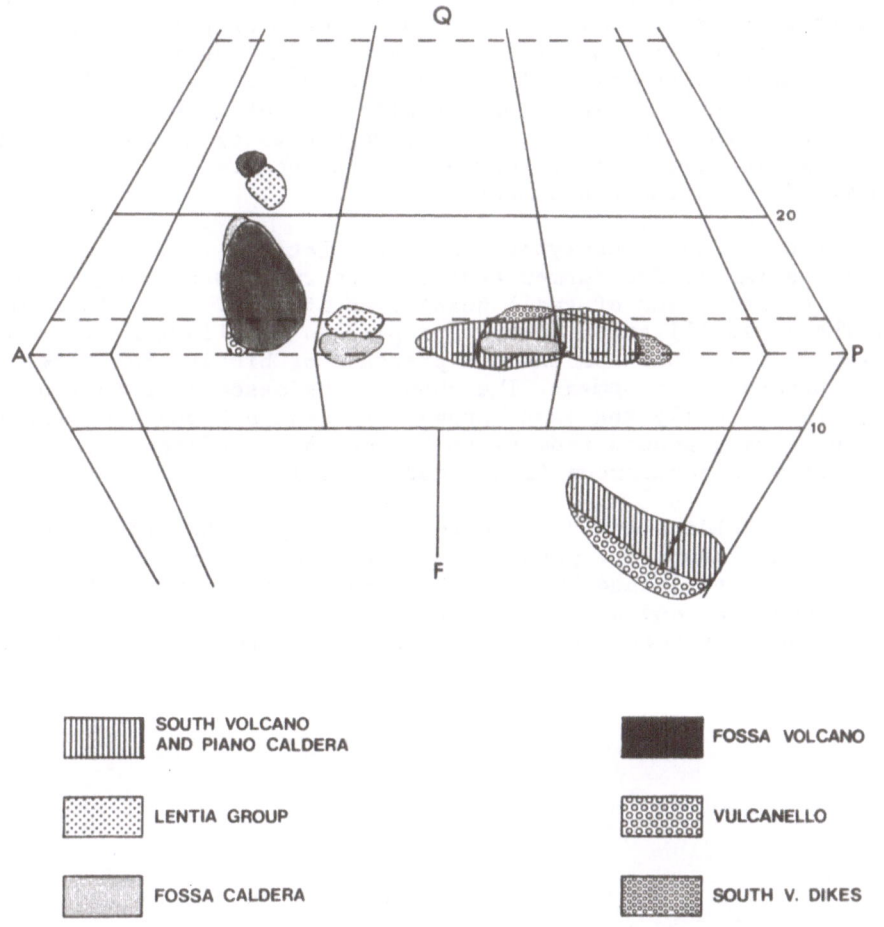

Fig. 3. QAPF double triangle of Streckeisen (1967) classification.

phenocrysts total about half of whole volume. They are represented by plagioclase (plag): 15÷41%; clinopyroxene (cpx): 3÷20%; olivine (ol): 2÷4%; opaques (op): 0.5÷1.5%. Fabric relationships indicate the following crystallization sequence: plag - cpx - ol, the same which has been obtained experimentally on an analogous system (8) operating at 2 Kb total pressure and H_2O undersaturated conditions. Groundmass is usually holocrystalline and it is composed basically of the same phenocrysts phases plus sanidine (san).

Piano Caldera rock-types are represented by: TB and TA (analogues with those outcropping at South Volcano); shoshonitic alkaline basalts (AB) and leucite tephrites (LTF). Among them the tephra

and scoriaceous lava blocks of Sommata Hill (AB-R1800) seem to represent the most primitive liquid of Vulcano eruptive complex. This
is indicated by the rock bulk composition (MgO number etc.), by the
low amount of phenocrysts (Table 2) and by the absence of cumulate
phases. Phenocrysts mineralogy recalls that of TB but the presence
of glass in the groundmass. LTF outcrops have again the same essential trachybasaltic mineralogy except the scattered presence of
leucite (lc) in the groundmass.

Lentia Group rock-types range from latites (LT) to alkali-
rhyolites (ARY). The former have a phenocrysts assemblage made of
cpx principally and of small quantities of plag, san, biot, ol and
op. The crystallization of cpx and plag on the liquidus is shown
by fabric relationships and the presence of biotite indicates a
water present environment. The phenocrysts assemblage (but ol and
biot) is basically the same through the Lentia Group rock-types;
its decreasing amount from LT to RY does not conflict with the hypothesis of a continuous fractionation series.

Fossa Caldera and Volcano rock-types span the widest compositional range over. The phenocrysts assemblage is always dominated
by cpx and plag joined by ol in trachybasalts, lc-tephrites and
some trachytes, and by san in more acidic trachytes and rhyolites.
The phenocrysts total amount is fairly moderate even for TB; the

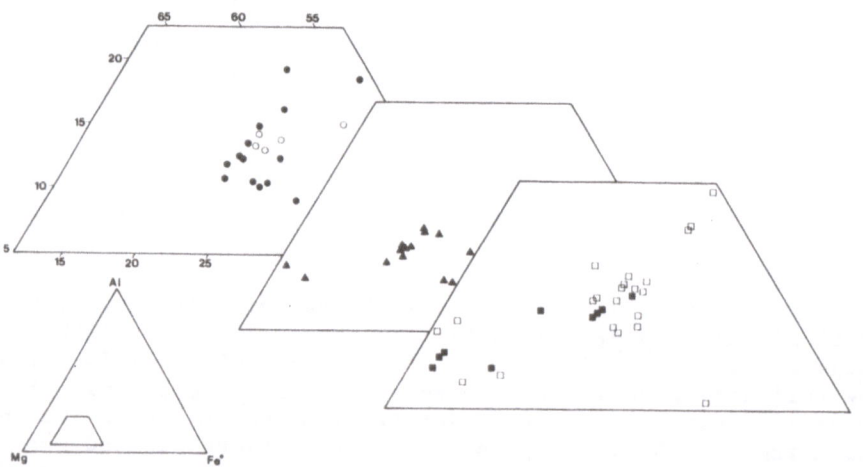

Fig. 4. Clinopyroxenes. Mg-Al-Fe°(Fe$_{tot}$+Mn) diagram. Symbols: dots,
South Volcano; circles, Piano Caldera; triangles, Lentia;
squares, Fossa Caldera; open squares, Fossa Volcano.

frequent inversion of plag/cpx ratios even for very similar compositions (i.e. N1340 and D1327) puts in evidence the rise of fresh
magma at the same time with the differentiated one, both with a
highly variable presence of volatile components. Vulcanello LTF
and TR present similar phenocrysts and groundmass patterns of the
already described Fossa analogues.

Microprobe mineralogy of the phenocrysts and of the groundmass dominant phases is presented in figs. 4-5-6.

Clinopyroxenes put in evidence chemical variations controlled
either by the environmental conditions than by the magmas compositions. The most of phenocrysts can be classified as augites. They
evolve during the late stages of crystallization (microphenocrysts
and groundmass microlites) towards fassaitic or salitic compositions as shown respectively by: i) increasingly high values of Σ
$(Fe^{2+}+Fe^{3+}+Mn)$ and Al at constant Ca level; ii) strong $(Fe^{2+}+Fe^{3+}+Mn)$
enrichment with falling values of Ca and Al. Both trends are shown
by cpx crystallizing from rock-types of the different volcanic u-
nits. The fassaitic trend is prevalent in the Piano Caldera TB and
can be recognized in some similar rock-types of South Volcano and

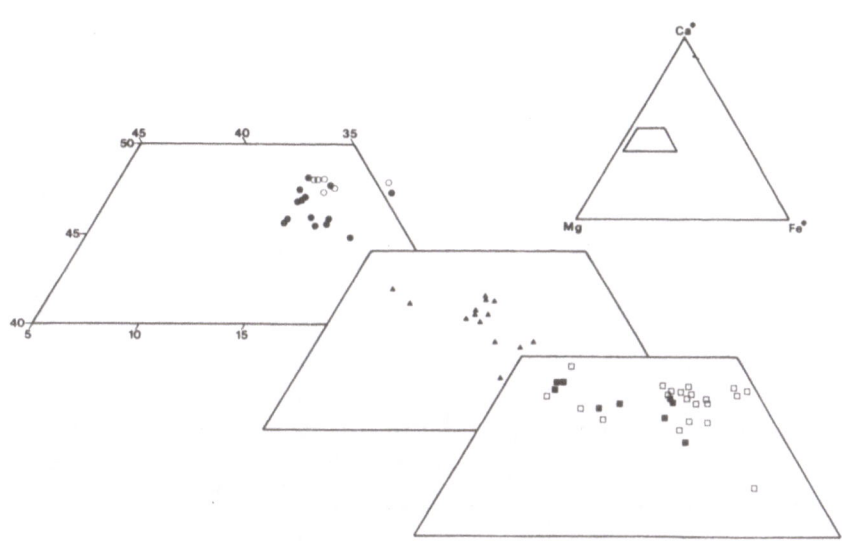

Fig. 5. Clinopyroxenes. Ca°(Ca+Na+K)-Mg-Fe°(Fe$_{tot}$+Mn) diagram.
        Symbols as in fig. 4.

Fossa Volcano; the salitic one is peculiar of Lentia LT and RY, and can be recognized in the more evolved rock-types: South Volcano TA and Fossa Volcano and Caldera TR.

Pigeonites s.s. can occur in scarce amounts in the groundmass of the most acidic types. Probe analyses from samples of Lentia RY and Fossa Volcano TR give a virtually identical composition, i.e.: En57; Fs38; Wo5.

Olivine phenocrysts and microphenocrysts (even if in small a-mounts) are ubiquitous in TB, TA, LTF, LT, RY and in the some mafic TR. Their compositions range from Fo71 (TB) to Fo53 (TR). In the

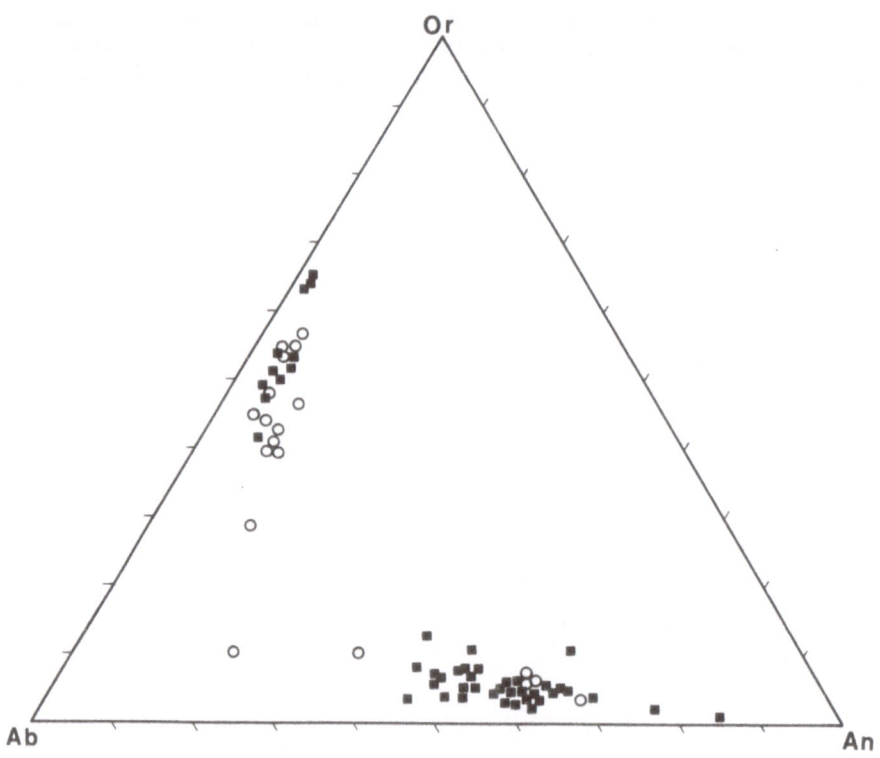

Fig. 6. Plagioclases and K-feldspars. An-Ab-Or diagram. Squares: phenocrysts; circles: groundmass microlites.

groundmass ol (Fo50) is confined into the more mafic rock-types.

Fe-Ti oxides are represented by ulvo-spinel s.s. The amount
of u.s.% is highly variable either among the different rock-types
than for different stages of crystallization. In the groundmass of
TB and the other mafic types it is around 50%; lower values(between
10 and 20%) are instead characteristic of TA, LT and of the most
acidic types.

Plagioclases phenocrysts range from An87 (plag cores of Piano
Caldera TB) to An40 (cores of Fossa Volcano TR) (fig. 6). Groundmass microlites exibit a comparable chemical variation from An67
(South Volcano TB) to An20 (Lentia RY). The composition of groundmass ternary feldspars (An21, Ab69, Or10) in such rhyolites is indicative of the most differentiated liquid which can be produced
by crystal fractionation in the shoshonitic suite (9).

K-feldspars in equilibrium with plag are progressively Ab-rich
as crystallization proceeds. TB have typically an Ab50 Or50 soda-
sanidine; TR range from Ab35 Or65 (phenocrysts) to Ab60 Or40
(groundmass microlites).

4. THE PETROLOGICAL MODEL

To get a comprehensive picture of magmatic evolution,a determi-
nistic model able to describe the fractionation relationships and
the magma T,P,H_2O paths was drawn up. This has been realized utili-
zing quantitatively the already given informations on chemistries
of rock-types and on compositions of dominant phases which crystallize from the different magmas batches during their way to the surface. Fractionation relationships have been elucidated using a modified version (10) of Nathan and Van Kirk (11) magma model which
simulates the crystal-liquid fractionation at atmospheric pressure
into a nine components anhydrous system. The liquid lines of descent
generated in such way have been taken as representative of fractionation processes under low pressure and water undersaturated conditions. In some istance they duplicate typical rock-types composition. Such relationships have been confirmed by mass-balance calculations. These have been performed using the computer program proposed by Nicholls and Stormer (12). Only solutions which gave values of ΣR^2 (sum of square residuals between calculated and actual
rock-types final compositions) less or around 1-1.5 have been accepted. In some cases, only the mass-balance calculations give
positive solutions. These are therefore taken as representative of
fractionation conditions under higher pressures and/or with signi-
ficative amounts of dissolved water in the melt. Some further
indications on the fractionation controlling conditions will be
given later on when discussing the estimates of P,T and % H_2O of
the melts in equilibrium with their respective phenocrysts. The

adopted scheme puts in evidence fractionation relationships among magmas of different volcanic units and inside the same volcanic unit. In the latter case this has been realized exploring fractionation relationships among compositions representative of the most primitive and of the most evolute magmas.

In fig. 7, liquidus temperatures obtained through the magma model calculations have been plotted against S.I. (100 MgO/MgO+ Fe_2O_3+FeO+Na_2O+K_2O). It appears that the reported compositions take part into a single trend. The change of slope around S.I.=20 can be related to the crystallization of calcic plagioclase which joins clinopyroxene close to the liquidus surface. Some gaps occur along the points alignment. However they do not seem linked to any large jump among confining compositions. Among them those with the highest S.I. ratios and highest liquidus temperatures are taken as

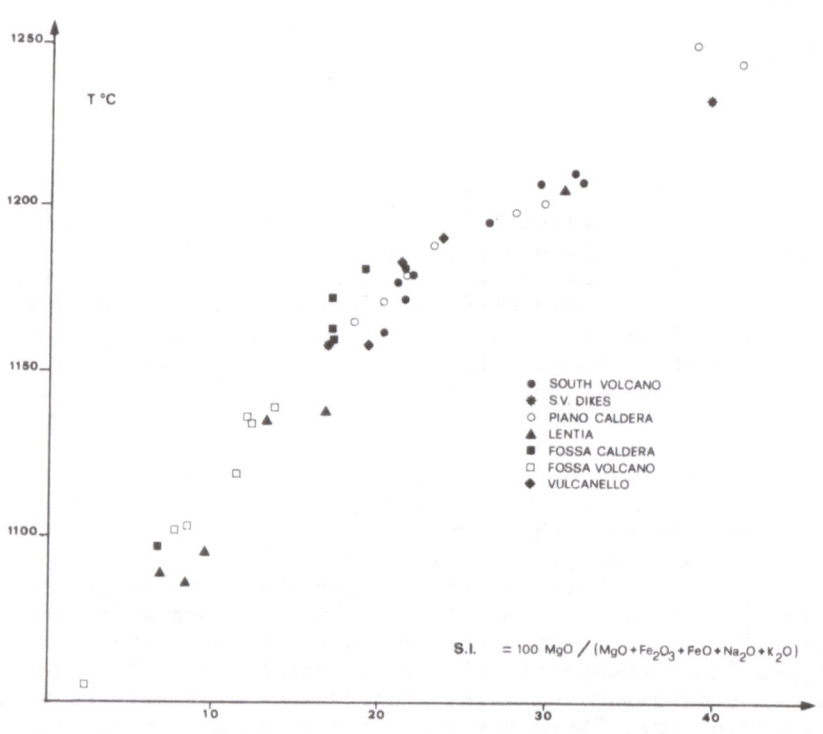

Fig. 7 . T°C-S.I. diagram. Liquidus temperatures have been obtained through the magma model calculations.

representative of the most primitive magmas which reach surface.
R1800 (AB) type has been used as parent composition for magma model and mass-balance calculations. In fig. 8 (AFM triangle), all
the compositions utilized for the fractionation calculations have
been projected. Into the diagram, R1800 and the rock-types taken
as representative of the most primitive and the most differentiate
compositions of the single volcanic units have been localized. The
liquid line of descent of R1800 reaches P1012 Piano Caldera TA and
D1200 Fossa Caldera TB compositions which,as confirmed by mass-
balance calculations, can be considered as differentiated melts
under low pressure conditions. For N1534 South Volcano TA, P1701
Lentia LT and V027 Vulcanello LTF, only the mass-balance program
gives positive solutions. Among the two sets of compositions,the
latter has lower Al_2O_3, and higher MgO and CaO values (Table 1).

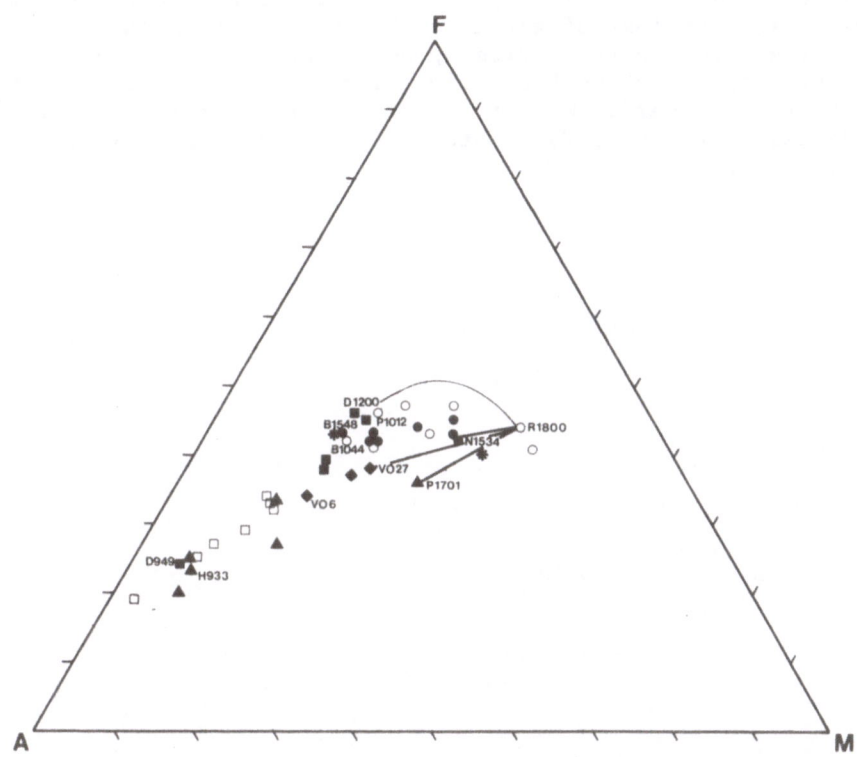

Fig. 8. AFM triangle. Fractionation relationships among the compositions of the different volcanic units are indicated.

Because the mass-balance calculations give for all compositions a
similar amount of fractionated solids, it appears that, in the
latter case, the more alkaline and mafic melts could be formed by
removal of a plagioclase rich solid.

On different AFM triangles (fig. 9), fractionation paths inside the single volcanic units are now illustrated. From any of the
projected compositions,liquid lines of descent as built up by the
magma model program are shown. Good solutions have been obtained
for South Volcano and in part for Piano Caldera rock-types. Mass-
balance calculations gave successfull results too. Similar attempts to link primitive compositions (TB,LTF) with typical differentiated ones (TR,RY) of Lentia, Fossa Caldera and Vulcanello
volcanic units are not verified by magma model lines of descent
(fig. 9) and even the mass-balance program gave too high ΣR^2 values
to be accepted. Eventual explanations of this behaviour could be
found supposing either: a number of fractionation steps (as suggested in case of Fossa Caldera and Volcano); or fractionation of
sanidine (anyway a typical phenocryst of the most differentiated
types only); or loss of alkalies through the magma chamber and
conduit walls when the system opened before or during eruptions.
To a certain extent each solution is possibly true; by working on
the third hypothesis, minima in the ΣR^2 values have been obtained
by depleting the Na_2O+K_2O content of the single parent compositions
by nearly 20% of the original amount.

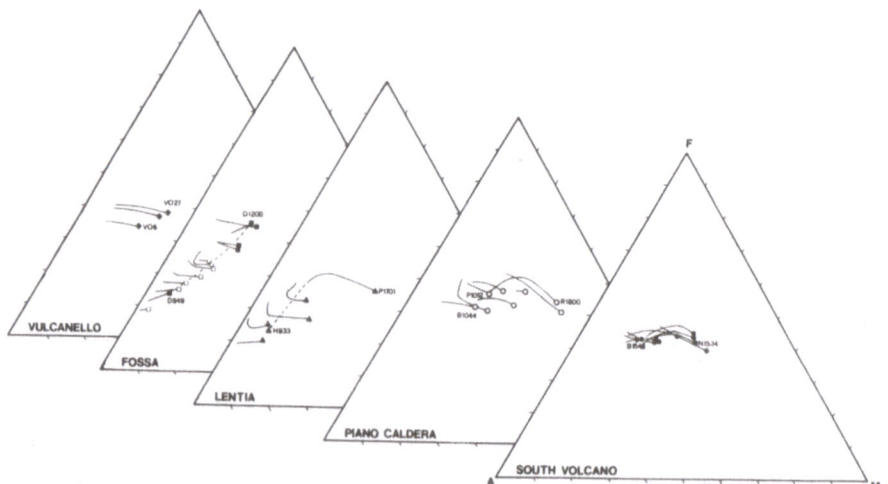

Fig. 9. AFM triangles. Liquid lines of descent of the single rock-
types are shown for each volcanic unit.

A simplified scheme of fractionations relationships is presented in fig. 10. According to it intermediate liquids (TB,TA,LT, LTF) can be generated by fractionation of approximately 50-65% AB parent magma. Greater amount (up to 80%) become necessary to generate the most acidic liquids (TR, RY).

An insight on enviromental conditions which occurred during magmatic evolution can be gathered by estimates of the intensive variables characterizing the early crystallization of phenocrysts

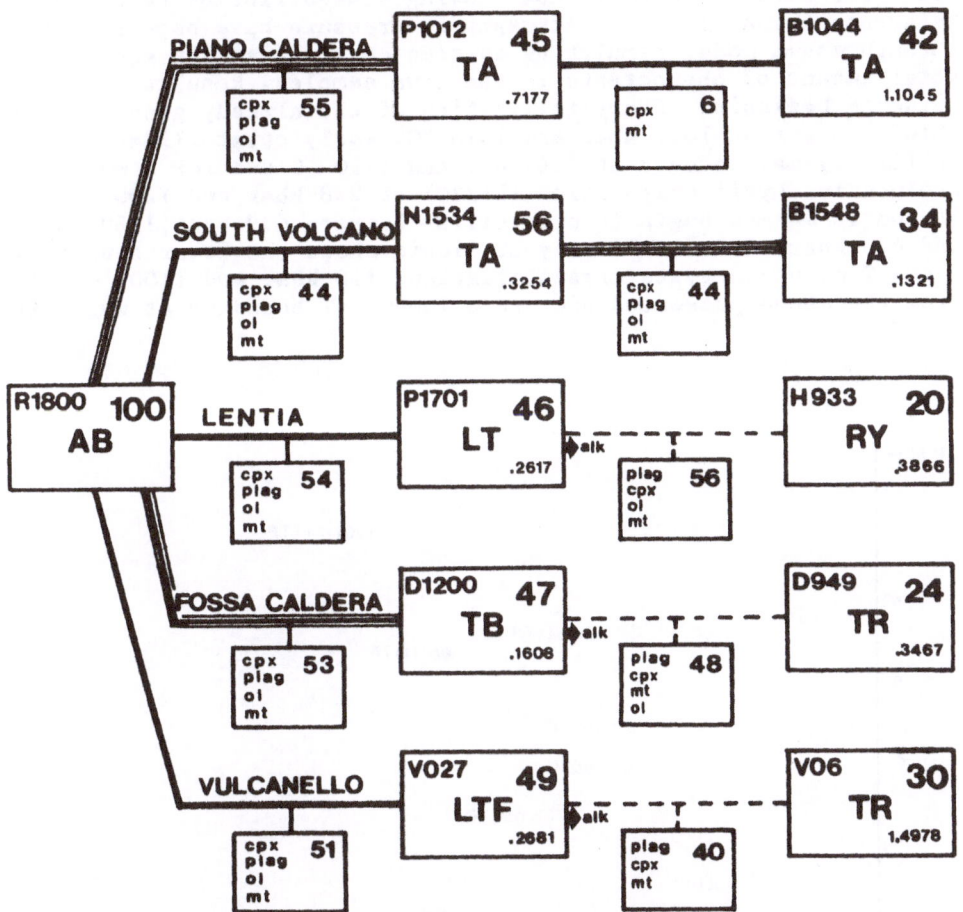

Fig. 10. Vulcano eruptive complex. Fractionation relationships as obtained by magma model+mass-balance (triple lines), mass-balance only (heavy lines), alk. modified mass-balance (dashed lines). Large boxes: rock types (center), amounts of fractionated melts (up right), ΣR^2 (down right); small boxes: amounts and mineral assemblages of removed solids.

assemblages in equilibrium with the respective liquids. Even if
fractionation and crystallization steps do not relate each other
necessarily,the latter give informations on magma pauses on its
way to the surface and put constraints for fractionation events.
The P and T equilibrium estimates of phenocrysts have been carried
out by thermodynamic calculations on ol,cpx,plag and san solid
solutions. Calculation procedures and method peculiarities have
been matter of detailed investigations (9),(13),(14),(15) and will
not repeated here. The results reported in fig. 11 refer to samples
having developed groundmass phases suitable for probe analyses and
well preserved phenocrysts not showing disequilibrium textures.
Reference temperatures at atmospheric pressure have been estimated
through magma model simulating systems crystallization equal to the
total amount of phenocrysts in the lava samples. Results in fig. 11
evidence beginning of crystallization of the already generated LTF
(G1023) magma at 10.1 kbar and 1310 °C. Early crystallization of
TR (L1344) melt occurs at 7.4 kbar and 1170 °C and for a more
evolved trachytic composition (L1220) at 2.8 kbar and 1200 °C.
TA and TB magmas begin to crystallize between 6.9 kbar;1260 °C
and 4.0 kbar;1150 °C. Finally LT liquid shows the lower equilibrium
P and T of phenocrysts crystallization: 1.0 kbar and 1100 °C.
From the above presented picture a lower P-T boundary at which the

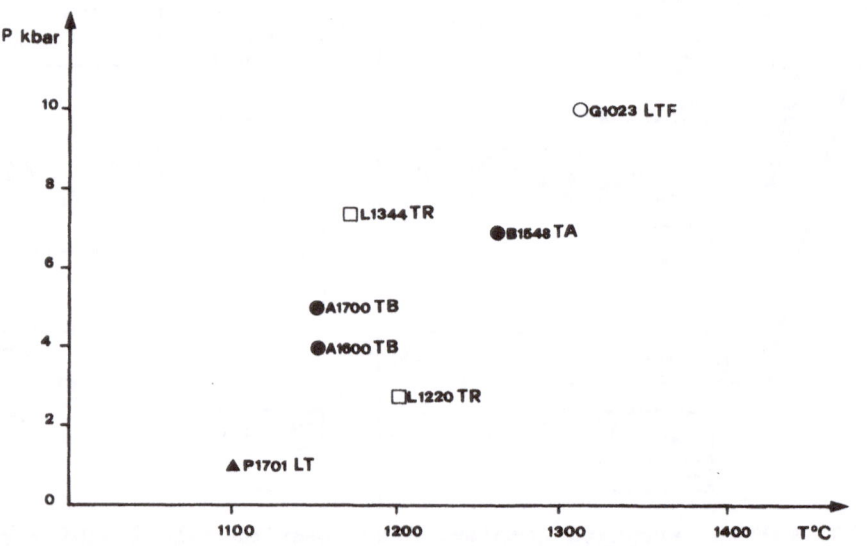

Fig. 11. Equilibrium pressures and temperatures of phenocrysts
          assemblages.Abbreviations same as in Table 1.

single magmas have been generated may be inferred. According to
this, the compositions of HK end member series (LTF-TR) appear to
be formed at higher pressures than those of LK shoshonitic one (TB,
TA,LT). The different behaviour of the two trachytic compositions
can be related either to differences in chemistry than to a different content of water dissolved into the melt.

In absence of hydrate phases the water content of magmas has
been deduced from the composition of <u>cpx</u> solid solutions for which
equilibrium P and T have been calculated. This has been realized
by calibration of X^{cpx}_{c+ts} against P and $\%H_2O$ in the melt for systems
in the range TB-LTF (8). The estimates can be related to the fractionation steps and to the extent of crystallization into differentiated liquids provided the systems behaved as closed. This seems
reasonable considering the relationships between the parent composition AB and the differentiated ones: LTF,TA,TB. The results of
calculations give the following water contents in the melts: LTF
(0.95%), TA (1.34%), TB (1.11%), LT (1.10%). Taking into account
the amounts of the removed solid to generate the above reported
types from AB (fig. 11) we obtain for this last one 0.46-0.52% of
water.

4. CONCLUSIONS

The approach followed here to take into account the various
aspects of the magmatic differentiation of Vulcano eruptive complex
gives a semplified scheme of a very complicated phenomenon. From
the results of fractionation models and from $P-T-H_2O^{liq}$ estimates
of the single differentiated liquids, we obtain the following
consistent indications:

1) Differentiated liquids of both HK (lc-tephritic) and LK
(shoshonitic) series can be generated by fractionation from a single parental magma of basaltic affinity.

2) Fractionation occurs at pressures in excess of: 10 kbar
(LTF) , 7 kbar (TA) , 4 kbar (TB) compositions and involves 50%
about of removed solids.

3) Water estimates for the corresponding melts at the beginning of their crystallization processes indicate an amount of water
in the parent magma close to 0.5%.

According to mass-balance calculations LT melts can be generated by AB parent magma. On the other hand, $P-T-H_2O^{liq}$ data indicate
a beginning of crystallization at very low pressure which is not
confirmed by magma model results. The most extreme differentiates
TR and RY may be eventually generated by lc-tephritic and latitic
systems which have to be considered open (at least for water and

alkalies) in the late stage of their evolution.

REFERENCES

1. F. Barberi, P. Gasparini, F. Innocenti, L. Villari: 1974,
    J. Geophys. Res. 78, pp. 5221-5232.
2. S.R. Carter and L. Civetta: 1977, Earth Planet. Sci. Lett.
    36,168.
3. L. Civetta and M. Cortini: 1979, CNR Geodin. 235, pp. 77-86.
4. J. Keller: 1980, Rend. SIMP, 36, 1, pp. 369-414.
5. M.F. Sheridan, T.C. Mayer and K.H. Wohletz: 1981, Mem. S.A.
    It., pp. 523-527.
6. A. Peccerillo and S.R. Taylor: 1976, Contrib. Miner. Petrol.
    58, pp. 63-81.
7. K.G. Cox, J.D. Bell and R.J. Pankhurst: 1979, The interpreta-
    tion of Igneous Rocks, London
8. D. Dolfi and R. Trigila: 1981, Min. Mag., in press.
9. I.S.E. Carmichael, F.J. Turner and J. Verhoogen: 1974, Igneous
    Petrology, McGraw-Hill Co.
10. R. Crescenzi, B. Giannetti and R. Trigila: 1981, Per. Min. in
    press.
11. H.D. Nathan and C.K. Van Kirk: 1978, J. of Petrol. 19, 1, pp.
    66-94.
12. J. Nicholls and J.C. Stormer Jr.: 1978, Computers and Geo-
    sciences 4, pp. 143-159.
13. J. Nicholls: 1977, in D.G. Fraser (ed.) Thermodynamics in Geo-
    logy, pp. 327-348.
14. D. Dolfi and R. Trigila: 1976, Per. Min. 35, 1-2-3, pp. 109-
    127.
15. I.S.E. Carmichael, J. Nicholls, F. Spera, B.J. Wood and S.A.
    Nelson: 1977, Phil. Trans. R. Soc. London A, 286, pp. 373-431.

MORPHOLOGY AND NETWORK PATTERNS OF MARTIAN VALLEYS

David C. Pieri

Jet Propulsion Laboratory, California Institute of
Technology, Pasadena, CA.

Martian valleys exhibit morphologies and network patterns
which do not provide compelling evidence of formation by
rainfall. Rather, cuspate amphitheater terminations and
sparse, open networks, often strongly structurally controlled,
argue for erosion by subsurface water.

Although martian valleys display a diverse mix of network
pattern and morphology, there are certain unifying
observations. Martian "valleys" are distinguished here from
"channels" by the absence in the former (Figure 2) of the
direct evidence of fluid erosion often found in the latter
(Figure 3) (1). There is no clear evidence of direct fluid
erosion (i.e. streamlined obstacles, interior channels, etc.)
yet observed within any martian valley interior. It is
possible that such features are too small to be seen in either
Mariner 9 or Viking images, although features as small as
100 m can be resolved in some Viking Orbiter images of
valleys. Valley walls are typically rugged and cliff-like,
sometimes displaying debris accumulations and talus deposits,
valley floors are generally flat. Mantling by materials of
eolian and volcanic origin is common. Some valleys display
cliff bench interior topography, similar in character and
scale to features which exist, for example, in the Grand
Canyon of the Colorado River, formed there in response to
differential resistance to erosion (2). This type of
morphology may be evidence of extensive layering and materials
contrast in the martian heavily cratered terrain as suggested
previously (3). The most consistent morphological
characteristic, however, is the presence of steep-walled,

477

A. Coradini and M. Fulchignoni (eds.), The Comparative Study of the Planets, 477–483.
Copyright © 1982 by D. Reidel Publishing Company.

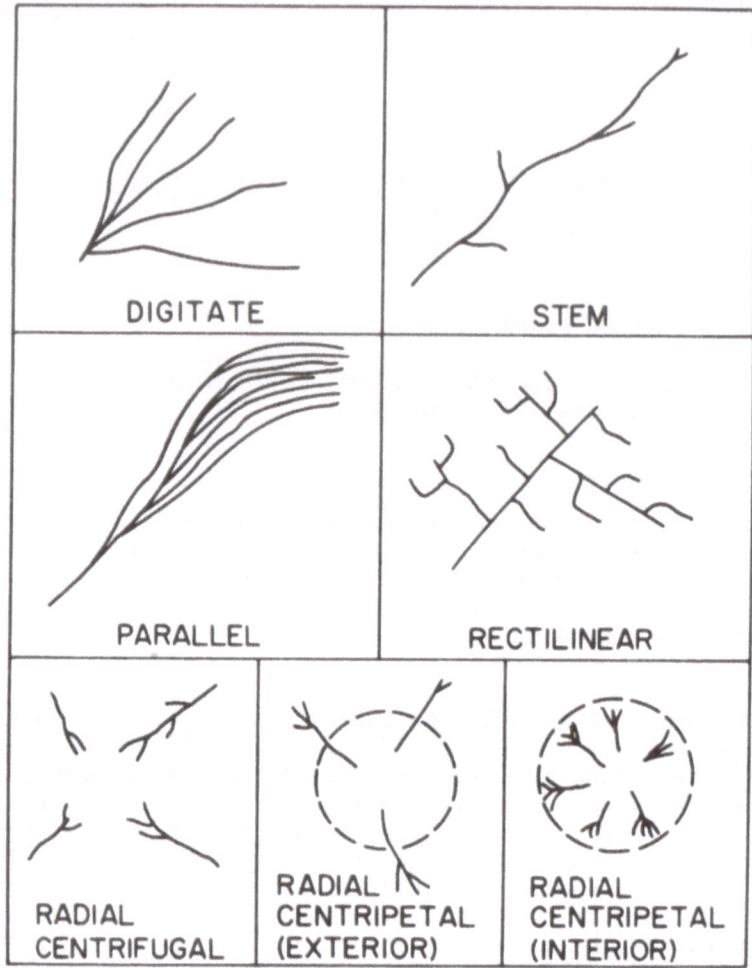

Figure 1. Sketches of valley planimetric patterns observed on Mars.

Figure 2. Parallel network near Claritas Fossae. This
network is expressed in a block of heavily fractured
ancient heavily cratered terrain. The upper reaches of
this network are intimately associated with the northe-
ast trending fractures and appear to be truncated by
them in several places. The network post-dates the east
west trending scarp at the bottom of the picture. Note
the long parallel valley sections and the strong coale
scence near the mouth, similar to digitate systems.The
strongly parallel arrangement of tributaries, lack of
evidence of capture across inter-valley divides, and
absence of deflection of small tributaries into larger
ones suggest formation other than by rainfall, although this system is often cited as evidence of pluvial
erosion (Masursky et al., 1977). Sun elevation is 35°,
illumination is from the right, and the viewing angle
is 13.5°.

Picture center: latitude - 43.8°, longitude 93.5°
Resolution: 350 meters
Viking Orbiter Image: 532A16

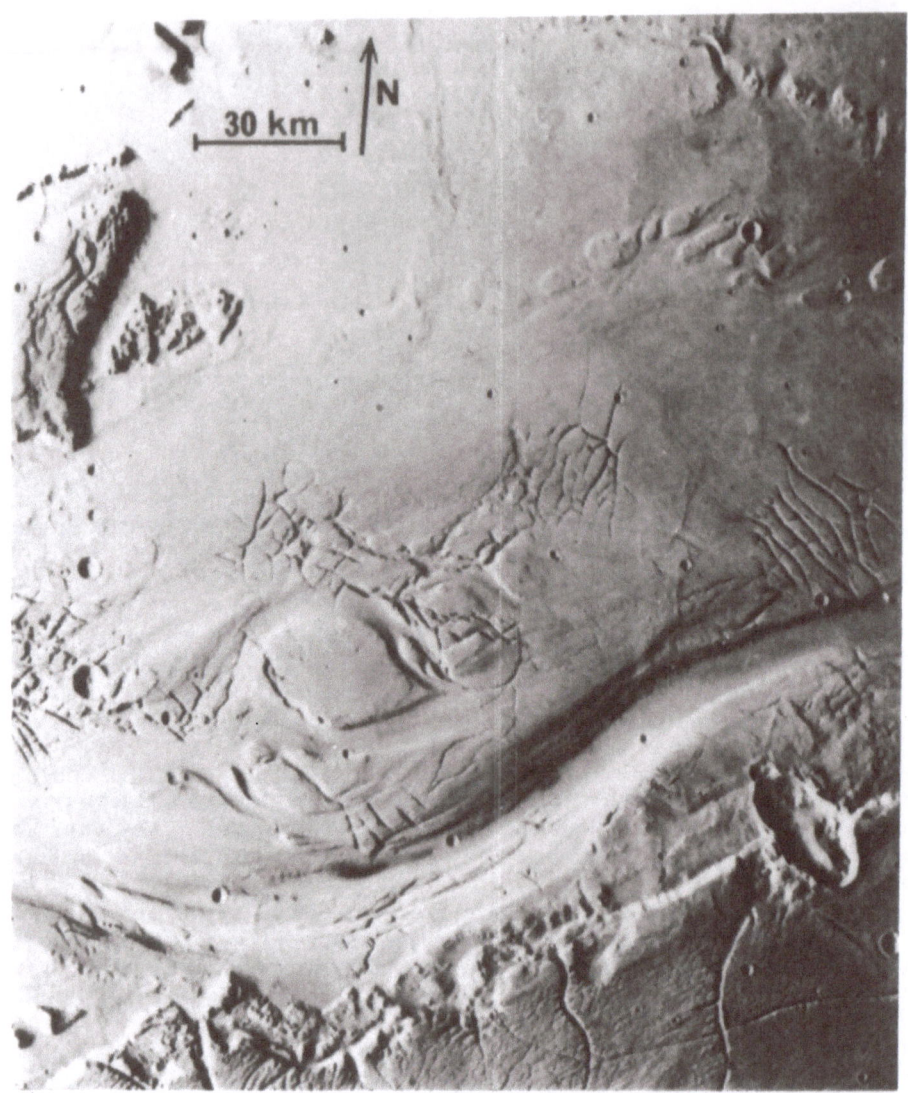

Figure 3. **Northern section of Kasei Vallis.** The main channel is
in the lower half of the scene. Note the streamlined shapes of
topographic projections and the parallel, sinuous grooves and
ridges, consistent with erosion by flow from the west. Illumination is from the left, sun elevation above the horizon is 25°,
and viewing angle is 13°.
Picture center: latitude 27.5°, longitude 69.3°
Resolution: 350 meters
Viking Orbiter Image: 519A29

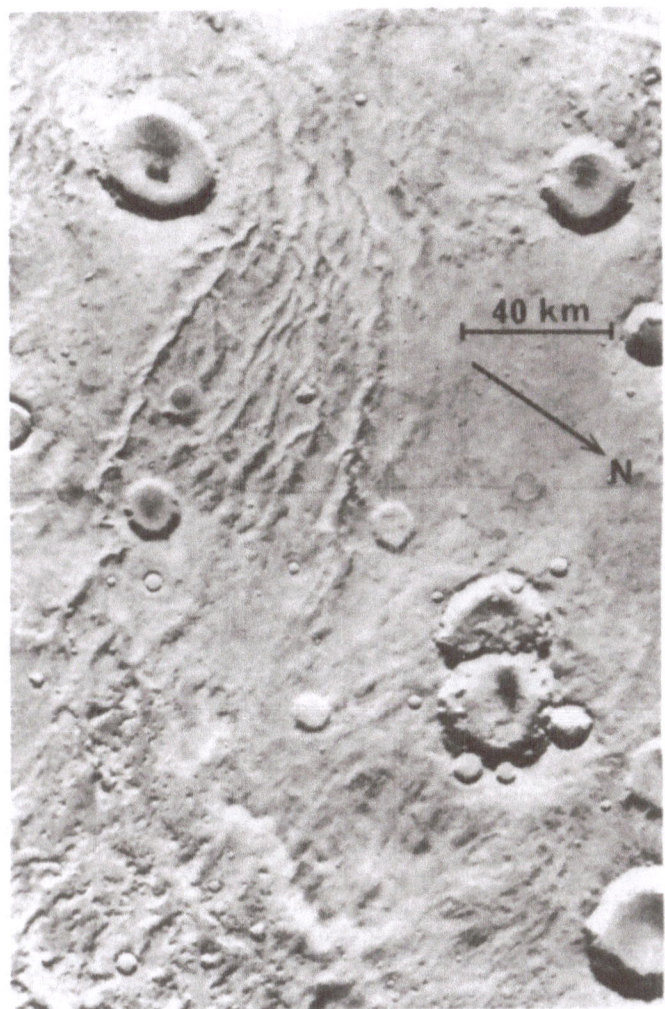

Figure 4. Parallel and digitate valley system in Thaumasia.
Amphitheater terminations and widening distal to the valley mouth
are observed. Many parallel networks are observed in this area
as part of a centrifugal assemblage of valley systems radial
to the chaotic terrain in the lower left. Notice the prominent
valley which appears to originate in the chaotic terrain. Sun
elevation is $24°$, illumination is from the bottom of the picture,
and the viewing angle is $14°$.

Picture center: latitude $-47°$, longitude $73°$
Resolution: 300 meters
Viking Orbiter Images: 535A23 and 25

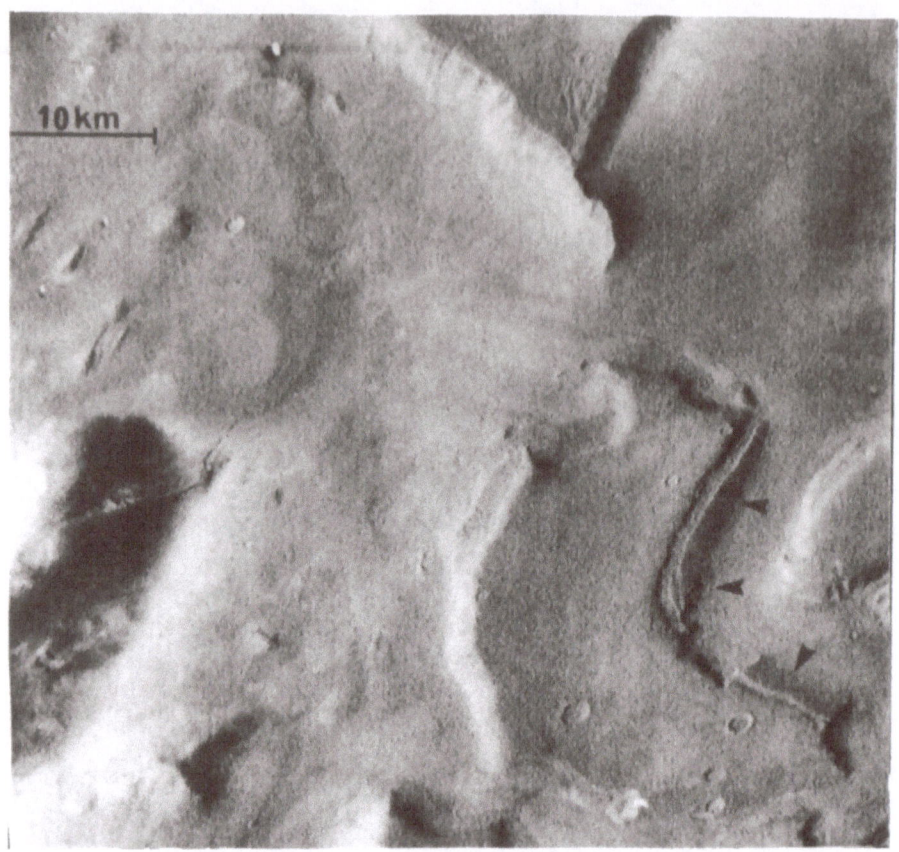

Figure 5. Example of wall degradation. Note how the block
marked by the arrow has moved down and away from the plateau
surface along the perimeter of the valley at lower right. Such
collapse, by undermining less competent substrata results in
valley widening and resurfacing, complicating the interpretation
of impact crater statistics with regard to determining valley
"age". Sun elevation is 38°, illumination is from the left,
and viewing angle is 41°.

Picture center: latitude 36.3°, longitude 345°
Resolution: 100 meters
Mariner 9 Image: DAS 10650904

cuspate terminations at the heads of the smallest tributary valleys (Figure 4). The existence of these steep-walled amphitheater (4, 5) terminations suggests headward extension ("sapping") by basal undermining and wall collapse (Figure 5), as in the predominant mode of headward extension for many terrestrial canyons.

Valley widths are variable between < 1 km to about 10 km. Lengths range from $\sim$ 5 km to about 10^3 km. This range in physical size alone implies a range of valley formational processes and probably also a sensitivity to materials differences between the host rock units.

Overall, network morphologies are variable from the rather spacially compact digitate and parallel networks to the generally mono-filamentous stem networks with few tributaries. This difference may also be attributable to contrasting modes of origin. Also, there is no compelling global correlation between network morphology and terrain type although a very few fresher and sparser appearing valley networks tend to occur in Lunae Planum type ridged-plains units (e.g. Nigal Vallis) while the more numerous degraded and somewhat more complex networks tend to occur in the heavily cratered terrain. Whether this dichotomy in occurrence and morphology is due to contrasting age or contrasting lithology is problematic.

The lack of clear evidence of fluid flow, the sparse network patterns and the vast undissected interfluves provides no compelling evidence for the formation of these features by rainfall. Rather, the prevalent amphitheater terminations, the steep cliff-like walls, the generally parallel network patterns, and the tendency to strongly follow structural controls, where they exist argues strongly for a near surface lithospheric source of water. Seepage runoff and basal sapping, perhaps in combination, may be the likely formative processes.

References. (1) Baker, V.R. and D.J. Milton, 1974, Erosion of catastrophic floods on Mars and Earth: Icarus, 23, p. 27-41; (2) Museum of Northern Arizona, 1974, Geology of the Grand Canyon, Museum of Northern Arizona and Grand Canyon Natural History Association; (3) Malin M.C., 1976, PhD. Thesis, California Institute of Technology, Pasadena, CA, 176 p.; (4) Pieri, D.C., 1979, Geomorphology of Martian Valleys, PhD. Thesis, Cornell University, Ithaca, N.Y., 280 p.; (5) Pieri, D.C., 1980, Martian Valleys: Morphology, Distribution, Age, and Origin, Science, 210, p. 895-897.

STRATIGRAPHIC RELATIONSHIPS AMONG THE UPPER LAYERS OF THE OUTER GALILEAN SATELLITES, INFERRED FROM THE INVESTIGATION OF THEIR RAY SYSTEMS

M.Poscolieri

Istituto di Astrofisica Spaziale, Rome,Italy

Abstract   Callisto and Ganymede are the outermost Jovian icy satellites: they exhibit similar low density and inferred internal structure, but their surface appearance and their albedo are markedly different. Callisto is very dark and uniformly cratered, whereas Ganymede is splitted in dark cratered terrains and relatively bright grooved terrains. This distinction is confirmed by rayed craters distribution and occurrence: Ganymede has many remarkable ray systems, Callisto exhibits mostly haloed craters. The different scenario between the two satellites is believed to be related to the stratigraphy and the thickness of the upper layers of their crust. The study of Ganymede's rayed craters distribution on grooved and cratered terrains reveals further dissimilarities. Dark rayed craters are present almost exclusively on cratered terrains and only in the small size range; bright rayed craters, on the other hand, occur in all size ranges and on both grooved and cratered regions. Supposing that a projectile excavates material from a depth of one tenth of the final crater size, we suggest the presence of a continuous layer, 4 ÷ 5 km thick, composed of a mixture of ice and rock under Ganymede's dark cratered areas. An analogous layer could be present on Callisto, but it should be ubiquitous and sufficiently thick to avoid the formation of prominent bright rayed craters.

1. OVERVIEW OF THE OUTER GALILEAN SATELLITES FEATURES
Strong differences in the global appearance of the icy

485

A. Coradini and M. Fulchignoni (eds.), The Comparative Study of the Planets, 485–494.
Copyright © 1982 by D. Reidel Publishing Company.

Galilean satellites have been clearly shown on the Voya
gers pictures (6,7). In fact, Europa exhibits a smooth
surface, cut only by shallow and linear features, with
a neglectable number of impact craters. Ganymede consists of two types of terrains: one brighter, grooved
and slightly cratered, the other darker and heavily cra
tered; Callisto, instead, appears very dark and almost
uniformly cratered, with two huge multiring basins.
Photometric studies of the inner Jovian satellites ha
ve confirmed these dissimilarities (1,2), giving for
their surfaces decreasing values of the general albedo
and of water content with increasing distance from the
planet. Relative measurements provided for Europa 98%
of water content and 0.68 of reflectivity, for Ganymede
90% of water and 0.44 of albedo, for Callisto 30 to 60%
of water and about 0.2 of albedo. These results might
look contradictory with the density values decreasing
from Europa (3 g/cm^3) to Ganymede (1.93 g/cm^3) and to
Callisto (1.79 g/cm3).
Such a scenario has been explained (3) hypothesizing
for Europa a silicatic bulk composition, except for a
thin crust made up of almost pure ice (according to the
model, 20 up to 100 km thick). Ganymede and Callisto,
instead, are thought to exhibit a similar internal struc
ture: a silicatic core, a nearly liquid water mantle,
and an upper crust consisting of ice and rocky material.
The differences in surface characteristics have been
interpreted in terms of a stronger or longer internal
activity for Ganymede with respect to Callisto, probably related to a higher amount of silicates.
So, Ganymede should have experienced a resurfacing period, witnessed by the presence of bright and grooved
regions, whereas Callisto, also because a thicker crust,
did not undergo a similar surface evolution. However,
dark cratered terrains on Ganymede are analogous to the
overall crust of Callisto.

2. RAYED CRATERS ANALYSIS TO INVESTIGATE GALILEAN
   SATELLITES STRATIGRAPHY
Taking into account the inferred structure of Ganymede
and Callisto we would gain some insight in the outer
stratigraphy of these satellites, studying the distribution and morphology of impact craters on their surface.
In particular, the analysis of the rayed craters size
frequency and geographical location is useful because
they represent very recent landforms, yet exhibiting a
clear pattern of linear features, which, at least partial
ly, appear to be primary ejecta. Thus, ray characteristics, together with the crater size values strictly related to the volume of material excavated and to the depth

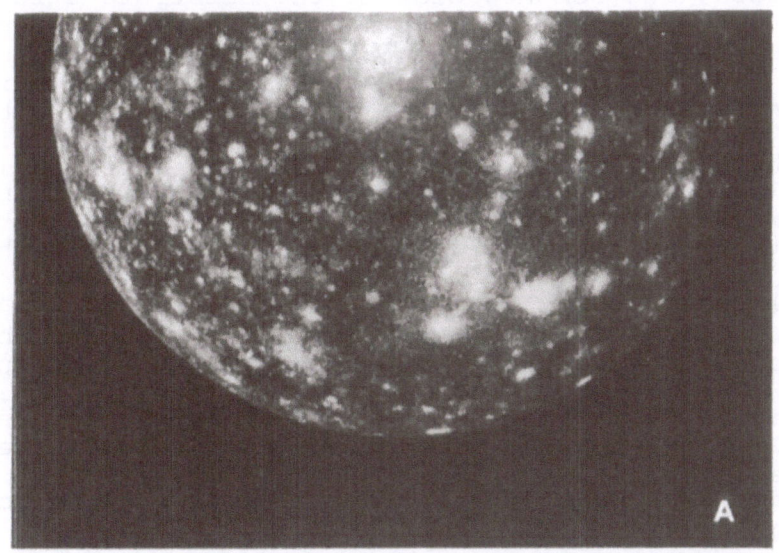

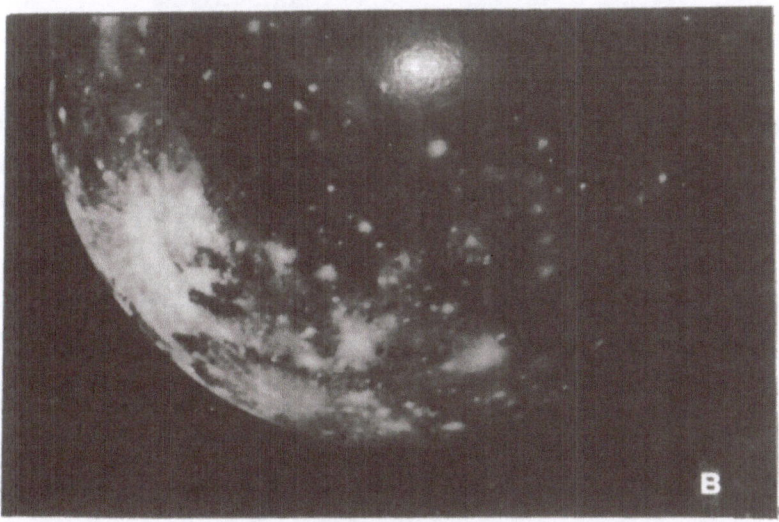

Fig. 1 - (a) Global view of Callisto's surface, sho
wing very few rayed craters. This picture was taken
by Voyager 2 at a range of 1,094,666 km. (b) Voyager 2 picture of Ganymede's surface taken at a range of 1,2 milion km. Note the presence of remarkable
bright rayed craters especially on the left side.

attained, can somehow clarify the substrate structure.
The Voyagers pictures have shown a sharp difference in
rayed craters distribution between Ganymede and Callisto. In fact, Ganymede has several large circular features with high albedo (see fig.1b), surrounded by a
halo and, often, by remarkable ray patterns. Similar
features are also present on Callisto's surface (see
fig.1a) but they are mostly haloed and without any significant bright streaks (5,8).
Furthermore, an interesting characteristic is the presence on both Ganymede and Callisto of very dark (almost
black) ray systems. They appear to be related to a central crater and are similar in shape to the bright ray
patterns. These dark landforms are not very frequent
and have been argued to have an impact origin (5,6,7).
Particularly Ganymede exhibits dark rayed craters (see
fig.2) mostly bright floored and dark haloed: they have
represented a main subject of this work and have been
regarded as a part of the overall rayed craters popula-
lation.

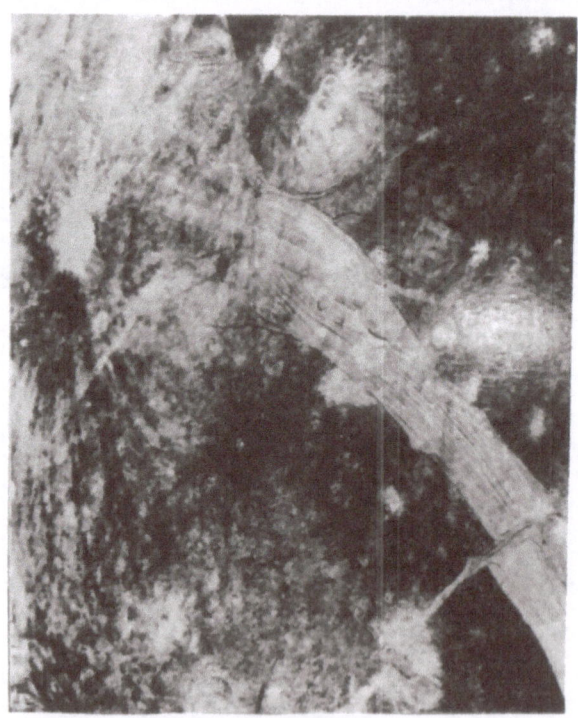

Fig. 2 - Voyager 2 picture of Ganymede's surface
taken at a range of 169,000 km. Note the presence
of prominent dark rayed craters.

So, the presence of both dark and bright rayed systems on the same satellite's surface, and their different di stribution between Ganymede and Callisto can put some constraint to the crustal model of these Galilean satellites.

3. DATA SET AND STATISTICAL APPROACH

Owing to the meaningless number of ray systems discovered on Callisto's surface, only Ganymede has been taken into account: all the impact craters exhibiting clear radial features (bright and dark) coming out of their rims, have been examined. Craters too close to the limb have been rejected because the oblique viewing angle prevented any satisfactory assessment of the rays trend. Moreover, particular care was taken in looking at areas near to the terminator, where brightness differences are not sufficiently enhanced: in this case pictures at lower resolution but with more contrast we-

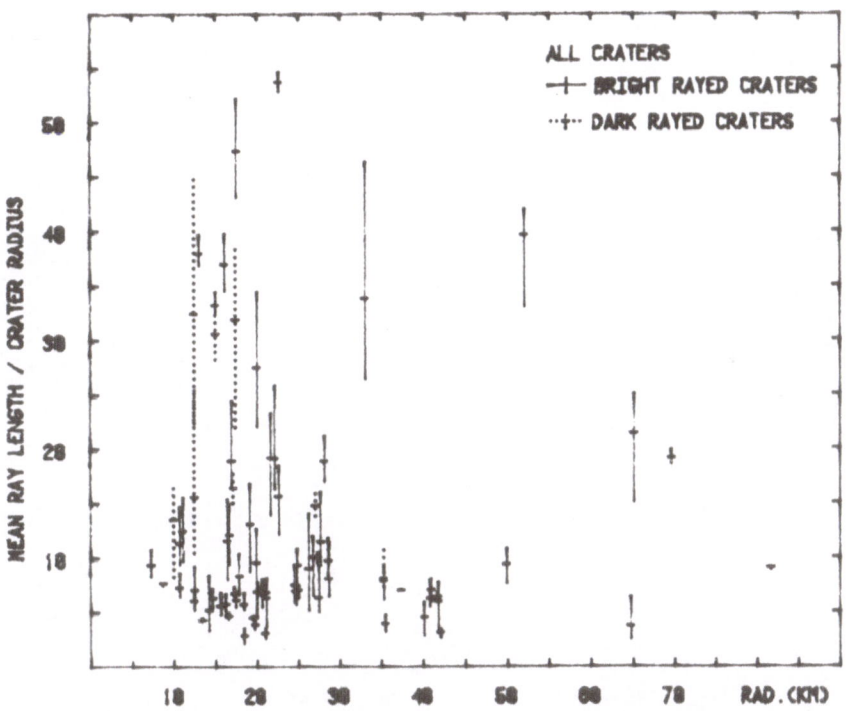

Fig. 3 - Plot showing the relations between crater radii and relative ray length expressed in terms of crater radii for all Ganymede's rayed craters

re considered for a safe estimate. The lengths of the
most prominent rays of each crater were then measured
putting their ends where the distinction between ray ma
terial and background was not evident.
  Once completed, Ganymede's data set included 76 rayed
craters, generally greater than 20 km in diameter. The
parameters chosen for each crater were size, geographi
cal location on the satellite's surface and length of
the most remarkable rays. In order to study the relationships between crater size and ray material distribution, a particular kind of plot has been used, showing crater radii values versus the ratios of average
ray length to crater radius (see fig.3).
On the plots each datum has been displayed, taking into account the three or four longest rays: the central
symbol represents their mean value, and the ends of the
bars the highest and lowest of them.
  Ganymede's rayed crater population has been splitted
in subsets, checking for color (bright or dark) and

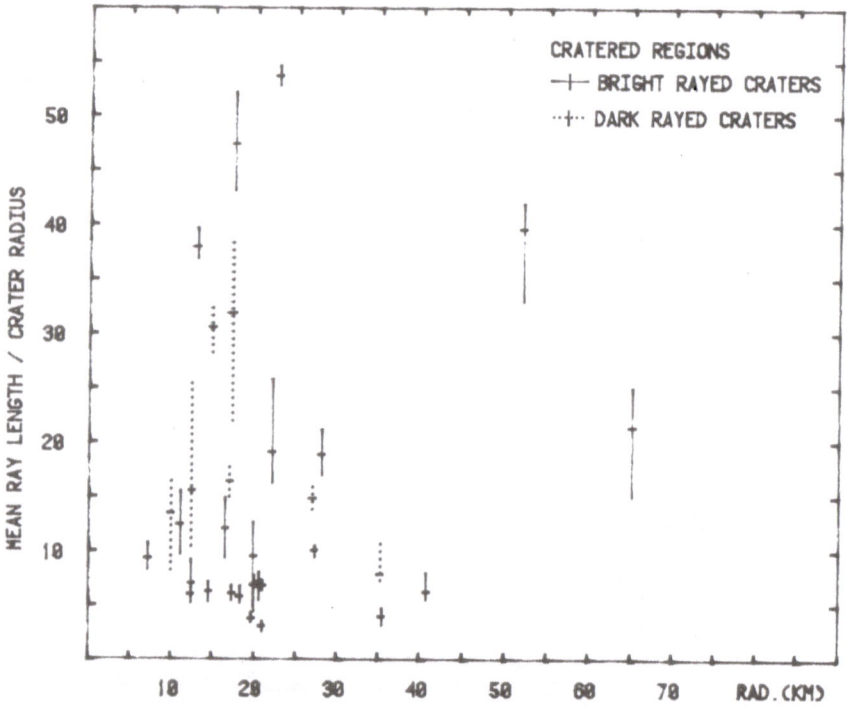

Fig. 4 - Plot analogous to that shown in fig. 3,
and relative to Ganymede's rayed craters lying
on dark cratered terrain

type of underlying terrain (grooved or cratered), and correspondent plots have been drawn (see figures 4 and 5).

4. RESULTS AND DISCUSSION

The plots displaying relative ray length versus crater radius give some insight in the ray formation and evolution process considered as a product of the final stage of the impact crater emplacement.

In particular, in the case of Ganymede it has been outlined (5) that all the obtained plots (see figures 3,4 and 5) show a general trend of ray length (in crater radii) decrease with increasing crater size. It could reflect the balance between ray fading (or erosion) rate and small craters production rate, regardless of the kind of underlying terrain (cratered or grooved).

Except for some very recent and bright rayed craters, a significant parameter influencing their appearance and distribution seems to be the type of substrate on which they formed. In fact, a strict relationship between ray material brightness and kind of underlying ter

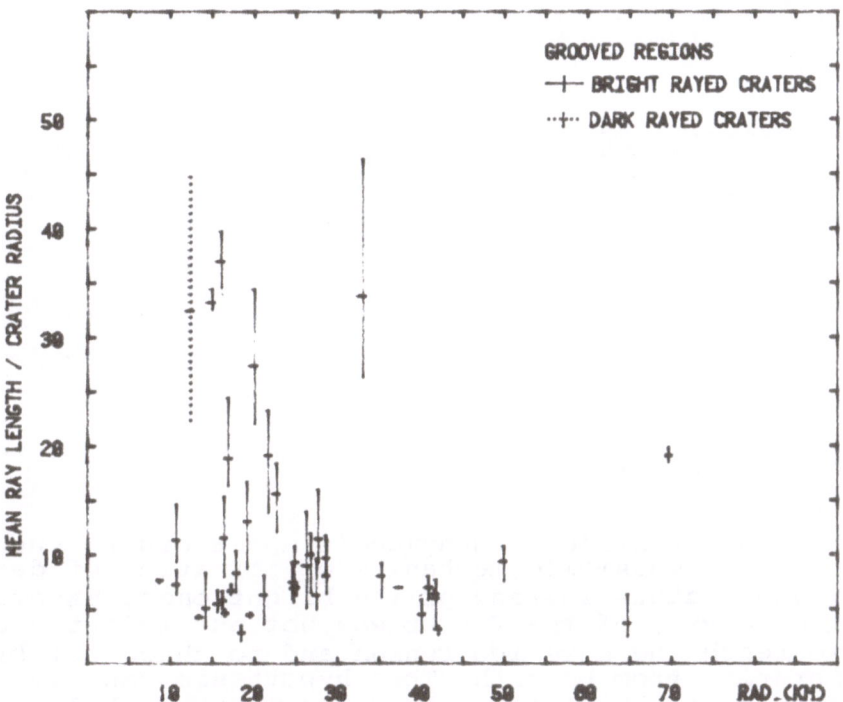

Fig. 5 - Same kind of plot as in fig.3 relative to Ganymede's rayed craters on the grooved terrains

rain has been found. Grooved terrains (see fig. 5)
exhibit a relative higher number of light ray systems
with respect to the heavily cratered regions; dark rayed
craters, on the other hand, (see fig. 4)are mostly present on cratered terrain, in the small size ranges, till
40 km in diameter (other two, a little larger , have so
me dark and some bright rays).
  On the basis of these results, and assuming that ray
material is generally fine-grained primary crater deposit, it is somehow possible to define the boundaries
and the structure of a dark upper layer compositionally
different from an underlying icy belt, evidenced on Ganymede's surface by the presence of the bright and groo
ved terrain.
  This dark layer should not be uniformly distributed
in the entire satellite's crust, but mostly under the
heavily cratered regions.
  Geographical location and size of the dark rayed craters   seem to confirm this hypothesis (see fig. 6):they

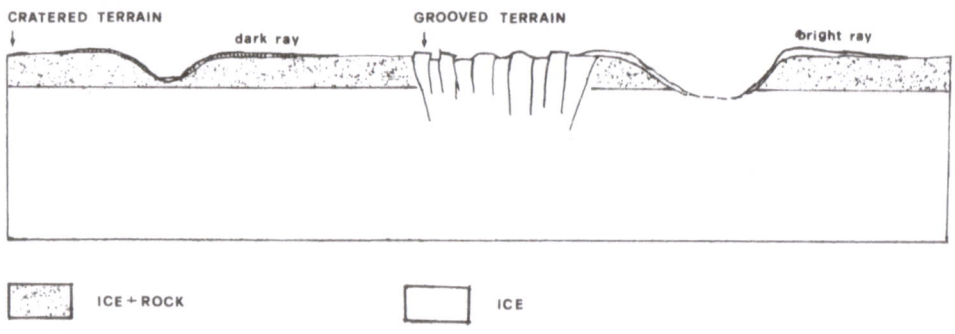

Fig. 6 - Profile of Ganymede's upper crust suggesting a possibile mechanism of formation of dark
rayed craters instead of the bright ones. Excavation cavity of the former was not sufficiently deep
to reach the pure ice region and to throw out bright
material from beneath. This hypothesis can give some
insight in the thickness of the upper dark layer pro
bably consisting of a mixture of rock and ice

represent craters emplaced mostly on heavily cratered
areas and not sufficiently large to excavate material
from a depth where only pure ice is present. Therefore,
dark ray and halo deposit should be the remnants of material coming from a relatively thin layer composed of
ice and rock.

Furthermore, following the assumption of Steven K.
Croft (4) that excavation cavity depth of a given crater would measure about 0.1 times the crater diameter,
a maximum value for the thickness of the upper dark
layer can be proposed: taking into account the largest
size found for the very dark rayed craters ($\sim$40 km) we
suggest a value of 4 $\div$ 5 km. The presence, however, of
two larger craters exhibiting a pattern of both dark
and bright rays, can mean that the inferred thickness
refers to an uniformly mixed outer layer and that the
boundary with the underlying icy regions is not really.
sharp, but probably made up of a quite complicate al-
ternance of icy and rocky thin layers.

5. CONCLUDING REMARKS

Cratered regions on Ganymede and Callisto are believed
to be analogously composed of a mixture of ice and rocky
material. Some insight on the thickness and the characteristics of this outer layer on both satellites can
be gained looking at the distribution of impact craters
as recent as rayed ones. Assuming that ray material is
mostly primary ejecta deposit, we can somehow put a
constraint to the different stratigraphic model of the
two Galilean satellites. In this framework particular
features like the dark rayed craters on Ganymede's sur
face have been found to be useful for their size and
location. In fact, they occur in the small size ranges
(mostly less than 40 km in diameter) and especially on
heavily cratered areas, whereas bright ray systems are
in all size ranges and on both cratered and grooved ter
rains. So, it can be suggested that dark ray deposits
on Ganymede represents material excavated from beneath
till 4 $\div$ 5 km of depth, supposing a ratio of about 1
to 10 between excavation cavity depth and correspondent
crater diameter (4). This value may be regarded as the
thickness of a layer representing the substrate of the
dark cratered terrain and overlying pure water ice material. Moreover, relative absence of remarkable bright
ray systems on Callisto's surface seems to confirm the
hypothesis that a thick layer of mixed-rock and ice is
ubiquitous in its crust.

ACKNOWLEDGEMENTS

The data base used for this research was acquired by
the author while he was visiting scientist at Lunar and
Planetary Institute.

REFERENCES

1) Clark,R.N.:1980,Icarus 44,pp.388-409

2) Clark,R.N. and McCord,T.B.:1980,Icarus 41,pp.323-339

3) Consolmagno,G.J. and Lewis,J.S.:1976,in "Jupiter",
   T. Geherels ed. (University of Arizona Press,Tuc-
   son),pp.1035-1051

4) Croft,S.K.:1980,Proc. Lunar Planet. Sci. Conf. 11th,
   pp.2347-2378

5) Poscolieri,M. and Schultz,P.H.:1980,Mem. S.A.iT.51,
   pp.359-377

6) Smith,B.A.,Soderblom,L.A.,Johnson,T.V.,Ingersoll,
   A.P.,Collins,S.A.,Shoemaker,E.M.,Hunt,G.E.,Masursky,
   H.,Carr,M.H.,Davies,M.E.,Cook II,A.F.,Boyce,J.,Da-
   nielson,G.E.,Owen,T.,Sagan,C.,Beebe,R.F.,Veverka,J.,
   Strom,R.G.,McCauley,J.F.,Morrison,D.,Briggs,G.A.,Suo
   mi,V.E.:1979,Science 204,pp.951-972

7) Smith,B.A.,Soderblom,L.A.,Beebe,R.F.,Boyce,J.,Briggs,
   G.A.,Carr,M.H.,Collins,S.A.,Cook II,A.F.,Danielson,
   G.E.,Davies,M.E.,Hunt,G.E.,Ingersoll,A.,Johnson,T.V.,
   Masursky,H.,McCauley,J.F.,Morrison,D.,Owen,T.,Sagan,
   C.,Shoemaker,E.M.,Suomi,V.E.,Veverka,J.:1979,Scien-
   ce 206,pp.927-950

8) Squyres,S.W. and Veverka,J.:1981,Icarus 46,pp.137-
   155

LIST OF SPEAKERS AND PARTICIPANTS

Anderson, D.M.	U.S.A.	State University of New York, Clemence Hall Buffalo, New York 14260 U.S.A.
Anderson, K.	U.S.A	Space Science Laboratory University of California Berkeley,California 94720 U.S.A.
Arvidson R.E.	U.S.A.	Earth Sciences/Geology Dept., Lindel and Sckinker Washington University St. Louis, Missouri 63130 U.S.A.
Baldea, I.	Rumania	Central Institute of Physics, Space Research Dept. P.O. Box MG-6 R 76900 Magurele-Bucharest Rumania
Barucci, M.A.	Italy	Osservatorio Astronomico di Collurania 64100 Teramo Italy
Bertola, F.	Italy	Osservatorio Astronomico Università di Padova 35100 Padova Italy
Binder, A.	U.S.A.	Institut fur Geophysik Neue Universitat 23 Kiel W. Germany

Blair, D.J.	U.S.A.	National Geographic Magazine Washington D.C. U.S.A.
Blanco, C.	Italy	Osservatorio Astrofisico Città Universitaria Catania Italy
Bodenheimer, P.	U.S.A.	Lick Observatory University of California Santa Cruz,CA 95064 U.S.A.
Bolle, H.J.	Austria	Institut fur Mineralogie und Geophysik Universitat Innsbruck Schopfstrasse 41 Innsbruck, Austria
Bonafede, M.	Italy	Istituto di Geofisica Università di Bologna Via Irnerio, 46 42126 Bologna Italy
Buedeler, W.	W.Germany	8153 Thalham/Obb W. Germany
Burchi, R.	Italy	Osservatorio Astronomico di Collurania 64100 Teramo Italy
Buser, W.	W.Germany	Institut fur Mineralogie Gievenbecker Weg 61 D - 4400 Munster W. Germany
Butterworth, P.S.	U.K.	Dept. of Astronomy University of Leicester Leicester LE 1 7RH United Kingdom
Caloi, V.	Italy	Istituto Astrofisica Spaziale del C.N.R. C.P. 67 - 00044 Frascati Roma Italy

Canovaro, F.	Italy	Via di Bravetta, 1 00100 Roma Italy
Carta, F.	Italy	Osservatorio Astronomico di Brera Via Brera, 28 20121 Milano Italy
Carusi, A.	Italy	Istituto Astrofisica Spaziale del C.N.R. V.le dell'Università,11 00185 Roma Italy
Casacchia, R.	Italy	Istituto Astrofisica Spaziale del C.N.R. V.le dell'Università,11 00185 Roma Italy
Castellani, V.	Italy	Istituto Astrofisica Spaziale del C.N.R. C.P. 67 00044 Frascati (Roma) Italy
Cembram, S.	Italy	Istituto Astrofisica Spaziale del C.N.R. V.le dell'Università,11 00185 Roma Italy
Cerroni, P.	Italy	School of M.A.P.S. University of Sussex Brighton BN1 9QH United Kingdom
Chlistowsky, F.	Italy	Osservatorio Astronomico di Brera Via Brera,28 20121 Milano Italy
Coradini, A.	Italy	Istituto Astrofisica Spaziale V.le dell'Università,11 00185 Roma Italy

Coradini, M.	Italy	Istituto Astrofisica Spaziale V.le dell'Università,11 00185 Roma Italy
Crescentini, L.	Italy	Istituto di Fisica Università di Roma P.le Aldo Moro,5 00185 Roma Italy
Dalaudier, F.	France	Centre National de la Recherche Scientifique Service d'Aeronomie B.P. No.3 91370 Verrieres le Buisson France
D'Alli, R.	U.S.A.	Arizona State University KAET TV 8 Tempe, Arizona 85287 U.S.A.
Dobrowolny, M.	Italy	Istituto plasma spazio Via G. Galilei, C.P. 27 00044 Frascati (Roma) Italy
Duncan, A.M.	U.K.	Dept. of Science Luton College of Higher Education Park Square Luton Bedfordshire LU1 3J United Kingdom
Dvorak, R.	Austria	Institut fur Astronomie Universitatsplatz 5 A - 8010 Graz Austria
Dziewonski, A.	U.S.A.	Hoffman Laboratory Harvard University 20 Oxford Street Cambridge, Mass. 02138 U.S.A.

Farinella, P.	Italy	Istituto di Matematica Università di Pisa Via Buonarroti, 2 56100 Pisa Italy
Federico, C.	Italy	Istituto Astrofisica Spaziale V.le dell'Università,11 00185 Roma Italy
Ferrini, F.	Italy	Istituto di Fisica Università di Pisa P.za Torricelli, 2 56100 Pisa Italy
Flamini, E.	Italy	Istituto Astrofisica Spaziale V.le dell'Università,11 00185 Roma Italy
Fofi, M.	Italy	Osservatorio Astronomico Via del Parco Mellini,84 00136 Roma Italy
Friolo, R.	Italy	AGIP Mineraria Via del Bergamino, 22 34139 Trieste Italy
Froeschle, C.	France	Observatoire de Nice Gran Corniche BP 252 06007 Nice Cedex France
Fulchignoni, M.	Italy	Istituto Astrofisica Spaziale V.le dell'Università,11 00185 Roma Italy
Genova, F.	France	Observatoire de Paris Section d'Astrophysique 92190 Meudon France

Gore, R.   U.S.A.   National Geographic
                                    Magazine
                                    Washington D.C. 20036
                                    U.S.A.

Greeley, R.   U.S.A.   Dept. of Geology
                                    Arizona State University
                                    Tempe, Arizona 85281
                                    U.S.A.

Guest, J.   U.K.   University of London
                                    Observatory
                                    Dept. of Physics and
                                    Astronomy
                                    Mill Hill Park
                                    London NW7 2QS
                                    United Kingdom

Hayli, A.   France   Observatoire de Lion
                                    Saint Genis
                                    Lavalle
                                    France

Hiller, K.   W.Germany   Institut fur Allgemeine
                                    und Angewandte Geologie
                                    der Universitat,
                                    Luisenstrasse 37
                                    D - 8000 München
                                    West Germany

Kaiser, T.   U.K.   Hicks Building
                                    Dept. of Physics
                                    University of Sheffield
                                    Sheffield 10
                                    United Kingdom

Karkoschka, E.   W.Germany   Nellingerstrasse 45C
                                    D - 7000 Stuttgart 75
                                    West Germany

Keller, U.   W.Germany   Max Planck Institut
                                    fur Aeronomie
                                    3411 Katlemburg Lindhau
                                    West Germany

Kilburn, C.   U.K.   University of London
                                    Observatory
                                    Dept. of Physics and
                                    Astronomy
                                    Mill Hill Park
                                    London NW7 2QS
                                    United Kingdom

Lanciano, P.	Italy	Istituto Astrofisica Spaziale V.le dell'Università,11 00185 Roma Italy
Leff, C.	U.S.A.	Dept. Of Earth and Planetary Sciences Washington University P.O. Box N. 1169 St.Louis, Missouri 63130 U.S.A.
Lepine, V.	France	C.N.R.S. Service d'Aeronomie BP No. 3 91370 Verrieres le Buisson France
Lopes, R.	Brazil	University of London Observatory Dept. of Physics and Astronomy Mill Hill Park London NW7 2QS United Kingdom
Magni, G.	Italy	Istituto Astrofisica Spaziale V.le dell'Università,11 00185 Roma Italy
Mariani, F.	Italy	Istituto di Fisica Università di Roma P.le Aldo Moro, 5 00185 Roma Italy
Martelli, G.	Italy	School of M.A.P.S. University of Sussex Brighton BN1 9QH United Kingdom
McConville, P.	U.K.	Dept. of Physics University of Sheffield Sheffield S3 7RH United Kingdom

Milani, A.	Italy	Istituto di Matematica Università di Pisa Via Buonarroti, 2 56100 Pisa Italy
Milano, L.	Italy	Osservatorio Astronomico di Capodimonte Via Moiariello, 16 80131 Napoli Italy
Mulargia, F.	Italy	Istituto di Geofisica Università di Bologna Via Irnerio, 46 40126 Bologna Italy
Nappi, G.	Italy	Centro di studio per la geologia termica del CNR Via Eudossiana, 18 00184 Roma Italy
Neukum, G.	W.Germany	Institut fur Allgemeine und Angewendte Geologie der Universitat Luisenstrasse 37 D - 8000 Munchen 2 West Germany
Nobili, A.	Italy	Istituto di Matematica Università di Pisa Via Buonarroti, 2 56100 Pisa Italy
Oberst, J.	W.Germany	Institut fur Mineralogie Gevenbecker Weg 61 D - 4400 Munster West Germany
Ott, U.	W.Germany	Max Planck Institut fur Chemie Saarstrasse 23 D - 6500 Mainz West Germany
Paterno', L.	Italy	Osservatorio Astrofisico Città Universitaria 95125 Catania Italy

Rogers, E.   U.S.A.   National Geographic
                                    Magazine
                                    Washington D.C. 20036
                                    U.S.A.

Romano, R.   Italy   Istituto Internazionale
                                    di Vulcanologia del CNR
                                    V.le Regina Margherita,6
                                    95123 Catania
                                    Italy

Rothwell, P.   U.K.   Dept. of Physics
                                    University Southampton
                                    Southampton SZ9 5NH
                                    United Kingdom

Roy, A.   U.K.   Dept.of Astronomy
                                    University of Glasgow
                                    Glasgow G12 8QW
                                    United Kingdom

Rosino, L.   Italy   Istituto di Astronomia
                                    Università di Padova
                                    Vicolo dell'Osservatorio,5
                                    55100 Padova
                                    Italy

Runcorn, S.,K.   U.K.   School of Physics
                                    The University
                                    Newcastle upon Tyne NE1 7RU
                                    United Kingdom

Sabadini, R.   Italy   Istituto di Geofisica
                                    Università di Bologna
                                    Via Irnerio 46
                                    40126 Bologna
                                    Italy

Salvatori, R.   Italy   Istituto di Astrofisica
                                    Spaziale
                                    V.le Università,11
                                    00185 Roma
                                    Italy

Scholl, H.   W.Germany   Astronomisches Rechen
                                    Institut
                                    Monch Hofstrasse 12-14
                                    6900 Heidelberg 1
                                    W. Germany

Paolicchi, P.	Italy	Osservatorio Astronomico di Merate Via Bianchi, 46 22055 Merate (Como) Italy
Pellarin, E.	Italy	Osservatorio Astronomico di Merate Via Bianchi, 46 22055 Merate (Como) Italy
Pellas, P.	France	Museum National d'Histoir Naturelle Laboratoire de Mineralo-gie 61, Rue de Buffon 75005 Paris France
Perozzi, E.	Italy	Istituto Astrofisica Spaziale V.le dell'Università,11 00185 Roma Italy
Pieri, D.	U.S.A.	Jet Propulsion Lab. 4800 Oak Grove Dr. Pasadena, California 91103 U.S.A.
Pohl, J.	W.Germany	Institut fur Allgemeine und Angewandte Geophysik Theresienstrasse 41 D - 8000 Munchen 2 West Germany
Poscolieri, M.	Italy	Istituto Astrofisica Spaziale V.le Università, 11 00185 Roma Italy
Pozio, S.	Italy	Istituto Astrofisica Spaziale V.le Università, 11 00185 Roma Italy

Silvestri, M.	Italy	Istituto di Fisica Università di Roma P.le Aldo Moro, 5 00185 Roma Italy
Skorve, J.	Norway	G.I.O.U. Oslo Norway
Smith, P.H.	U.S.A.	Lunar and Planetology Laboratory University of Arizona Tucson, Arizona 85721 U.S.A.
Steitz, W.	W.Germany	Max Planck Institut fur Chemie Saarstrasse 23 D - 6500 Mainz West Germany
Theilen-Willige, B.	W.Germany	Huttenweg 8 Clausthal-Zellerfeld West Germany
Tozer, D.	U.K.	School of Physics The University Newcastle upon Tyne NE1 7RU United Kingdom
Trigila, R.	Italy	Istituto di Fisica Università di Roma P.le Aldo Moro, 5 00185 Roma Italy
Tulunay, Y.	Turkey	Ortadogu Teknik Universitesi Inonu Bulvari Ankara Turkey
Turner, G.	U.K.	Dept. of Physics University of Sheffield Sheffield S3 7RH United Kingdom
Valsecchi, G.	Italy	Istituto Astrofisica Spaziale V.le Università,11 00185 Roma - Italy

Wanke, H.L.   W.Germany Max Planck Institut fur
                                          Chemie
                                          Saarstrasse 23
                                          D - 6500 Mainz
                                          West Germany

Weidenschilling, S.J. U.S.A.   Planetary Science
                                          Institute
                                          2030 E. Speedway
                                          Suite 201
                                          Tucson, Arizona 85719
                                          U.S.A.

Wise, D.U.   U.S.A.   Dept. of Geology
                                          Univ. of Massachusetts
                                          Amherst, Mass.01002
                                          U.S.A.

Woodhouse, J.H.   U.S.A.   Harvard University
                                          Hoffman Laboratory
                                          20 Oxford Street
                                          Cambridge, Mass.02138
                                          U.S.A.

Zappalà, V.   Italy   Osservatorio Astronomico
                                          Pino Torinese
                                          10025 Pino Torinese (TO)
                                          Italy

INDEX OF AUTHORS

INDEX OF SUBJECTS AND NAMES